AF896001

Free-Ranging Dogs and Wildlife Conservation

Free-Ranging Dogs and Wildlife Conservation

EDITED BY

Matthew E. Gompper

University of Missouri, USA

OXFORD
UNIVERSITY PRESS

Free-Ranging Dogs and Wildlife Conservation. Edited by Matthew E. Gompper
© Oxford University Press 2014. Published 2014 by Oxford University Press.

OXFORD
UNIVERSITY PRESS

Great Clarendon Street, Oxford, OX2 6DP,
United Kingdom

Oxford University Press is a department of the University of Oxford.
It furthers the University's objective of excellence in research, scholarship,
and education by publishing worldwide. Oxford is a registered trade mark of
Oxford University Press in the UK and in certain other countries

© Oxford University Press 2014

The moral rights of the author have been asserted

First Edition published in 2014

All rights reserved. No part of this publication may be reproduced, stored in a retrieval system, or transmitted, in any form or by any means, without the prior permission in writing of Oxford University Press, or as expressly permitted by law, by licence or under terms agreed with the appropriate reprographics rights organization. Enquiries concerning reproduction outside the scope of the above should be sent to the Rights Department, Oxford University Press, at the address above

You must not circulate this work in any other form
and you must impose this same condition on any acquirer

Published in the United States of America by Oxford University Press
198 Madison Avenue, New York, NY 10016, United States of America

British Library Cataloguing in Publication Data
Data available

Library of Congress Control Number: 2013944190

ISBN 978–0–19–966321–7

Links to third party websites are provided by Oxford in good faith and for information only. Oxford disclaims any responsibility for the materials contained in any third party website referenced in this work.

Foreword

Jako is my family dog, a Munsterländer born to be a hunter and frustrated by a family of non-hunters. When we walk in the woods, he often disappears to follow his nose on the track of a deer or a wild boar, but he always comes back after a while. When he wanders alone in the wood, is he a free-ranging dog? Hard to say, as the definition is open to many interpretations. His movements are indeed not restrained by any physical impediment but by a strong psychological leash that keeps him tight to his master. Is this enough to put him safely in the category of non-free-ranging dogs? In the infinite diversity of the dog world, where we can find all possible degrees of stronger and weaker links to humans, and all levels of freedom to roam unrestrained near or far from humans, where do we set the threshold?

Wild, stray, sylvatic, feral, village, unrestrained, are just some of the many labels used to define a huge variety of ecotypes of dogs that share a fundamental ecological feature: they are free to wander where they want and follow the occasional lure. They are free, temporarily or permanently, from the control of a human who dictates their times, movements, and lifestyle. The many categories of free-ranging dogs mostly refer to their lifestyle and the degree of their dependence and social relationship with humans. Some of them may still have a human boss, their 'owner,' to whom they are attached and keen to return more or less frequently, but most of them have no human companion. This last category is highly heterogeneous and includes dogs that had and lost a human companion, and dogs that never had a social connection with humans. For those of the first group, their dependence on humans is often reduced to a faint image of a former symbiotic relationship; sometimes the loss of a social bond with humans is suffered as an unhealed wound, driving the dog to a never-ending search for a new human companion. But sometimes the loss is fully turned into a new life of canine-only relationships. Those of the second group, the largest majority of free-ranging dogs, never had a social connection to humans. However, no matter how detached they are from humans, they still depend on us for food and shelter. Only a tiny minority of them, the true feral dogs, are back in the woods living, as they can, as intruders into the natural community and the existing balance of predators and prey.

There are at least two very different ways of looking at the phenomenon of free-ranging dogs. On one side, they have been with us since the early days of the domestication of the wolf-like ancestor from which all dogs originate. It is likely, in fact, that the transition from a wild to a tame animal under strict human control passed through a long phase of reciprocal understanding and adaptation, when the dog's ancestors were necessarily hanging around human settlements and were gradually losing fear of humans. Thus, the condition of free-ranging is all-natural and common for dogs; they learned to exploit the rich niche in the vicinity of humans, taking advantage of their protection and food remains. In this sense, village dogs (i.e., those living in and around human villages and towns without a direct social connection with humans) are the quintessential dog, the original and best example of the man–dog relationship for most of our common history.

On the other side, the dominating Western culture sees dogs only as faithful companions and helpers in a large number of human activities (hunting, defense, herding livestock, combat, etc.). We tend to see dogs as a product of the human capacity to bend nature to our needs and, as such, we like to think that dogs are acceptable only as far as they are under our full control. A free-ranging dog is an element of disturbance to our orderly view of nature and therefore becomes a problem for society. This is an understandable and legitimate position but it shows how limited our cultural system is when we try to build a broader view of the ecological processes around us. Stray and village dogs of many tropical and subtropical areas, so often in poor condition, shy, and elusive, are perceived as the paradigm of a degraded and contemptible environment and human condition. They are the symptoms and contributors of a situation to be corrected: to our Western, narrow view, free-ranging dogs are not good dogs, they are a disturbance to be eliminated.

The two views of free-ranging dogs are not in antithesis, however, and there is all to be gained by trying to integrate them into a more comprehensive view of the multifaceted entity that we call dogs. In doing so we can get much better insights into dogs' evolution, behavior, sociality, and ecological flexibility and, more importantly, we may be able to establish a more respectful relationship with them. It seems to me that the fondest love for our best friend when he is at home or working with us is just too little for a species that has shown the most incredible capacity for thriving on human societies across all ages and cultures.

This first book on free-ranging dogs is timely and excellent in bringing together a vast amount of information on many aspects of the role of free-ranging dogs in our wild and domestic settings. There is no doubt that the very same features of dogs' biology that made them the most successful material to shape as a tool to extend man's senses into the environment, that is morphological and behavioral plasticity, are the same that preoccupy us most when dogs enter the inner circle of our interests: dogs as predators of livestock and wild prey, as carriers of diseases and zoonosis, as potential predators of humans, as competitors for resources needed by other protected wild predators, as cause of hybridization with wild canids. All are serious threats to be concerned about and these threats justify our attempts at mitigation.

The extraordinary capacity of dogs to live inside and at the fringes of human settlements in a variety of ecotypes makes them also extremely powerful in bringing those threats directly to the core of the human circle. More recently, to the serious problems caused by free-ranging dogs to human health as carriers of several pathogens, a suite of concerns related to conservation issues have been added: dogs, together with domestic cats, are the most numerous predators on Earth and their potential impact is devastating. There can be no compromise when this impact threatens wild species of our natural heritage. No matter how broad our view is of dogs as 'natural' animals, it is our responsibility to manage and restrict the niche in which dogs thrive: our own human environment.

All these issues and many more are accounted for in this book: its goal is daunting as the subject cannot be reduced to a few definitions and generalizations. The complex array of ecotypes, each defined in different terms depending on the local ecological conditions and the human social, economic, and cultural contexts, is hardly summarized by just one or a few conceptual schemes and, conversely, a long list of different anecdotes and case studies would remain without a unifying vision. This book succeeds in achieving the best possible compromise between specificity and generalization, and it contains important clues to address the key questions on managing free-ranging dogs.

I hope that this book will not only be read by those who are professionally engaged in managing dog populations either in urban or natural settings, but also by those who enjoy a close relationship with their dog. It will teach them that there is much more to a dog than just love for its master.

Luigi Boitani
Department of Biology and Biotechnologies
University La Sapienza, Rome

Preface

Matthew E. Gompper

The initial seed for this book was planted approximately a decade ago when I was teaching an undergraduate-level course in conservation biology at the University of Missouri. At the time, I had several graduate students working on wildlife conservation and management projects, and their work always seemed to be influenced directly or indirectly by the presence of dogs. Dogs were chasing, killing, or used to hunt their study organisms. Dogs were observed to harbor pathogens that were also problems for their study organisms. Dogs were being photographed at remote sites where students were conducting wildlife surveys. Then came the conservation biology class, during which we examined the fall and rise of the rare black-footed ferret (*Mustela nigripes*).

In brief, throughout much of 1981 the prevailing wisdom regarding the black-footed ferret, a charismatic small carnivore endemic to the grasslands of North America, was that the species was extinct. Extensive searches had not identified any remaining wild populations, and the sole captive population had died out several years earlier. However, on September 26 of that year, this prevailing wisdom was proven wrong when a dog killed a black-footed ferret near Meeteetse, Wyoming (Miller et al., 1996). The dead animal was discovered by the dog's owner. Thereafter, a wild population of black-footed ferrets was discovered, renewed captive breeding efforts ensued, multiple reintroductions of the progeny occurred, and now the black-footed ferret is viewed by many as recovering (Grenier et al., 2007).

The role of a dog in the rediscovery of the black-footed ferret is well known among North American conservationists and wildlife managers, and because people often have positive views of dogs, we know a great deal about this particular dog (Figure P.1). We know it was a Blue-Heeler mixed breed animal and that its name was Shep (Miller et al., 1996). Eminent conservationist Archie Carr (1986) met the dog and noted that Shep was " . . . very tractable, and blithely unconcerned with the hoopla stirred up by his routine vigilance." The dog became a bit of a star within conservation circles, and the dog's name worked its way into numerous articles for the general public. Thirty years after the rediscovery of the black-footed ferret, Dean Biggins penned the introduction to a journal volume focused on recent research on this now well-studied species, and acknowledged the many individuals who contributed so much to black-footed ferret conservation. This included Shep: "Finally, it also seems fitting to pay tribute to Shep, the dog owned by John and Lucille Hogg, who by killing an individual ferret near Meeteetse in 1981 gave life to the species (Biggins, 2012)." The general message of this tale is one that is humorously positive: Shep did what many researchers could not in rediscovering black-footed ferrets. Strange as it may seem, Shep put a human face on the tale.

Figure P.1 Wyoming Game and Fish Department employee Dennie Hammer poses with Shep, the dog that killed a black-footed ferret (shown in the lower panel; now *Mustela nigripes* specimen MSB 107934, USGS Biological Surveys Collection, Museum of Southwestern Biology, Albuquerque, New Mexico, USA). At the time the black-footed ferret was thought extinct. When Shep's owner reported the dead ferret, Hammer was part of the investigating team that identified an extant black-footed ferret population, which subsequently led to the recovery of the species. Photos courtesy of Dennie Hammer and Steven Raniszewski. Specimen information courtesy of Cindy Ramotnik, US Geological Survey.

I began to ponder our perception of Shep and the broader importance of such perceptions. What has, to my knowledge, never been explicitly stated is that Shep, a relatively large free-ranging dog, killed one of the world's last black-footed ferrets (although Carr, for his part, did appear to recognize the conflicted role of Shep). The fact that we thought black-footed ferrets were extinct is irrelevant. We do not know exactly how many black-footed ferrets persisted when the population was rediscovered, as censuses could only start at that point and so the ferret population size data for 1982 is incomplete (Forrest et al., 1988). However we do have census data for 1983, and if we assume the 1982 population was similar to the 1983 population, then the black-footed ferret that Shep killed (an adult male now labeled as MSB 107934 and housed in the Biological Surveys Collection of the University of New Mexico's Museum of Southwestern Biology in Albuquerque) represented approximately 10% of the remaining global adult male population (6 adult males plus 10 unknown sex adult individuals; Forrest et al., 1988). And while Shep's interaction with a black-footed ferret might have been unexpected, it was not the first time the dog had interacted with wildlife, as Shep's owners initially "heard the commotion and thought Shep had crossed paths with yet another porcupine" (Miller et al., 1996).

Reading about Shep, while simultaneously recognizing that dogs were influencing the work of my students, clarified the need to think about the ecological role of dogs. Interactions between dogs and wildlife are likely the norm, whether the dogs are owned or un-owned, free-ranging or restrained. With the exception of indoor or fully confined dogs, most dogs interact directly or indirectly with wildlife on a regular basis. These interactions range from direct agonistic interactions such as dogs killing (or being killed by) wildlife, to interactions that at first glimpse are seemingly benign, such as dogs disturbing wildlife by instilling a perceived risk of predation, even if the true risk is negligible. This book aims to focus and refine our understanding of how dogs interact with wildlife. A great deal has been written about dogs from the perspectives of the human–dog bond, veterinary medicine, and public health. Generally these perspectives focus on companion animals, with less attention given to free-ranging dogs (although, of course, companion animals can also be free-ranging), and still less to the free-ranging dog populations inhabiting the rural and less developed regions of the world where they are most likely to interact with wildlife.

Yet, over the past two decades, researchers working with wildlife have become increasingly attuned to the fact that dogs are not necessarily benign players in the environment. With this recognition has come concern among applied ecologists, animal conservationists, and wildlife managers that the presence and seemingly great numbers of dogs, whether owned or un-owned, might have far reaching effects on the outcome of wildlife management programs. While an increasing number of research papers have addressed components of this issue, the topic has not been broadly synthesized. This lack of a holistic attempt to address the interactions of dogs and wildlife is potentially problematic because of the virtually ubiquitous nature of dog populations, and the varied mechanisms by which dogs may influence wildlife. Many scientists (and indeed, the broader public) are likely aware of some issues associated with dogs, but unaware of other concerns. Yet studies of wildlife, and sometimes direct studies of the dogs themselves, have identified a diversity of concerns on almost every continent. Examples include dogs killing wildlife and so contributing to declines of species of conservation concern, dogs acting as vectors of pathogens that result in epidemics in wildlife, dogs acting as the prey base for large carnivores and in turn underpinning conflicts between large carnivores and people, dogs competing with native species for limited habitat and resources, and dogs hybridizing with wild species of canids.

The influence of dogs may, however, be more nuanced than the string of putatively negative interactions noted above. For instance, in some settings dogs may, perhaps, offset the loss of the large carnivorous species that once were abundant. In this capacity dogs may act to control invasive species and thereby actually aid wildlife conservation efforts. In addition, when used as a tool (e.g., livestock guard dogs, detection dogs, hunting dogs) dogs may facilitate wildlife conservation efforts, help to minimize human–wildlife conflicts, or allow humans to better access the resource that wildlife sometimes represents. As such, these dogs—which when 'working' are used in a free-ranging context—may have a mix of positive and negative influences on wildlife that are not widely recognized outside limited research circles. Furthermore, some free-ranging dog populations may even represent worthy conservation targets in and of themselves. Thus, in planning this book, I attempted to identify authors who would pen chapters that looked not solely at the agonistic and negative interactions of dogs and wildlife, but also at nuanced interactions that working dogs have with wildlife, and even at the possible positive role that free-ranging dogs may sometimes play.

This book could not have come together without the enthusiastic participation of the authors of each chapter. Thanks are also deserved for the numerous individuals who assisted with the review process, including Michael Alvard, Arman Ardalan, Arne Arnberger, Jerry Belant, Aniruddha Belsare, Dean Biggins, Juliet Clutton-Brock, Karen DeMatteo, Al Glen, Inger Hansen, Anne Hoylman, Roland Kays, Greger Larson, John Linnell, Grainne Maguire, Darcy Morey, Charlie Nilon, Teet Otstavel, Heidi Parker, Kent Redford, Sarah Reed, Ben Sacks, Stephanie Schuttler, James St Clair, Todd Steury, Cat Urbigkit, and Lisette Waits. The enthusiastic, friendly, and knowledgeable guidance and support of Ian Sherman, Lucy Nash, and Helen Eaton made this project enjoyable. Finally, I owe thanks to Anne, Alex, and Isabel for their consistent encouragement, indulgence, and fond humoring during a project that seemed to integrate itself all too much into our family.

References

Biggins, D.E. (2012). Information on black-footed ferret biology collected within the framework of ferret conservation. *Western North American Naturalist*, **72**, 129–133.

Carr, A., III. (1986). Black-footed ferret: Introduction. *Great Plains Naturalist Memoirs*, **8**, 1–7.

Forrest, S.C., Biggins, D.E., Richardson, L., Clark, T.W., Campbell, T.M. III, Fagerstone, K.A. and Thorne, E.T. (1988). Population attributes for the black-footed ferret (*Mustela nigripes*) at Meeteetse, Wyoming, 1981–1985. *Journal of Mammalogy*, **69**, 261–273.

Grenier, M.B., McDonald, D.B., and Buskirk, S.W. (2007). Rapid population growth of a critically endangered carnivore. *Science*, **317**, 779.

Miller B., Reading R. and Forest S. (1996). *Prairie Night: Black-footed Ferrets and the Recovery of Endangered Species.* Smithsonian Institution Press, Washington, D.C.

Contents

Contributors

Vidya Athreya Wildlife Conservation Society—India, 1669, 31st cross, 16th main. Banashankari 2nd stage, Bangalore, 560070, India

Luigi Boitani Department of Biology and Biotechnologies, Sapienza Università di Roma, Viale dell'Università 32, Roma 00185, Italy

Adam R. Boyko College of Veterinary Medicine, Cornell University, Ithaca, New York, United States of America

Ryan H. Boyko School of Public Health, Yale University, New Haven, Connecticut, United States of America

James R.A. Butler CSIRO Ecosystem Sciences, GPO Box 2583, Brisbane, QLD 4001, Australia

Rob Critchlow Boyd Orr Centre for Population and Ecosystem Health, Institute of Biodiversity, Animal Health and Comparative Medicine, College of Medical, Veterinary and Life Sciences, University of Glasgow, Glasgow, United Kingdom

Christopher R. Dickman School of Biological Sciences, A08, University of Sydney, NSW 2006, Australia

Jorge Echegaray Conservation and Evolutionary Genetics Group, Estación Biológica de Doñana (EBD-CSIC), Avda. Américo Vespucio s/n, 41092 Seville, Spain

Thomas M. Gehring Department of Biology, Central Michigan University, Mt. Pleasant, Michigan, 48859, United States of America

Matthew E. Gompper Department of Fisheries and Wildlife Sciences, School of Natural Resources, University of Missouri, Columbia, Missouri, 65211, United States of America

Aimee Hurt Working Dogs for Conservation, 52 Eustis Road, Three Forks, Montana, 59752, United States of America

Darryn L. Knobel Department of Veterinary Tropical Diseases, Faculty of Veterinary Science, University of Pretoria, P/Bag X04, Onderstepoort 0110, South Africa

Jeremy Koster Department of Anthropology, University of Cincinnati, Cincinnati, Ohio, 45221–0380, United States of America

Jean-Marc Landry Institute for the Promotion and Research on Guarding Animals, Martigny, Switzerland

Michael Lavelle National Wildlife Research Center, USDA/APHIS/Wildlife Services, Fort Collins, Colorado, 80521, United States of America

Tiziana Lembo Boyd Orr Centre for Population and Ecosystem Health, Institute of Biodiversity, Animal Health and Comparative Medicine, College of Medical, Veterinary and Life Sciences, University of Glasgow, Glasgow, United Kingdom

Jennifer A. Leonard Conservation and Evolutionary Genetics Group, Estación Biológica de Doñana (EBD-CSIC), Avda. Américo Vespucio s/n, 41092 Seville, Spain

Nicolas Lescureux Norwegian Institute for Nature Research, PO Box 5685 Sluppen, NO-7485 Trondheim, Norway

Mike Letnic School of Biological, Earth and Environmental Sciences, University of New South Wales, Sydney, New South Wales, 2052, Australia

John D.C. Linnell Norwegian Institute for Nature Research, PO Box 5685 Sluppen, NO-7485 Trondheim, Norway

Laurie Marker Cheetah Conservation Fund, Otjiwarongo, Namibia

Adam McKeown CSIRO Ecosystem Sciences, Tropical Forest Research Centre, CSIRO Atherton, PO Box 780, Atherton, QLD 4883, Australia

Kelly K. Miller Centre for Integrative Ecology, School of Life and Environmental Sciences, Deakin University, 221 Burwood Highway, Burwood, Victoria, 3125, Australia

Damian Morrant School of Marine and Tropical Biology, James Cook University, PO Box 6811, Cairns, QLD 4870, Australia

Andrew Noss 5217 NW 59th Lane, Gainesville, Florida, 32653, United States of America

Ettore Randi Laboratory of Genetics, Istituto Superiore per la Protezione e la Ricerca Ambientale (ISPRA), Via Cà Fornacetta, 9, 40064 Ozzano dell'Emilia (Bologna), Italy

Ngaio L. Richards Working Dogs for Conservation, 52 Eustis Road, Three Forks, Montana, 59752, United States of America

Euan G. Ritchie School of Life and Environmental Sciences, Deakin University, 221 Burwood Highway, Burwood, Victoria, 3125, Australia

Eduardo A. Silva-Rodriguez Departamento de Ecología y Biodiversidad, Facultad de Ecología y Recursos Naturales, Universidad Andrés Bello, Santiago, Chile

Theodore Stankowich Department of Biological Sciences, California State University, Long Beach, 1250 Bellflower Blvd, Long Beach, California, 90840, United States of America

Abi Tamim Vanak Centre for Biodiversity and Conservation, Ashoka Trust for Research in Ecology and the Environment, Bangalore 560 064, India

Kurt VerCauteren National Wildlife Research Center, USDA/APHIS/Wildlife Services, Fort Collins, Colorado, 80521, United States of America

Carles Vilà Conservation and Evolutionary Genetics Group, Estación Biológica de Doñana (EBD-CSIC), Avda. Américo Vespucio s/n, 41092 Seville, Spain

Michael A. Weston Centre for Integrative Ecology, School of Life and Environmental Sciences, Faculty of Science, Engineering and the Built Environment, Deakin University, 221 Burwood Highway, Burwood, Victoria 3125, Australia

Deborah A. (Smith) Woollett Working Dogs for Conservation, 52 Eustis Road, Three Forks, Montana, 59752, United States of America

Introduction: outlining the ecological influences of a subsidized, domesticated predator

Matthew E. Gompper

How we see dogs versus how wildlife sees dogs

We have mixed feelings about dogs (*Canis familiaris*). On the one hand we view dogs in the positive light of man's best friend. Examples of flamboyant prose describing the dog abound. In George Graham Vest's (1870) *Eulogy of the Dog*, the dog is "[T]he one absolutely unselfish friend that man can have in this selfish world, the one that never deserts him, the one that never proves ungrateful or treacherous ...." Such wordsmithing implies a comfortable state of camaraderie and symbiotic association. In other words, dogs are domestic companion animals. They are the seemingly welcome commensals of humans. And as a result of this commensal affiliation with humans, the population sizes and distributions of dogs mirror those of humans. We directly or indirectly provide food and shelter, such that even where dogs are not owned or cared for, dogs are still strongly associated with human communities, and across the globe the number of dogs one finds in an area is predicated on the number of humans.

On the other hand, we recognize the negative aspects associated with dogs. Dogs have the potential to vector pathogens to humans (most notoriously, the rabies virus), have the potential to become predators of livestock and farm animals, and may represent a direct risk to humans. Such risks are deeply embedded in the human language. The phrase 'hair of the dog (that bit you),' referring to the drinking of alcohol as a hangover remedy, has been in use for over 400 years, and traces to a purported remedy for the bite of a rabid dog (OED, 2010). Shakespeare's 'Dogges of Warre' from *Julius Caesar* has come to represent the uncontrolled savagery and bloodshed of conflict, while the expression 'dog-eat-dog' (dating from the 1700s) is indicative of such ruthless competitiveness that people are willing to harm each other in order to succeed. Guard dogs are used to protect people, premises, and livestock precisely because they are inherently threatening to outsiders. Visitors wishing to enter a rural Mongolian household may announce their presence by calling out 'Nokhoi khor!' (literal translation: Hold your dog!) even if no dog is known to be present. Dogs represent risks to humans and our personal spheres of influence, and recognition of these risks is deeply engraved in our language and cultures.

The balance between these positive and negative perspectives differs greatly among societies and among individuals within societies (e.g., Knobel et al., 2008; Ruiz-Izaguirre and Eilers, 2012). Yet because we generally view dogs through a human or utilitarian lens, we often fail to recognize that there are other players in the ecological communities where both dogs and humans reside, and that these players may view the dog not as something distinct and aside from native wildlife species, but rather as a highly interactive member of the vertebrate community on every continent except Antarctica (e.g., Butler et al., 2004; Fritts and Paul, 1989; Harris, 1981; Lacerda et al., 2009; Letnic et al., 2012;

Free-Ranging Dogs and Wildlife Conservation. Edited by Matthew E. Gompper
© Oxford University Press 2014. Published 2014 by Oxford University Press.

Vanak et al., 2009). When viewed in this latter context, we begin to realize that dogs represent both *domesticated predators* (that is, a population derived from a carnivorous ancestor, the wolf (*C. lupus*), and selected upon for a commensal and relatively less-agonistic relationship with humans) and *subsidized predators* (that is, a population of predators whose densities are above what one might normally expect if human-derived resources were unavailable). And, as such, these predators might have strong impacts on surrounding communities. Furthermore, because dogs tend to be found where humans are found, dogs may strongly influence how humans interact with wildlife (Figure 0.1).

Dogs have the potential to interact with wildlife in a multitude of ways, and as a result their influences on wildlife are complex and nuanced. These influences are likely to change as a function of the regional population density of dogs, the extent to which these dogs are constrained by humans, and the extent to which they are directly cared for (fed, sheltered, vaccinated). Yet we do not fully understand how wildlife 'sees' dogs. At one extreme, dogs may be viewed as phenotypic variants of wolves. Wildlife may respond to the presence of dogs or, indeed, to even the perception of the presence of dogs, much as it might to the presence or the risk represented by the possible presence of wolves or similar large canids. Alternatively, wildlife could hypothetically view dogs as something else entirely: as a rather benign member of the community that is *perceived* to represent little if any risk. It is critical, however, to realize that a perception of risk (whether low or high) held by wildlife may not correlate closely with the actual risk dogs may represent (again, low or high).

Such hypothetical extremes in how wildlife may visualize dogs represent a continuum, and the various taxa that comprise an animal community likely view dogs differently, much like they might view any species differently. Yet the evidence available to date, put forth in the chapters of this volume, suggest that in most cases (the possible exception being animal communities that have evolved in independence of terrestrial predators, such as those inhabiting remote oceanic islands) wildlife is more likely to perceive a dog as an analog of a wolf than as a benign and relatively non-interactive member of the community. The presence of dogs triggers strong and diverse responses by wildlife. This is most obvious among prey species, but the presence of dogs can also elicit responses by other carnivores, perhaps due to a perceived competitive dynamic or even because dogs may be perceived as potential prey.

Generally the size and dietary habits of an organism are indicative of its energetic needs. Large wild canid species, such as wolves and coyotes (*C. latrans*), have large caloric requirements and thus have larger home ranges to obtain these requirements. Such large home range sizes are generally associated with low population densities. In contrast, small canid species, such as foxes, have lower caloric needs and thus smaller home range sizes and higher population densities. While the sizes of dog populations are highly variable, especially when specific breeds are considered (toy breeds such as Chihuahuas may weigh just 2 kg, while large breeds such as Great Danes may exceed 50 kg), if one considers dog populations with a mean weight of *ca.* 16–20 kg (a typical weight for a village dog; Bekoff et al., 1981; Vanak and Gompper,

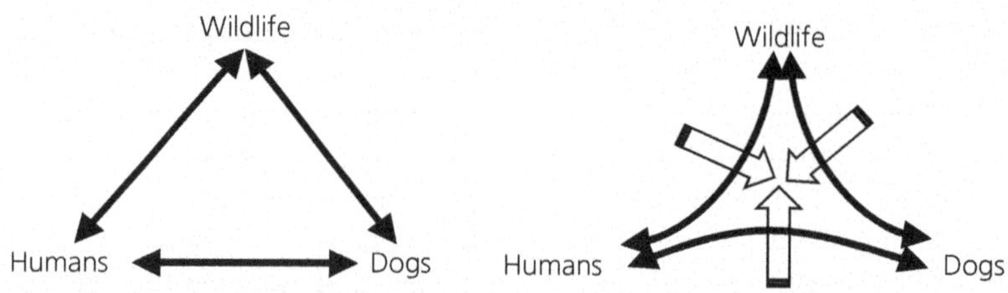

Figure 0.1 Schematic representations of the interactions of humans, dogs, and wildlife. (Left) A large body of research has focused on human–dog interactions (bottom arrow) and human–wildlife interactions (left arrow). Relatively little attention has been paid to dog–wildlife interactions (right arrow), and the simultaneous interaction of the three groups is rarely considered. (Right) When considered together we recognize that any of the three populations influence the interactions of the other two. For example, dogs may strongly influence human–wildlife interactions. Therefore the shape of the interaction for any two components of the system is altered (or, in this figure, bent).

2009), one might expect densities of dogs to exist at levels comparable to those observed for mid-sized and large wild canid species. Yet this is not the case. Whether or not a dog is owned, most dogs gain nearly all their caloric needs from human-derived foods; that is, the food that humans directly and indirectly provide. Therefore the per capita reliance of any single dog on wildlife or on the resources that wildlife simultaneously seeks is likely quite low. As a direct result of this access to human-derived foods, local dog population densities can be remarkably high (e.g. 468 dogs per km^2 in landscapes surrounding Philippine villages (Childs et al., 1998); 330 dogs per km^2 on Navajo reservation lands in Arizona, USA (Daniels and Bekoff, 1989); 195 dogs per km^2 in rural Sri Lanka (Matter et al., 2000)). Even less extreme levels (e.g. 6–21 dogs per km^2 in rural Kenya (Kitala et al., 2001); 1–16 dogs per km^2 in rural Chile (Acosta-Jammett et al., 2010)) represent population densities far higher than typically occur for similar-sized wild canids. For example, population densities of coyotes range from 0.02 to 0.44 per km^2 and densities of wolves range from 0.005 to 0.04 per km^2 (Carbone and Gittleman, 2002).

It is the combined effect of these large population sizes and the free-ranging behavior of the individuals that comprise these populations that is potentially problematic for wildlife. Because there are so many dogs, the presumably small per capita impact sums to a collective impact on other species that may be dramatic. Thus, a critical issue is how dog densities come to be so great in areas where wildlife co-occurs. That is, how do the subsidies humans provide for dogs result in high densities of dogs even in areas where such subsidies are not provided?

The concept of the subsidized predator

The abundance of an organism is limited by one of three processes: predator pressures, pathogen pressures, or resource availability. When resources are the key limiting factor, increases in resource availability will result in an augmented carrying capacity. Thus, the density of the population may increase to the new carrying capacity or it may remain below it because of the other two limiting processes. In this framework, predator populations are no different than those of any other organism, and are often limited by the availability of critical resources such as food or shelter. The concept of the subsidized predator (Soulé, 1988) derives from the idea that atypical resource inputs facilitate the maintenance of densities of a predator population that are higher than would occur in the absence of these resources. Thus a predator subsidy is a type of ecological subsidy: a flux of organisms, energy, or materials across ecosystem boundaries (Power et al., 2004).

While resource subsidies occasionally are documented in putatively natural systems (e.g. Rose and Polis, 1998), they are widespread in human-dominated systems. Ravens (*Corvus corax*) and glaucous gulls (*Larus hyperboreus*) feed on human refuse, increase in population density, and then negatively influence desert tortoise (*Gopherus agassizii*) ecology and the reproductive success of tundra nesting birds, respectively (Boarman, 2003; Liebezeit et al., 2009; Webb et al., 2004; Weiser and Powell, 2010). Predatory beetles subsidized by agricultural crops influence insect herbivore populations (Rand and Louda, 2006) and mountain lions (*Puma concolor*) subsidized by domestic cattle (*Bos taurus*) influence bighorn sheep (*Ovis canadensis*) demographics (Rominger et al., 2004). In each of these cases the subsidies provided by humans to the predatory species are accidental; the predator is simply adapting to the resource that humans produce for other purposes.

The ecology of systems with subsidized predators can be complex, as the subsidies insulate the predator against density-dependent fluctuations due to decreases in native prey availability. For example, golden eagles (*Aquila chrysaetos*) on the California Channel Islands were able to persist and to negatively influence island fox (*Urocyon littoralis*) numbers because introduced pigs (*Sus scrofa*) provided a readily available food source. Prior to pig introductions, golden eagles were unable to persist on the islands as there simply was not enough prey biomass (including island foxes) to support eagles. Once eagles could persist on the island, however, they directly drove declines in foxes, which indirectly triggered reorganization of the island food webs (Roemer et al., 2002, 2009).

In systems more directly linked to human-derived materials (i.e., refuse dumps, crops and livestock, buildings), such subsidies may also immunize predators against some density-independent limiting factors, such as extreme weather events, that might normally reduce population sizes because of lack of access to food or refugia.

The importance of such subsidies would be spatially limited to the site of the subsidy if not for the fact that predator populations may traverse habitat edges and move away from the location of the subsidy, thereby increasing their abundance in areas where subsidies are limited and increasing their impact on resident prey species in adjacent natural habitats (spillover predation; Rand et al., 2006). Prey species in recipient habitats will be especially vulnerable if their reproductive rates are low, if attack rates are high, if the distance that predators move is great, and if predator mortality rates in the natural habitats are low (Holt and Hochberg, 2001; Rand et al., 2006). Similarly, if the subsidized predator also acts as a prey species, the result is what might be termed a 'secondary predator subsidy,' which can indirectly enhance the density and role of the secondary predator species in the natural habitat, as well as create spillback effects in the anthropogenic site (e.g., Butler et al., Chapter 5).

Furthermore, subsidies that enhance predator densities also influence intraspecific contact rates. Both the subsidies themselves, and the altered contact rates, can mediate changes in the local parasite community (Monello and Gompper, 2010, 2011; Wright and Gompper, 2005) and increase opportunity for transmission of these parasites to sympatric species (Acosta-Jamett et al., 2011; Knobel et al., Chapter 6). Such increases in intraspecific contact rates as a result of the provision of subsidies are not unique to predators. However, subsidized predators are likely to travel greater distances than non-predatory species and travel further into natural habitats.

A fundamental difference between the anthropogenic subsidies provided to native species, such as golden eagles and ravens, and subsidies provided to domestic carnivores (both dogs and cats, *Felis catus*) is that the latter are often targeted. That is, humans provide food, shelter, and health care with the specific goal of benefiting these species (Gompper, Chapter 1). Furthermore, while the subsidies to native predatory taxa are spatially limited, the spatial distribution of subsidies for dogs mirrors the spatial distribution of people and is therefore a global issue. While populations of dogs may vary in the level of subsidies they receive, the end results are nonetheless enhanced densities of dogs wherever people are found.

If the extent and importance of dog–wildlife interactions is a function of the density of dogs, then the subsidies provided to dogs should be considered the ultimate mediator of such interactions. What are those potential interactions and how might subsidies play a role? In a very simplistic sense, we can assess dogs as potential predators on wildlife, as potential prey for native predators, as potential competitors for resources (with other carnivorous taxa) and for mates (with sympatric *Canis* populations), and as potential reservoirs or vectors of pathogens that might influence wildlife. Each of these themes is developed in depth in the subsequent chapters of this book. The importance of each of these potential interactions is a function of the local population density of dogs (in turn, a function of the availability of subsidies) and how dogs are managed (i.e., the extent to which dog movement into habitats that also contain wildlife is permitted).

In addition, because dogs live where people live, and because dogs can influence what wildlife species are found in an area and how those species behave, dogs influence human–wildlife interactions (Figure 0.1). The extent of this influence is likely a function of the density of dogs in an area and thus the subsidies that underpin that density. For example, through secondary predator subsidies, high densities of dogs may attract and support locally high densities of still larger predators, such as wolves or leopards (*Panthera pardus*), which may broaden human–wildlife conflicts. Similarly, high dog densities beget high reservoir potential for pathogens that can also influence other species of wildlife. If that wildlife in turn facilitates the transmission of the pathogen back into susceptible domestic animals (spillback to dogs or other species) or even into people, wildlife may receive the brunt of the blame without recognition that the underlying problem is facilitated by high dog densities and the subsidies that support these dog populations.

The structure of this book

This book was conceived to explore dog–wildlife interactions, to better understand how subsidies influence such interactions, to understand the theoretical and applied implications of the interactions for both wildlife and humans, and to offer research and management suggestions pertaining to both the subsidies and the interactions. As the editor of the volume, I solicited the insights of authors who work extensively with dogs in diverse settings and who are knowledgeable and capable of addressing the ecological issues associated with dogs. Such issues, of course, include the direct and indirect interactions that dogs have with wildlife. However, because dog population densities are also a function of human population densities, I felt it important to also attempt to task authors to, where possible, examine how dogs influence human–wildlife interactions as well as address the human dimensions of the issues raised in the book. Ultimately, how humans perceive dog–wildlife interactions will be fundamental to attempts to manage the extent to which dogs are allowed to range across landscapes and interact with wildlife.

To capture these issues, the initial chapter addresses the broad issue of who these dogs are. In Chapter 1 (Gompper) I set a framework for the book by assessing global patterns in the distribution and abundance of dogs, how we define the 'types' of dogs that might be relevant to wildlife, what we know about the demography and population ecology of dogs, as well as the seemingly simple yet ultimately tricky questions pertaining to the origin and taxonomic classification of dogs. This latter issue is particularly interesting and relevant once we recognize that in some parts of the world dogs, like people, are relatively recent arrivals, with colonization events occurring just a few hundred or a few thousand years ago. In other parts of the world, particularly in Northern Africa, Europe, Asia, and North America, dogs have a long regional history. The variable period of co-occurrence between dogs and wildlife suggests the possibility of greater impacts of dogs in regions of the planet where they are relative newcomers.

This opening chapter is followed by others examining how dogs directly or indirectly influence wildlife as predators, as prey, as competitors, and as pathogen vectors. In each chapter the authors not only review the available materials, but also delve into the more complicated nuances that such a synthesis allows. Chapters 2 (Ritchie et al.) and 3 (Vanak et al.), for example, examine how dogs interact with prey species and with other predators, respectively. With regard to prey, we know that dogs occasionally kill wildlife, but unfortunately the scale and importance of this is often unclear. Ritchie et al. examine the extent to which dogs kill wildlife, and the broader potential importance of the topic. They also examine situations where dogs have become the apex predators in the system and therefore act as trophic regulators, driving the structure and biomass of prey communities. Through such examinations we gain an understanding of whether and how dogs may have ecosystem-scale influences. Vanak et al. examine the competitive dynamic that exists between dogs and other carnivores. The chapter assesses whether, how, and why dogs directly (i.e., killing, chasing) and indirectly (competing for food) interact with native species of carnivores, and the implications of these interactions for the conservation of these native carnivore species.

While many interactions between dogs and wildlife are straightforward in principle, the role of disturbance is far more nuanced. The mere presence of a dog can affect individuals of other species in subtle ways that involve the disruption of behavioral or physiological states, despite the dog having no or limited serious interest in the putative prey. In other words, there is a perceived risk despite a lack of true risk. As noted in Chapter 4 (Weston and Stankowich) relatively little is known about the importance of such patterns of disturbance by dogs. Yet, given the abundance of dogs in many seemingly natural landscapes, disturbance events are likely to occur repeatedly and play an important role in determining the population dynamics or even the persistence of the native populations of interest.

Chapter 5 (Butler et al.) examines the role of the dog as a prey species. In landscapes where large carnivores persist and dog population densities are high, dogs can become an important food source for these predators. If the large predator population increases because of the availability of dogs (i.e., a secondary predator subsidy occurs), or if large predators are attracted to locales with high dog densities, human–carnivore conflicts may be underpinned by the subsidies humans provide to dogs. These conflicts may reflect the perception of a direct risk to humans or they may reflect the emotional and economic impacts associated with the loss of valued companion and working dogs.

Aside from direct predator–prey interactions with wildlife, the issue of pathogen transmission from free-ranging dogs to wildlife is the issue biologists and the broader public are perhaps most attuned to when considering how dogs may influence native wildlife populations. Dogs and wild carnivores share many of the same pathogens. Because dogs occur at high densities, these pathogens are generally enzootic (that is, maintained in the dog population), and the likelihood of pathogen transmission from the dog population to wild carnivore populations is increased. High profile cases of cross-species pathogen transmission in the Serengeti and Ethiopia (Randall et al., 1994; Roelke-Parker et al., 1996) and the long history of rabies transmission from dog reservoirs to humans (Steele and Fernandez, 1991) have sensitized people to the potential for dog populations to act as pathogen reservoirs. In Chapter 6 (Knobel et al.), the role of dogs as pathogen reservoirs and vectors is assessed in depth. The important pathogens are reviewed and the important ecological, epidemiological, and management issues are addressed. In particular, lessons learned from recent management actions in Africa are discussed, as the lack of vaccination programs for dogs in many parts of the work suggests that cross-species transmission events are likely to regularly occur across the globe.

Two chapters examine genetic themes, one from the perspective of wildlife and the other from the perspective of dogs. In Chapter 7 (Leonard et al.), the issue of dog hybridization with wild taxa is addressed. The presence of dogs in landscapes where other *Canis* species are present at low densities may result in mating events between dogs and wild canids. In natural settings, where population densities of native *Canis* species are high, the likelihood of hybridization is low. But as opportunities for same-species pairings decrease, and opportunities for cross-species mating increase, the likelihood of hybridization also increases. The extent to which this is a significant concern is a function of the likelihood of the introgression of genetic materials from dogs to native canids. These events can potentially place the wild canids at risk both because of the loss of genetic integrity and because of the loss of the political protections afforded to 'pure' native taxa.

A very different genetic issue is addressed in Chapter 8 (Boyko and Boyko). Given that some dog populations may have a long history of isolation resulting in unique phenotypic and genotypic attributes, are some of these populations deserving of a conservation focus much like unique populations of wildlife are deemed worthy of conservation efforts? This thorny issue has, for example, received extensive attention for a small handful of populations such as Australian dingoes. But the diverse populations of free-ranging dogs that are found elsewhere across the globe are rarely examined closely in an attempt to discern the extent of their relative isolation and uniqueness. By reviewing the genetic structure of dog populations, Boyko and Boyko shed further light on the history of dog radiations, and indirectly on dog–wildlife interactions, facilitating recognition of the basis for the distinctiveness of dog populations and perhaps even dog conservation efforts.

Following these contributions on dog–wildlife interactions and dog genetic structure are several chapters focusing on the working dog; that is, dogs that are either bred or trained for a specific purpose. There are numerous types of working dogs, including police and military dogs, herding dogs, sled dogs, search and rescue dogs, and dogs that assist disabled people. All of these dogs have the potential to interact with wildlife. However, most relevant in the context of this volume are dogs that are specifically used by people to detect or respond to wildlife. These include livestock guard dogs, which are used to mediate the potential conflicts that arise when livestock and wildlife co-occur. Chapter 9 (VerCauteren et al.) reviews this topic and also examines how these dogs might be better used so as to minimize impacts on non-target wildlife species while simultaneously maximizing the ability to mediate conservation conflicts. Chapter 10 (Woollett et al.) examines the use of detection dogs in conservation studies. Largely as a function of their discriminating sense of smell, dogs can be trained to find evidence of particular species, and as a result are increasingly used as a seemingly non-invasive and efficient tool for surveying rare taxa. Yet such studies are sometimes conducted without fully recognizing the potential limitations or even drawbacks of the tool, including the importance of the dog handler as well as the potential for dogs to negatively influence the very species the field workers are hoping to detect and protect. Finally, Chapter 11 (Koster and Noss) focuses on hunting dogs. Dogs have been used across the globe for millennia as a tool for increasing the success of hunters. Yet our understanding of the ecology of hunting dogs is, at best, limited.

In Chapter 12 (Miller et al.) the focus is on the human dimension of dog–wildlife interactions. That is, to what extent do humans view dogs as interactors with wildlife, do humans see such interactions as a concern, and to what extent do humans see these interactions as deriving from the very resources that humans openly and incidentally provide for dogs? Furthermore, do humans recognize that the dogs that surround them influence how people themselves view wildlife? It is striking how little work has been conducted on these questions. Yet, addressing the topic is fundamentally important for initiating any informed, appropriate, and feasible management strategies that might be deemed necessary to minimize the impacts that dogs have on wildlife.

In each chapter the authors make an effort to identify potential approaches and the practicalities of solving identified problems or, where dogs are discussed in a beneficial context, enhancing their value. Ultimately, the goal is to help readers identify and address important practical issues as well as to discern topics that might be important, but for which research is lacking. The hope is that such efforts, and indeed this book as a whole, will aid those who need to understand and manage the interactions of dogs and wildlife and will set the stage for future studies of these interactions.

References

Acosta-Jamett, G., Cleaveland, S., Cunninghamm, A.A., and Bronsvoort, B.M. (2010). Demography of domestic dogs in rural and urban areas of the Coquimbo region of Chile and implications for disease transmission. *Preventive Veterinary Medicine*, **94**, 272–81.

Acosta-Jamett, G., Chalmers, W.S., Cunningham, A.A., Cleaveland, S., Handel, I.G., and Bronsvoort, B.M. (2011). Urban domestic dog populations as a source of canine distemper virus for wild carnivores in the Coquimbo region of Chile. *Veterinary Microbiology*, **152**, 247–57.

Bekoff, M., Diamond, J., and Mitton, J.B. (1981). Life-history patterns and sociality in canids: body size, reproduction, and behavior. *Oecologia*, **50**, 386–90.

Boarman, W.I. (2003). Managing a subsidized predator population: reducing common raven predation on desert tortoises. *Environmental Management*, **32**, 205–17.

Butler, J.R.A., du Toit, J.T., and Bingham, J. (2004). Free-ranging domestic dogs as predators and prey in rural Zimbabwe: threats of competition and disease to large carnivores. *Biological Conservation*, **115**, 369–78.

Carbone, C. and Gittleman, J.L. (2002). A common rule for the scaling of carnivore density. *Science*, **295**, 2273–6.

Childs, J.E., Robinson, L.E., Sadek, R., Madden, A., Miranda, M.E., and Miranda, N.L. (1998). Density estimates of rural dog populations and an assessment of marking methods during a rabies vaccination campaign in the Philippines. *Preventive Veterinary Medicine*, **33**, 207–18.

Daniels, T.J. and Bekoff, M. (1989). Population and social biology of free-ranging dogs, *Canis familiaris*. *Journal of Mammalogy*, **70**, 754–62.

Fritts, S.H. and Paul, W.J. (1989). Interactions of wolves and dogs in Minnesota. *Wildlife Society Bulletin*, **17**, 121–3.

Harris, S. (1981). An estimation of the number of foxes (*Vulpes vulpes*) in the city of Bristol, and some possible factors affecting their distribution. *Journal of Applied Ecology*, **18**, 455–65.

Holt, R.D. and Hochberg, M.E. (2001). Indirect interactions, community modules and biological control: a theoretical perspective. In E. Wajnberg, J.K. Scott, and P.C. Quimby, Eds, *Evaluating Indirect Ecological Effects of Biological Control*. CAB International, Wallingford, pp. 13–37.

Kitala, P., McDermott, J., Kyule, M., Gathuma, J., Perry, B., and Wandeler, A. (2001). Dog ecology and demography information to support the planning of rabies control in Machakos District, Kenya. *Acta Tropical*, **78**, 217–30.

Knobel, D.L., Laurenson, M.K., Kazwala, R.R., and Cleaveland, S. (2008). Development of an item scale to assess attitudes towards domestic dogs in the United Republic of Tanzania. *Anthrozoös*, **21**, 285–95.

Lacerda, A.C.R., Tomas, W.M., and Marinho-Filho, J. (2009). Domestic dogs as an edge effect in the Brasilia National Park, Brazil: interactions with native mammals. *Animal Conservation*, **12**, 477–87.

Letnic, M., Ritchie, E.G., and Dickman, C.R. (2012). Top predators as biodiversity regulators: the dingo *Canis lupus dingo* as a case study. *Biological Reviews*, **87**, 390–419.

Liebezeit, J.R., Kendall, S.J., Brown, S., Johnson, C.B., Martin, P., Mcdonald, T.L., Payer, D.C., Rea, C.L., Streever, B., Wildman, A.M., and Zack, S. (2009). Influence of human development and predators on nest survival of tundra birds, Arctic Coastal Plain, Alaska. *Ecological Applications*, **19**, 1628–44.

Matter, H.C., Wandeler, A.I., Neuenschwander, B.E., Harischandrad, L.P.A., and Meslin, F.X. (2000). Study of the dog population and the rabies control activities in the Mirigama area of Sri Lanka. *Acta Tropical*, **75**, 95–108.

Monello, R.J. and Gompper, M.E. (2010). Differential effects of experimental increases in sociality on ectoparasites of free-ranging raccoons. *Journal of Animal Ecology*, **79**, 602–9.

Monello, R.J. and Gompper, M.E. (2011). Effects of resource availability and social aggregation on the species richness of raccoon endoparasite infracommunities. *Oikos*, **120**, 1427–33.

OED. (2010). *Oxford English Dictionary*, 3rd edition. Online version (March 2012). Oxford University Press, Oxford <http://www.oed.com>.

Power, M.E., Vanni, M.J., Stapp, P.T., and Polis, G.A. (2004). Subsidy effects on managed ecosystems: implications for sustainable harvest, conservation and control. In G.A. Polis, M.E. Power, and G.R. Huxel, Eds, *Food Webs at the Landscape Level*. University of Chicago Press, Chicago, pp. 387–409.

Rand, T.A. and Louda, S.M. (2006). Spillover of agriculturally subsidized predators as a potential threat to native insect herbivores in fragmented landscapes. *Conservation Biology*, **20**, 1720–29.

Rand, T.A., Tylianakis, J.M., and Tscharntke, T. (2006). Spillover edge effects: the dispersal of agriculturally subsidized insect natural enemies into adjacent natural habitats. *Ecology Letters*, **9**, 603–14.

Randall, D.A., Williams, S.D., Kuzmin, I.V., Rupprecht, C.E, Tallents, L.A., Tefera, Z., Argaw, K., Shiferaw, F., Knobel, D.L., Sillero-Zubiri, C., and Laurenson, M.K. (1994). Rabies in endangered Ethiopian wolves. *Emerging Infectious Diseases*, **10**, 2214–17.

Roelke-Parker, M.E., Munson, L., Packer, C., Kock, R., Cleaveland, S., Carpenter, M., O'Brien, S.J., Pospischil, A., Hofmann-Lehmann, R., Lutz, H., Mwamengele, G.L., Mgasa, M.N., Machange, G.A., Summers, B.A., and Appel, M.J. (1996). A canine distemper virus epidemic in Serengeti lions (*Panthera leo*). *Nature*, **379**, 441–5.

Roemer, G.W., Donlan, C.J., and Courchamp, F. (2002). Golden eagles, feral pigs and insular carnivores: how exotic species turn native predators into prey. *Proceedings of the National Academy of Sciences, USA*, **99**, 791–96.

Roemer, G.W., Gompper, M.E., and Van Valkenburgh, B. (2009). The ecological role of the mammalian mesocarnivore. *Bioscience*, **59**, 165–73.

Rominger, E.M., Whitlaw, H.A., Weybright, D.L., Dunn, W.C., and Ballard, W.B. (2004). The influence of mountain lion predation on bighorn sheep translocations. *Journal of Wildlife Management*, **68**, 993–9.

Rose, M.D. and Polis, G.A. (1998). The distribution and abundance of coyotes: the effects of allochthonous food subsidies from the sea. *Ecology*, **79**, 998–1007.

Ruiz-Izaguirre, E. and Eilers, C.H.A.M. (2012). Perceptions of village dogs by villagers and tourists in the coastal region of rural Oaxaca, Mexico. *Anthrozoös*, **25**, 75–91.

Soulé, M. E. (1988). Reconstructed dynamics of rapid extinctions of chaparral-requiring birds in urban habitat islands. *Conservation Biology*, **2**, 75–92.

Steele, J.H. and Fernandez, P.J. (1991). History of rabies and global aspects. In G.M. Baer, Ed., *The Natural History of Rabies*, 2nd edn. CRC Press, Boca Raton, pp. 1–26.

Vanak, A.T. and Gompper, M.E. (2009). Dietary niche separation between sympatric free-ranging domestic dogs and Indian foxes in central India. *Journal of Mammalogy*, **90**, 1058–65.

Vanak, A.T., Thaker, M., and Gompper, M.E. (2009). Experimental examination of behavioural interactions between free-ranging wild and domestic canids. *Behavioral Ecology and Sociobiology*, **64**, 279–87.

Vest, G.G. (1870). Eulogy of the Dog. *U.S. Congress, Senate, Congressional Record*, 101st Congress, 2nd session, S4823-4824. (Written version of original oral presentation given in the court case, Burden v. Hornsby, September 23, 1870, Johnson County Court of Common Pleas, Warrensburg, Missouri, USA).

Webb, W.C., Boarman, W.I., and Rotenberry, J.T. (2004). Common raven juvenile survivorship in a human augmented landscape. *Condor*, **106**, 517–28.

Weiser, E.L. and Powell, A.N. (2010). Does garbage in the diet improve reproductive output of glaucous gulls? *Condor*, **112**, 530–8.

Wright, A.N. and Gompper, M.E. (2005). Altered parasite assemblages in raccoons in response to manipulated resource availability. *Oecologia*, **144**, 148–56.

CHAPTER 1

The dog–human–wildlife interface: assessing the scope of the problem

Matthew E.Gompper

1.1 Introduction

Dogs are ubiquitous organisms on our planet, occurring nearly everywhere that humans occur and in some places where humans are practically absent. Understanding the ecological role and impact of dogs requires an appreciation of the nature of the dog (that is, what is a dog and what is the global geographic history of the dog?) and an evaluation of numbers and distributions of dogs that occur on the landscape. By recognizing what a dog is, we gain a sense of the variability inherent in the dog populations that occur across the globe. We also come to recognize the limits to the common use of the term 'domestic dog' to define these animals. A broad assessment of the size of dog populations, as well as the basic population biology of dogs, facilitates understanding where and how dogs live, and where dog populations are most likely to interact with wildlife populations. To collectively address these issues, in this chapter I review dog evolutionary history, dog population sizes across the globe, and dog demographics. The goal is to provide a baseline for discussions regarding the interactions of dogs, humans, and wildlife.

1.1.1 What makes a dog domestic?

The dog (*Canis familiaris*; see Box 1.1 for an overview of dog taxonomic nomenclature) is an evolutionary lineage of mammal that is derived from the wolf (*C. lupus*). We term the dog a 'domestic' or 'domesticated' animal based on its association with humans and on the role that humans presumably played in the origin and rise of this lineage. Yet defining what we mean by domestication is nuanced, complex, and contentious, with the definition of such terms varying as a function of the disciplinary background of the scholar and how it maps along the "hoary nature/culture dichotomy" (Russell, 2002). While the process of domestication may initially involve taming (on the part of humans) or habituation (by the animal), neither alone results in domestication. Rather, from a biological (Russell's nature) perspective, domestication is a relationship between people and animals that results in morphological, physiological, and behavioral adaptations or changes in the animal population through human modification of breeding and isolation from the ancestral gene pool. This manipulation of reproductive success in the lineage is deliberate and is conducted because it offers humans some utilitarian value (Bökönyi, 1969; Clutton-Brock, 1994; Price, 2002; Russell, 2002).

While defining domestication based on control of reproduction and isolation of a lineage is likely important, in the context of the wolf–dog dichotomy such a definition by itself is not entirely satisfactory as it suggests that people living 10–20,000 years ago deliberately attempted to manipulate wolves prior to the very notion of animal domestication (dogs were almost certainly the earliest domestic animal; Clutton-Brock, 2012). Furthermore, the focus on reproductive control and phenotype manipulation leaves little room for understanding the mutualistic relationship that also benefits the dog. Dogs are likely the world's most common member of the mammalian order Carnivora (see Section 1.2). As a whole they have clearly benefited from their association

Free-Ranging Dogs and Wildlife Conservation. Edited by Matthew E. Gompper
© Oxford University Press 2014. Published 2014 by Oxford University Press.

with humans. Thus, an alternative biological view of animal domestication is based on a symbiotic model (O'Connor, 1997; Russell, 2002; Zeuner, 1963). Early domestication of dogs builds on wolves scavenging human refuse (and perhaps humans scavenging wolf kills?) resulting in a commensal relationship and ultimately genetic and phenotypic change. It is worth noting that the focus on control of breeding and on the symbiotic relationship need not be entirely exclusive. While the latter may predate the former, both processes currently occur, although more so for some kinds of dogs than others.

On the opposite end of the nature/culture continuum is a social definition of domestication that is based on the integration of animals into the human socioeconomic realm by converting free-living animals into property (Ducos, 1989). Once this occurs, both human and animal behaviors change and artificial selection becomes intensified. Driscoll and Macdonald (2010) note that the timing of the appearance of small dogs in the Neolithic Middle East matches the timing of the human transition from living as hunter-gatherers to living in agricultural settlements, which in turn results in drastic changes in human society, including material wealth accumulation and social inequity. Domestic animals represent material wealth (Borgerhoff Mulder et al., 2009), and as Driscoll and Macdonald speculate, dogs are likely to have become intrinsically valuable (for instance as guard animals, as agents for the removal of detritus, and as aids in hunting) and also valuable as status symbols of property and wealth that could be passed between generations.

Thus the dog can be defined as a domestic animal based on some combination of human manipulation of their reproduction, human selection on their genotypes and phenotypes, their commensal interactions with humans, and their role in the culture of humans. Because there are diverse kinds of dogs, it is important to consider multiple definitions of domestication simultaneously. On one extreme are distinct phenotypes (breeds)[1] derived from line breeding. These dogs fall squarely within both the biological and cultural definitions of domestication. On the other extreme are feral populations of dogs with a history of independence from humans. These populations may be relatively young, such as the feral dogs of the Galapagos Islands whose origin dates to the mid 1800s (Barnett, 1986; Phillips et al., 2012), or have a timeline measurable in millennia, as in the Australian dingo. The phenotypes of these dogs have been shaped by their past interactions with people, but they no longer neatly fit some of the definitions of domestication. They are, in a sense, non-domestic dogs. But this diminishing of domestication does not result in reversion to a typical wolf phenotype. They clearly remain dogs. Thus a consideration of the ecological effects of dogs must define the dog quite broadly to include the owned and closely monitored members of breeds, the feral dogs whose interactions with humans are minimal, and those dogs that occupy the vast middle ground between these extremes: the owned dogs who spend some or all of their lives beyond the boundaries prescribed by their owners, as well as the un-owned dogs that inhabit human-modified landscapes.

Finally, tameness (again a relative term) and characteristics considered to have resulted from long-term artificial selection may be generated quickly through intense natural or artificial selection on a small set of behaviors (e.g., the flight response). The effects of this selection may be independent of the canid species (Trut et al., 2009). This suggests that many of the distinct phenotypic characteristics that separate a dog and its wild progenitor may be quantitatively and qualitatively superficial. As Wayne and vonHoldt (2012) report, much of the diversity of dog breeds derives from the fixation of discrete mutations that have large effects in individual lineages, which are then crossed to various breed groupings. This transfer of mutations across the dog evolutionary tree leads to the appearance of high phenotypic diversity that in actuality reflects a small number of important genes.

1.1.2 A brief history of the dog

Three issues are perceived as central controversies in understanding dog evolution: the timing of

[1] Clutton-Brock (2012, p. 3; see also Miklósi 2007, p. 33) defines breeds as groups of animals that have been bred by humans to possess uniform characteristics that are passed down through generations and that distinguish the groups from other animals of the same species. Note that by emphasizing the human role, such a definition excludes free-ranging dingoes, New Guinea singing dogs, and other primitive or wild dog populations.

Box 1.1 The taxonomy of the dog

The taxonomy of dogs is fraught with confusion. On the one hand, dogs are derived from wolves (*Canis lupus*), and their relatively recent evolutionary origin and their ability to hybridize with wolves in some circumstances has led some researchers to consider dogs a subspecies or variety of wolf (e.g., *C. l. familiaris*; *C. l. dingo*; *C. l. hallstromii*). On the other hand, dogs are unmistakably phenotypically distinct from wolves, and even where wolves and dogs are sympatric, hybridization is relatively uncommon (Leonard et al., Chapter 7), suggesting that the Latin binomial *Canis familiaris* is appropriate. After all, if wolves and dogs were indeed distinct subspecies, then hybridization should be the norm where the populations overlap. This is not the case, indicating that there is reproductive isolation between the two populations. Thus, to suggest that dogs are a subspecies of wolf is problematic. As Coppinger and Coppinger (2001, p. 282) note, "Dogs are not wolves, no matter what you call them." Clutton-Brock (2012, p. 5) astutely writes that classifying dogs as wolves and not including domestication as an evolutionary process that has lead to the production of new species is ". . . for cultural and historical reasons rather than for biological ones. . . ."

A difficulty in accepting or rejecting the argument I make in the previous paragraph is that it ignores the rules that govern the scientific naming of animal taxa. These rules, compiled by the International Commission on Zoological Nomenclature (ICZN), regulate how names are correctly established and which names have priority in cases of conflict among names. In this context, the naming of domestic animals is particularly complicated. Many wild species share the same Latin binomial with their domestic derivative (e.g., *Oryctolagus cuniculus* for the wild and the domestic rabbit). However, 16 mammals, including the dog, have names that are distinct from those designated for their wild ancestral counterparts. For most of these the names of the domestic taxa either pre-date or are contemporary with the names of their wild congeners (Gentry et al., 1996, 2004; Groves, 1995). *Canis familiaris* and *C. lupus* were both described by Linnaeus (1758), and *C. familiaris* is the type specimen for the genus. Until a recent ruling by the ICZN (Opinion 2027) a strong argument could have been put forth that the name of the wolf was *C. familiaris lupus*.

Thankfully Opinion 2027 has clarified this issue, such that the wolf is now indeed *C. lupus* (Gentry et al., 2004). Gentry et al. (2004) have also proposed to fix the specific name of the dog as *C. familiaris*, and Clutton-Brock (2012) provides a further discussion of this issue. Presumably this could justify designation of Australian dingoes and New Guinea singing dogs as *C. f. dingo* and *C. f. hallstromii*, respectively. However, to date, Gentry et al.'s proposal has not been acted on, which is unfortunate as the taxonomic treatment of dogs as a subspecies of the wolf engenders confusion, even among those with a strong working knowledge of taxonomic nomenclature.

the split from wolves, the location of the origin of dogs, and the more recent source and age of specific breeds. The dog is derived from the wolf, with nodal dates for the origin of dogs placed between *ca*. 15,000 years before present (ybp) and 135,000 ybp. The extraordinary difference in these dates derives from the study approach (molecular dating or fossil specimens) and from assessments of what constitutes the fossil remains of a dog. That is, when are the examined remains dogs and when are they wolves? Answering such a question depends in part on one's focal definition of domestication. Early, or 'proto,' dogs may not be fully osteologically or genetically distinguishable from the ancestral wolves from which they derived, and the transition from wolf to dog may have taken millennia (Ducos, 1989; Dimitrijević and Vuković, In press).

The earliest definitively dog-like remains date to between 14,000 and 17,000 ybp (Benecke, 1987; Musil, 2000; Larson et al., 2012; Napierala and Uerpmann, 2012; Pionnier-Capitan et al., 2011; Sablin and Khlopachev, 2002). However, recent studies of Paleolithic canid skulls from Siberia, Belgium, and the Czech Republic suggest a considerably earlier origin—perhaps as early as 33,000 ybp (Germonpré et al., 2009, 2012; Ovodov et al., 2011), although these " . . . canids most probably [were] both 'proto' or incipient dogs that did not persist long enough to found enduring lineages . . . " (Ovodov et al., 2011). The quantitative and qualitative extent to which these specimens of putative Paleolithic proto-dogs are truly distinct from sympatric specimens of wolves is contentious (Crockford and Kuzmin, 2012; Germonpré et al., 2013; Morey, 2010).

Nonetheless, whether or not the Paleolithic specimens represent dogs, by about 14,000 ybp canids existed that were distinct from wolves and were associated with humans or sites used by humans. These specimens underpin dogs as the earliest known domestic organisms (Clutton-Brock, 2012) with origins in Europe, the Middle East, or elsewhere in Asia. However, it is not clear that these animals were actually domestic based on the criteria given earlier, since the deliberate role of humans in breeding these animals, in driving artificial selection, or in treating these animals as property is not evidenced. Thus they might be better thought of as proto-dogs or proto-domesticated wolves (Galibert et al., 2011; Miklósi, 2007). Such caveats also speak to the interpretative difficulties that arise when multiple nuanced definitions of domestication are available.

Older dates for the origin of dogs derive from molecular studies, and most prominently Vilà et al.'s (1997) examination of mitochondrial DNA (mtDNA) sequence data. This study concluded that mtDNA sequence divergence within dogs, and between dogs and wolves, supported a lineage split of >100,000 ybp, and perhaps as early as 135,000 ybp. Such dates are early enough to open up the possibility of dog associations with both modern humans (*Homo sapiens*) and Neanderthals (*H. neanderthalensis*). Perhaps more intriguing than the exact purported timing of the origin was the observation that several clades of dog lineages were interspersed among the examined wolves, suggesting that dogs derived from multiple wolf lineages. This later finding has been upheld in follow-up genetic studies (Savolainen et al., 2002; Verginelli et al., 2005), although Savolainen et al. suggest a more recent origin: *ca.* 15,000 or 40,000 ybp. The latter two estimates are a function of whether the analyses allowed for a single origin or multiple origins of dogs, and Savolainen et al. recommend the younger date as more reasonable, especially given its agreement with the available fossil evidence.

1.1.3 The source and spread of dogs

The locality of the rise of dogs is not definitively known; support exists for an origin in eastern Asia (Olsen and Olsen, 1977; Pang et al., 2009; Savolainen et al., 2002) or western Asia and Europe (Verginelli et al., 2005; vonHoldt et al., 2010), or with contributions from the wolf genome in each of these regions (Wayne and vonHoldt, 2012). In the New World, archeological evidence confirms the presence of dogs by 9,000–10,000 ybp (Grayson et al., 1988). MtDNA sequences isolated from ancient dog remains from the New World indicate that native American dogs, which have now been largely replaced by dogs brought by European colonists (Castroviejo-Fisher et al., 2011), originated from multiple Old World lineages of dogs that accompanied late Pleistocene humans across the Bering Strait *ca.* 12,000–14,000 ybp (Leonard et al., 2002). Thus, independent of exact dates of origin, dogs existed across Europe, Asia, the Middle East, and North America by 10,000 ybp (Figure 1.1).

Beyond the Middle East, the dog record for the rest of Africa as well as for South America is sparser and more recent. Dogs occur in northern Africa 5,000–6,500 ybp, but dogs appear in the faunal record of sub-Saharan Africa only within the past two millennia (Clutton-Brock, 1993, 2012; Larson et al., 2012; van Sittert and Swart, 2003) (Figure 1.1). In Central and South America, the earliest archeological materials date to 5,600 ybp (MacNeish and Viera, 1983; Larson et al., 2012), and Prates et al. (2010) propose that the spread of dogs in South America occurred mainly among Andean societies (especially from 3,500 ybp) and that the introduction of dogs into egalitarian hunter-gatherer societies in the Southern Cone occurred only around 1,000 ybp, when these societies increased their long-distance social contacts. In parts of Amazonian South America there is linguistic and anthropologic evidence that suggests dogs appeared only in the past several centuries (Koster, 2009). In Australia, dogs arrived *ca.* 3,500–5,000 ybp (Savolainen et al., 2004) and, unlike in most other parts of the world, returned to a state of nondependence on humans. This lack of association of Australian dingoes with humans has also resulted in considerable taxonomic contention (Box 1.2).

Finally, islands with permanent human inhabitants have dogs as well. However, in many locations the arrival time of these dog populations is quite recent (<200—1,000 years ago; e.g.: Galapagos Islands (Barnett, 1986; Phillips et al., 2012); Hawaii (Burney et al., 2001); New Zealand (Towns and Daugherty,

Figure 1.1 World map in which the approximate maximal range of gray wolves (*Canis lupus*) is shaded in gray. Circles represent regions where confidently dated remains of dogs have been described in at least one archeological site. Circles are divided into eight segments, each of which represents 1,500 ybp. Completely filled circles represent remains older than 10,500 ybp. Figure modified from Larson et al. (2012) and used with permission.

Box 1.2 Defining the dingo

Dogs arrived in Australia between 3,500 and 5,000 ybp (Savolainen et al., 2004), approximately 45,000 years after humans first colonized the continent (Hudjashov et al., 2011). Faced with a naïve native fauna, and perhaps with release from limiting pathogens, one might hypothesize that the dog population increased at the rate of its biotic potential and colonized the entire Australian mainland within several centuries (much like red foxes, *Vulpes vulpes*, which colonized virtually the entire continent within a century). The arrival of the dog may have contributed to the loss of the top Australian predators, the thylacine, *Thylacinus cynocephalus*, and the Tasmanian devil, *Sarcophilus harrisii*, and perhaps led to a larger-scale reorganization of the indigenous animal communities (Baird, 1991; Brown, 2006; Johnson and Wroe, 2003; Letnic et al., 2012; Ritchie et al., Chapter 2).

While these 'feral' dogs were occasionally tamed and kept as pets by indigenous Australians, with a few possible exceptions they were not re-domesticated or the subject of artificial selection (Smith and Litchfield, 2009). Thus the population of dogs found across Australia evolved a relatively distinct phenotype. However, the brief period of isolation of these dogs has not been nearly long enough to facilitate reproductive isolation from dogs that have arrived on the island within the past several centuries, and thus gene flow between the original dog population (which have come to be known as dingoes, although the origin of this name is recent; Ryan, 1964) and these later dogs is high and there is an extensive literature associated with the difficulties of delineating dingoes, non-dingo feral dogs, and their hybrids (e.g., Elledge et al., 2008; Glen, 2010; Jones, 2009).

The history of the Australian dingo, the quagmire associated with differentiating differing degrees of 'dingo-ness,' and the potential role of the dingo to act as the apex mammalian predator in Australia (Ritchie et al., Chapter 2) have collectively resulted in a series of contentious management quandaries (e.g., Allen et al., 2011a; Claridge and Hunt, 2008; Glen, 2010, 2012; Letnic et al., 2011). At their core, these quandaries involve the quasi-philosophical issues of whether a feral dog population that has been on the landscape for several millennia should be considered native and thus treated as a protected, or even threatened, taxon (for which hybridization with non-dingo dogs should be viewed as compromising dingo purity), or whether it is still an exotic species and should thus be treated as a pest.

continued

Box 1.2 *Continued*

Finally, it is worth noting that the term 'dingo' has been appropriated for other dog populations. Corbett (2004) has used the term to refer to dog populations in 11 other countries, from India to the Philippines. Most prominently, the term is sometimes used when referring to the New Guinea singing dog, a feral dog of the New Guinea highlands with a phenotype similar to the Australian dingo. The singing dog population is apparently down to just a few hundred animals in the wild, and the same issues of hybridization that plague identification of the historic Australian dingo population also occur on the island of New Guinea.

1994); Tasmania (Boyce, 2006)). Similarly, in locales such as Madagascar where the dog archeological record is unclear, the recent human arrival suggests a maximum age of dog colonization of <2,500 ybp. These recent dates of dog arrival in parts of South America, in southern Africa, and on many oceanic islands indicate that the history of dog–wildlife interactions in these locales is far more recent than in Northern Africa, Europe, Asia, and North America. As such the influence of dogs on wildlife might be greater in these more recently colonized regions.

Did the diversity of these early dog populations resemble the dog diversity we see today? Early dog populations might have been phenotypically distinct or uniform in local regions, but quite variable across landscapes. Coppinger and Coppinger (2001) have suggested the 'village dog' found across much of the globe today may reflect what these early dog populations looked like, with populations evolving through founder effects, genetic drift, and natural selection (rather than through intensive artificial selection) into populations that differed in shape, pelage, and behavior (Boyko and Boyko, Chapter 8). While dog keeping may have occurred, only more recently might humans have begun to practice strong artificial selection to deliberately augment particular characteristics (following Bökönyi's (1969) division of domestication into two phases: animal keeping and animal breeding). Distinct breeds indicative of artificial selection likely existed in western Asia and Egypt by 3,000–4,000 ybp, and in Europe by 2,000 ybp (Clutton-Brock, 1999; Sablin and Khlopachev, 2002). However, the extent to which extant breeds reflect such early origins is contentious; with a few notable exceptions virtually all currently recognized breeds arose within the past several centuries (Larson et al., 2012).

1.2 The dog is the most common carnivore

The size of the global dog population is rarely estimated and those estimates that are available are typically generated through unclear methodologies. Wandeler et al.'s (1993) estimate of 500 million dogs, and Coppinger and Coppinger's (2002) estimate of a global population of 400 million dogs are widely cited. However, neither source provides details on the calculation of these values. Large-scale estimates of dog abundance are typically derived from either estimates of human:dog ratios, which in turn are gained through diverse survey methods, or through estimates provided by manufacturers of pet products, veterinary organizations, and shelters or animal welfare organizations. Recently Hughes and Macdonald (2013) used a combination of such estimates, made at either continental-level or country-level scales, to extrapolate a global estimate of 700 million dogs.

A difficulty with such estimates is that they do not fully address regional variance in dog numbers that derives from differences in the likelihood of dog ownership within rural and urban landscapes, and from the presence of un-owned dogs. For example, one estimate of the number of owned dogs in the United States is 78.2 million (American Pet Products Association 2011–2012 National Pet Owners Survey). Given an estimated July 1, 2011 United States human population of 311.6 million (source: United States Census Bureau, Population Division; document NST-EST2011-01), a human:dog ratio of 3.98:1 is extrapolated. Extrapolating such metrics to a global scale (7 billion people and a 4:1 human:dog ratio equates to 1.75 billion dogs) would be flawed, as the United States survey excluded un-owned

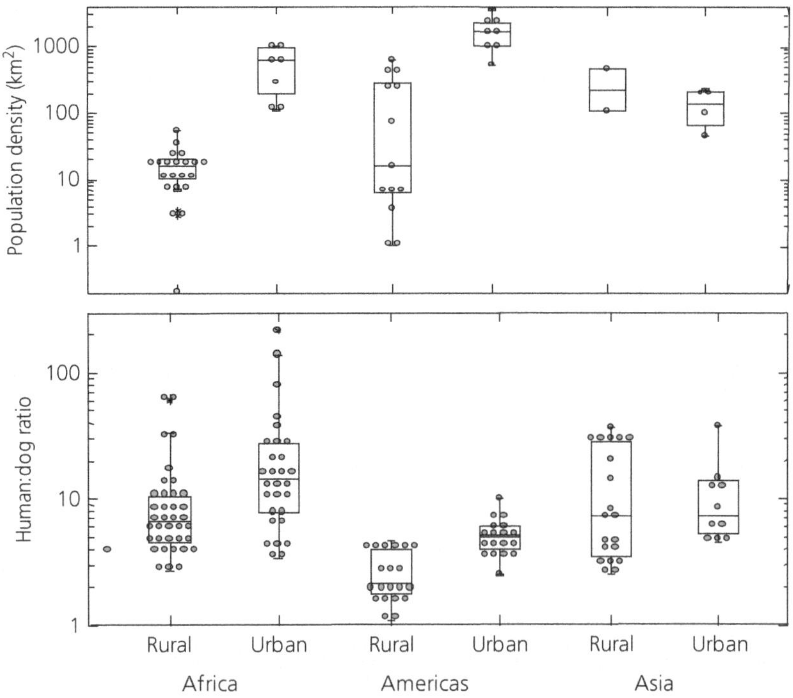

Figure 1.2 Human:dog ratios (n = 135 locations) and dog population density estimates (n = 55), subdivided by global region and by urban and rural locales. Note log scale and scarcity of density estimates from Asia. Box plot defines median and quartile ranges. Europe and Australia are excluded because of lack of applicable studies.

dogs and thereby overestimated the human:dog ratio. More importantly, extrapolating across geographically far-flung locales fails to account for regional differences in human:dog ratios. These values may differ because of societal norms and dog management strategies, as well as differences in the local carrying capacity of dogs. Since dogs are often entirely dependent on resources provided or derived from humans (Vanak and Gompper, 2009a, b), variance in human densities, waste disposal practices, and regional resource availability result in large differences in human:dog ratios and in dog population densities in urban versus rural areas (Figure 1.2). Although in some countries urban areas have lower ratios, overall the ratios are lower in rural landscapes. Ratios are also lower in the Americas compared to Asia and Africa.

By using more spatially realistic estimates of human:dog ratios we can extrapolate to 0.987 billion dogs (Box 1.3). Aside from a shockingly large estimate, which like that of Hughes and Macdonald (2013) is only a coarse number in need of further refinement, what can we take from this value and the human:dog ratios used to generate them? First, dogs are by far the planet's most abundant member of the taxonomic order Carnivora.[2] The biomass necessary to support this many animals is substantial. Coppinger and Coppinger (2001) worked with a decade-old estimate of 52 million United States dogs and coarsely estimated that because these animals subsist primarily on commercial pet food, they collectively have the caloric needs equivalent to that of approximately 26 million people. More precisely, Sallander et al. (2010) showed that the daily energetic requirements (in kilojoules/day, kJ/d)

[2] A caveat here is that many of the difficult issues associated with estimating dog populations also apply to domestic cats, *Felis catus*. In some countries cat populations may be larger than those of dogs (e.g., Perrin, 2009; FEDIAF, 2010) although many other regionally focused studies have found dogs to be more abundant than cats (Chommel et al., 1988; Kongkaew et al., 2004; Alves et al., 2005; Alie et al., 2007; Cardinal et al., 2007; Westgarth et al., 2007; Knobel et al., 2008; Durr et al., 2009; ACAC, 2010).

of dogs equals $554BW^{0.66}$. Thus a 20 kg dog requires about 4,000 kJ/d. Energetic requirements of humans vary by age, body size, and activity levels. For 18–20 year olds, a moderately active 55 kg woman requires about 10,000 kJ/d, and a moderately active 70 kg male requires about 13,000 kJ/d (FAO, 2001). Given that dogs in developed countries are fed primarily from commercially-produced foods that are similar to those of people, the agricultural production requirements to support dogs equate to about 30–40% of that required to support humans. In much of the world, however, dogs are not fed commercially prepared foods. Therefore, if even a small proportion of the dietary needs of these animals are derived directly from wildlife, or derived from materials that in the absence of dogs would be available to wildlife, then the impact on wildlife populations is likely large. Furthermore, in a disease context such large population estimates indicate a tremendous capacity for dogs to act as the reservoir for pathogens that may influence other species of wildlife (Knobel et al., Chapter 6).

Box 1.3 Extrapolating dog population sizes

Estimates of dog population sizes are often derived from human household surveys, or less commonly direct surveys of dogs, resulting in human:dog ratios that are then extrapolated over broader regions. These values can vary considerably as a function of survey methodology (for instance, were un-owned dogs included?) and where the survey was conducted (Figure 1.2). Nonetheless, they facilitate a coarse understanding of the densities and numbers of dogs that exist in particular regions and also allow extrapolation to country, continental, or global scales. In Table 1.1 human population estimates (in millions) and urban–rural proportions are derived from 2010 United Nations estimates (United Nations, 2011). For countries within which studies of human:dog ratios have been conducted, these local estimates were used to extrapolate rural and urban dog densities at a country scale. In some cases, only country-wide estimates were available. Collectively, these ratios (excluding those that were conducted country wide without consideration of rural and urban differences) were applied to countries where ratios or dog population size estimates were not available. Countries were categorized into geographical regions (following United Nations, 2011) and ratios available from countries within each region were applied to those countries from the region lacking human:dog ratio estimates. For regions in which no human:dog estimates were available, ratios were extrapolated from surrounding geographic regions.

Table 1.1 Global dog population estimates subdivided by country and geographic region. Human population estimates and urban–rural population proportions are 2010 United Nations estimates (United Nations, 2011). Dog population estimates and human:dog ratios derive from either country-wide population estimates that were used to calculate ratios, or from published and unpublished studies that estimate ratios, which were then used to extrapolate rural, urban, and total dog populations. For countries where no ratio or dog population estimates were available, mean ratio estimates for the region were used to extrapolate population sizes. Where no ratios were available for a region, mean ratios from adjacent regions and countries were used to extrapolate population sizes. Data sources are given in the last column and associated footnotes. Human and dog population estimates are given in 1000s.

Region country	Human population	% Urban	Rural ratio	Urban ratio	Derived rural ratio	Derived urban ratio	Rural dog population	Urban dog population	Total dog population	Source
Eastern Africa										
Burundi	8,383	11			6.29	22.46	1,186.8	44.1	1,230.9	
Comoros	735	28			6.29	22.46	84.2	9.8	94.0	
Djibouti	889	76			6.29	22.46	33.9	32.3	66.3	

continued

Box 1.3 *Continued*

Table 1.1 *Continued*

Region country	Human population	% Urban	Rural ratio	Urban ratio	Derived rural ratio	Derived urban ratio	Rural dog population	Urban dog population	Total dog population	Source
Eritrea	5,254	22			6.29	22.46	651.9	55.3	707.2	
Ethiopia	82,950	17	3.85	14.3			17,882.7	986.1	18,868.8	1
Kenya	40,513	22	9.02	6.7			3,503.3	1,330.3	4,833.6	2
Madagascar	20,714	30		4.4	6.29		2,306.4	1,412.3	3,718.8	3
Malawi	14,901	20	3.8	3.4			3,137.1	876.5	4,013.6	4
Mauritius	1,299	42			6.29	22.46	119.8	24.2	144.1	
Mayotte	204	50			6.29	22.46	16.2	4.5	20.8	
Mozambique	23,391	38			6.29	22.46	2,306.9	395.0	2,701.9	
Réunion	846	94			6.29	22.46	8.1	35.3	43.4	
Rwanda	10,624	19			6.29	22.46	1,368.8	89.7	1,458.6	
Somalia	9,331	37			6.29	22.46	935.1	153.4	1,088.5	
Uganda	33,425	13			6.29	22.46	4,625.6	193.1	4,818.7	
United Rep. Tanzania	44,841	26	9.36	56.5			3,545.1	206.3	3,751.5	5
Zambia	13,089	36	6.7	45			1,250.3	104.7	1,355.0	6
Zimbabwe	12,571	38	4.99	26.9			1,561.9	177.6	1,739.5	7
Mean			6.29	22.46						
Middle Africa										
Angola	19,082	59			5.75	13.18	1,360.6	854.2	2,214.8	
Cameroon	19,599	58	5.75	4.66			1,431.6	2,439.4	3,870.9	8
Central African Rep.	4,401	39			5.75	13.18	466.9	130.2	597.1	
Chad	11,227	28		21.7	5.75		1,405.8	144.9	1,550.7	9
Congo	4,043	62			5.75	13.18	267.2	190.2	457.4	
Dem. Rep. Congo	65,966	35			5.75	13.18	7,457.0	1,751.8	9,208.8	
Equatorial Guinea	700	40			5.75	13.18	73.0	21.2	94.3	
Gabon	1,505	86			5.75	13.18	36.6	98.2	134.8	
Sao Tome & Principe	165	62			5.75	13.18	10.9	7.8	18.7	
Mean			5.75	13.18						

continued

Box 1.3 *Continued*

Table 1.1 *Continued*

Region country	Human population	% Urban	Rural ratio	Urban ratio	Derived rural ratio	Derived urban ratio	Rural dog population	Urban dog population	Total dog population	Source
Northern Africa										
Algeria	35,468	66			3.84	9.83	3,140.4	2,381.4	5,521.8	
Egypt	81,121	43			3.84	9.83	12,041.4	3,548.5	15,589.9	
Libya	6,355	78			3.84	9.83	364.1	504.3	868.4	
Morocco	31,951	58			3.84	9.83	3,494.6	1,885.2	5,379.8	
Sudan	43,552	40			3.84	9.83	6,805.0	1,772.2	8,577.2	
Tunisia	10,481	67	3.84	9.83			900.7	714.4	1,615.1	10
Western Sahara	531	82			3.84	9.83	24.9	44.3	69.2	
Mean			3.84	9.83						
Southern Africa										
Botswana	2,007	61			9.95	15.2	78.7	80.5	159.2	
Lesotho	2,171	27			9.95	15.2	159.3	38.6	197.8	
Namibia	2,283	38			9.95	15.2	142.3	57.1	199.3	
South Africa	50,133	62	9.95	15.2			1,914.6	2,044.9	3,959.5	11
Swaziland	1,186	21			9.95	15.2	94.2	16.4	110.6	
Mean			9.95	15.2						
Western Africa										
Benin	8,850	42			32.9	24.65	156.0	150.8	306.8	
Burkina Faso	16,469	26			32.9	24.65	370.4	173.7	544.1	
Cape Verde	496	61			32.9	24.65	5.9	12.3	18.2	
Côte d'Ivoire	19,738	51			32.9	24.65	294.0	408.4	702.3	
Gambia	1,728	58			32.9	24.65	22.1	40.7	62.7	
Ghana	24,392	51			32.9	24.65	363.3	504.7	867.9	
Guinea	9,982	35			32.9	24.65	197.2	141.7	338.9	
Guinea-Bissau	1,515	30			32.9	24.65	32.2	18.4	50.7	
Liberia	3,994	48			32.9	24.65	63.1	77.8	140.9	
Mali	15,370	36			32.9	24.65	299.0	224.5	523.5	
Mauritania	3,460	41			32.9	24.65	62.0	57.5	119.6	
Niger	15,512	17			32.9	24.65	391.3	107.0	498.3	
Nigeria	158,423	50	32.9	35.3			2,407.6	2,244.0	4,651.6	12
Senegal	12,434	42			32.9	24.65	219.2	211.9	431.1	

continued

Box 1.3 *Continued*

Table 1.1 *Continued*

Region country	Human population	% Urban	Rural ratio	Urban ratio	Derived rural ratio	Derived urban ratio	Rural dog population	Urban dog population	Total dog population	Source
Sierra Leone	5,868	38		14	32.9		110.6	159.3	269.9	13
Togo	6,028	43			32.9	24.65	104.4	105.2	209.6	
Mean			32.9	24.65						
Central Asia										
Kazakhstan	16,026	59			14.2	8.1	462.7	1,167.3	1,630.0	
Kyrgyzstan	5,334	35			14.2	8.1	244.2	230.5	474.6	
Tajikistan	6,879	26			14.2	8.1	358.5	220.8	579.3	
Turkmenistan	5,042	50			14.2	8.1	177.5	311.2	488.8	
Uzbekistan	27,445	36			14.2	8.1	1,237.0	1,219.8	2,456.7	
Derived			14.2	8.1						14
East Asia										
China	1,341,335	47		37.5	2.9		245,140.5	16,811.4	261,951.9	15
China, Hong Kong	7,053	100			2.9	25.7	0.0	274.2	274.2	
China, Macao	544	100			2.9	25.7	0.0	21.1	21.1	
Dem. Peo. Rep. Korea	24,346	60			2.9	25.7	3,358.1	567.8	3,925.9	
Japan	126,536	67			10.3	10.3	4,054.1	8,231.0	12,285.0	
Mongolia	2,756	62	2.9			25.7	361.1	66.4	427.6	16
Rep. Korea	48,184	83			2.9	25.7	2,824.6	1,554.6	4,379.2	
Other areas (Taiwan)	23,216	n/a		13.95	2.9				1,664.2	17
Mean			2.9	25.7						
South-eastern Asia										
Brunei Darussalam	399	76			3.9	6.9	24.6	43.9	68.5	
Cambodia	14,138	20	3.1	4.46			3,648.5	634.0	4,282.5	18
Indonesia	239,871	44	2.5			6.9	53,731.1	15,296.1	69,027.2	19
Lao Peo. Dem. Rep	6,201	33			3.9	6.9	1,065.3	296.6	1,361.9	
Malaysia	28,401	72			3.9	6.9	2,039.0	2,963.6	5,002.6	
Myanmar	47,963	34			3.9	6.9	8,116.8	2,363.4	10,480.2	

continued

Box 1.3 *Continued*

Table 1.1 *Continued*

Region country	Human population	% Urban	Rural ratio	Urban ratio	Derived rural ratio	Derived urban ratio	Rural dog population	Urban dog population	Total dog population	Source
Philippines	93,261	49	5.35			6.9	8,890.3	6,622.9	15,513.2	20
Singapore	5,086	100			3.9	6.9	0.0	737.1	737.1	
Thailand	69,122	34	4.6	9.4			9,917.5	2,500.2	12,417.7	21
Timor-Leste	1,124	28			3.9	6.9	207.5	45.6	253.1	
Viet Nam	87,848	30			3.9	6.9	15,767.6	3,819.5	19,587.1	
Mean			3.9	6.9						
Southern Asia										
Afghanistan	31,412	23			19.97	7.85	1,211.2	920.4	2,131.5	
Bangladesh	148,692	28			19.97	7.85	5,361.0	5,303.7	10,664.6	
Bhutan	726	35			19.97	7.85	23.6	32.4	56.0	
India	1,224,614	30	30.67	12			27,950.1	30,615.4	58,565.5	22
Iran	73,974	71	20.5			7.85	1,046.5	6,690.6	7,737.1	23
Maldives	316	40			19.97	7.85	9.5	16.1	25.6	
Nepal	29,959	19		5.29	19.97		1,215.2	1,076.0	2,291.2	24
Pakistan	173,593	36			19.97	7.85	5,563.3	7,961.0	13,524.3	
Sri Lanka	20,860	14	8.73	6.25			2,054.9	467.3	2,522.2	25
Mean			19.97	7.85						
Western Asia										
Armenia	3,092	64			14.2	8.1	78.4	244.3	322.7	
Azerbaijan	9,188	52			14.2	8.1	310.6	589.8	900.4	
Bahrain	1,262	89			14.2	8.1	9.8	138.7	148.4	
Cyprus	1,104	70			14.2	8.1	23.3	95.4	118.7	
Georgia	4,352	53			14.2	8.1	144.0	284.8	428.8	
Iraq	31,672	66			14.2	8.1	758.3	2,580.7	3,339.0	
Israel	7,418	92			14.2	8.1	41.8	842.5	884.3	
Jordan	6,187	79			14.2	8.1	91.5	603.4	694.9	
Kuwait	2,737	98			14.2	8.1	3.9	331.1	335.0	
Lebanon	4,228	87			14.2	8.1	38.7	454.1	492.8	
Occ. Palestinian Terr.	4,039	74			14.2	8.1	74.0	369.0	442.9	
Oman	2,782	73			14.2	8.1	52.9	250.7	303.6	
Qatar	1,759	96			14.2	8.1	5.0	208.5	213.4	
Saudi Arabia	27,448	82			14.2	8.1	347.9	2,778.7	3,126.6	
Syria	20,411	56			14.2	8.1	632.5	1,411.1	2,043.6	

continued

Box 1.3 *Continued*

Table 1.1 *Continued*

Region country	Human population	% Urban	Rural ratio	Urban ratio	Derived rural ratio	Derived urban ratio	Rural dog population	Urban dog population	Total dog population	Source
Turkey	72,752	70			14.2	8.1	1,537.0	6,287.2	7,824.2	
United Arab Emirates	7,512	84			14.2	8.1	84.6	779.0	863.7	
Yemen	24,053	32			14.2	8.1	1,151.8	950.2	2,102.1	
Derived			14.2	8.1						26
Eastern Europe										
Belarus	9,595	75			8.4	8.4	285.6	856.7	1,142.3	
Bulgaria	7,494	71			8.4	8.4	258.7	633.4	892.1	
Czech Republic	10,493	74			3.33	3.33	819.5	2,332.5	3,152.0	27
Hungary	9,984	68			3.50	3.50	913.9	1,942.1	2,856.0	27
Poland	38,277	61			5.24	5.24	2,851.3	4,459.7	7,311.0	27
Republic of Moldova	3,573	47			8.4	8.4	225.4	199.9	425.4	
Romania	21,486	57			5.16	5.16	1,791.4	2,374.6	4,166.0	27
Russian Federation	142,958	73			11.42	11.42	3,380.4	9,139.6	12,520.0	27
Slovakia	5,462	55			21.85	21.85	112.5	137.5	250.0	27
Ukraine	45,448	69			8.4	8.4	1,677.2	3,733.2	5,410.5	
Mean of Derived					8.41	8.41				28
Northern Europe										
Channel Islands	153	31			7.9	7.9	13.4	6.0	19.4	
Denmark	5,550	87			9.57	9.57	75.4	504.6	580.0	27
Estonia	1,341	69			7.68	7.68	54.1	120.5	174.6	27
Finland	5,365	85			8.24	8.24	97.7	553.4	651.0	27
Iceland	320	93			7.9	7.9	2.8	37.7	40.5	
Ireland	4,470	62			4.3	4.3	395.0	644.5	1,039.5	29
Latvia	2,252	68			8.35	8.35	86.3	183.5	269.8	27
Lithuania	3,324	67			4.45	4.45	246.3	500.0	746.3	27
Norway	4,883	79			10.80	10.80	94.9	357.1	452.0	27
Sweden	9,380	85			10.93	10.93	128.7	729.5	858.2	30
United Kingdom	62,036	80			6.55	6.55	1,894.2	7,576.9	9,471.1	31
Mean of Derived					7.87	7.87				32

continued

Box 1.3 *Continued*

Table 1.1 *Continued*

Region country	Human population	% Urban	Rural ratio	Urban ratio	Derived rural ratio	Derived urban ratio	Rural dog population	Urban dog population	Total dog population	Source
Southern Europe										
Albania	3,204	52			9	9	170.9	185.1	356.0	
Bosnia & Herzegovina	3,760	49			9	9	213.1	204.7	417.8	
Croatia	4,403	58			9	9	205.5	283.7	489.2	
Greece	11,359	61			17.08	17.08	259.4	405.7	665.0	27
Italy	60,551	68			4.3	4.3	4,506.1	9,575.5	14,081.6	33
Malta	417	95			9	9	2.3	44.0	46.3	
Montenegro	631	61			9	9	27.3	42.8	70.1	
Portugal	10,676	61			5.50	5.50	756.6	1,183.4	1,940.0	27
Serbia	9,856	56			9	9	481.8	613.3	1,095.1	
Slovenia	2,030	50			8.46	8.46	120.0	120.0	240.0	27
Spain	46,077	77			9.76	9.76	1,085.6	3,634.4	4,720.0	27
TFYR Macedonia	2,061	59			9	9	93.9	135.1	229.0	
Mean of Derived					9.02	9.02				34
Western Europe										
Austria	8,394	68			13.72	13.72	195.8	416.2	612.0	27
Belgium	10,712	97			8.05	8.05	39.9	1,290.8	1,330.7	27
France	62,787	85			8.27	8.27	1,139.3	6,455.8	7,595.0	27
Germany	82,302	74			15.53	15.53	1,378.0	3,922.0	5,300.0	27
Luxembourg	507	85			12.3	12.3	6.2	35.0	41.2	
Netherlands	16,613	83			11.13	11.13	253.8	1,239.2	1,493.0	27
Switzerland	7,664	74			17.22	17.22	115.7	329.3	445.0	27
Mean of Derived					12.32	12.32				35
Caribbean										
Aruba	107	47			3.2	3.9	17.7	12.9	30.6	
Bahamas	343	84		2.5	3.2		17.2	115.2	132.4	36
Barbados	273	44			3.2	3.9	47.8	30.8	78.6	
Cuba	11,258	75			3.2	3.9	879.5	2,165.0	3,044.5	
Dominican Rep.	9,927	69			3.2	3.9	961.7	1,756.3	2,718.0	

continued

Box 1.3 *Continued*

Table 1.1 *Continued*

Region country	Human population	% Urban	Rural ratio	Urban ratio	Derived rural ratio	Derived urban ratio	Rural dog population	Urban dog population	Total dog population	Source
Grenada	104	39			3.2	3.9	19.8	10.4	30.2	
Guadeloupe	461	98			3.2	3.9	2.9	115.8	118.7	
Haiti	9,993	52			3.2	3.9	1,499.0	1,332.4	2,831.4	
Jamaica	2,741	52			3.2	3.9	411.2	365.5	776.6	
Martinique	406	89			3.2	3.9	14.0	92.7	106.6	
Netherlands Antilles	201	93			3.2	3.9	4.4	47.9	52.3	
Puerto Rico	3,749	99			3.2	3.9	11.7	951.7	963.4	
Saint Lucia	174	28			3.2	3.9	39.2	12.5	51.6	
St. Vincent & the Grenadines	109	49			3.2	3.9	17.4	13.7	31.1	
Trinidad and Tobago	1,341	14			3.2	3.9	360.4	48.1	408.5	
US Virgin Islands	109	95			3.2	3.9	1.7	26.6	28.3	
Mean				3.9						37
Derived			3.2							38
Central America										
Belize	312	52			3.2	4.74	46.8	34.2	81.0	
Costa Rica	4,659	64			3.2	4.74	524.1	629.1	1,153.2	
El Salvador	6,193	64			3.2	4.74	696.7	836.2	1,532.9	
Guatemala	14,389	49			3.2	4.74	2,293.2	1,487.5	3,780.7	
Honduras	7,601	52			3.2	4.74	1,140.2	833.9	1,974.0	
Mexico	113,423	78	3.2	4.74			7,797.8	18,664.5	26,462.4	39
Nicaragua	5,788	57			3.2	4.74	777.8	696.0	1,473.8	
Panama	3,517	75			3.2	4.74	274.8	556.5	831.3	
Mean			3.2	4.74						
South America										
Argentina	40,412	92	1.69			6.34	1,918.7	5,863.5	7,782.1	40
Bolivia	9,930	67	1.5	4.6			2,184.6	1,446.3	3,630.9	41
Brazil	194,946	87	4.7	4.74			5,392.1	35,751.1	41,143.2	42
Chile	17,114	89	1.7	5.16			1,107.4	2,951.8	4,059.2	43
Colombia	46,295	75			2.50	6.34	4,635.1	5,475.8	10,110.9	

continued

Box 1.3 *Continued*

Table 1.1 *Continued*

Region country	Human population	% Urban	Rural ratio	Urban ratio	Derived rural ratio	Derived urban ratio	Rural dog population	Urban dog population	Total dog population	Source
Ecuador	14,465	67		7.2	2.50		1,911.7	1,346.0	3,257.7	44
French Guiana	231	76			2.50	6.34	22.2	27.7	49.9	
Guyana	754	29			2.50	6.34	214.4	34.5	248.9	
Paraguay	6,455	61			2.50	6.34	1,008.2	621.0	1,629.2	
Peru	29,077	77	2.9	10			2,306.1	2,238.9	4,545.0	45
Suriname	525	69			2.50	6.34	65.2	57.1	122.3	
Uruguay	3,369	92			2.50	6.34	107.9	488.8	596.8	
Venezuela	28,980	93			2.50	6.34	812.4	4,250.5	5,062.9	
Mean			2.50	6.34						
Northern America										
Canada	34,017	81			5.60	5.60	1,153.4	4,917.3	6,070.8	46
United States	310,384	82			3.97	3.97	14,076.0	64,124.0	78,200.0	47
Australia/New Zealand										
Australia	22,268	89				6.4	6.4	382.7	3,096.6	48
New Zealand	4,368	86				6.4	6.4	95.6	587.0	
Derived for New Zealand						6.4	6.4			49
Melanesia										
Fiji	861	52			5.14	6.67	80.3	67.2	147.5	
New Caledonia	251	57			5.14	6.67	21.0	21.5	42.4	
Papua New Guinea	6,858	13			5.14	6.67	1,159.9	133.8	1,293.7	
Solomon Islands	538	19			5.14	6.67	84.7	15.3	100.1	
Vanuatu	240	26			5.14	6.67	34.5	9.4	43.9	
Derived			5.14	6.67						50

continued

Box 1.3 *Continued*

Table 1.1 *Continued*

Region country	Human population	% Urban	Rural ratio	Urban ratio	Derived rural ratio	Derived urban ratio	Rural dog population	Urban dog population	Total dog population	Source
Micronesia										
Guam	180	93			5.144	6.67	2.4	25.1	27.6	
Micronesia	111	23			5.144	6.67	16.6	3.8	20.4	
Derived			5.14	6.67						50
Polynesia										
French Polynesia	271	51			5.14	6.67	25.8	20.7	46.6	
Samoa	183	20			5.14	6.67	28.5	5.5	34.0	
Tonga	104	23			5.14	6.67	15.6	3.6	19.2	
Derived			5.14	6.67						50
Total							**585,916.1**	**399,333.2**	**986,913.5**	

Sources: (1) Shiferaw et al., 1998; Atickem, 2003; Atickem et al., 2009; (2) Kitala et al., 2001; Woodroffe and Donnelly, 2011; (3) Ratsitorahina et al., 2009; (4) Edelsten, 1995; (5) Cleaveland et al., 2001; Knobel et al., 2008; Kaare et al., 2009; (6) DeBalogh et al., 1993; (7) Hill, 1985; Brooks, 1990; Butler and Bingham, 2000; (8) Awah-Ndukum, 2003; (9) Durr et al., 2009; Mindekem et al., 2005; Kayali et al., 2003; (10) Matter et al., 1998; Yousseff et al., 1998; Seghaier et al., 1999; (11) Rautenbach et al., 1991; McCrindle et al., 1999; Bishop, 2001; (12) Ezeokoli et al., 1984; Faleke, 2003; Oboegbulem and Nwakonobi, 1989; Aiyedun and Olugasa, 2012; Idachaba, 2009; El-Yuguda, et al. 2007; Ezeolkoli and Umoh, 1987; (13) Suluku et al., 2012; (14) Derived from means of Southern Asia and Eastern Europe; (15) Wang et al., 2006 as cited in Hu et al., 2009; (16) Buuveibaatar et al., 2009; Young, pers. comm.; (17) Tung et al., 2010; (18) Ly et al., 2009; Lunney et al., 2012; (19) Bingham, 2001; (20) Robinson et al., 1996; Childs et al., 1998; Estrada et al., 2001; (21) Sagarasaeranee et al., 2001; Singhchai, 2001; Kongkaew et al., 2004; (22) Sudarshan et al., 2001, 2006; Belsare pers. comm.; (23) Gavgani et al., 2002; (24) Joshi and Bogel, 1988; Bogel and Joshi, 1990; Kato et al., 2003; (25) Matter et al., 2000; Wandeler et al., 1988; (26) Derived from means of Southern Asia and Eastern Europe; (27) Ratios back-calculated from FEDIAF, 2010; (28) Mean of Eastern Europe back-calculated values applied to Belarus, Bulgaria, Moldova, and Ukraine; (29) Ratios back-calculated from Downes et al., 2009; (30) ratios back-calculated from Egenvall, 1999; (31) Ratios back-calculated from Asher et al., 2011; (32) Mean of Northern Europe back-calculated values applied to Channel Islands, Iceland; (33) Ratios back-calculated from Slater, 2008; (34) Mean of Southern Europe back-calculated values applied to Albania, Bosnia, Croatia, Malta, Montenegro, Serbia, and Macedonia; (35) Mean of Western Europe back-calculated values applied to Luxembourg; (36) Fielding et al., 2005; (37) includes 5.36 in urban Dominica; Alie et al., 2007; (38) Derived from the means of Central America, South America, and the back-calculated value from the United States; (39) Fishbein et al., 1992; Orihuela and Solano, 1995; Flores-Ibarra and Estrella-Valenzuela, 2004; Ortega-Pacheco et al., 2007; Romero-Lopez et al., 2008; Ruiz-Izaguirre and Eilers, 2012; (40) Cardinal et al., 2007; Gürtler et al., 2007; (41) Fiorello et al., 2006; Suzuki et al., 2008; (42) Dias et al., 2004; Moreira et al., 2004; Alves et al., 2005; Andrade et al., 2008; Nunes et al., 2008; Dantas-Torres et al., 2010; (43) Morales et al., 2009; Acosta-Jamett et al., 2010; Silva-Rodriguez and Sieving, 2012; (44) Beran and Frith, 1988; (45) Chomel et al., 1988; Reithinger et al., 2003; (46) Perrin, 2009; (47) Ratios back-calculated from APPA, 2012; (48) Ratios back-calculated from ACAC, 2010; (49) Derived from Australia and applied to New Zealand; (50) Derived from the means of South-Eastern Asia and Australia and applied to Melanesia, Micronesia, and Polynesia.

Collectively this exercise results in an estimate of 0.987 billion dogs. Several important caveats must be noted. First, the vast majority of human:dog ratios and dog population size estimates used in Table 1.1 are based on household surveys and so focus on owned dogs. Estimates of numbers of un-owned dogs are rarely available; in some regions the un-owned are proportionately uncommon, but in other regions they make up a significant portion of the population. For instance, Ratsitorahina et al. (2009) examined the dog population of six municipal subdivisions in Antananarivo, Madagascar, and calculated that un-owned dogs represented a mean of 11.3% (range = 3.3–16.9%) of the total dog population. If approximately 10% of the global dog population is un-owned, the estimate in Table 1.1 may be biased downwards by 100 million animals.

Second, robust assessments of human:dog rations or of dog population sizes are severely lacking in many parts of the world, and thus estimates of dog population sizes are derived from regions that may not be comparable. Furthermore, the methodologies used to generate many country-wide estimates are often unavailable. Seemingly slight changes to the ratios applied in many regions could significantly alter global population estimates, especially for populous countries.

Second, the global population is not dominated by urban dogs. Rather, the summed rural and urban dog population estimates suggest that approximately 60% of the global dog population is found in rural settings (human populations are approximately 51% rural based on United Nations data). The per capita abundance of rural dogs is particularly striking in some regions. Reflecting much of Latin America, for example, Acosta-Jamet et al. (2010) report that human:dog ratios decrease from about 5–6:1 to 1–2:1 as one moves along an urban to rural gradient. Similar patterns have been observed in other studies (e.g., Knobel et al., 2008; Vos and Turan, 1996), although this is not a universal pattern (Table 1.1).

Are dog populations growing? If human:dog ratios are deterministically related to human population growth, then dog populations are expanding wherever human populations are growing. However, few countries have a national system for monitoring dog population size, and as such the data on population growth rates are typically collected either by trade and veterinary organizations or by individual researchers. For individual researchers, data on the demographics of dogs is usually collected with a primary goal of better understanding pathogen epidemiology. Such data can then be used to characterize the population growth of dogs. For example, Acosta-Jamett et al. (2010) calculated annual growth rates of 20, 19, and 9% for owned dog populations inhabiting Chilean cities, towns, and rural settings, respectively. Such estimates, of course, represent a single point in time. Rarely do multiple estimates exist for a region that would facilitate an understanding of temporal variation in growth rates. In rural Zimbabwe, Butler and Bingham (2000) estimated the annual growth rate of dogs on communal rural lands to be 6.5%, more than double the human growth rate during that period, and an increase of 1–2 percentage points over estimates generated by past researchers.

Alternatively, the regular collection of survey data by veterinary organizations or pet product industry groups may provide population estimates, although such estimates are likely only applicable to owned populations. The work of such groups depends on having reliable and current data on dog populations (Clancy and Rowan, 2003) and can ultimately be used for trend analysis. For instance, in the United States, such data are collected primarily by the American Veterinary Medical Association (AVMA) and the American Pet Products Association (APPA). Similar organizations publish data for other countries (e.g., the Australian Companion Animal Council Inc., the European Pet Food Industry Federation, the Pet Food Manufacturers' Association). The APPA has conducted surveys biennially since 1988, and the AVMA has published survey results every five years since 1983, and these data can be used to assess longer term trends in population growth (Table 1.2) although the data may exclude un-owned dogs. Furthermore, the accuracy of the methodologies used to estimate densities (Clancy and Rowan, 2003), and sometimes the methodologies themselves, are not always clear. Nonetheless, such data reveal interesting patterns. The United States data, for instance, indicates that the dog population was stable from 1987 to 1996 despite increased human population size, but then increased at a rate of 3–4% annually. This increase resulted in a decline in the human:dog ratio from approximately 5.1:1 to 4.0:1. The causes of this increase are unclear. An inverse pattern has occurred in Australia, where the owned dog population has declined by approximately 15% between 1998 and 2009 despite human population growth, resulting in an increase in the human:dog ratio from 4.7:1 to 6.4:1 (Table 1.2).

In Zimbabwe, most dogs are free-ranging and supplement the protein-poor food they directly receive from their owners by scavenging waste food and human feces (Butler and Bingham, 2000). In the United States and Australia most dogs subsist on commercially prepared foods and are less likely to be free-ranging. Contrasting these populations suggests that, without active management of either dog numbers or of the availability of human-derived foods, dog populations will track human populations in a deterministic fashion. However, where dogs are directly fed, economic and social considerations may partially insulate the growth rates of dog populations from those of humans.

Table 1.2 United States (1987–2012) and Australia dog and human population estimates in millions based on American Veterinary Medical Association (1987–2007), American Pet Products Association (2012), and Australian Companion Animal Council (1994–2009) data. Percentage growth of dog population is calculated relative to the previous survey. Average annual growth rate is since 1987, or since 1996 for the United States dog population and since 1994 for the Australian dog population.

Year	Dog population size	Percent change	Average annual growth (since 1987)	Average annual growth (since 1996)	Average annual growth (since 1994)	Human population Size[1]	Human:dog ratio
United States Dog Population Size							
1987	52.4					245.9	4.69
1991	52.5	0.002	0.000			255.8	4.87
1996	52.9	0.008	0.001			269.4	5.09
2001	61.6	0.164	0.015	0.033		285.5	4.64
2007	72	0.169	0.019	0.033		302.3	4.20
2012	78.2	0.086	0.020	0.037		313.1	4.00
Australia Dog Population Size							
1994	3.78					17.9	4.74
1998	4	0.058			0.029	18.7	4.70
2002	3.97	–0.007			0.006	19.6	4.93
2005	3.75	–0.055			–0.001	20.4	5.44
2007	3.69	–0.016			–0.002	21.1	5.72
2009	3.41	–0.076			–0.007	21.9	6.42

[1] Human population from United Nations (2011) except 2012 which is from United States Census Bureau.

1.3 How has the dog–human–wildlife interface developed?

Rural environments are where dog–wildlife interactions are most likely to occur. However, not all rural dogs are similar. The likelihood and types of interactions that may occur between dogs and wildlife are a function of the collective space use of the individual dogs that make up a population. For example, Vanak and Gompper (2010) observed that free-ranging dogs in rural Maharashtra, India had small home ranges that overlapped only partially with the much larger home ranges of Indian foxes, *V. bengalensis*, despite the foxes being only approximately 10% of the body mass of the dogs. Yet, because there were so many dogs in the region, the collective impact of the dog population was potentially extensive because each fox home range was also overlapped by that of many dogs.

The use of space by individual dogs, as well as associated characteristics such as the social behavior of dogs, varies with the quantity and distribution of the dietary subsidies provided directly and indirectly by humans (Vanak and Gompper, 2009b). For example, free-ranging dogs that lived in Italian villages and those that inhabited an adjacent more rural region were both heavily dependent on human-derived foods (Boitani et al., 1995; MacDonald and Carr, 1995). However, the former population, although in loose territorial social groups, were usually observed as solitary individuals, while the latter more rural dogs were members of highly territorial packs and were rarely observed alone. Thus, where a dog lives and how it is able to make use of human subsidies are important predictors of dog behavioral and foraging ecology and ultimately of dog–wildlife interactions. As such, many efforts have been made to categorize dogs (Boitani et al., 1995; Daniels and Bekoff, 1989; Green and Gipson, 1994; MacDonald and Carr, 1995; Nesbitt, 1975; Vanak and Gompper, 2009b; Ritchie et al., Chapter 2). Common delineations of dogs include: owned and un-owned dogs; rural, village, pariah,[3] and farm dogs; stray, sylvatic, wild, and feral dogs; dingoes; pet dogs; and working dogs. However, the fluid and non-exclusive nature of these categories is often problematic. It is quite possible for a dog to be an owned village pariah dog or a stray farm dog. It is also possible for individual dogs to shift between categories (e.g., Boitani et al., 1995). The nuanced difference between feral dogs and dingoes has also been particularly troublesome (Box 1.2). Furthermore, dogs from all of these categories can be free-ranging; indeed, a common misconception is that ownership of an animal reduces its free-ranging behavior. Yet the size of a mammal's home range is typically determined by the body mass of the individual, the distribution of resources (for females), or the distribution of females (for males) (Gompper and Gittleman, 1991; Sandell, 1989). Therefore, unless ownership is associated with constraining the movement of the animal, either through physical barriers or through behavioral training, owned dogs and un-owned dogs likely have similar home range sizes (although perhaps with differing intensities of use) unless ownership is associated with feeding, which is not always the case (e.g., Saluku et al., 2012).

Thus, a critical issue when attempting to categorize dogs is the extent of their reliance on human-derived foods. As a modification to categories put forth by Vanak and Gompper (2009b) in their examination of the interactions of dogs and free-ranging carnivores, a useful framework for categorizing the potential of dogs to interact with wildlife can be made based on where the animal lives (that is, to what extent is the animal sympatric with wildlife), and the reliance on human-derived foods (extensive vs. independent). In this context, urban (owned or un-owned) and fully constrained dogs are highly reliant on human-derived foods and have limited opportunities to interact with wildlife even if the latter group inhabits rural areas. Although these dogs can have effects on wildlife when they accompany humans into natural areas or when they contribute to the pathogen reservoir status of the broader dog population (Banks and Bryant, 2007; Koster, 2008; Lenth et al., 2008; Reed and Merenlender, 2011; Sterl et al., 2008; Steven et al., 2011), their per capita daily likelihood of interacting with wildlife is low.

In contrast, unrestrained dogs in rural locales are more likely to interact with wildlife, especially if the foods provided to them are energetically insufficient such that they must range more broadly to scavenge or hunt. Furthermore, rural owned dogs are more likely to be free-ranging than are urban dogs (Figure 1.3), and the health and nutritional status of these dogs can strongly influence not only the likelihood of interactions with wildlife but also the type of interactions (Silva-Rodríguez and Sieving, 2011). The space use of rural free-ranging dogs (owned or un-owned) commonly facilitates contact with wildlife, especially when the human habitations these dogs are associated with border wildlife reserves or natural areas (Butler et al., 2004; Srbek-Araujo and Chiarello, 2008; Vanak and Gompper, 2010). Indeed, it is at the rural–wildland interface where

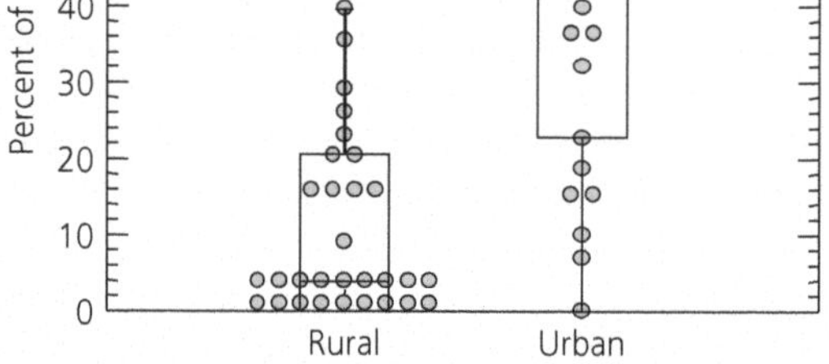

Figure 1.3 Percentage of owned dogs categorized as restricted in their ability to move beyond boundaries set by an owner. Data points represent sites ($n = 58$), subdivided into rural ($n = 32$) and urban ($n = 26$) settings. Box plot defines median and quartile ranges.

[3] The terms 'pariah dog' refers to a subset of free-ranging village dogs of the Indian subcontinent that are considered to represent an ancient or primitive dog morphotype. The morphologic, geographic, and evolutionary bases for the designation have not been well defined and are in need of focused study.

dog–wildlife interactions peak. Many of the most broadly publicized interactions of dogs and wildlife involve rural free-ranging dogs (e.g., Serengeti, Ethiopia, Chile, Mongolia), and rural dogs are also likely to negatively influence the conservation of large native carnivores; if the latter view dogs as food or as competitors and therefore kill them, they can then be perceived as a risk to people (Goodrich et al., 2011; Butler et al., Chapter 5).

The traditional 'village' dog sits between the urban and rural populations with respect to interactions with wildlife. Because the term 'village' is vague (a village may include just a few hundred homes or tens of thousands), this category of dogs is also difficult to define. In general, these dogs are unconfined, but are closely associated with particular human habitations (independent of whether they are owned). In larger villages, these dogs are entirely dependent on human-derived foods and rarely leave the village (MacDonald and Carr, 1995). However, in small villages the opportunity to interact with wildlife, or to interact with rural dogs who themselves interact with wildlife, increases. In these settings one may find a mix of individual dogs, some of whom have very small home ranges and others who may range widely (Vanak and Gompper, 2010). For instance, Meek (1999) observed that half of a set of monitored dogs in an Australian village had home ranges of <3 ha, while the other half had home ranges of >900 ha, and traveled 8–30 km on forays that averaged 26 hours to hunt in areas where eastern grey kangaroos (*Macropus giganteus*) and swamp wallabies (*Wallabia bicolor*) were abundant.

Two categories of dogs that do not neatly fit within the urban–rural framework are feral dogs (including dingoes), and working dogs that are specifically bred and trained to interact with wildlife. The former group represents those dog populations that are independent of human subsidies.[4] With the important exception of Australia, such populations are quite rare. Feral dog populations are limited by the occurrence of larger carnivores and human persecution (Boitani et al., 1995; Butler et al., 2004). Yet in the absence of larger predators and human persecution, dogs are sometimes able to make a living entirely independent of human-derived foods. For instance, Kruuk and Snell (1981) detail the foraging ecology of Galapagos Island feral dogs that were entirely independent of humans. These animals fed predominantly on marine iguanas (*Amblyrhynchus cristatus*) as well as sea lions (*Zalophus californianus*) and fur seals (*Arctocephalus australis*). Marine iguanas are unable to cope with the risk presented by dogs because of their failure to appropriately recognize the risks inherent in dogs (Rödl et al., 2007). The presence of naïve prey may also have underpinned the ability of feral dogs to thrive in Australia following their initial colonization, although several studies have shown that native fauna now show appropriate avoidance behavior to dingo signs (e.g., Carthey and Banks, 2012; Parsons and Blumstein, 2010).

Working dogs differ from other categories of dogs in that some have been bred and/or trained specifically to interact with wildlife. When considering dog–wildlife interactions, the two most notable members of this category are hunting dogs (Koster and Noss, Chapter 11), which have been selected or trained by humans to kill particular types of animals or to facilitate the hunting success rates of humans, and livestock guard dogs (VerCauteren et al., Chapter 9), which accompany domestic ungulates so as to minimize predation by native carnivores. While other working dog breeds, such as herding dogs, have the potential to interact with wildlife, they are not specifically kept by humans with this goal in mind.

1.4 The demographics and ownership of free-ranging dogs

Given our ability to categorize dogs as urban and rural, albeit with the imprecision inherent in such categories, how do the demographics of these categories differ? Furthermore, what proportion of the human population 'manages' these dogs? Surprisingly, there have been few attempts to thoroughly review the demographics of dog populations or to understand the distribution of dogs among human households. I compiled a dataset comprised of

[4] Some authors (e.g., Boitani et al., 1995) use the term *feral* when referring to dogs that are not *intentionally* fed or sheltered. Here I focus on populations that are able to persist entirely independent of human subsidies.

demographic and ownership parameters from 225 study areas derived from 109 sources (Table 1.3). Some of these dataset entries represent surveys at a country-wide spatial scale and are limited to a single parameter, but most represent the results of studies conducted within small rural villages or urban household surveys. I used this dataset to identify broad demographic and ownership patterns that characterize dogs.

For each study I attempted to extract information on the locality, the ownership patterns, demographic information, and management information. In some cases, authors report details for some measures for multiple localities, but give a single mean value for other parameters. In these cases, where the same parameter estimate is repeated across several sites examined by a source, only a single value is applied in quantitative analyses used to calculate summary characterizations of parameter estimations and distributions.

The global dog demographic may be summarized in brief: at a global scale, dog populations are comprised of young animals that are unlikely to be sterilized. The percentage of females that breed is relatively high and litter size is relatively large. However, most dogs are male and mortality rates are also high. While the proportion of households that own dogs varies considerably across the globe, a relatively large portion of the owned dog community is concentrated among a relatively small portion of the human community. These patterns are detailed below.

1.4.1 Dog demographics

The mean age of 37 dog populations was 2.8 ± 1.2 years of age, with a mean of 34.4 ± 17.4% of dog populations comprised of individuals ≤1 yr of age. Neither of the parameters differed significantly for rural versus urban populations (mean age: Mann Whitney $U = 108, p = .361$; % ≤1 yr: $U = 167, p = 0.862$) or at broader continental or Old World versus New World spatial scales. Where sterilization data was available, rates were low (Figure 1.4). Mean and median rates of sterilization were 11.6 and 6.2%, respectively. All rates above 10% were from localities in the developed world, although there were also localities such as Sweden (5%, Egenvall et al., 1999) with low rates of sterilization affirming that cultural norms strongly influence rates of dog sterilization (Orihuela and Solano, 1995).

Across studies, reproduction within the past year occurred for 48.0% (± 10.8%; $n = 10$) of the adult female population. The data were too sparse to assess global differences among urban and rural studies, but data collected by Acosta-Jamett et al. (2010) in Chile using household surveys of urban (46%), town (38%), and rural (42%) dog populations suggest no clear trend. The mean litter size for the studies from 15 localities was 4.9 ± 1.2. Interpreting litter sizes is difficult because this value is strongly size and age-related. For example, the litter size of dogs in Chile peaks at 4–5 years of age (Acosta-Jamett et al., 2010). Larger dogs also tend to have larger litter sizes resulting in curvilinear size*age interactions in litter size (Borge et al., 2011).

Sex ratios are strongly male biased. For 85 study sites, the mean [± s.d.] percentage of the population that was male was 62.6 ± 11.5, and only six studies reported <50% males (Figure 1.4). Sex ratio biases were slightly, but not statistically significantly, different between urban (mean = 60.6; $n = 35$) and rural (64.8; $n = 45$) settings ($U = 968, p = 0.08$). This sex ratio bias may indicate a preference for male dogs inherent in patriarchal societies (Polgreen, 2009), or the more pragmatic selection for male guard dogs and selection against female dogs in rural environments. The latter occurs when rural residents actively exclude female dogs from small villages or settlements to reduce the likelihood of population growth (Acosta-Jamett et al., 2010; Daniels and Bekoff, 1989; Laurensen et al., 1997; Woodroffe and Donnelly, 2011). However, caution must be used in interpreting sex ratios among adult dogs, as pup sex ratios are also male biased in both free-ranging and owned dogs of recognized breeds (Boitani et al., 2005; Lyngset and Lyngset, 1970; Pal, 2001). Indeed, one might conjecture that these biases in pup sex ratios are influenced by artificial selection against female dogs.

Given a relationship between age and reproductive success, as well as the observation that a large percentage of dogs is ≤1 yr of age, it is important to recognize that mortality rates are often high, and in many populations most pups do not survive to adulthood. In an examination of data from 10,810

Table 1.3 Compiled demographic characteristics of dog populations based on surveys and field studies conducted across the globe. Data are delineated by country and by setting (U = urban; R = rural; T = town; C = country-wide; nd = no data). HH = household. DOHH = dog-owning household.

Country	Setting	Human:dog	% DOHH	dogs/ HH	dogs/ DOHH	Density/ km^2	%<1 yr	mean age	% male	% restricted	% sterilized	% females bred/yr	mean litter size	% 1st yr mortality	Source
Argentina	R	1.87	94	2.6	2.68		22	1.8	85.3						1
Argentina	U		54			2,114	11.6	4							2
Argentina	U		75			3,575	26.2	3							2
Argentina	U		27.9	1.36							0.17				3
Argentina	U		35.7	1.95							0.37				3
Argentina	R	1.5	97	2.5	2.5			2	81						4
Australia	C	6.4													5
Australia	R	3.79	56	0.88	1.6										6
Austria	C	13.72													7
Bahamas	U	2.5	52	0.94	1.8			2.6	53.8	57	36	41	4.73		8
Belgium	C	8.05													7
Bolivia	R	1.5	100	4	4			3.5	60				4.1	73	9
Bolivia	U	4.6	77.2	1.4	1.8		21.7		58	14.4					10
Brazil	U	4.98		1.2			57	1.7	52.8	39.8					11
Brazil	U	4	52.5	0.92	1.6					60.7					12
Brazil	U	5.6					32.5								13
Brazil	R	4.7		1.23			41	2.4	70.7	0					14
Brazil	U	5.14													15
Brazil	U	4		0.97			90		52	70					16
Brazil	R		93.9	1.6	2.4	6.3				21.8					17
Cambodia	U	4.46	56.4	1.2	2.1										18
Cambodia	R	3.1	75	1.7	2.3										19
Cameroon	U	4.66	70.5	1.57	2.22		52.1	1.75	56.3	35.5			4.7		20
Cameroon	R	5.75	70.5	1.95	2.76		52.1	1.75	56.3	35.5			4.7		20
Canada	C	5.6	32.3	0.45	1.38		7	5.9			69				21

continued

Table 1.3 *Continued*

Country	Setting	Human:dog	% DOHH	dogs/ HH	dogs/ DOHH	Density/ km^2	%<1 yr	mean age	% male	% restricted	% sterilized	% females bred/yr	mean litter size	% 1st yr mortality	Source
Chad	U	21.7	28	0.4	1.43	1,049	22	3.4	77						22
Chile	U	5.2	49	0.8	1.7	2,380		3	56	49	2	63.2	5	19.4	23
Chile	U	6.2	49	0.7	1.5	1,509		3	56	49	2	63.2	5	19.4	23
Chile	T	5.3	63	0.8	1.6	119		3	74	26	7	47.6	5	45.4	23
Chile	T	4.5	63	0.9	1.5	1,544		3	74	26	7	47.6	5	45.4	23
Chile	T	2.3	63	1.4	1.9	311		3	74	26	7	47.6	5	45.4	23
Chile	R	1.3	89	2.5	2.7	7.2		3	83	15	3	55.1	5	20	23
Chile	R	1.1	89	2.8	3	1.3		3	83	15	3	55.1	5	20	23
Chile	R	2	89	2.1	2.2	1		3	83	15	3	55.1	5	20	23
Chile	R	2.1	89	1.8	2.3	15.9		3	83	15	3	55.1	5	20	23
Chile	U	4.1		0.95				4.6	62						24
Chile	R	2	85.6		1.9	3.6	12.7		92	4.7					25
Chile	R		94.4		2.7	7.3			88	15.7	0				25
China	U	37.5													26
China	R		53.7	0.54	1.05										26
Czech Rep.	C	3.33													7
Denmark	C	9.57													7
Dominica	U	5.36	38.6	0.61	1.59					56					27
Dominica	U		29.5	0.47	1.59			3.5	59.5	70.3	8.5	26	7	18	28
Ecuador	U	7.2		0.59		1,792	32		61				4.1		29
Estonia	C	7.68													7
Ethiopia	U	14.3				290	17	2	83						30
Ethiopia	R	4.6				16	26	2	63						30
Ethiopia	R	3.1	100	2.1	2.1	7				0					31
Finland	C	8.24													7

continued

Table 1.3 *Continued*

Country	Setting	Human:dog	% DOHH	dogs/ HH	dogs/ DOHH	Density/ km^2	%<1 yr	mean age	% male	% restricted	% sterilized	% females bred/yr	mean litter size	% 1st yr mortality	Source
France	C	8.27													7
Germany	C	15.53													7
Greece	C	17.08													7
Hungary	C	3.50													7
India	U					185			57.8				5.03	82	32
India	U											47.5			33
India	U	12	12.3		1.3										34
India	C	36.2	16.9		1.1										35
India	U						9.2		58.4				4.6		36
India	R	28													37
India	R	31													37
India	R	36													37
India	R	31													37
India	R	30													37
India	R	28													37
Indonesia	R	2.5													38
Iran	R	20.5	48	1.28	2.6										39
Ireland	C	4.3	38.9	0.63	1.54						47.3				40
Italy	P		33	0.46	1.4			4.75	62	88.7	20		2.6		41
Italy	C	4.3													42
Japan	C	10.3													43
Kenya	U	6.7	53	1.1	2.1	110.1	50.2	1.9	55.2	10	9.1	54	4.7	52	44
Kenya	R	14.8	53	1	2	10.3	50.2	1.9	51.9	10	9.1	54	4.7	52	44
Kenya	R	8.2	81	2.2	2.7	2.8	50.2	1.9	57.8	10	9.1	54	4.7	52	44
Kenya	R	9.4	58	0.8	2.2	20.8	50.2	1.9	65.2	10	9.1	54	4.7	52	44

continued

Table 1.3 *Continued*

Country	Setting	Human:dog	% DOHH	dogs/ HH	dogs/ DOHH	Density/ km²	%<1 yr	mean age	% male	% restricted	% sterilized	% females bred/yr	mean litter size	% 1st yr mortality	Source
Kenya	R	5	71	1.5	2.1	16.1	50.2	1.9	61.6	10	9.1	54	4.7	52	44
Kenya	R	7.7	74	1.3	1.8	11.5	50.2	1.9	61.6	10	9.1	54	4.7	52	44
Kenya	R					3.4			71.2				5.06		45
Kenya	R					0.21			77.7				4.94		45
Korea	U		21												46
Latvia	C	8.35													7
Lithuania	C	4.45													7
Madagascar	U	4.4	64.8	1.6	2.5		38.8		60	18.6		50.9			47
Malawi	U	3.4								7					48
Malawi	R	3.8								1					48
Mexico	R	4			1.9				84	0					49
Mexico	R	4			1.6				51	29					49
Mexico	R	2			1.8				60	16					49
Mexico	U	4	56.4						66.5	62.4					50
Mexico	U	6	42.8						60.7	62					50
Mexico	U	6	49.3						59.3	60					50
Mexico	U					534			61.5						51
Mexico	U					936			66.7						51
Mexico	R	2.6	77.3	2.7	3.5			3	52						52
Mexico	R	4.3	69	1.6	2.3			3	67						52
Mexico	R	4	74.8	1.9	2.5			4	58						52
Mexico	U	4.3	53.6	0.95	1.78				60	38	5				53
Mexico	U	3.4	72.8	1.2	1.6	1,163		3.1	48.8		3.1	37.5			54
Mexico	R	4.6	63.6	1.5	1.6	75		2.5	58.5		1.8	57.3			54

continued

Table 1.3 *Continued*

Country	Setting	Human:dog	% DOHH	dogs/ HH	dogs/ DOHH	Density/ km²	%<1 yr	mean age	% male	% restricted	% sterilized	% females bred/yr	mean litter size	% 1st yr mortality	Source
Mexico	R	1.7	71.7	2	2.9	288		2.5	58.5		1.8	57.3			54
Mexico	R	2.2	65.5	1.5	2.3	390		2.5	58.5		1.8	57.3			54
Mexico	R	2.63	84.9	1.78	2.09	625	51.5	1	67.9	55	7.8				55
Mexico	U		47				27	2	65						56
Mongolia	R									83					57
Mongolia	R	3.25													57
Mongolia	R	2.8													57
Mongolia	R	3.2													57
Mongolia	R	4.46													57
Namibia	R		79	1.3						3					58
Nepal	U	5.88	90						59						59
Nepal	U	4.7													60
Nether-lands	C	11.13													7
Nigeria	U	17.7													61
Nigeria	R	10.3													61
Nigeria	U	7.9		0.1						32					62
Nigeria	U									16.6					63
Nigeria	U	21	38.2	0.5	1.3	131	57		48	0					64
Nigeria	R	59	13.4	0.18	1.3	11	45		36	0					64
Nigeria	R	62	18.2	0.23	1.3	17	46		45	0					64
Nigeria	R	33	29.6	0.37	1.3	11	47		49	0					64
Nigeria	U	139		0.07					53						65
Nigeria	U	31													66
Nigeria	U	4.1		2.07			28.1		54.9	22.7					67

continued

Table 1.3 *Continued*

Country	Setting	Human:dog	% DOHH	dogs/ HH	dogs/ DOHH	Density/ km^2	%<1 yr	mean age	% male	% restricted	% sterilized	% females bred/yr	mean litter size	% 1st yr mortality	Source
Nigeria	R	3.2		1.33			26.8		78.6	19.4					67
Nigeria	U	27													68
Nigeria	R	30													68
Norway	C												5.4		69
Norway	C	10.8													7
Peru	U	10					38		58						70
Peru	R	2.9	64	1.16	1.8										71
Philippines	R	3.8	69	1.4	2.1	468.4		1	51	15					72
Philippines	R	6.9	39.8	0.71	1.8		66	1.2	37	0					73
Philippines	R			2.6					54	5					74
Philippines	S			2.3					50	39					74
Philippines	U			3.2					54	57					74
Poland	C	5.24													7
Portugal	C	5.5													7
Romania	C	5.16													7
Russian Federation	C	11.42													7
Sierra Leone	U	14								56					75
Slovakia	C	21.85													7
Slovenia	C	8.46													7
South Africa	R						21		63	23					76
South Africa	U	12.4		0.42											77
South Africa	U	6.2		0.8											78

continued

Table 1.3 *Continued*

Country	Setting	Human:dog	% DOHH	dogs/ HH	dogs/ DOHH	Density/ km^2	%<1 yr	mean age	% male	% restricted	% sterilized	% females bred/yr	mean litter size	% 1st yr mortality	Source
South Africa	U	27		0.2											78
South Africa	R	8.8		0.8											78
South Africa	R	11.1													79
Spain	C	9.76													7
Sri Lanka	nd	5	85.5	0.93	1.1										80
Sri Lanka	R	4.6	57	0.83	1.45	108	21.6	3.5	73.6	39.6					81
Sri Lanka	R		49												82
Sri Lanka	S	9.1		0.62											83
Sri Lanka	U	6.25		0.97			24								83
Sri Lanka	R	7.7		0.68											83
Sri Lanka	R	14.3		0.45											83
Sri Lanka	R	8.3		0.61											83
Sudan	R					24.3									84
Sweden	C	10.93		0.02	1.36			5.7	50.3		5.4				85
Sweden	U		8.5												85
Sweden	R		27.1												85
Switzer-land	C	17.22													7
Taiwan	U		22.9	0.37	1.63				57	79.1	20.3				86
Taiwan	U	13.95	13.4	0.2	1.5	45.9									87
Taiwan	U		31						58	68.3	46				88
Tanzania	R	7.3		1.1											89
Tanzania	R	7.5		2.14											89

continued

Table 1.3 *Continued*

Country	Setting	Human:dog	% DOHH	dogs/ HH	dogs/ DOHH	Density/ km^2	%<1 yr	mean age	% male	% restricted	% sterilized	% females bred/yr	mean litter size	% 1st yr mortality	Source
Tanzania	U	15.4	16.8	0.37											90
Tanzania	U	8.3	14.1	0.37											90
Tanzania	U	11.1	14.7	0.28											90
Tanzania	U	11	10.3	0.32											90
Tanzania	U	78.6	8.1	0.11											90
Tanzania	U	214.6	2.5	0.03											90
Tanzania	R	6.6	22	0.83											90
Tanzania	R	17.5	12.7	0.26											90
Tanzania	R	5.3	33.8	0.71											90
Tanzania	R	8.3	18.1	0.43											90
Tanzania	R	11.7	11.5	0.32											90
Tanzania	R	14	19.4	0.43											90
Tanzania	R	6.02		1.28											91
Thailand	R	4.6	54	0.9	1.7				66.5	26					92
Thailand	U	14.5	19.7	0.39	1.99										93
Thailand	U	5.2	45.8	1	2.18										93
Thailand	U	8.5													94
Tunisia	C				1.63										95
Tunisia	R	5.14	79.5	1.29	1.63		26	2.8	68.8						96
Tunisia	R	3.69	98.1	2.06	2.09		27	3.1	72.8						96
Tunisia	U	15.6	23	0.42	1.79	916			67.2						97
Tunisia	U	10.1		0.7		990	30	2.7	66						98
Tunisia	U	3.8		2.4		625	30	2.7	66						98
Tunisia	R	2.7		2.7		20	30	2.7	66						98
Turkey	S		12.5	0.14	1.16				87	86.3					99

continued

Table 1.3 *Continued*

Country	Setting	Human:dog	% DOHH	dogs/ HH	dogs/ DOHH	Density/ km²	%<1 yr	mean age	% male	% restricted	% sterilized	% females bred/yr	mean litter size	% 1st yr mortality	Source
Turkey	U		13.2	0.15	1.16				87	61.6					99
Turkey	U					231									99
Turkey	U					101									99
Turkey	U		19.1	0.24	1.27				87	61.6					99
Turkey	U		9.4	0.11	1.12				87	61.6					99
Turkey	U		4.3	0.04	1				87	61.6					99
Turkey	U		0.1	0.01	1				87	61.6					99
Turkey	R		45.7	0.54	1.19				87	9.1					99
Turkey	U				1.22				87	61.6					99
Turkey	S				1.21				87	61.6					99
Turkey	U				1.09				87	61.6					99
Turkey	U				1.14				87	61.6					99
Turkey	U				1.22				87	61.6					99
UK	C	6.55	23.9	0.35	1.47										100
UK	R		24	0.25	1.3										101
USA	R					228			80						51
USA	R					431			75						51
USA	C								51				7.6		102
USA	C		36.1	1.6							70				103
USA	C	3.97													104
USA	U		50	0.71	1.42					85					105
Zambia	U	45													106
Zambia	R	6.7													106
Zimbabwe	U	37.8	13.5	0.19	1.43										107
Zimbabwe	U	16													108

continued

Table 1.3 *Continued*

Country	Setting	Human:dog	% DOHH	dogs/ HH	dogs/ DOHH	Density/ km^2	%<1 yr	mean age	% male	% restricted	% sterilized	% females bred/yr	mean litter size	% 1st yr mortality	Source
Zimbabwe	R	4.5													108
Zimbabwe	R	9.1	45.5	1.5		17	40.8	2	55.6	0			4.6	71.8	109
Zimbabwe	R	3.7	64.5	1.5		35.2	40.8	2	52.6	0			4.6	71.8	109
Zimbabwe	R	5.6	46.5	1.5		53.3	40.8	2	55.6	0			4.6	71.8	109
Zimbabwe	R	5	61.8	1.5		26.1	40.8	2	55.6	0			4.6	71.8	109
Zimbabwe	R	3.7	72.2	1.5		17.5	40.8	2	52.6	0			4.6	71.8	109
Zimbabwe	R	4.3	59	1.5		8.3	40.8	2	50	0			4.6	71.8	109
Zimbabwe	R	4	78	1.5		11.5	40.8	2	66.7	0			4.6	71.8	109

Source: (1) Cardinal et al., 2006, 2007; (2) Rupel et al., 2003; (3) Zumpano et al., 2011; (4) Gurtler et al., 2007; (5) ACAC, 2010; (6) Meehan et al., 1999; (7) FEDIAF, 2010; (8) Fielding, 2010; Fielding et al., 2005; Fielding and Mather, 2000; Fielding and Plumridge, 2005; (9) Fiorello et al., 2006; (10) Suzuki et al., 2008; (11) Moreira et al., 2004; (12) Alves et al., 2005; (13) Andrade et al., 2008; (14) Dantas-Torres et al., 2010; (15) Dias et al., 2004; (16) Nunes et al., 2008; (17) Torres and Prado, 2010; (18) Lunney et al., 2012; (19) Ly et al., 2009; (20) Awah-Ndukum, 2003; (21) Perrin, 2009; (22) Durr et al., 2009; Mindekem et al., 2005; Kayali et al., 2003; (23) Acosta-Jamett et al., 2010, 2011; (24) Morales et al., 2009; (25) Silva-Rodriguez et al., 2010; Silva-Rodriguez and Sieving, 2011, 2012; (26) Wang et al., 2006 as cited in Hu et al., 2009; (27) Alie et al., 2007; (28) Davis et al., 2007; (29) Beran and Frith, 1988; (30) Laurensen et al., 1997; Shiferaw et al., 1998; (31) Atickem, 2003; Atickem et al., 2009; (32) Pal, 2001; (33) Reece et al., 2008; Reece and Chawla, 2006; (34) Sudarshan et al., 2001; (35) Sudarshan et al., 2006; (36) Totten et al., 2010a,b; (37) Belsare, pers. comm.; (38) Bingham, 2001; (39) Gavgani et al., 2002; (40) Downes et al., 2009; (41) Slater et al., 2008; DiNardo et al., 2007; (42) Slater, 2008; (43) Takahashi-Omoe and Omoe, 2012; (44) Kitala et al., 2001; (45) Woodroffe and Donnelly, 2011; (46) Podberscek, 2009; (47) Ratsitorahina et al., 2009; (48) Edelsten, 1995; (49) Ruiz-Izaguirre and Eilers, 2012; (50) Romero-Lopez et al., 2008; (51) Daniels and Bekoff, 1989; (52) Fishbein et al., 1992; (53) Flores-Ibarra and Estrella-Valenzuela, 2004; (54) Ortega-Pacheco et al., 2007; (55) Orihuela and Solano, 1995; (56) Eng et al., 1993; (57) Buuveibaatar et al., 2009; Young, pers. comm.; (58) Sorin and Mvula, 2001 in Mettler, 2003; (59) Joshi and Bogel, 1988; Bogel and Joshi, 1990; (60) Kato et al., 2003; (61) Ezeokoli et al., 1984; (62) Faleke, 2003; (63) Awoyomi et al., 2007; (64) Oboegbulem and Nwakonobi, 1989; (65) Aiyedun and Olugasa, 2012; (66) Idachaba, 2009; (67) El-Yuguda et al., 2007; (68) Ezeolkoli and Umoh, 1987; (69) Borge et al., 2011; (70) Chomel et al., 1988; (71) Reithinger et al., 2003; (72) Childs et al., 1998; Robinson et al., 1996; (73) Estrada et al., 2001; (74) Beran, 1982; (75) Suluku et al., 2012; (76) van Sittert et al., 2010; (77) McCrindle et al., 1999; (78) Bishop, 2001; (79) Rautenbach et al., 1991; (80) Perera et al., 2007; (81) Matter et al., 2000; (82) Matibag et al., 2009; (83) Wandeler et al., 1988; (84) Hassan et al., 2009; (85) Egenvall et al., 1999; (86) Hsu et al., 2003; (87) Tung et al., 2010; (88) Weng et al., 2006; (89) Kaare et al., 2009; (90) Knobel et al., 2008; (91) Cleaveland et al., 2001; (92) Kongkaew et al., 2004; (93) Sagarasaeranee et al., 2001; (94) Singhchai, 2001; (95) Touihri et al., 2009; (96) Yousseff et al., 1998; (97) Seghaier et al., 1999; (98) Matter et al., 1998; (99) Vos and Turan, 1996; (100) Asher et al., 2011; (101) Westgarth et al., 2007; (102) New et al., 2004; (103) Clancy and Rowan, 2003; (104) APPA, 2012; (105) McCollum et al., 2012; (106) DeBalogh et al., 1993; (107) Hill, 1985; (108) Brooks, 1990; (109) Butler and Bingham, 2000.

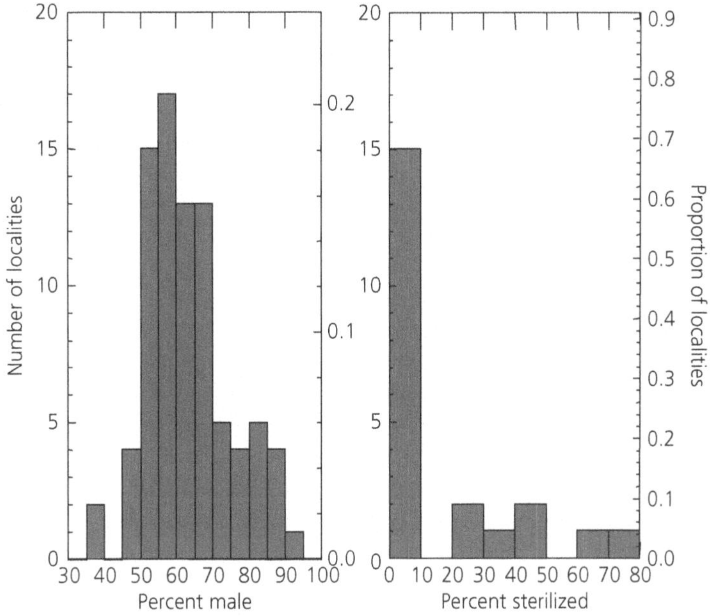

Figure 1.4 Distributions of dog sex ratios (percentage of the population comprised of male dogs) for 85 study populations, and of the proportions of surgically sterilized adult dogs from 22 study populations. Sex ratios are strongly male biased and the percentage of adult dogs that have been spayed or neutered is typically low.

litters (58,339 puppies) of known-breed dogs in Norway, Tønnessen et al. (2012) found that 13.3% of litters had some mortality of pups, and 3.6% of pups (excluding still-born pups) died within a week of birth. Thereafter, mortality rates decline sharply; 99% of pups alive at 8 days survive to 8 weeks of age. Treating the Norway data as a baseline for survival potential given modern veterinary care and sufficient nutrient availability, we can contrast these estimates of mortality to those from owned and un-owned populations across the planet. For eight studies that provided estimates of mortality among dogs ≤1 yr of age, mean mortality was 47.7%, but with high variance (s.d. = 26.4; range = 18–82%). Even the lowest rate of mortality (18% from dogs inhabiting urban Dominica in the Caribbean; Davis et al., 2007) is 4–5 times greater than that in Norway. In other urban and rural populations (e.g., Fiorello et al., 2006; Pal, 2001) rates are far higher. Much of this mortality is likely associated with disease (although see Boitani et al., 1995). For example, Fiorello et al. (2006) examined the demographics of rural dogs in Bolivia and calculated that 73% of pups died. When questioned about the cause of these deaths, dog owners mentioned diarrhea (48%), mange (44%), worms (26%), and vomiting (22%). While these may be manifestations of nutritional stressors, the enzootic nature of micro and macroparasites (Knobel et al., Chapter 6) suggest that disease is likely a fundamental driver of pup mortality. After dogs reach a year of age, age-specific mortality rates can decline considerably (Acosta-Jamett et al., 2010) and human associated mortality (both direct killing of dogs and accidental killing, primarily via vehicle impact) is increasingly important. Nonetheless, it is clear that we need further insights into the dynamics of parasites in free-ranging dog populations.

1.4.2 Might disease limit dog populations?

Despite high mortality and turnover rates, dog populations are apparently growing in many parts of the world (see Section 1.2). The role of disease in limiting dog populations is not clear, as we don't know if the cause-specific mortality that occurs in dog populations (and especially in free-ranging dog populations) is compensatory. That is, if dog population size is determined in a density dependent fashion by food availability, then the mortality due to disease may substitute (compensate) for mortality due to competition for resources. In such situations the disease does not limit the density of the population and therefore the reduction of disease-related mortality will not result in population growth because the population density is set by food availability, irrespective of morbidity and mortality caused by disease. Alternatively, disease may limit dog populations in a density dependent fashion, irrespective of and below the nutritionally-set carrying capacity of the local environment. If that is the case, then removal of disease from the dog population should result in an increase in the dog population to a level set by some other limiting factor, such as food or societal tolerances.

The question of whether disease limits dog populations has immense applied importance in the context of managing dog–wildlife interactions and even for managing human–wildlife interactions in situations where dogs mediate their likelihood and context. Dog populations can be problematic for wildlife managers and conservationists because of the risk of pathogen transmission to other carnivore species (the analogy in dog–human interactions is the risk of the transmission of the rabies virus). As a result, there has been intensive recent work in the wildlife conservation realm on how best to reduce these impacts by reducing transmission opportunities between the reservoir dog population and the susceptible wildlife population. The primary strategies available include the culling of dogs, animal birth control (ABC; spay and neuter campaigns), the vaccinating of dogs, and the vaccination of susceptible wildlife. These strategies are nonexclusive and different management goals and local societal norms influence the approach used. Culling of dogs to benefit wildlife has rarely been used on a large scale (but is occasionally used for other species: e.g., López et al., 2009), and while ABC programs are used in some urban environments to reduce the likelihood of human rabies (e.g., Totton et al., 2010a, b), their use to address wildlife management goals is minimal due to the logistic costs, the spatial scale over which such a management approach must be taken, as well as cultural attitudes (Laurenson et al., 2005). In contrast, vaccination-based programs to reduce transmission are increasingly common for managing the disease risk inherent in dog–carnivore interactions (Knobel et al., Chapter 6). The most prominent uses of these approaches include the vaccination of dogs for rabies and canine distemper in Tanzania and Ethiopia to reduce spillover of these diseases to large carnivore species (in particular, African wild dogs *Lycaon pictus*; lions *Panthera leo*; and Ethiopian wolves *C. simensis*; Haydon et al., 2006; Vial et al., 2006).

These programs have shown that through dog or wildlife vaccination campaigns, reducing dog-derived disease mortality events in target wildlife taxa is possible if a wide cordon solitaire (a barrier created through vaccination or culling that prevents a pathogen from spreading through a population) can be put in place. But how such campaigns influence dog populations is not entirely clear. While Laurenson et al. (2005) indicate that preliminary unpublished data suggest no increase in population growth rates in Tanzania, they also call for further assessment of this issue. If vaccination of dogs reduces the role that pathogens may have in limiting dog populations, then an indirect result of such programs might be a larger dog population. This in turn might impact wildlife in other non-disease related ways, might enhance the ability of the dog population to act as a reservoir for other parasites that were not the target of vaccination but might also influence wildlife, and might increase the risk of an epidemic should fiscal, logistic, or political constraints result in an interruption of vaccination campaigns.

1.4.3 Who owns dogs?

The long history of the study of rabies and the need to predict rabies incidence has led to a strong reliance on the use of human:dog ratios. Unfortunately, this ratio is relatively uninformative with regards to the variance in dog ownership within a population.[5] While information facilitating the calculation

[5] The very concept of ownership itself can vary across regions. For example, in some regions such as rural India (A. Belsare, pers. comm.) surveys easily identify 'reference persons' associated with many dogs. These persons may provide resources to the dog, can handle the dog, and have a bond with the animal. If asked, however, these reference persons will deny ownership of the dog.

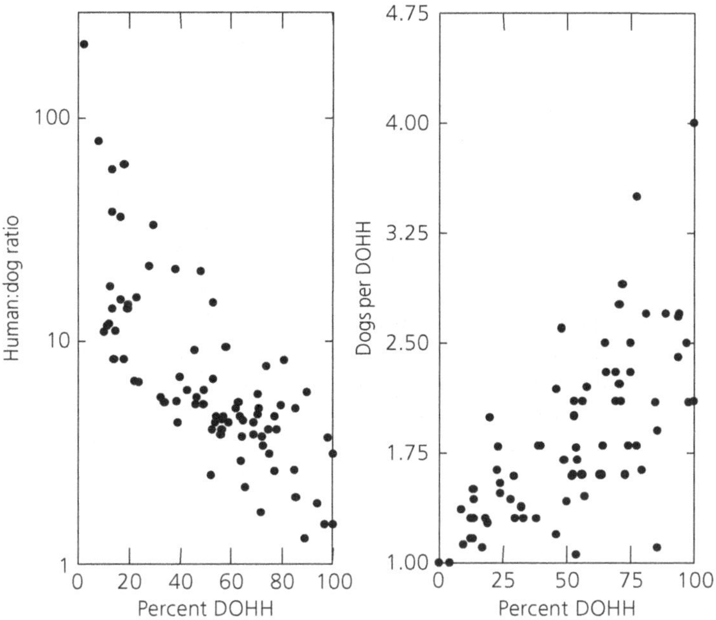

Figure 1.5 Observed relationships between the percentage of households in a locality that are dog-owning (DOHHs) and (left) the human:dog ratio on a log scale ($n = 80$; r = 0.767) and (right) the number of dogs per DOHH at the site ($n = 72$; r = 0.697).

of measures, such as the percentage of household that own dogs or the number of dogs per dog-owning household (DOHH), is often collected and either reported directly or reported in such a way that these measures can be calculated, the insights to be gained from these ownership parameters are rarely discussed in a holistic fashion. This is somewhat unfortunate, as while human:dog ratios and the percentage of households that own dogs are strongly correlated (Figure 1.5), the latter measure provides insights into the aggregation of dogs amongst the human population. That is, insight is gained into the proportion of human households that are directly responsible for subsidizing the dog population. Those households that own dogs can represent <10% of the broader household population, or alternatively dog ownership may approach 100% in some locations.

Furthermore, there is also a positive relationship between the percentage of households that own dogs and the mean number of dogs per DOHH. While the mean number of dogs in rural DOHHs is approximately 40% greater than the number in urban DOHHs (2.12 and 1.55, respectively; U = 1099; $p < 0.001$), in both settings the number of dogs per DOHH increases with the percentage of the population that maintains dogs. At sites where very few households maintain dogs, those households that do have relatively few dogs. In contrast, DOHHs situated in locales where DOHHs are more common tend to have a greater number of dogs. This may in part be due to the role of hunting and livestock guard dogs in rural environments.

The percentage of households that own dogs is an indication of the aggregation of dogs that can be contrasted across locales (Figure 1.5), but further insights may be gained when more precise data on the distribution of dog number per DOHH is provided. Among DOHHs there is considerable variance in the number of dogs owned. By graphing the distribution of DOHH ownership intensity (the number of dogs per DOHH) we observe a negative binomial distribution (Figure 1.6). Most DOHHs have a low number of dogs (i.e., 1–2 dogs), but a small proportion of households maintain many dogs. Such patterns are relatively common in ecology, and recognizing them often provides predictive power and management insights. For instance, the distribution pattern of dogs among DOHHs is strikingly similar to the epidemiological patterns observed in the intensity of parasitism of individual hosts by macroparasites: most hosts harbor few

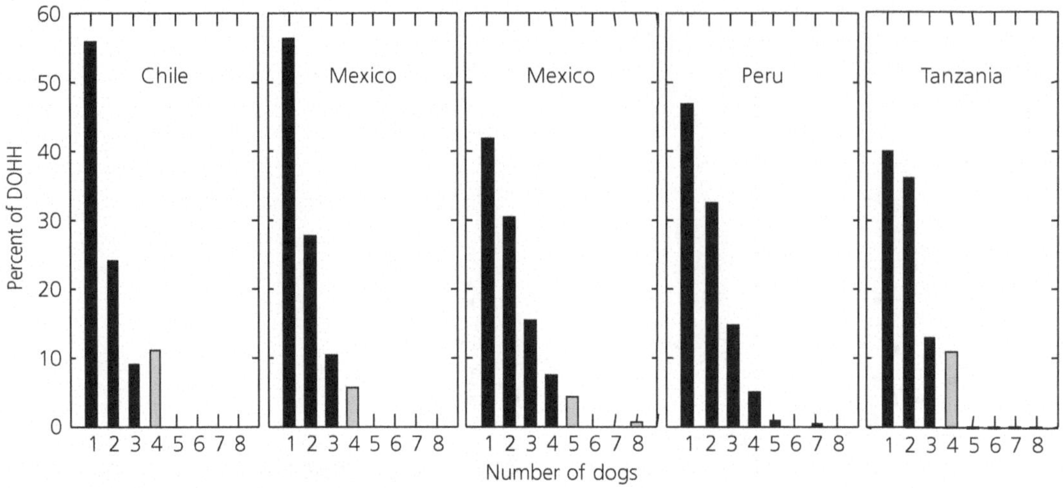

Figure 1.6 Distribution of dogs among dog-owning households (DOHHs) for five study populations. Most DOHHs (>70 in each locale) have 1–2 dogs, but a smaller portion of DOHHs have a larger number of dogs and thus underpin a disproportionally large percentage of the dog population. Gray bars represent cases for which the numbers of dogs is ≥ the indicated number of dogs. Sources for the data, left to right: Acosta-Jammett (2010); Flores-Ibarra and Estrella-Valenzuela (2004); Orihuela and Solano (1995); Reithinger et al. (2003); Knobel et al. (2008).

parasites while a few support many (Wilson et al., 2001). Recognizing such patterns facilitates opportunities for discerning the factors that underpin the extent of an individual's parasite burden (e.g., Monello and Gompper, 2007) or, in the current context, the likelihood of a household containing a particular abundance of dogs.

Recognition of heterogeneities in the abundance of dogs across DOHHs may also provide important opportunities in managing dogs and impacts of dogs on wildlife. For example, for dog populations such as those graphed in Figure 1.6, we can calculate the contribution of households with different numbers of dogs to the broader dog population. In rural Peru, Reithinger et al. (2003) report a complete dataset on the number of dogs per 326 DOHHs (total dog population = 594). The DOHHs with 3 (48 households with 144 dogs), 4 (16, 64), 5 (3, 15), and 7 (1, 7) dogs collectively represent 68 DOHHs (20.8%) and collectively account for 230 (38.7%) of the total dog population. In contrast, the 152 DOHHs (46.6%) with a single dog account for 25.5% of the total dog population. Thus management efforts that target the smaller group of households with three or more dogs will have a greater impact (that is, reach a greater portion of the dogs) than efforts that target the larger group of households that have just a single dog.

1.5 Future research needs

In considering the relevance of our understanding of the evolution, population size, and demography of dogs in the context of dog–wildlife interactions, four principal knowledge voids are apparent. First, a great deal of attention has gone into dating and placing the origin of dogs. Less consideration has been paid to the implications of these dates for understanding dog–wildlife interactions. For instance, dogs have only recently (within the past 1,500–2,000 yrs) colonized sub-Saharan Africa, parts of Amazonian South America (<1,000 ybp), Madagascar (<2,500 ybp), and Australia (<5,000 ybp), and in many island systems colonization events are likely even more recent. The time since colonization, in combination with the structure of the native predator community, may mediate how wildlife communities respond to the now nearly ubiquitous presence of dogs.

Second, while the estimates of regional and global dog numbers presented here are more precise than those values previously put forth, they remain coarse estimates. To make such values of greater use (vis-à-vis extrapolating the environmental impacts of dogs and understanding spatial variance in how dogs are likely to interact with wildlife) further refinement is greatly needed. In addition, while data on the demographics of free-ranging dog populations exist, they have rarely been put to use in such a way that the population dynamics of dogs are modeled.

Third, high densities of dogs inhabiting the human–wildland interface may act as reservoirs for multihost pathogens that have the potential to be transmitted to wild carnivore species. As a result, there has been extensive effort focusing on vaccinating dogs and the wildlife species of concern to mitigate this risk. Yet there is a need to better understand whether these diseases play a role in limiting dog populations and, if so, whether vaccination campaigns may result in higher dog population densities that might influence wildlife in related ways (such as via other parasites) or altogether different ways (such as by directly interacting with the species of interest).

Finally, data on heterogeneities in the distribution of dogs among households has not been well explored. If relatively small proportions of dog-owning households subsidize disproportionately large numbers of dogs, efforts to identify these dog owners should be made. Furthermore, targeted efforts to work with these households may offer novel opportunities for managing dogs. Working with these households on issues related to dog reproduction, pathogen load, nutrition, and roaming limitations may provide large gains for reducing problematic dog–wildlife interactions because these households have access to and support a proportionately large number of dogs.

Acknowledgments

Aniruddha Belsare, Juliet Clutton-Brock, Anne Hoylman, and Darcy Morey provided comments on earlier versions of this chapter. Their input is greatly appreciated.

References

ACAC (2010). *Contribution of the Pet Care Industry to the Australian Economy.* 7th edn. Australian Companion Animal Council, East Kew, Victoria.

Acosta-Jamett, G., Chalmers, W.S.K., Cunningham, A.A., Cleaveland, S., Handel, I.G., and Bronsvoort, B.M. de C. (2011). Urban domestic dog populations as a source of canine distemper virus for wild carnivores in the Coquimbo region of Chile. *Veterinary Microbiology*, **152**, 247–57.

Acosta-Jamett, G., Cleaveland, S., Cunninghamm, A.A., and Bronsvoort, B.M. (2010). Demography of domestic dogs in rural and urban areas of the Coquimbo region of Chile and implications for disease transmission. *Preventive Veterinary Medicine*, **94**, 272–81.

Aiyedun, J.O. and Olugasa, B.O. (2012). Use of aerial photograph to enhance dog population census in Ilorin, Nigeria. *Sokoto Journal of Veterinary Sciences*, **10**, 22–7.

Alie, K., Davis, B.W., Fielding, W.J., and Maldonado, F.G. (2007). Attitudes towards dogs and other "pets" in Roseau, Dominica. *Anthrozoös*, **20**, 143–54.

Allen, B.L., Engeman, R.M., and Allen, L.R. (2011a). Wild dogma: an examination of recent "evidence" for dingo regulation of invasive mesopredator release in Australia. *Current Zoology*, **57**, 568–83.

Allen, B.L., Engeman, R.M., and Allen, L.R. (2011b). Wild dogma II: the role and implications of wild dogma for wild dog management in Australia. *Current Zoology*, **57**, 737–40.

Alves, M.C., Matos, M.R., Reichmann, M. de L., and Dominiguez, M.H. (2005). Estimation of the dog and cat population in the State of São Paulo. *Revista de Saúde Pública*, **39**, 891—97.

Andrade, A.M., Queiroz, L.H., Perri, S.H.V., and Nunes, C.M. (2008). Estudo descritivo da estrutura populacional canina da área urbana de Araçatuba, São Paulo, Brasil, no período de 1994 a 2004. *Cadernos de Saúde Pública*, **24**, 927–32.

APPA (2012). 2011/2012 APPA National Pet Owners Survey. American Pet Product Association, Greenwich, Connecticut.

Asher, L.Buckland, E.L., Phylactopoulos, C.I., Whiting, M.C., Abeyesinghe, S.M., and Wathes, C.M. (2011). Estimation of the number and demographics of companion dogs in the UK. *BMC Veterinary Research*, **7**, 74.

Atickem, A. (2003). The ecology of domestic dog (*Canis familiaris*) and its effect on the survival of the Ethiopian wolf (*Canis simensis*) in the Web Valley of the Bale Mountains National Park, Ethiopia. MS thesis, Addis Ababa University.

Atickem, A., Bekele, A., and Williams, S.D. (2009). Competition between domestic dogs and Ethiopian wolf (*Canis simensis*) in the Bale Mountains National Park, Ethiopia. *African Journal of Ecology*, **48**, 401–7.

Awah-Ndukum, J. (2003). Ecological aspects of dogs in to rabies control and public health significance in Northwest Cameroon. *Journal of the Cameroon Academy of Sciences*, **3**, 25–31.

Awoyomi, O.J., Adeyemi, I.G., and Awoyomi, F.S.O. (2007). Socioeconomic factors with non-vaccination of dogs against rabies in Ibadan, Nigeria. *Nigerian Veterinary Journal*, **28** (3), 59–63.

Baird, R.F. (1991). The dingo as a possible factor in the disappearance of *Gallinula morlierii* from the Australian mainland. *Emu*, **91**, 121–22.

Banks, P.B. and Bryant, J.V. (2007). Four-legged friend or foe? Dog walking displaces native birds from natural areas. *Biology Letters*, **3**, 611–13.

Barnett, B.D. (1986) Eradication and control of feral and free-ranging dogs in the Galapagos Islands. *Proceedings of the Twelfth Vertebrate Pest Conference*, **8**, 358–68.

Benecke, N. (1987). Studies on early dog remains from northern Europe. *Journal of Archaeological Science*, **14**, 31–49.

Beran, G.W. (1982). Ecology of dogs in the central Philippines in relation to rabies control efforts. *Comparative Immunology, Microbiology and Infectious Diseases*, **5**, 265–70.

Beran, G.W. and Frith, M. (1988). Domestic animal rabies control: an overview. *Review of Infectious Diseases*, **10**, Suppl 4, S672–77.

Bingham, J. (2001). Rabies in Flores, Indonesia: comparisons and contrasts to rabies control in Africa. In *Proceedings of the Sixth Southern and Eastern African Rabies Group Meeting*, Lilongwe, Malawi. pp. 127–30.

Bishop, G. (2001). Rabies in South Africa. In *Proceedings of the Sixth Southern and Eastern African Rabies Group Meeting*, Lilongwe, Malawi. pp. 45–49.

Boitani, L., Francisci, F., Ciucci, P., and Andreoli, G. (1995). Population biology and ecology of feral dogs in central Italy. In J. Serpell, Ed., *The Domestic Dog: Its Evolution, Behavior and Interactions with People*. Cambridge University Press, Cambridge, pp. 217–44.

Borge, K.S., Tønnessen, R., Nødtvedt, A., and Indrebø, A. (2011). Litter size at birth in purebred dogs: a retrospective study of 224 breeds. *Theriogenology*, **75**, 911–19.

Borgerhoff Mulder, M., Bowles, S., Hertz, T., Bell, A., Beise, J., Clark, G., Fazzio, I., Gurven, M., Hill, K., Hooper, P.L., Irons, W., Kaplan, H., Leonetti, D., Low, B., Marlowe, F., McElreath, R., Naidu, S., Nolin, D., Piraino, P., Quinlan, R., Schniter, E., Sear, R., Shenk, M., Smith, E.A., von Rueden, C., and Wiessner, P. (2009). Intergenerational wealth transmission and the dynamics of inequality in small-scale societies. *Science*, **326**, 682–88.

Boyce, J. (2006). Canine revolution: the social and environmental impact of the introduction of the dog to Tasmania. *Environmental History*, **11**, 102–29.

Brooks, R. (1990). Survey of the dog population of Zimbabwe and its level of rabies vaccination. *Veterinary Record*, **127**, 592—96.

Brown, O.J.F. (2006). Tasmanian devil (*Sarcophilus harrisii*) extinction on the Australian mainland in the mid-Holocene: multicausality and ENSO intensification. *Alcheringa*, **30 (S1)**, 49–57.

Burney, D.A., James, H.F., Burney, L.P., Olson, S.L., Kikuchi, W., Wagner, W. L., Burney, M., McCloskey, D., Kikuchi, D., Grady, F. V., Gage, R., and Nishek, R. (2001). Fossil evidence for a diverse biota from Kaua'i and its transformation since human arrival. *Ecological Monographs*, **71**, 615–41.

Butler, J.R.A. and Bingham, J. (2000). Demography and dog–human relationships of the dog population in Zimbabwean communal lands. *Veterinary Record*, **147**, 442–46.

Butler, J.R.A., du Toit, J.T., and Bingham, J. (2004). Free-ranging domestic dogs (*Canis familiaris*) as predators and prey in rural Zimbabwe: threats of competition and disease to large wild carnivores. *Biological Conservation*, **115**, 369–78.

Buuveibaatar, B., Young, J.K., and Fine, A.E. (2009). Mongolian saiga in Sharga Nature Reserve: are domestic dogs a threat to saiga? *Mongolian Journal of Biological Sciences*, **7**, 37–43.

Bögel, K. and Joshi, D.D. (1990). Accessibility of dog populations for rabies control in Kathmandu valley, Nepal. *Bulletin of the World Health Organization*, **68**, 611–17.

Bökönyi, S. (1969). Archaeological problems and methods of recognizing animal domestication. In P.J. Ucko and G.W. Dímbleby, Eds., *The Domestication and Exploitation of Plants and Animals*. Duckworth, London, pp. 219–29.

Cardinal, M.V., Castañera, M.B., Lauricella, M.A., Cecere, M.C., Ceballos, L.A., Vazquez–Prokopec, G.M., Kitron, U., and Gürtler, R.E. (2006). A prospective study of the effects of sustained vector surveillance on *Trypanosoma cruzi* infection of dogs and cats in rural northwestern Argentina. *American Journal of Tropical Medicine and Hygiene*, **75**, 753–61.

Cardinal, M.V., Lauricella, M.A., Marcet, P.L., Orozco, M.M., Kitron, U., and Gürtler, R.E. (2007). Impact of community–based vector control on house infestation and *Trypanosoma cruzi* infection in *Triatoma infestans*, dogs and cats in the Argentine Chaco. *Acta Tropica*, **103**, 201–11.

Carthey, A.J.R. and Banks, P.B. (2012). When does an alien become a native species? A vulnerable native mammal recognizes and responds to its long–term alien predator. *PLoS ONE*, **7**, e31804.

Castroviejo–Fisher, S., Skoglund, P., Valadez, R., Vilà, C., and Leonard, J.A. (2011). Vanishing native American dog lineages. *BMC Evolutionary Biology*, **11**, 73.

Childs, J.E., Robinson, L.E., Sadek, R., Madden, A., Miranda, M.E., and Miranda, N.L. (1998). Density estimates of rural dog populations and an assessment of marking methods during a rabies vaccination campaign in the Philippines. *Preventive Veterinary Medicine*, **33**, 207–18.

Chomel, B., Chappuis, G., Bullon, F., Cardenas, E., David de Beublain, T., Lombard, M., and Giambruno, E. (1988). Mass vaccination campaign against rabies: Are dogs correctly protected? The Peruvian experience. *Reviews of Infectious Diseases*, **10**, Suppl. 4, S697–S702.

Clancy, E.A. and Rowan, A.N. (2003). Companion animal demographics in the United States: a historical perspective. In D.J. Salem and A.N. Rowan, Eds., *The State of the Animals II: 2003*. Humane Society of the United States, Washington D.C., pp. 9–26.

Claridge, A.W. and Hunt, R. (2008). Evaluating the role of the dingo as a trophic regulator: additional practical suggestions. *Ecological Management and Restoration*, **9**, 116–19.

Cleaveland, S., Kaare, M., Tiringa, P., and Mlengeya, T. (2001). A dog rabies vaccination campaign in rural Africa: impact on the incidence of animal rabies and human bite injuries. In *Proceedings of the Sixth Southern and Eastern African Rabies Group Meeting*, Lilongwe, Malawi. pp. 114–26.

Clutton-Brock, J. (1993). The spread of domestic animals in Africa. In T. Shaw, P. Sinclair, B., Andah and A. Okpoko, Eds., *The Archaeology of African Foods, Metals and Towns*. Routledge, London, pp. 61–70.

Clutton-Brock, J. (1994). The unnatural world: behavioural aspects of humans and animals in the process of domestication. In A. Manning and J. A. Serpell, Eds., *Animals and Human Society*: Changing Perspectives. Routledge, London, pp. 23–35.

Clutton-Brock, J. (1999). *A Natural History of Domesticated Mammals*, 2nd edn. Cambridge University Press, Cambridge.

Clutton-Brock, J. (2012). *Animals as Domesticates: A World View Through History*. Michigan State University Press, East Lansing.

Coppinger, R. and Coppinger, L. (2001). *Dogs: A Startling New Understanding of Canine Origin, Behavior and Evolution*. Scribner, New York.

Corbett, L K. (2004). Dingo (*Canis lupus dingo*). In C. Sillero–Zubiri, M. Hoffmann and D.W. Macdonald, Eds., *Canids: Foxes, Wolves, Jackals and Dogs. Status Survey and Conservation Action Plan*. IUCN/SSC Canid Specialist Group, Gland and Cambridge, pp. 223–30.

Crockford, S.J. and Kuzmin, Y.V. (2012). Comments on Germonpré et al., Journal of Archaeological Science 36, 2009 "Fossil dogs and wolves from Palaeolithic sites in Belgium, the Ukraine and Russia: osteometry, ancient DNA and stable isotopes," and Germonpré, Lázkičková–Galetová, and Sablin, Journal of Archaeological Science 39, 2012 "Palaeolithic dog skulls at the Gravettian Předmostí site, the Czech Republic." *Journal of Archaeological Science*, **39**, 2797–801.

Daniels, T.J. and Bekoff, M. (1989). Population and social biology of free ranging dogs, *Canis familiaris*. *Journal of Mammalogy*, **70**, 754–62.

Dantas–Torres, F., Paiva–Cavalcanti, M., Figueiredo, L.A., Melo, M.F., Silva, F.J., Silva, A.L., Almeida, E.L., and Brandão–Filho, S.P. (2010). Cutaneous and visceral leishmaniosis in dogs from a rural community in northeastern Brazil. *Veterinary Parasitology*, **170**, 313–17.

Davis, B.W., Alie, K., Fielding, W.J., Morters, M., and Galindo, F. (2007). Preliminary observations on the characteristics of the owned dog population in Roseau, Dominica. *Journal of Applied Animal Welfare Science*, **10**, 141–51.

De Balogh, K.K., Wandeler, A.I., and Meslin, F.X. (1993). A dog ecology study in an urban and a semi-rural area of Zambia. *Onderstepoort Journal of Veterinary Research*, **60**, 437–43.

Di Nardo, A., Candeloro, L., Budke, C.M., and Slater, M.R. (2007). Modeling the effect of sterilization rate on owned dog population size in central Italy. *Preventive Veterinary Medicine*, **82**, 308–13.

Dias, R.A., Garcia, R.C., Silva, D.F., Amaku, M., Ferreira Neto, J.S., and Ferreira, F. (2004). Estimate of the owned canine and feline populations in urban area in Brazil. *Revista de Saúde Pública*, **38**, 565–70.

Dimitrijević, V. and Vuković, S. (In press). Was the dog locally domesticated in the Danube Gorges? Morphometric study of dog cranial remains from four Mesolithic–Early Neolithic archaeological sites by comparison with contemporary wolves. *International Journal of Osteoarchaeology*.

Downes, M., Canty, M.J., and More, S.J. (2009). Demography of the pet dog and cat population on the island of Ireland and human factors influencing pet ownership. *Preventive Veterinary Medicine*, **92**, 140–49.

Driscoll, C. and Macdonald, D. (2010). Top dogs: wolf domestication and wealth. *Journal of Biology*, **9**, 10.

Ducos, P. (1989). Defining domestication: a clarification. In J. Clutton–Brock, Ed., *The Walking Larder: Patterns of Domestication, Pastoralism, and Predation*. Unwin Hyman, London, pp. 28–30.

Durr, S., Mindekem, R., Kaninga, Y., Doumagoum Moto, D., Meltzer, M.I., Vounatsou, P., and Ziinsstag, J. (2009). Effectiveness of dog rabies vaccination programmes: comparison of owner–charged and free vaccination campaigns. *Epidemiology and Infection*, **137**, 1558–67.

Edelsten, R.M. (1995). Epidemiology and control of rabies in Malawi. *Tropical Animal Health Production*, **27**, 155–63.

Egenvall, A., Hedhammar, A., Bonnett, B.N., and Olson, P. (1999). Survey of the Swedish dog population: age, gender, breed, location and enrollment in animal insurance. *Acta Veterinaria Scandinavica*, **40**, 231–40.

Elledge, A.E., Allen, L.R., Carlsson, B.–L., Wilton, A.N., and Leung, L.K.–P. (2008). An evaluation of genetic analyses, skull morphology and visual appearance for assessing dingo purity: implications for dingo conservation. *Wildlife Research*, **35**, 812–20.

El-Yuguda, A.D., Baba, A.A., and Baba, S.S.A. (2007). Dog population structure and cases of rabies among dog bite victims in urban and rural areas of Borno State, Nigeria. *Tropical Veterinarian*, **25**, 34–40.

Eng, T.R., Fishbein, D.B., Talamante, H.E., Hall, D.B., Chavez, G.F., Dobbins, J.G., Muro, F.J., Bustos, J.L., de los Angeles Ricardy, M., Munguia, A., Carrasco, J., Robles, A.R., and Baer, G.M. (1993). Urban epizootic of rabies in Mexico: epidemiology and impact of animal bite injuries. *Bulletin of the World Health Organization*, **71**, 615–24.

Estrada, R., Vos, A., De Leon, R., and Mueller, T. (2001). Field trial with oral vaccination of dogs against rabies in the Philippines. *BMC Infectious Diseases*, **1**, 23–30.

Ezeokoli, C.D., Ogunkoya, A.B., Beran, G.W., Belino, E.D., and Eze, E.U. (1984). The epidemiology of rabies in Kaduna State: results of a survey. *Bulletin of Animal Health and Production in Africa*, **32**, 378–84.

Ezeokoli, C.D. and Umoh, J.U. (1987). Epidemiology of rabies in northern Nigeria. *Transactions of the Royal Society of Tropical Medicine and Hygiene*, **81**, 268–72.

Faleke, O.O. (2003). Studies on dog population and its implication for rabies control. *Nigerian Journal of Animal Production*, **30**, 242–45.

FAO (2001). *Human Energy Requirements*. Report of a Joint FAO/WHO/UNU Expert Consultation, Food and Nutrition Technical Report Series, Rome.

FEDIAF (2010). *The European Pet Food Industry: Facts and Figures 2010*. European Pet Food Industry Federation, Brussels.

Fielding, W. J. (2010). Dog breeding in New Providence, The Bahamas and its potential impact on the roaming dog population I: planned and accidental. *Journal of Applied Animal Welfare Science*, **13**, 1–13.

Fielding, W. J. and Mather, J. (2000). Stray dogs in an island community: a case study from New Providence, The Bahamas. *Journal of Applied Animal Welfare Science*, **3**, 305–19.

Fielding, W. J., Mather, J., and Issacs, M. (2005). *Potcakes: Dog Ownership in New Providence, The Bahamas*. Purdue University Press, West Lafayette, Indiana.

Fielding, W. J. and Plumridge, S. J. (2005). Characteristics of owned dogs on the island of New Providence, The Bahamas. *Journal of Applied Animal Welfare Science*, **8**, 245–60.

Fiorello, C.V., Noss, A.J., and Deem, S.L. (2006). Demography, hunting ecology, and pathogen exposure of domestic dogs in the Isoso of Bolivia. *Conservation Biology*, **20**, 762–71.

Fishbein, D.B., Frontini, M.G., Dobbins, J.G., Flores Collins, E., Quiroz Huerta, G., Gamez Rodriguez, J.J., Woo-Ming, B., Garza Ramos, J., Belotto, A.J., Balderas Torres, J.M., Yenne, K.M., Linhart, S.B., and Baer, G.M. (1992). Prevention of canine rabies in rural Mexico: an epidemiologic study of vaccination campaigns. *American Journal of Tropical Medicine and Hygiene*, **47**, 317–27.

Flores-Ibarra, M. and Estrella-Valenzuela, G. (2004). Canine ecology and socioeconomic factors associated with dogs unvaccinated against rabies in a Mexican city across the US–Mexico border. *Preventive Veterinary Medicine*, **62**, 79–87.

Galibert, F., Quignon, P., Hitte, C., and Andre, C. (2011). Toward understanding dog evolutionary and domestication history. *Comptes Rendus Biologies*, **334**, 190–96.

Gavgani, A.S.M., Mohite, H., Edrissian, G.H., Mohebali, M., and Davies, C.R. (2002). Domestic dog ownership in Iran is a risk factor for human infection with *Leishmania infantum*. *American Journal of Tropical Medicine and Hygiene*, **67**, 511–15.

Gentry, A., Clutton-Brock, J., and Groves, C.P. (1996). Proposed conservation of usage of 15 mammal specific names based on wild species which are antedated by or contemporary with those based on domestic animals. *Bulletin of Zoological Nomenclature*, **53**, 28–37.

Gentry, A., Clutton-Brock, J., and Groves, C.P. (2004). The naming of wild animal species and their domestic derivatives. *Journal of Archaeological Science*, **31**, 645–51.

Germonpré, M., Lázničková-Galetová, M., and Sablin, M.V. (2012). Palaeolithic dog skulls at the Gravettian Předmostí site, the Czech Republic. *Journal of Archaeological Science*, **39**, 184–202.

Germonpré, M., Sablin, M.V., Després, V.R., Hofreiter, M., Lázničková-Galetová, M., Stevens, R.E., and Stiller, M. (2013). Palaeolithic dogs and the early domestication of the wolf: a reply to the comments of Crockford and Kuzmin (2012). *Journal of Archaeological Science*, **40**, 786–92.

Germonpré, M., Sablin, M.V., Stevens, R.E., Hedges, R.E.M., Hofreiter, M., Stiller, M., and Després, V.R. (2009). Fossil dogs and wolves from Palaeolithic sites in Belgium, the Ukraine and Russia: osteometry, ancient DNA and stable isotopes. *Journal of Archaeological Science*, **36**, 473–90.

Glen, A.S. (2010). Hybridisation between dingoes and domestic dogs: A comment on Jones (2009). *Australian Mammalogy*, **32**, 76–7.

Glen, A.S. (2012). Enough dogma: seeking the middle ground on the role of dingoes. *Current Zoology*, **58**, 189–210.

Gompper, M.E. and Gittleman, J.L. (1991). Home range scaling: intraspecific and comparative patterns. *Oecologia*, **87**, 343–48.

Goodrich, J.M., Seryodkin, I.V., Miquelle, D.G., and Bereznuk, S.I. (2011). Conflicts between Amur (Siberian) tigers and humans in the Russian Far East. *Biological Conservation*, **144**, 584–92.

Grayson, D.K., Parmalee, P.W., Lyman, R.L., and Mead, J.I. (1988). Danger Cave, Last Supper Cave, and Hanging

Rock Shelter: the faunas. *Anthropological Papers of the American Museum of Natural History*, **66** (**part 1**), 1–130.

Green, J.S. and Gipson, P.S. (1994). Feral dogs. In S.E. Hygnstrom, R.M. Timm and G.E. Larson, Eds., *The Handbook: Prevention and Control of Wildlife Damage*. University of Nebraska, Lincoln, pp. 1–7.

Groves, C.P. (1995). On the nomenclature of domestic animals. *Bulletin of Zoological Nomenclature*, **52**, 137–41.

Gürtler, R.E., Cecere, M.C., Lauricella, M.A., Cardinal, M.V., Kitron, U., and Cohen, J.E. (2007). Domestic dogs and cats as sources of *Trypanosoma cruzi* infection in rural northwestern Argentina. *Parasitology*, **134**, 1–14.

Hassan, M.M., Osman, O.F., El-Raba'a, F.M.A., Schallig, H.D.F.H., and Elnaiem, D.-E.A. (2009). Role of the domestic dog as a reservoir host of *Leishmania donovani* in eastern Sudan. *Parasites and Vectors*, **2**, 26.

Haydon, D.T., Randall, D.A., Matthews, L., Knobel, D.L., Tallents, L.A., Gravenor, M.B., Williams, S.D., Pollinger, J.P., Cleaveland, S., Woolhouse, M.E., Sillero-Zubiri, C., Marino, J., Macdonald, D.W., and Laurenson, M.K. (2006). Low-coverage vaccination strategies for the conservation of endangered species. *Nature*, **443**, 692–95.

Hill, F.W.G. (1985). A survey of the animal population of four high density suburbs in Harare. *Zimbabwe Veterinary Journal*, **16** (3/4), 31–6.

Hsu, Y., Severinghaus, L.L., and Serpell, J.A. (2003). Dog keeping in Taiwan: its contribution to the problem of free-roaming dogs. *Journal of Applied Animal Welfare Science*, **6**, 1–23.

Hu, R., Tang, Q., Tang, J., and Fooks, A.R. (2009). Rabies in China: an update. *Vector-Borne and Zoonotic Diseases*, **9**, 1–11.

Hudjashov, G., Kivisild, T., Underhill, P.A., Endicott, P., Sanchez, J.J., Lin, A.A., Shen, P., Oefner, P., Renfrew, C., and Villems, R. (2007). Revealing the prehistoric settlement of Australia by Y chromosome and mtDNA analysis. *Proceedings of the National Academy of Sciences, USA*, **104**, 8726–30.

Hughes, J. and Macdonald, D.W. (2013). A review of the interactions between free-roaming domestic dogs and wildlife, *Biological Conservation*, **157**, 341–51.

Idachaba, S.E. (2009). Status of canine vaccination and the prevalence of rabies in humans and dogs in Plateau State, Nigeria 1998–2007. MS thesis, University of Pretoria.

Johnson C. N. and Wroe, S. (2003). Causes of extinction of vertebrates during the Holocene of mainland Australia: arrival of the dingo, or human impact? *The Holocene*, **13**, 941–8.

Jones, E. (2009). Hybridisation between the dingo *Canis lupus dingo* and the domestic dog *Canis lupus familiaris* in Victoria: a critical review. *Australian Mammology*, **31**, 1–7.

Joshi, D.D. and Bögel, K. (1988). Role of lesser developed nations in rabies research. *Reviews of Infectious Diseases*, **10**, Suppl. 4, S600–3.

Kaare, M., Lembo, T., Hampson, K., Ernest, E., Estes, A., Mentzel, C., and Cleaveland, S. (2009). Rabies control in rural Africa: evaluating strategies for effective domestic dog vaccination. *Vaccine*, **27**, 152—60.

Kato, M., Yamamoto, H., Inukai, Y., and Kira, S. (2003). Survey of the stray dog population and the health education program on the prevention of dog bites and dog-acquired infections: a comparative study in Nepal and Okayama Prefecture, Japan. *Acta Medica Okayama*, **57**, 261–66.

Kayali, U., Mindekem, R., Yémadji, N., Vounatsou, P., Kaninga, Y., Ndoutamia, A.G., and Zinsstag, J. (2003). Coverage of pilot parenteral vaccination campaign against canine rabies in N'Djaména, Chad. *Bulletin of the World Health Organization*, **81**, 739–44.

Kitala, P., McDermott, J., Kyule, M., Gathuma, J., Perry, B., and Wandeler, A. (2001). Dog ecology and demography information to support the planning of rabies control in Machakos District, Kenya. *Acta Tropica*, **78**, 217—30.

Knobel, D.L., Laurenson, M.K., Kazwala, R.R., Boden, L.A., and Cleaveland, S. (2008). A cross-sectional study of factors associated with dog ownership in Tanzania. *BMC Veterinary Research*, **4**, 5.

Kongkaew, W., Coleman, P., Pfeiffer, D.U., Antarasena, C., and Thiptara, A. (2004). Vaccination coverage and epidemiological parameters of the owned-dog population in Thungsong District, Thailand. *Preventive Veterinary Medicine*, **65**, 105–15.

Koster, J. M. (2008). The impact of hunting with dogs on wildlife harvests in the Bosawas Reserve, Nicaragua. *Environmental Conservation*, **35**, 211–20.

Koster, J. M. (2009). Hunting dogs in the lowland Neotropics. *Journal of Anthropological Research*, **65**, 575–610.

Kruuk, H. and Snell, H. (1981). Prey selection by feral dogs from a population of marine iguanas (*Amblyrhynchus cristatus*). *Journal of Applied Ecology*, **18**, 197–204.

Larson, G., Karlsson, E. K., Perri, A., Webster, M. T., Ho, S .Y., Peters, J., Stahl, P.W., Piper, P.J., Lingaas, F., Fredholm, M., Comstock, K.E., Modiano, J.F., Schelling, C., Agoulnik, A.I., Leegwater, P.A., Dobney, K., Vigne, J.D., Vilà, C., Andersson, L., and Lindblad-Toh, K. (2012). Rethinking dog domestication by integrating genetics, archeology, and biogeography. *Proceedings of the National Academy of Sciences, USA*, **109**, 8878–83.

Laurenson, M.K., Mlengeya, T., Shiferaw, F., and Cleaveland S. (2005). Approaches to disease control in domestic canids for the conservation of endangered wild carnivores. In S. Cleaveland, W.B. Karesh, D. Kock, P.J. Nyhus, L. Starr and A. Yang, Eds., *Conservation and*

Development Interventions at the Wildlife/Livestock Interface: Implications for Wildlife, Livestock and Human Health. IUCN Publications Services Unit, Gland and Cambridge, pp. 141–6.

Laurenson, M.K., Shiferaw, F., and Sillero-Zubiri, C. (1997). Rabies as a threat to the Ethiopian wolf (Canis simensis). In *Proceedings of the Southern and Eastern African Rabies Group (SEARG) Meeting*, Nairobi. Fondation Merieux, France, pp. 97–103.

Lenth, B.E., Knight, R.L., and Brennan, M.E. 2008. The effects of dogs on wildlife communities. *Natural Areas Journal*, **28**, 218–27.

Leonard, J.A., Wayne, R.K., Wheeler, J., Valadez, R., Guillén, S., and Vilà, C. (2002). Ancient DNA evidence for Old World origin of New World dogs. *Science*, **298**, 1613–6.

Letnic, M., Crowther, M.S., Dickman, C.R., and Ritchie, E. (2011). Demonising the dingo: how much wild dogma is enough. *Current Zoology*, **57**, 668–70.

Letnic, M., Fillios, M., and Crowther, M.S. (2012). Could direct killing by larger dingoes have caused the extinction of the thylacine from mainland Australia? *PLoS ONE*, **7**, e34877.

Linnaeus, C. (1758). *Systema Naturae*, Edn. 10 vol. 1. Salvii, Holmiae.

Lunney, M., Fèvrea, S.J.S., Stiles, E., Ly, S., San, S., and Vong, S. (2012). Knowledge, attitudes and practices of rabies prevention and dog bite injuries in urban and peri-urban provinces in Cambodia, 2009. *International Health*, **4**, 4–9.

Ly, S., Buchy, P., Heng, N.Y., Ong, S., Chhor, N., Bourhy, H., and Vong, S. (2009). Rabies situation in Cambodia. *PLoS Neglected Tropical Diseases*, **3**(9), e511.

Lyngset, A. and Lyngset, O. (1970). Kullstørrelse hos hund [Litter size in the dog]. *Nordisk Veterinaermedicin*, **22**, 186–91.

López, G., López-Parra, M., Fernández, L., Martínez-Granados, C., Martínez, F., Meli, M.L., Gil-Sánchez, J.M., Viqueira, N., Díaz-Portero, M.A., Cadenas, R., Lutz, H., Vargas, A., and Simón, M.A. (2009). Management measures to control a feline leukemia virus outbreak in the endangered Iberian lynx. *Animal Conservation*, **12**, 173–82.

MacNeish, R.S. and Vierra, R.K. (1983). The Preceramic way of life in the Thorn Forest Riverine Ecozone. In R.S. MacNeish, R.K. Vierra, A. Nelken-Terner, R. Lurie and A.G.Cook, Eds., *Prehistory of the Ayacucho Basin, Peru, Vol.4: The Preceramic Way of Life*. University of Michigan Press, Ann Arbor, pp. 48–129.

Macdonald, D.W. and Carr, G.M. (1995). Variation in dog society: between resource dispersion and social flux. In J. Serpell, Ed., *The Domestic Dog: Its Evolution, Behavior and Interactions with People*. Cambridge University Press, Cambridge, pp. 199–216.

Matibag, G.C., Ohbayashi, Y., Kanda, K., Yamashina, H., Bandula Kumara, W.R., Perera, I.N.G., De Silva, D.D.N., Gunawardena, G.S.P. De S., Jayasinghe, A., Ditangco, R.A., and Tamashiro, H. (2009). A pilot study on the usefulness of information and education campaign materials in enhancing the knowledge, attitude and practice on rabies in rural Sri Lanka. *Journal of Infection in Developing Countries*, **3**(1), 55–64.

Matter, H.C., Schumacher, C.L., Kharmachi, H., Hammami, S., Tlatli, A., Jemli, J., Mrabet, L., Meslin, F.-X., Aubert, M.F.A., Neuenschwander, B., and El Hicheri, K. (1998). Field evaluation of two bait delivery systems for the oral immunization of dogs against rabies in Tunisia. *Vaccine*, **16**, 657–65.

Matter, H.C., Wandeler, A.I., Neuenschwander, B.E., Harischandra, L.P., and Meslin, F.X. (2000). Study of the dog population and the rabies control activities in the Mirigama area of Sri Lanka. *Acta Tropica*, **75**, 95–108.

McCollum, A.M., Blanton, J.D., Holman, R.C., Callinan, L.S., Baty, S., Phillips, R., Callahan, M., Levy, C., Komatsu, K., Sunenshine, R., Bergman, D.L., and Rupprecht, C.E. (2012). Community survey after rabies outbreaks, Flagstaff, Arizona, USA. *Emerging Infectious Diseases*, **18**, 932–8.

McCrindle, C.M.E., Gallant, J., Cornelius, S.T., and Schoeman, H.S. (1999). Changing roles of dogs in urban African society: a South African perspective. *Anthrozoös*, **12**, 157–161.

Meehan, B., Jones, R., and Vincent, A. (1999). Gulu-kula: dogs in Anbarra society, Arnhem Land. *Aboriginal History*, **23**, 83–106.

Meek, P.D. (1999). The movement, roaming behaviour and home range of free-roaming domestic dogs, *Canis lupus familiaris*, in coastal New South Wales. *Wildlife Research*, **26**, 847–55.

Mettler, F. (2003). Some aspects of the rabies epizootic in Namibia. In *Proceedings of the Seventh Southern and Eastern African Rabies Group Meeting*. Ezulwini, Swaziland, pp. 43–46.

Miklósi, A. (2007). *Dog Behaviour, Evolution, and Cognition*. Oxford University Press, Oxford.

Mindekem, R., Kayali, U., Yemadji, N., Ndoutamia, A.G., and Zinsstag, J. (2005). Impact of canine demography on rabies transmission in N'djamena, Chad. *Médecine Tropicale: Revue du Corps de Santé Colonial*, **65**, 53–8.

Monello, R.J. and Gompper, M.E. (2007). Biotic and abiotic predictors of tick (*Dermacentor variabilis*) abundance and engorgement on free-ranging raccoons (Procyon lotor). *Parasitology*, **134**, 2053–62.

Morales, M.A., Varas, C., and Ibarra, L. (2009). Caracterización demográfica de la población de perros de Viña del Mar, Chile. *Archivos de Medicina Veterinaria*, **41**, 89–95.

Moreira, Jr., E.D., Mendes de Souzam, V.M., Sreenivasan, M., Nascimento, E.G., and Pontes de Carvalho, L. (2004). Assessment of an optimized dog-culling program in the dynamics of canine *Leishmania* transmission. *Veterinary Parasitology*, **122**, 245–52.

Morey, D.F. (2010). *Dogs: Domestication and the Development of a Social Bond*. Cambridge University Press, Cambridge.

Musil, R. (2000). Evidence for the domestication of wolves in central European Magdalenian sites. In S.J. Crockford, Ed., *Dogs Through Time: An Archaeological Perspective*. British Archaeological Reports, Oxford, pp. 21–8.

Napierala, H. and Uerpmann, H.P. (2012). A "new" palaeolithic dog from central Europe. *International Journal of Osteoarchaeology*, **22**, 127–37

Nesbitt, W.H. (1975). Ecology of a feral dog pack on a wildlife refuge. In M.W. Fox, Ed., *The Wild Canids: Their Systematics, Behavioral Ecology and Evolution*. Van Nostrand Reinhold, New York, pp. 391–5.

New, J.C., Kelch, W.J., Hutchison, J.M., Salman, M.D., King, M., Scarlett, J.M., and Kass, P.H. (2004). Birth and death rate estimates of cats and dogs in U.S. households and related factors. *Journal of Applied Animal Welfare Science*, **7**, 229–41.

Nunes, C.M., Lima, V.M.F., Paula, H.B., Perri, S.H.V., Andrade, A.M., Dias, F.E.F., and Burattini, M.N. (2008). Dog culling and replacement in an area endemic for visceral leishmaniasis in Brazil. *Veterinary Parasitology*, **153**, 19–23.

Oboegbulem, S.I. and Nwakonobi, I.E. (1989). Population density and ecology of dogs in Nigeria: a pilot study. *Revue Scientifique et Technique*, **8**, 733–45.

Olsen, S.J. and Olsen, J.W. (1977). The Chinese wolf, ancestor of New World dogs. *Science*, **197**, 533–5.

Orihuela, T. A. and Solano, V. J. (1995). Demographics of the owned dog population in Miacatlan, Mor. Mexico. *Anthrozoös*, **8**, 171–5.

Ortega-Pacheco, A., Rodriguez-Buenfil, C., Bolio-Gonzalez, M. N., Sauri-Arceo, C. H., Jiménez-Coello, M., and Forsberg, C. L. (2007). A survey of dog populations in urban and rural areas of Yucatan, Mexico. *Anthrozoös*, **20**, 261–74.

Ovodov, N.D., Crockford, S.J., Kuzmin, Y.V., Higham, T.F.G., Hodgins, G.W.L., and van der Plicht, J. (2011). A 33000-year-old incipient dog from the Altai Mountains of Siberia: evidence of the earliest domestication disrupted by the last glacial maximum. *PLoS ONE*, **6**, e22821.

O'Connor, T.P. (1997). Working at relationships: another look at animal domestication. *Antiquity*, **71**, 149–56.

Pal, S.K. (2001). Population ecology of free-ranging urban dogs in West Bengal, India. *Acta Theriologica*, **46**, 69–78.

Pang, J.-F., Kluetsch, C., Zou, X.-J., Zhang, A., Luo, L.-Y., Angleby, H., Ardalan, A., Ekström, C., Sköllermo, A., Lundeberg, J., Matsumura, S., Leitner, T., Zhang, Y.-P., and Savolainen, P. (2009). mtDNA data indicate a single origin for dogs south of Yangtze River, less than 16300 years ago, from numerous wolves. *Molecular Biology and Evolution*, **26**, 2849–64.

Parsons, M.H. and Blumstein, D.T. (2010). Familiarity breeds contempt: kangaroos persistently avoid areas with experimentally deployed dingo scents. *PLoS ONE*, **5**, e10403.

Perera, I.N.G., Pinidiyapathirage, M.J., and Wickremasinghe, A.R. (2007). Rabies vaccination of domesticated dogs in the Central Province of Sri Lanka. *Ceylon Medical Journal*, **52**, 69–70.

Perrin, T. (2009). The Business of Urban Animals Survey: the facts and statistics on companion animals in Canada. *Canadian Veterinary Journal*, **50**, 48–52.

Phillips, R.B., Wiedenfeld, D.A., and Snell, H.L. (2012). Current status of alien vertebrates in the Galápagos Islands: invasion history, distribution, and potential impacts. *Biological Invasions*, **14**, 461–80

Pionnier-Capitan, M., Bemilli, C., Bodu, P., Célérier, G., Ferrié, J.G., Fosse, P., Garcia, M., and Vigne, J.D. (2011). New evidence for upper Palaeolithic small domestic dogs in South Western Europe. *Journal of Archaeological Science*, **38**, 2123–40.

Podberscek, A.L. (2009). Good to pet and eat: the keeping and consuming of dogs and cats in South Korea. *Journal of Social Issues*, **65**, 615–32.

Polgreen, L. (2009). Matchmaking in India—canine division. *New York Times*, August 18, 2009, page A5.

Prates, L., Prevosti, F., and Berón, M. (2010). First records of prehispanic dogs in southern South America (Pampa-Patagonia, Argentina). *Current Anthropology*, **51**, 273–80.

Price, E.O. (2002) *Animal Domestication and Behavior*. CABI Publishing, New York.

Ratsitorahina, M., Rasambainarivo, J.H., Raharimanana, S., Rakotonandrasana, H., Andriamiarisoa, M.P., Rakalomanana, F.A., and Richard, V. (2009). Dog ecology and demography in Antananarivo, 2007. *BMC Veterinary Research*, **5**, 21.

Rautenbach, G.H., Boomker, J., and de Villiers, I.L. (1991). A descriptive study of the canine population in a rural town in southern Africa. *Journal of the South African Veterinary Association*, **62**, 158–62.

Reece, J.F. and Chawla, S.K. (2006). Control of rabies in Jaipur, India, by the sterilisation and vaccination of neighbourhood dogs. *Veterinary Record*, **159**, 379–83.

Reece, J.F., Chawla, S.K., Hiby, E.F., and Hiby, L.A. (2008). Fecundity and longevity of free-roaming dogs in Jaipur, India. *BMC Veterinary Research*, **4**, 6.

Reed, S.E. and Merenlender, A.M. (2011). Effects of management of domestic dogs and recreation on carnivores in protected areas in Northern California. *Conservation Biology*, **25**, 504–13.

Reithinger, R., Canales Espinoza, J., Llanos-Cuentas, A., and Davies, C.R. (2003). Domestic dog ownership: a risk factor for human infection with *Leishmania* (*Viannia*) species. *Transactions of the Royal Society of Tropical Medicine and Hygiene*, **97**, 141–5.

Robinson, L.E., Miranda, M.E., Miranda, N.L., and Childs, J.E. (1996). Evaluation of a canine rabies vaccination campaign and characterization of owned-dog populations in the Philippines. *Southeast Asian Journal of Tropical Medicine and Public Health*, **27**, 250–6.

Rödl, T., Berger, S., Romero, L.M., and Wikelski, M. (2007). Tameness and stress physiology in a predator-naive island species confronted with novel predation threat. *Proceedings of the Royal Society, Series B*, **274**, 577–82.

Romero-Lopez, J.A., Jaramillo-Arango, C.J., Martinez-Maya, J.J., Alvarez Peralta, E., and Terrones, C.R. (2008). Study of the population structure of dogs in a political district in Mexico City. *Journal of Animal and Veterinary Advances*, **7**, 1352–7.

Rubel, D., Zunino, G., Santillán, G., and Wisnivesky, C. (2003). Epidemiology of *Toxocara canis* in the dog population from two areas of different socioeconomic status, Greater Buenos Aires, Argentina. *Veterinary Parasitology*, **115**, 275–86

Ruiz-Izaguirre, E. and Eilers, C.H.A.M. (2012). Perceptions of village dogs by villagers and tourists in the coastal region of rural Oaxaca, Mexico. *Anthrozoös*, **25**, 75–91.

Russell, N. (2002). The wild side of animal domestication. *Society & Animals*, **10**, 285–302.

Ryan, J. S. (1964). Plotting an isogloss: the location and types of aboriginal names for native dog in New South Wales. *Oceania*, **35**, 111–23.

Sablin, M.V. and Khlopachev, G.A. (2002). The earliest ice age dogs: evidence from Eliseevichi 11. *Current Anthropology*, **43**, 795–9.

Sagarasaeranee, P., Puanghat, A., Kasempimolparn, S., and Khawplod, P. (2001). Efficacy of oral rabies vaccine in dogs in Thailand. In B. Dodet, F.-X. Meslin, and E. Heseltine, Eds., *Rabies Control in Asia*. John Libbey Eurotext, Paris, pp. 101–4.

Sallander, M., Hagberg, M., Hedhammar, Å., Rundgren, M., and Lindberg, J.E. (2010). Energy-intake and activity risk factors for owner-perceived obesity in a defined population of Swedish dogs. *Preventive Veterinary Medicine*, **96**, 132–41.

Sandell, M. (1989). The mating tactics and spacing patterns of solitary carnivores. In J.L. Gittleman, Ed., *Carnivore Behavior, Ecology and Evolution*. Cornell University Press, Ithaca, pp. 164–82.

Savolainen, P., Leitner, T., Wilton, A.N., Matisoo-Smith, E., and Lundeberg, J. (2004). A detailed picture of the origin of the Australian dingo, obtained from the study of mitochondrial DNA. *Proceedings of the National Academy of Sciences, USA*, **101**, 12387–90.

Savolainen, P., Zhang, Y.P., Luo, J., Lundeberg, J., and Leitner, T. (2002). Genetic evidence for an East Asian origin of domestic dogs. *Science*, **298**, 1610–3.

Seghaier, C., Cliquet, F., Hammami, S., Aouina, T., Tlatli, A., and Aubert, M. (1999). Rabies mass vaccination campaigns in Tunisia: are vaccinated dogs correctly immunized? *American Journal of Tropical Medicine and Hygiene*, **61**, 879–84.

Shiferaw, F., Laurenson, K., and Sillero Zubiri, C. (1998). Disease and dogs as threats to the endangered Ethiopian Wolf (*Canis simensis*). *Journal of the Ethiopian Veterinary Association*, **2**(1), 15–27.

Silva-Rodriguez, E.A., Ortega-Solis, G.R., and Jimenez, J.E. (2010). Conservation and ecological implications of the use of space by chilla foxes and free-ranging dogs in a human-dominated landscape in southern Chile. *Austral Ecology*, **35**, 765–77.

Silva-Rodríguez, E.A. and Sieving, K.E. (2011). Influence of care of domestic carnivores on their predation on vertebrates. *Conservation Biology*, **25**, 808–15.

Silva-Rodríguez, E.A. and Sieving, K.E. (2012). Domestic dogs shape the landscape-scale distribution of a threatened forest ungulate. *Biological Conservation*, **150**, 103–10.

Singhchai, C. (2001). Dog rabies control in Bangkok metropolitan area: why has rabies not been eliminated from Bangkok? In B. Dodet, F.-X. Meslin, and E. Heseltine, Eds., *Rabies Control in Asia*. John Libbey Eurotext, Paris, pp. 91–93.

Slater, M.R., Di Nardo, A., Pediconi, O., Dalla Villa, P., Candeloro, L., Alessandrini, B., and Del Papa, S. (2008). Free-roaming dogs and cats in central Italy: public perceptions of the problem. *Preventive Veterinary Medicine*, **84**, 27–47.

Slater, M.R., Di Nardo, A., Pediconi, O., Villa, P.D., Candeloro, L., Alessandrini, B., and Del Papa, S. (2008) Cat and dog ownership and management patterns in central Italy. *Preventive Veterinary Medicine*, **85**, 267–94.

Smith, B.P. and Litchfield, C.A. (2009). A review of the relationship between indigenous Australians, dingoes (*Canis dingo*) and domestic dogs (*Canis familiaris*). *Anthrozoös*, **22**, 111–28.

Srbek-Araujo, A. and Chiarello, A. (2008). Domestic dogs in Atlantic forest preserves of south-eastern Brazil: a camera-trapping study on patterns of entrance and site occupancy rates. *Brazilian Journal of Biology*, **68**, 771–9.

Sterl, P., Brandenburg, C., and Arnberger, A. (2008). Visitors' awareness and assessment of recreational disturbance of wildlife in the Donau-Auen National Park. *Journal for Nature Conservation*, **16**, 135–45.

Steven, R., Pickering, C., and Castley, J.G. (2011). A review of the impacts of nature based recreation on birds. *Journal of Environmental Management*, **92**, 2287–94.

Sudarshan, M.K., Mahendra, B.J., Madhusudana, S.N., Ashwoath Narayana, D.H., Rahman, A., Rao, N.S.X-, Meslin, F., Lobo, D., Ravikumar, K., and Gangaboraiah (2006). An epidemiological study of animal bites in India: results of a WHO sponsored national multi-centric rabies surveys. *Journal of Communicable Diseases*, **38**, 32–9.

Sudarshan, M.K., Mahendra, B.J., and Narayan, D.H. (2001) A community survey of dog bites, anti-rabies treatment, rabies and dog population management in Bangalore city. *Journal of Communicable Diseases*, **33**, 245–51.

Suluku, R., Abu-Bakarr, I., Johnny, J., and Jonsyn-Ellis F. (2012). Post-war demographic and ecological survey of dog populations and their human relationships in Sierra Leone (a case study of urban Freetown). *Science Journal of Agricultural Research & Management*, **2012**, sjarm–282, doi: 10.7237/sjarm/28.

Suzuki, K., Pereira, J.A., Frías, L.A., López, R., Mutinelli, L.E., and Pons, E.R. (2008). Rabies-vaccination coverage and profiles of the owned-dog population in Santa Cruz de la Sierra, Bolivia. *Zoonoses and Public Health*, **55**, 177–83.

Takahashi-Omoe, H. and Omoe, K. (2012). Social environment and control status of companion animal-borne zoonoses in Japan. *Animals*, **2**, 38–54.

Torres, P.C. and Prado, P.I. (2010). Domestic dogs in a fragmented landscape in the Brazilian Atlantic Forest: abundance, habitat use and caring by owners. *Brazilian Journal of Biology*, **70**, 987–94.

Totton, S.C., Wandeler, A.I., Zinsstag, J., Bauch, C.T., Ribble, C.S., Rossett, R.C., and McEwen, S.A. (2010a). Stray dog population demographics in Jodhpur, India following a population control/rabies vaccination program. *Preventive Veterinary Medicine*, **97**, 51–7.

Totton, S.C., Wandeler, A.I., Gartley, C.J., Kachhawaha, S., Suman, M., Ribble, C.S., Rosatte, R.C., and McEwen, S.A. (2010b). Assessing reproductive patterns and disorders in free-ranging dogs in Jodhpur, India to optimize a population control program. *Theriogenology*, **74**, 1115–20.

Touihri, L., Zaouia, I., Elhili, K., Dellagi, K., and Bahloul, C. (2009). Evaluation of mass vaccination campaign coverage against rabies in dogs in Tunisia. *Zoonoses and Public Health*, **58**, 110–18.

Towns, D.R. and Daugherty, C.H. (1994). Patterns of range contractions and extinctions in the New Zealand herpetofauna following human colonization. *New Zealand Journal of Zoology*, **21**, 325–39.

Trut, L., Oskina, I., and Kharlamova, A. (2009). Animal evolution during domestication: the domesticated fox as a model. *Bioessays*, **31**, 349–60.

Tung, M.-C., Fei, C.-Y., Chiang, J.-T., Chou, C.-H., Yeh, L.-S., Liao, C.-Y., Su, Y.-C., Chang, J.-C., and Tung, K.-C. (2010), Surveys of dog populations in Taiwan from 1999 to 2009. *Journal of the Chinese Society of Animal Science*. **39**, 175–88.

Tønnessen. R., Sverdrup Borge, K., Nødtvedt, A., and Indrebø, A. (2012). Canine perinatal mortality: a cohort study of 224 breeds. *Theriogenology*, **77**, 1788–1801.

United Nations (2011). *World Population Prospects: The 2010 Revision*, CD-ROM Edition. United Nations, Department of Economic and Social Affairs, Population Division <http://www.un.org/popin/data.html>.

Vanak, A.T. and Gompper, M.E. (2009a). Dietary niche separation between sympatric free-ranging domestic dogs and Indian foxes in central India. *Journal of Mammalogy*, **90**, 1058–65.

Vanak, A.T. and Gompper, M.E. (2009b). Dogs as carnivores: their role and function in intraguild competition. *Mammal Review*, **39**, 265–83.

Vanak, A.T. and Gompper, M.E. (2010). Interference competition at the landscape level: the effect of free-ranging dogs on a native mesocarnivore. *Journal of Applied Ecology*, **47**, 1225–32.

van Sittert, L. and Swart, S. (2003). *Canis familiaris*: A dog history of South Africa. *South African Historical Journal*, **48**, 138–73.

Verginelli, F., Capelli, C., Coia, V., Musiani, M., Falchetti, M., Ottini, L., Palmirotta, R., Tagliacozzo, A., De Grossi Mazzorin, I., and Mariani-Costantini, R. (2005). Mitochondrial DNA from prehistoric canids highlights relationships between dogs and South-East European wolves. *Molecular Biology and Evolution*, **22**, 2541–51.

Vial, F., Cleaveland, S., Rasmussen, G., and Haydon, D.T. (2006). Development of vaccination strategies for the management of rabies in African wild dogs. *Biological Conservation*, **131**, 180–92.

Vilà, C., Savolainen, P., Maldonado, J.E., Amorim, I.R., Rice, J.E., Honeycutt, R.L., Crandall, K.A., Lundeberg, J., and Wayne, R.K. (1997). Multiple and ancient origins of the domestic dog. *Science*, **276**, 1687–89.

vonHoldt, B.M., Pollinger, J.P., Lohmueller, K.E., Han, E.J., Parker, H.G., Quignon, P., Degenhardt, J.D., Boyko, A.R., Earl, D.A., Auton, A., Reynolds, A., Bryc, K., Brisbin, A., Knowles, J.C., Mosher, D.S., Spady, T.C., Elkahloun, A., Geffen, E., Pilot, M., Jedrzejewski, W., Greco, C., Randi, E., Bannasch, D., Wilton, A., Shearman, J., Musiani, M., Cargill, M., Jones, P.G., Qian, Z., Huang, W., Ding, Z.-L., Zhang, Y.-P., Bustamante, C.D., Ostrander, E.A., Novembre, J., and Wayne, R.K. (2010). Genomewide SNP and haplotype analyses reveal a rich history underlying dog domestication. *Nature*, **464**, 898–903.

Vos, A. and Turan, B. (1996). A descriptive and estimative study of the dog population in Istanbul, Turkey. *Rabies Bulletin Europe*, **20(2)**, 11–14.

Wandeler, A.I., Budde, A., Capt, S., Kappeler, A., and Matter, H. (1988). Dog ecology and dog rabies control. *Reviews of Infectious Diseases*, **10**(Suppl. 4), S684–8.

Wandeler, A.I., Matter, H.C., Kappeler, A., and Budde, A. (1993). The ecology of canine rabies: A selective review. *Revue Scientifique el Technique de l'Office International des Epizooties*, **12**, 51–71.

Wayne, R.K. and vonHoldt, B.M. (2012). Evolutionary genomics of dog domestication. *Mammal Genome*, **23**, 3–18.

Weng, H.-Y., Kass, P.H., Hart, L.A., and Chomel, B.B. (2006). Risk factors for unsuccessful dog ownership: an epidemiologic study in Taiwan. *Preventive Veterinary Medicine*, **77**, 82–95.

Wilson, K., Bjornstad, O.N., Dobson, A.P., Merler, S., Poglayen, G., Randolph, S.E., Read, A.F., and Skorping, A. (2002). Heterogeneities in macroparasite infections: patterns and processes. In P.J. Hudson, A. Rizzoli, B. T. Grenfell, H. Heesterbeek and A.P. Dobson, Eds., *The Ecology of Wildlife Diseases*. Oxford University Press, pp. 6–44.

Woodroffe, R. and Donnelly, C.A. (2011). Risk of contact between endangered African wild dogs *Lycaon pictus* and domestic dogs: opportunities for pathogen transmission. *Journal of Applied Ecology*, **48**, 1345–54.

Youssef, S.B., Matter, H.C., Schumacher, C.L., Kharmachi, H., Jemli, J., Mrabet, L., Ghargi, M., Hammami, S., Hicheri, K.E., Aubert, M.F.A., and Meslin, F.X. (1998). Field evaluation of a dog owner, participation-based, bait delivery system for the oral immunization of dogs against rabies in Tunisia. *American Journal of Tropical Medicine and Hygiene*, **58**, 835–45.

Zeuner, F.E. (1963). *A History of Domesticated Animals*. Harper & Row, New York.

Zumpano, R., Tortosa, A., and Degregorio, O.J. (2011). Estimación del impacto de la esterilización en el índice de crecimiento de la población de caninos. *Revista de Investigaciones Veterinarias del Perú*, **22**, 336–41.

CHAPTER 2

Dogs as predators and trophic regulators

Euan G. Ritchie, Christopher R. Dickman, Mike Letnic, and Abi Tamim Vanak

2.1 Introduction

There is increasing interest in the ecology of dogs, the world's most abundant carnivore (estimated 700 million–1 billion worldwide) (Hughes and Macdonald, 2013; Silva-Rodríguez and Sieving, 2012; Vanak and Gompper, 2009a; Gompper, Chapter 1). Like other large-bodied predators, dogs have important functional roles in structuring and maintaining ecological communities (Letnic et al., 2012; Ritchie and Johnson, 2009; Vanak and Gompper, 2009a), but they are also a regular source of conflict with humans due to their negative effects on people and their enterprises, for instance through their roles in disease transmission and the killing of livestock and wildlife (Baker et al., 2008; Hughes and Macdonald, 2013; King et al., 2012; Ritchie et al., 2012; Van Bommel and Johnson, 2012; Young et al., 2011). Indeed, some of the negative effects of maintaining dogs in the landscape have led to considerable recent debate about their management, most notably for dingoes, whose structuring role in Australian ecosystems may be comparable to that of wolves (*Canis lupus*) in North American settings (Allen et al., 2011a; Houston et al., 2010b, 2013; Fleming et al., 2012; Letnic et al., 2011a; Mech, 2012; Johnson and Ritchie, 2013).

Despite their controversial effects, global ubiquity, and significant ecological roles, dogs and their ecological impacts remain surprisingly poorly understood. In particular, we know little about how dogs influence other species across environmental gradients (from wild to urban) and how their influence varies between different 'types' of dogs (see Box 2.1). This lack of information is concerning, as in some parts of the world dogs, and in particular free-ranging dogs, have declined substantially due to direct human persecution and active management programs, whereas in other parts they have increased considerably, benefiting from human food and shelter subsidies (Gompper and Vanak, 2008; Ritchie et al., 2012; Vanak and Gompper, 2009b).

To better understand the consequences of changes in the distribution and abundance of dogs we require information on the roles dogs have in ecosystems, and how these vary both within and between types of dogs and different environmental contexts. Such information is also critical for scenarios where efforts are necessary to effectively manage or conserve free-ranging dogs. With this in mind, we review what is known about the predation pressure that is exerted by dogs and how this may influence trophic patterns of communities. We have deliberately chosen not to focus extensively on the theory regarding predators (including dogs) and their effects, as this is covered in detail by other recent reviews (Letnic et al., 2012; Prugh et al., 2009; Ritchie and Johnson, 2009; Ritchie et al., 2012; Vanak and Gompper, 2009a); instead we present an overview of recent work concerning the ecological roles of dogs as predators.

We based this review mainly on the primary literature that examines patterns of predation by dogs. We also used anecdotal reports from the popular press and other media to determine the range of species that are preyed upon by dogs and the extent of predation pressure. We identified several factors that contribute to the extent of predation pressure by dogs on wildlife. These range from population

Free-Ranging Dogs and Wildlife Conservation. Edited by Matthew E. Gompper
© Oxford University Press 2014. Published 2014 by Oxford University Press.

Box 2.1 A problem of definitions: ecological functions versus the identities of predators

Before we can define and understand the ecological roles of dogs, we must be clear about what we actually mean by a dog. This may seem obvious, as most people would have no problem identifying a dog from either a direct observation or from a picture, but in fact the issue of definition is far more complex. Importantly, the way dogs are defined directly influences the way they are managed (Claridge and Hunt, 2008; Glen, 2010). Uncertainty as to what defines a dog exists in two contexts: (1) taxonomic and (2) at the level of association and dependence on humans. Such difficulties in definitions are evident by examining the dingo, a naturalized canid of mainland Australia. Dingoes are referred to as invasive by some and native by others, having arrived in Australia less than 5,000 years ago (Savolainen et al., 2004; but see Oskarsson et al., 2012). Like large canids elsewhere (Gottelli et al., 1994), dingoes have undergone hybridization to varying degrees with dogs introduced to Australia over the past several centuries (Radford et al., 2012; Savolainen et al., 2004). 'Pure' dingoes are therefore regarded as rare in many parts of the continent and dingoes, feral dogs, and their hybrids are all frequently referred to collectively as wild dogs (Letnic et al., 2012), and managed similarly as pests, with the ultimate goal being extermination. In the absence of combined genetic and ecological information about 'dingoes,' it is difficult to ascertain whether previous studies (e.g., Johnson and Vanderwal, 2009; Johnson et al., 2007; Letnic et al., 2009b; Wallach et al., 2009, 2010) actually provide information on dingoes only, or dingoes, feral dogs, and their hybrids. If it is the latter, then it is difficult to ascertain the extent to which the ecology of these dog types differs according to their genetics (Claridge and Hunt, 2008).

Coupled with this problem of taxonomic and genetic identity is the variation in association and dependency of dingoes with humans. A review by Vanak and Gompper (2009a) places dogs under one of six categories: owned dogs, urban free-ranging dogs, rural free-ranging dogs, village dogs, feral dogs and wild dogs (e.g., dingoes, feral dogs, and their hybrids).

1. Owned dogs: Dogs that are owned and restricted in movement to a prescribed outdoor or indoor area. Although the potential for these dogs to interact with wildlife is limited, they can nonetheless have an effect on wildlife when they accompany humans into natural areas or if their unvaccinated status enhances the disease reservoir competency of the broader dog population (Banks and Bryant, 2007; Fiorello et al., 2006; Koster, 2008; Lenth et al., 2008).
2. Urban free-ranging dogs: Dogs that are not owned by humans, but are commensals, subsisting on garbage and other human-derived material (HDM) as their primary food source (Beck, 1975). They usually do not come into contact with wildlife, except in urban parks (Banks and Bryant, 2007; Lenth et al., 2008).
3. Rural free-ranging dogs: Dogs that are owned or peripherally associated with human habitations, but are not confined to a prescribed outdoor area. These include (but are not limited to) 'stray' dogs and owned farm and grazing companion dogs whose daily activity pattern may involve ranging that can bring them into contact with wildlife, especially when human habitations border wildlife reserves or other natural areas (Butler et al., 2004; Vanak, 2008).
4. Village dogs: Unconfined dogs that are associated with human habitations in rural environments, but rarely leave the immediate vicinity of the village (Macdonald and Carr, 1995; Vanak, 2008).
5. Feral dogs: Dogs that are completely wild and independent of human-derived food sources (Green and Gipson, 1994; Nesbitt, 1975).
6. Wild dogs: Dingoes and their hybrids in South-east Asia and Australasia that have a long history of independence from humans and are no longer considered domesticated (Corbett, 1995; Sillero-Zubiri et al., 2004).

Yet with the exception of urban free-ranging dogs, dingoes and their hybrids could actually fit all of these categories, as they are sometimes owned by people in urban and rural environments, they occur on the fringes of rural properties and aboriginal communities, and they also occur as completely wild populations that are entirely independent of humans (Hamilton, 1972; Letnic et al., 2012; Smith and Litchfield, 2009). These observations illustrate the difficulty in categorizing dogs, and raise questions about their ecological roles and to what degree we can generalize about dogs. With such classification complexity in mind, we assert that regardless of the name ascribed to a dog type and the categorization of where and how it lives, it is most important to focus on the ecological function of dogs within the spatial and temporal context of where they have been studied.

density and ranging behavior to the diversity and size of prey species available, and the presence and abundance of other sympatric and potentially competing predators.

2.2 Dog diet: influence of location and ranging behavior

Dogs are generalist and opportunistic predators, and their flexibility in this regard is illustrated in Figure 2.1. Between global regions, dog diet is highly variable and changes according to the resources that are locally available. In some cases the diets of dogs at a local scale may be dominated by one or two food types (e.g., human-derived materials (HDM) and vegetation in India or mammals in Zimbabwe; Figure 2.1). However, this does not imply that dogs are specialized with respect to their diets, but simply that they are able to capitalize on different food sources where and when these foods are abundant. This dietary plasticity is a major contributing factor to the success of dogs and their ability to survive in a diverse range of environments. In general, the diets of free-ranging dogs tend to contain a high proportion of vegetation (including fruits), HDM, and mammals, with birds, reptiles, amphibians, fish, and insects being less common prey items. As a comparison, the diet of dingoes contains a very high percentage of mammals and birds. Diet may be specialized further within a particular food class. As an example, in a comprehensive review of the diet of dingoes across Australia (Corbett, 1995), which included 12,802 diet samples collected between 1966 and 1986, 72% of prey identified were mammal. This mammal prey category was further subdivided, and comprised 20.3% large mammals, 29.7% small mammals, and 50% medium-sized mammals. This shows that dingoes have high consumption rates of medium-sized mammals.

The impact of dogs on wildlife is not only dependent on the size of their potential prey, but also, and perhaps more importantly, on the mean encounter rate between dogs and wildlife. It is clear that some categories of dogs (such as feral, wild dog, and rural free-ranging) can potentially have greater impacts on wildlife than others (such as urban or village dogs). The relationship between ranging behavior, location, and diet was explored by Vanak and Gompper (2009a). They found that all examples of urban dogs had a limited range and diets that were entirely human-dependent (Figure 2.1). However, as ranging behavior increased and the location of

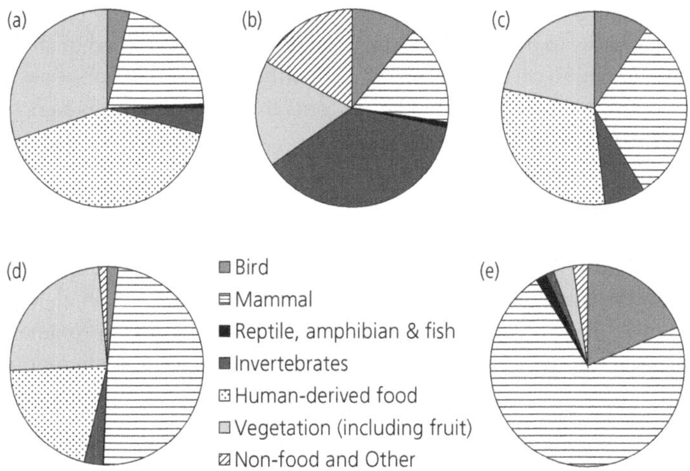

Figure 2.1 A comparison of rural dog diets derived from five studies on four continents. (a) Free-ranging dogs in India (Vanak and Gompper, 2009b); (b) Free-ranging dogs in Brazil (Campos et al., 2007); (c) Free-ranging dogs in Chile (Silva-Rodríguez et al., 2010); (d) Free-ranging dogs in Zimbabwe (Butler et al., 2004); (e) Dingoes in Australia (Corbett, 1995).

dogs became more rural, their diets become more opportunistic and less human-dependent. For example, five of eight (~63%) feral dog populations were almost entirely dependent on wild-caught food. Vanak and Gompper (2009a) concluded that the diet of dogs was closely linked to their location and ranging behavior, such that as dogs ranged farther into natural areas, they were more likely to have an impact as predators on wildlife.

2.3 Predation by dogs and its effects

In environments where predator guilds are largely intact and relatively diverse (e.g., parts of Asia, Africa, and North America), dogs often assume the roles of smaller-bodied and subordinate mesopredators, due to the presence of larger (apex) predators such as wolves and large cats (*Puma concolor* and *Panthera* spp.) (Butler et al., 2004; Dalerum et al., 2009; Ritchie and Johnson, 2009). However, when the predator guild is severely depleted or absent altogether, dogs can function as the top-predator. For example, in Australia, the ~15–20 kg dingo is similar in body size to the New World coyotes (*C. latrans*), and not considerably larger than Old World golden jackals (including the subspecies now considered a wolf species; Rueness et al., 2011) and side-striped jackals (*C. aureus* and *C. adustus*, respectively). Dingoes are considered the apex terrestrial predator due to the early extinction of much larger-bodied native carnivores, such as the marsupial lion (*Thylacoleo carnifex*) (Johnson, 2006; Wroe et al., 2005). Indeed, the situation in Australia reflects an overall trend whereby, due to the dramatic reduction or extinction of many larger predators around the globe (Estes et al., 2011), dogs now fulfill the role of apex predators in many environments. In addition to these influences, supplementation of the diet of dogs by humans may serve to exacerbate (through hyper-predation and diet switching) or potentially reduce the impact on wildlife by dogs (Silva-Rodríguez and Sieving, 2011, 2012; Vanak and Gompper, 2009a).

Dogs may suppress prey by killing them (lethal) or through instilling fear (non-lethal), which may cause changes in prey behavior, physiology, and habitat use (Clinchy et al., 2013; Ritchie and Johnson, 2009). Here, we explore these lethal and non-lethal effects of dogs in the context of their position in the carnivore guild, their population size, and their dependence on humans. We also draw a distinction between individual and population-level effects of dogs on prey. For example, a change in the behavior or the death of a prey animal has obvious effects on that individual, but if the animal would not have survived or its death allowed increased survival or reproduction of surviving members of its population, it is possible that no population-level effects would occur. That is, while changes in prey demography due to predation by dogs may alter population growth rates (e.g., because prey of different age or sex classes may differ in reproductive output), more generally a simplified framework for considering the risk that dogs represent to prey populations is that for predation to influence prey population size, mortality must be additive to existing causes of mortality rather than compensatory to those sources.

Examples of the impact of dogs on wildlife, according to their categorization and local context, are further explored in Table 2.1 below. What emerges from these studies is that dogs have the capacity to impact a range of wildlife species (often mammals and birds), via direct predation of individuals as well as through harassment and disturbance that results in lowered breeding success. It is also apparent that dogs, whether owned, un-owned, or wild, and whether restrained or free-ranging, may significantly impact other species across landscape gradients from urban settings to rural and wild habitats.

2.3.1 Direct killing

Dogs are capable of killing significant numbers of individuals of species, across a range of taxa and body sizes (see Vanak et al., Chapter 3; Young et al., 2011). For example, approximately 10,000 saiga antelope (*Saiga tatarica*) were reportedly killed annually by dogs in Kazakhstan (Sludskii, 1962). However, the extent of predation pressure exerted by dogs on prey populations varies considerably across studies. In some cases, no or low evidence of predation is found (e.g., Lowry and McArthur, 1978; Scott and Causey, 1973) or it is not possible to verify if dogs were the primary predators or scavengers (e.g., Bergeron and Pierre, 1981). Nonetheless,

Table 2.1 Examples of the impacts of dogs on prey according to dog categorization and habitat.

Type of dog	Location of study	Habitat	Major prey	Effects on prey	Reference
Owned	Australia	Suburban forest patches	None: human-provided	Reduced bird abundance and richness	(Banks and Bryant, 2007)
Urban free-ranging	New Zealand	Coastal	None: human-provided	Reduced shorebird breeding success	(Lord et al., 2001)
Rural free-ranging	Malaysia	Plantations	Wildlife	Reduced abundance	(Azhar et al., 2013)
Rural free-ranging or village	Australia	Forest and heath	Medium-large macropods	Harassment and killing; reduced abundance	(Meek, 1999)
Feral	New Zealand	Forest	Kiwi	Reduced population	(Taborsky, 1988)
Wild	Australia	Forest	Medium-large macropods	Elevated (compensatory) reproduction and reduced population size	(Robertshaw and Harden, 1986)

several studies have shown that dogs can have significant localized impacts that lead to decreases in some prey populations (Barnett and Rudd, 1983; Genovesi and Dupre, 2000; Iverson, 1978; Kruuk, 1972; Taborsky, 1988; Azhar et al., 2013).

Dogs may also limit species not only by killing adults, but also by affecting reproductive success. A study of the nesting success of freshwater crocodiles (*Crocodylus johnstoni*) in northern Australia (Somaweera et al., 2011) demonstrated that 72% of nests were opened by predators, and dingoes were responsible for 98% of these disturbances in one part of the study region and 54% in the other. Dogs are one of the main contributors to a declining kid/female ratio in mountain gazelles *Gazella gazella* in Israel (Manor and Saltz, 2004), and in a study of chiru (*Pantholops hodgsonii*), dogs killed 19 mostly young individuals that were malnourished and impeded by deep snow near a highway in Qinghai, China (Schaller, 1998). Similarly, dogs also have been reported to chase young argali *Ovis ammon* (Fedosonko and Blank, 2005) which, through increased energy expenditure and elevated stress levels, has the capacity to negatively impact argali condition, growth, and survival. However, the occurrence and severity of such effects remains to be determined for most species.

Elevated levels of predation such as those above may have severe consequences for some prey populations. In the Caicos Islands, an initial estimated population of 5,500 rock iguanas (*Cyclura carinata*) was nearly extirpated due to predation by both dogs and cats, *Felis catus* (Iverson, 1978), while in New Zealand a single dog was estimated to have killed at least 500 North Island brown kiwi (*Apteryx mantelli*) in a population of just 900 (Taborsky, 1988). Populations of rare species are likely to be particularly susceptible and less able to cope with dog predation, as small and isolated populations may not be able to recover due to insufficient reproduction and immigration that could compensate for dog-induced mortality.

Elevated levels of predation occur in another situation that can potentially exacerbate the negative impact of dogs on prey populations still further. In some situations the rate of killing by dogs greatly exceeds that necessary to meet the needs of the predator for food, and so prey carcasses frequently are not consumed after being killed. This has been termed 'surplus killing' (Kruuk, 1972). Two key reasons for surplus killing of prey by dogs include: (1) ineffective predator responses on the behalf of prey, particularly for species that have not coevolved with dogs and hence interactions with dogs are novel encounters for which prey are unlikely to exhibit effective anti-predator responses; and, (2) dogs have the ability to defend multiple kills due to their tendency to hunt and live in groups, which allows hunting to continue despite sufficient food often being available from an initial kill (Short et al., 2002). Short et al. (2002) noted in their study that they found no instances of surplus killing by cats, whereas it was a common feature of hunting by both red foxes (*Vulpes vulpes*) and dingoes. However, it remains conjectural whether this difference in hunting behavior between cats, foxes, and dingoes is widespread and consistent. Indeed, it is important to note that dogs do not always exhibit surplus killing, even for easily accessible prey (Kruuk and Snell, 1981).

Where surplus killing does occur the impacts of dogs on wildlife can be strong. In the example of dog predation on kiwis noted above, Taborsky (1988) commented that the presumed killing of 500 birds over a period of just 6 weeks would have met the energetic requirements of the single dog many times over. He also reported finding carcasses of kiwi that had been bitten but not eaten. In another example, Shepherd (1981) reported the killing of 83 red kangaroos (*Macropus rufus*) near a water hole over a 7-week period by a group of just 5 dingoes. Dingoes are known to target smaller individuals (juveniles and females) of sexually size-dimorphic prey, such as macropods (Grigg et al., 1989), as has also been found for wolves and their prey (Stahler et al., 2006). In Shepherd's (1981) study, 96% of the kangaroos killed were juveniles but, regardless of their size, their combined mass would have been more than enough to support the food requirements of the dingoes; indeed, many animals had been killed and not consumed at all, while others had been consumed only in part. Given that dogs often target juveniles and females, it is possible that relatively few individuals may be able to impact and suppress populations (Ritchie and Johnson, 2009). Wild and feral dogs in many parts of the world are notorious for their disproportionately intense attacks on livestock (e.g., Mech and Boitani, 2003; Short et al., 2002). In these situations prey are often aggregated, unable to run far owing to their confinement in a paddock, and may show inappropriate responses to the presence of dogs. These factors may combine to elicit continued killing behavior that ceases only when no further prey are conspicuous (Short et al., 2002).

2.3.2 Non-lethal effects

In a camera trapping study in Chile, Silva-Rodríguez and Sieving (2012) found that the probability of dog attacks (>85%) on a forest ungulate, pudu (*Pudu puda*), and the lethality of these attacks (50%), was high. These attacks are presumed to have invoked fear of dogs in pudu and explain why pudu distribution models are best explained by the probability of dog presence. In another study, Banks and Bryant (2007) investigated the responses of birds in suburban woodland sites to the presence of dogs. They found that the simple act of a person walking with a dog on a leash caused a 41% reduction in numbers of individual birds and a 35% reduction in species richness compared with control sites where no walking occurred; people walking alone induced less than half the disturbance compared with when they were walking a dog. This study showed further that dog-induced disturbance to birds was similar in areas where dog-walking occurred regularly compared to areas where it was not allowed, thus suggesting that birds did not become habituated to dog presence and could be at risk of long-term population declines.

Limited work has investigated the indirect effects of generalized disturbance by dogs (Weston and Stankowich, Chapter 4). It is clear from studies such as that by Lord et al. (2001), which found that off-leash dog walking disturbs shorebirds on their breeding nests, that the potential for indirect impacts on reproductive success is high if the prey species perceives dogs as a predation risk. Such effects have the capacity to alter the trophic structure of communities by causing lower-order predators (e.g., shorebirds) to avoid areas of habitat through fear (Laundre et al., 2001), which in turn may benefit smaller prey and/or the competitors of those species affected.

2.4 Human facilitation of dog predation of wildlife

The role of humans in mediating dog predation of other species is a critical consideration when attempting to manage dog populations and for conserving biodiversity. In some cases, ready supplies of food for dogs may serve to reduce impacts. Silva-Rodríguez and Sieving (2011) found that dogs preyed on most endemic and threatened mammals in their study region in Chile, but that the probability of dogs eating these prey was higher for poorly fed than adequately fed dogs. While these results suggest that feeding dogs may be an easy way to reduce their impacts on wildlife, there may be an unexpected and long-term negative consequence of such actions. By feeding dogs, which inevitably leads to sustaining dog populations at levels beyond their natural carrying capacity, humans may assist in maintaining constant and high predation rates on prey, which may become compounded during times when dogs have reduced access to human food and switch their diets to hunt native animals (Butler et al., 2004; Daniels and Bekoff, 1989).

In addition to human food subsidies for dogs, but far less understood, are the effects that habitat modification may play in mediating dog access to, and capture of, prey. Silva-Rodríguez et al. (2010) observed that the impact of dogs on pudu appeared to be exacerbated by roads that allowed dogs to increase their access to protected areas of forest. Habitat change (e.g., more roads, fragmentation, increased edge habitats) may serve to increase dog predation and threaten biodiversity by providing dogs with easier and more open access to hunt in otherwise complex habitats (Paschoal et al., 2012; Torres and Prado, 2010).

2.5 Ecosystem-wide effects of dogs

The non-consumptive effects of dogs on prey species are not restricted to their primary prey and competitors. For example, in the Strzelecki Desert of Australia, dingoes suppress the abundances of smaller invasive red foxes and in turn provide the dusky hopping mouse (*Notomys fuscus*) with refuge from predation by foxes (Letnic et al., 2009a). In areas where dingoes are present, hopping mice are less likely to occur in predator scats, are more abundant and forage less apprehensively (Figure 2.2; Letnic and Dworjanyn, 2011). These findings provide evidence that dingoes, through their suppression of fox populations, create a safer environment for hopping mice where the frequency of fatal encounters with foxes is reduced and the non-consumptive effects of foxes are lower.

The direct killing by dogs, or the fear dogs induce, may have strong negative effects on species depending on the environmental context and the identities of species comprising the faunal assemblage. The effects of dogs, however, go beyond just their direct predatory effects on individual species, and may extend to whole communities and ecosystems. At this broader ecosystem scale, the ecological effects of dogs become more complex and may include both negative and positive effects on the abundances and phenotypes of other species. Positive effects for some species can arise because dogs suppress the abundances and shape the activity patterns of smaller carnivores and herbivores. This suppression of smaller carnivores and herbivores by dogs can result in increases in the abundance and biomass of vegetation and small prey. Thus dogs can induce ecosystem-wide trophic cascades and indirectly facilitate increases in the abundances of species at lower trophic levels within the same food chain.

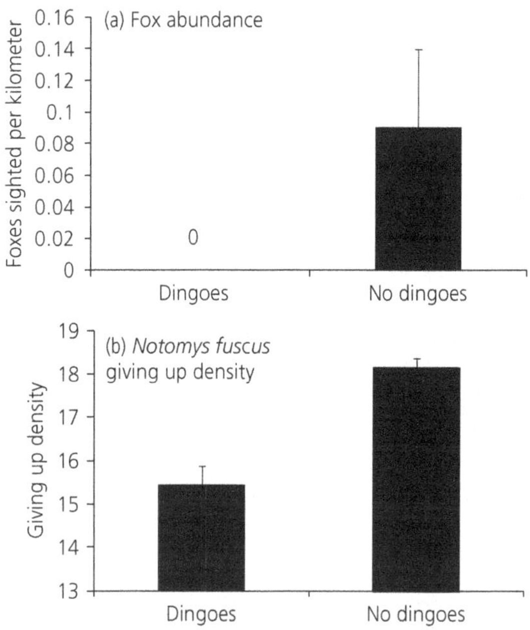

Figure 2.2 The presence of dingoes may alter the occurrence of smaller carnivore species as well as those taxa fed on by these smaller predators. (a) The abundance of red foxes, *Vulpes vulpes*, in areas where dingoes were present and where dingoes were rare. (b) The giving up density of dusky hopping mice, *Notomys fuscus*, determined using foraging trays where dingoes were present and rare. Lower giving up densities in the presence of dingoes indicate that *N. fuscus* foraged less apprehensively. Redrawn from Letnic and Dworjanyn (2011). Reproduced with permission from John Wiley & Sons.

For example, in Australia, recent research has shown that dingoes, through their suppression of

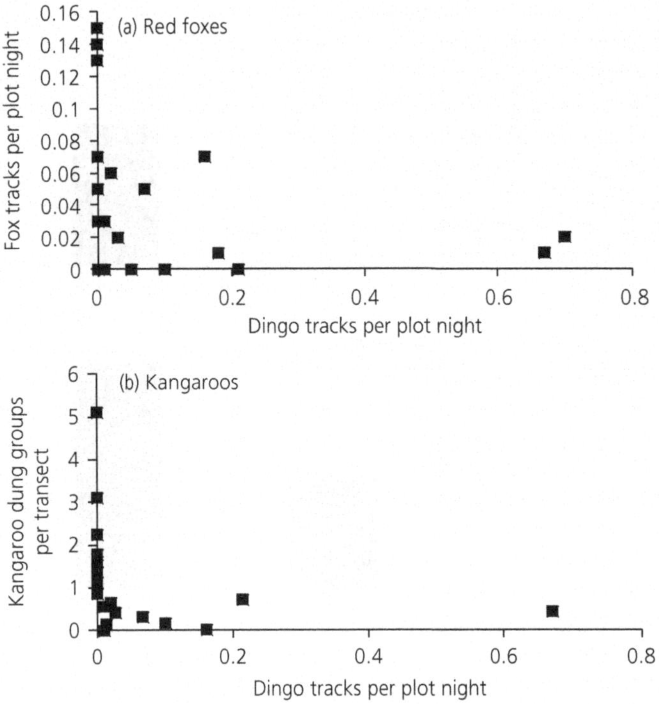

Figure 2.3 Dogs often have strong, nonlinear effects on the abundances of their competitors and prey. For example, plots of abundance indices of dingoes versus those of (a) red foxes, *Vulpes vulpes*, and (b) kangaroos, *Macropus* spp., in desert regions of Australia display threshold relationships which suggest that even low density populations of dingoes have strong suppressive effects on the abundances of mesopredators and herbivores. Redrawn from (a) Letnic et al. (2011b) and (b) Letnic and Crowther 2013. Reproduced with permission from John Wiley & Sons.

smaller predators and herbivores (Figure 2.3), induce ecosystem-wide trophic cascades with a net positive effect for biodiversity conservation (Letnic et al., 2012) (Figure 2.4). Within the last 200 years, two novel invasive predators have been introduced to the continent: the red fox and the domestic cat (Johnson, 2006). These species have been implicated in a large number (>20 species) of native mammal extinctions (Johnson, 2006). Comparisons of ecosystem attributes in areas where dingoes are actively controlled or not controlled (Figure 2.5) indicate that the impacts of these mesopredators are moderated by the presence or absence of dingoes (Johnson et al., 2007; Letnic et al., 2009b). Where studied, dingoes appear to suppress both cats and foxes, through the direct killing of individuals, but also by dingoes causing these subordinate predators to avoid them both spatially and temporally (Brook et al., 2012; Johnson and VanDerWal, 2009; Kennedy et al., 2012; Letnic and Dworjanyn, 2011). In addition, dingoes have been shown to be very effective at suppressing herbivore populations, contributing to more available food and shelter for native prey (Letnic et al., 2009b; Pople et al., 2000; Wallach et al., 2010). Together, the limiting effects of dingoes on cat, fox, and herbivore populations have been shown to protect smaller native prey, with native mammals in particular achieving higher diversity and abundance where dingoes are common (Letnic et al., 2012; Ritchie and Johnson, 2009; Ritchie et al., 2012; Wallach and O'Neill, 2009; Wallach et al., 2009, 2010).

Despite the benefits for biodiversity of dogs such as the dingo, in most other cases dogs appear to have significant detrimental effects. In Europe, for example, the common practice of dog walking may impact the European Nightjar *Caprimulgus europaeus* (Langston et al., 2007) and, as mentioned above, Banks and Bryant (2007) showed that bird diversity was reduced by more than one third and abundance by ~40% in woodland areas where dogs are walked. Where dogs are human-subsidized it would appear they frequently have negative effects.

Many studies, including a large number of those mentioned above, report the impact of dogs on prey species as measured and inferred primarily through

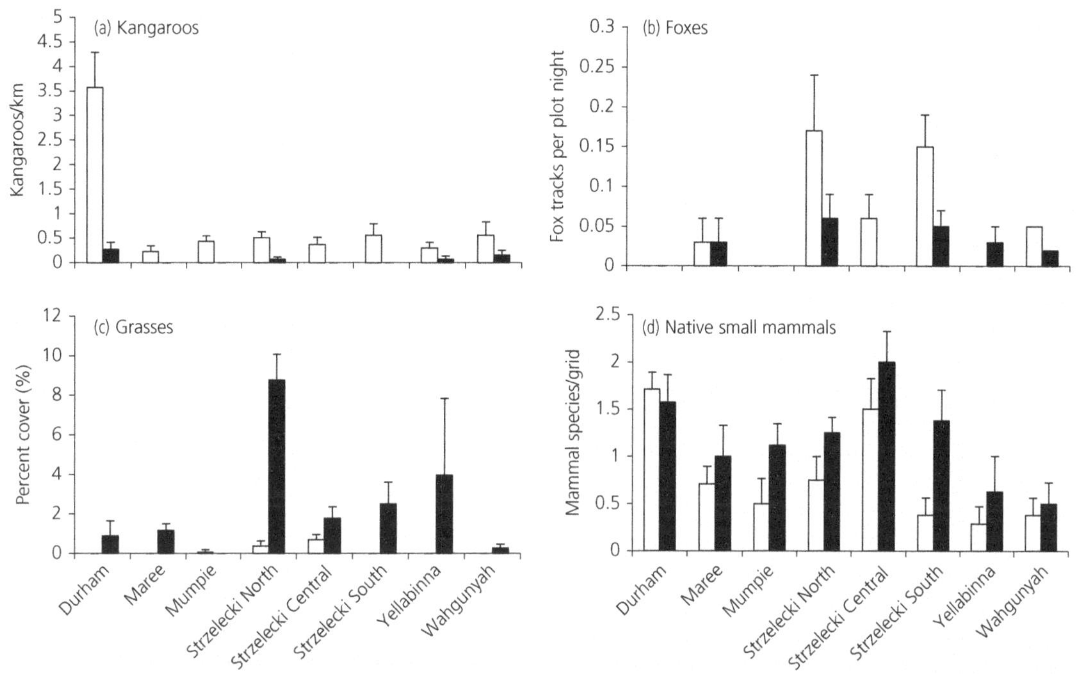

Figure 2.4 Evidence that dingoes induce a community-wide trophic cascade is provided by a comparison of the abundances of kangaroos, red foxes, and grasses and the species richness of small mammals at eight paired study sites situated on either side of the dingo fence in arid Australia (Letnic et al., 2009b). Within each pair of sites, dingoes were common in one (black bars) and rare in the other (white bars). Kangaroos and foxes were more abundant in the absence of dingoes. Grasses, the preferred forage of kangaroos, were more abundant in the presence of dingoes. Small mammals are subject to high rates of predation by foxes. Small mammal communities were more species rich in the presence of dingoes.

dog–wildlife habitat associations and dog diet (Hughes and Macdonald, 2013; Vanak and Gompper, 2009a; Young et al., 2011). With few exceptions what remains largely unknown and should be of concern is:

1. Whether dog predation on wildlife is additive (total annual mortality rate that is greater than what would occur without the predation) or compensatory (a population's total mortality remains unchanged because the other, presumably natural, causes of mortality, such as intraspecific competition for food, decrease to compensate for reduced density caused by dogs' predation; see Ritchie and Johnson, 2009).
2. Whether dogs are the principal drivers of decline for particular species, or whether they are contributing to the final decline or extinction of populations and species only because they have already been heavily impacted by other causal agents, such as habitat loss and disease. This is, in essence, a formulation of Caughley's contrast of small population versus declining population paradigms (Caughley, 1994).

There is no question dogs can pose a severe risk to species that are already threatened due to their low population sizes, but to assess the true impacts of dogs on other species, and hence to ensure appropriate biodiversity conservation and management, there is an urgent need for studies that address these two points.

2.6 Future research

From our review it is clear that dogs, spanning both wild and urban environments, interact with and affect biodiversity through a variety of important pathways. Most importantly, this occurs by dogs acting as predators or competitors within communities. In doing so, in some cases, they may contribute to the decline of already rare and threatened

Figure 2.5 A dingo left to hang on Australia's dingo barrier fence. This sight is common across much of arid Australia, with conflict occurring between cattle and sheep production and dingoes. Dingoes are known to prey on livestock, sometimes causing significant economic damage (photo courtesy of Mike Letnic).

species, or indeed of common species. But in other instances, conversely, through their suppression of other predators (including invasive species) and herbivores, dogs may help to protect and promote biodiversity, maintaining the resilience of ecosystems. It is therefore difficult to generalize about the trophic roles and ecological functions of dogs, as they are environmentally and temporally context dependent.

Humans have important roles in either facilitating or reducing the effects of dogs on other wildlife. Through anthropogenic changes to habitat and the supply of food provided to dogs, humans may make areas more or less suitable for dogs, with concomitant effects on wildlife. However, despite dogs being the most widespread and abundant carnivores worldwide, there are surprisingly few studies of their ecology. Indeed, the impacts of cats are far better known (Fitzgerald and Turner, 2000; Loss et al., 2013; Medina et al., 2011; Woinarski et al., 2011), with this species being listed in the 100 worst invasive species globally (Lowe et al., 2004). Given the difficulty of working on cats, due to their largely solitary and cryptic nature and their generally lower abundance than dogs globally, it is surprising we know so little about dogs, including basic information such as their diets, hunting behavior, and whether they impact native species to the same or greater degrees as do other non-native species. There are many key questions that remain to be answered in regards to dogs, and below we suggest what we consider to be urgent research priorities.

First, determine more broadly the importance of differences in dog categories (Box 2.1). How do such differences influence the functional roles of dogs within similar environments? Further, how do the behaviors of these different categories of dogs differ? The dingo provides an excellent model system in which to examine these questions, given the large area and range of environments in which it occurs, and its complex Holocene and recent history. Determining the answer to these questions has strong implications for how we should manage the dingo and other dog populations worldwide.

Second, more work needs to be done comparing dog populations from across the world, in both similar and different environments. Do dogs in South America, North America, Asia, Europe, Australia, and Africa behave in similar ways? Much could be learned from such comparisons and the ensuing knowledge would aid a global synthesis about dogs and their roles as trophic regulators. At present we have too few studies to do this.

Third, more experimental work on dogs (dog removals or dog additions) is needed to establish their effects, as the majority of work to date has been largely observational and correlative.

Finally, a greater focus on dog behavior is required, rather than just focusing on the effects of dogs in relation to their abundance. Dogs may affect other species in sublethal ways as these species respond to perceived predation risk. These indirect and subtle effects are only just beginning to be appreciated. Given the densities of dogs in many parts of the world, their perception as a predatory risk by other members of the animal community suggests that they may alter community structure even when not having a significant direct predatory effect.

References

Allen, B.L., Engeman, R.M., and Allen, L.R. (2011a). Wild dogma II: the role and implications of wild dogma for wild dog management in Australia. *Current Zoology*, **57**, 737–40.

Allen, B.L., Engeman, R.M., and Allen, L.R. (2011b). Wild dogma: an examination of recent "evidence" for dingo regulation of invasive mesopredator release in Australia. *Current Zoology*, **57**, 568–83.

Allen, B.L., Fleming, P.J.S., Allen, L.R., Engeman, R.M., Ballard, G., and Leung, L.K.P. (2013). As clear as mud: a critical review of evidence for the ecological roles of Australian dingoes. *Biological Conservation*, **159**, 158–174.

Azhar, B., Lindenmayer, D., Wood, J., Fischer, J., Manning, A., McElhinny, C., and Zakaria, M. (2013). Contribution of illegal hunting, culling of pest species, road accidents and feral dogs to biodiversity loss in established oil-palm landscapes. *Wildlife Research*, **40**, 1–9.

Baker, P.J., Boitani, L., Harris, S., Saunders, G., and White, P.C.L. (2008). Terrestrial carnivores and human food production: Impact and management. *Mammal Review*, **38**, 123–66.

Banks, P.B. and Bryant, J.V. (2007). Four-legged friend or foe? Dog walking displaces native birds from natural areas. *Biology Letters*, **3**, 611–13.

Barnett, B.D. and Rudd, R.L. (1983). Feral dogs of the Galapagos Islands: impact and control. *International Journal on Studies on Animal Problems*, **4**, 44–58.

Beck, A.M. (1975). The ecology of "feral" and free-roving dogs in Baltimore. In M.W. Fox, Ed., *The Wild Canids: their Systematics, Behavioral Ecology and Evolution*. Van Nostrand Reinhold Company, New York, pp. 380–90.

Bergeron, J.-M. and Pierre, D. (1981). Le regime alimentaire du coyote (*Canis latrans*) et du chien errant (*C. familiaris*) dans le sud du Quebec. *Canadian Field-Naturalist*, **95**, 172–7.

Brook, L.A., Johnson, C.N., and Ritchie, E.G. (2012). Effects of predator control on behaviour of an apex predator and indirect consequences for mesopredator suppression. *Journal of Applied Ecology*, **49**, 1278–86.

Butler, J.R.A., Du Toit, J.T., and Bingham, J. (2004). Free-ranging domestic dogs (*Canis familiaris*) as predators and prey in rural Zimbabwe: Threats of competition and disease to large wild carnivores. *Biological Conservation*, **115**, 369–78.

Campos, C.B., Esteves, C.F., Ferraz, K.M.P.M.B, Crawshaw, Jr., P.G., and Verdade, L.M. (2007). Diet of free-ranging cats and dogs in a suburban and rural environment, south-eastern Brazil. *Journal of Zoology*, **273**, 14–20.

Caughley, G. (1994). Directions in conservation biology. *Journal of Animal Ecology*, **63**, 215–44.

Claridge, A.W. and Hunt, R. (2008). Evaluating the role of the dingo as a trophic regulator: Additional practical suggestions. *Ecological Management & Restoration*, **9**, 116–9.

Clinchy, M., Sheriff, M.J., and Zanette, L.Y. (2013). Predator-induced stress and the ecology of fear. *Functional Ecology*, **27**, 56–65.

Corbett, L.K. (1995). *The Dingo in Australia and Asia*. Cornell University Press, Ithaca, New York.

Dalerum, F., Cameron, E.Z., Kunkel, K., and Somers, M.J. (2009). Diversity and depletions in continental carnivore guilds: implications for prioritizing global carnivore conservation. *Biology Letters*, **5**, 35–8.

Daniels, T.J. and Bekoff, M. (1989). Population and social biology of free-ranging dogs, *Canis familiaris*. *Journal of Mammalogy*, **70**, 754–62.

Estes, J.A., Terborgh, J., Brashares, J.S., Power, M.E., Berger, J., Bond, W.J., Carpenter, S.R., Essington, T.E., Holt, R.D., Jackson, J.B.C., Marquis, R.J., Oksanen, L., Oksanen, T., Paine, R.T., Pikitch, E.K., Ripple, W.J., Sandin, S.A., Scheffer, M., Schoener, T.W., Shurin, J.B., Sinclair, A.R.E., Soulé, M.E., Virtanen, R., and Wardle, D.A. (2011). Trophic downgrading of planet earth. *Science*, **333**, 301–6.

Fedosonko, A.K. and Blank, D.A. (2005). *Ovis ammon*. *Mammalian Species*, **773**, 1–15.

Fiorello, C.V., Noss, A.J., and Deem, S.L. (2006). Demography, hunting ecology, and pathogen exposure of domestic dogs in the Isoso of Bolivia. *Conservation Biology*, **20**, 762–71.

Fitzgerald, B.M. and Turner, D.C. (2000). Hunting behaviour of domestic cats and their impact on prey populations. In D.C. Turner and P. Bateson, Eds., *The Domestic Cat: The Biology of its Behaviour*. Cambridge University Press, Cambridge, pp. 151–75.

Fleming, P.J.S., Allen, B.L., and Ballard, G.A. (2012). Seven considerations about dingoes as biodiversity engineers: The socioecological niches of dogs in Australia. *Australian Mammalogy*, **34**, 119–23.

Genovesi, P. and Dupre, E. (2000). Strategia nazionale di conservazione del Lupo (*Canis lupus*): indagine sulla presenza e la gestione dei cani vaganti in Italia. *Biologia e Conservazione della Fauna*, **104**, 1–33.

Glen, A.S. (2010). Hybridisation between dingoes and domestic dogs: a comment on Jones (2009). *Australian Mammalogy*, **32**, 76–7.

Gompper, M.E. and Vanak, A.T. (2008). Subsidized predators, landscapes of fear and disarticulated carnivore communities. *Animal Conservation*, **11**, 13–14.

Gottelli, D., Sillero-Zubiri, C., Applebaum, G.D., Roy, M.S., Girman, D.J., Garcia-Moreno, J., Ostrander, E.A., and Wayne, R.K. (1994). Molecular genetics of the most endangered canid: The Ethiopian wolf *Canis simensis*. *Molecular Ecology*, **3**, 301–12.

Green, J.S. and Gipson, P.S. (1994). Feral Dogs. In S.E. Hygnstrom, R.M. Timm and G.E. Larson, Eds., *The Handbook: Prevention and Control of Wildlife Damage*. University of Nebraska, Lincoln, pp. 1–7.

Grigg, G., Hume, I., and Jarman, P. (1989). *Kangaroos, Wallabies, Rat-Kangaroos*. Surrey Beatty & Sons, Sydney.

Hamilton, A. (1972). Aboriginal man's best friend? *Mankind*, **8**, 287–95.

Houston, M.J., Bruskotter, J.T., and Fan, D. (2010). Attitudes toward wolves in the United States and Canada: A content analysis of the print news media, 1999–2008. *Human Dimensions of Wildlife*, **15**, 389–403.

Hughes, J. and Macdonald, D.W. (2013). A review of the interactions between free-roaming domestic dogs and wildlife. *Biological Conservation*, **157**, 341–51.

Iverson, J.B. (1978). The impact of feral cats and dogs on populations of the West Indian rock iguana, *Cyclura carinata*. *Biological Conservation*, **14**, 63–73.

Johnson, C.N. (2006). *Australia's Mammal Extinctions: A 50000 Year History*. Cambridge University Press, Cambridge.

Johnson, C.N., Isaac, J.L., and Fisher, D.O. (2007). Rarity of a top predator triggers continent-wide collapse of mammal prey: Dingoes and marsupials in Australia. *Proceedings of the Royal Society Biological Sciences Series B*, **274**, 341–46.

Johnson, C.N. and Ritchie, E.G. (2013). The dingo and biodiversity conservation: response to Fleming *et al.* (2012). *Australian Mammalogy*, **35**, 8-14.

Johnson, C.N. and VanDerWal, J. (2009). Evidence that dingoes limit abundance of a mesopredator in eastern Australian forests. *Journal of Applied Ecology*, **46**, 641–46.

Kennedy, M., Phillips, B., Legge, S., Murphy, S., and Faulkner, R. (2012). Do dingoes suppress the activity of cats in northern Australia?*Austral Ecology*, **37**, 134–9.

King, J.S., Brown, G.K., Jenkins, D.J., Ellis, J.T., Fleming, P.J.S., Windsor, P.A., and Šlapeta, J. (2012). Oocysts and high seroprevalence of *Neospora caninum* in dogs living in remote Aboriginal communities and wild dogs in Australia. *Veterinary Parasitology*, **187**, 85–92.

Koster, J. (2008). The impact of hunting with dogs on wildlife harvests in the Bosawas Reserve, Nicaragua. *Environmental Conservation*, **35**, 211–20.

Kruuk, H. (1972). Surplus killing by carnivores. *Journal of Zoology*, **166**, 233–44.

Kruuk, H. and Snell, H. (1981). Prey selection by feral dogs from a population of marine iguanas *Amblyrhynchus cristatus*. *Journal of Applied Ecology*, **18**, 109–16.

Langston, R.H.W., Liley, D., Murison, G., Woodfield, E., and Clarke, R.T. (2007). What effects do walkers and dogs have on the distribution and productivity of breeding European Nightjar *Caprimulgus europaeus*? *Ibis*, **149**, 27–36.

Laundre, J.W., Hernandez, L., and Altendorf, K.B. (2001). Wolves, elk, and bison: Reestablishing the "landscape of fear" in Yellowstone National Park, USA. *Canadian Journal of Zoology*, **79**, 1401–09.

Lenth, B.E., Knight, R.L., and Brennan, M.E. (2008). The effects of dogs on wildlife communities. *The Natural Areas Journal*, **28**, 218–27.

Letnic, M. and Crowther, M.S. (2013). Patterns in the abundance of kangaroo populations in arid Australia are consistent with the exploitation ecosystems hypothesis. *Oikos*, **122**, 761-69.

Letnic, M. and Dworjanyn, S.A. (2011). Does a top predator reduce the predatory impact of an invasive mesopredator on an endangered rodent? *Ecography*, **34**, 827–35.

Letnic, M., Crowther, M.S., Dickman, C.R., and Ritchie, E.G. (2011a). Demonising the dingo: how much wild dogma is enough? *Current Zoology*, **57**, 668–70.

Letnic, M., Crowther, M.S., and Koch, F. (2009a). Does a top-predator provide an endangered rodent with refuge from an invasive mesopredator? *Animal Conservation*, **12**, 302–12.

Letnic, M., Koch, F., Gordon, C., Crowther, M.S., and Dickman, C.R. (2009b). Keystone effects of an alien top-predator stem extinctions of native mammals. *Proceedings of the Royal Society B-Biological Sciences*, **276**, 3249–56.

Letnic, M., Greenville, A., Denny, E., Dickman, C.R., Tischler, M., Gordon, C., and Koch, F. (2011b). Does a top predator suppress the abundance of an invasive mesopredator at a continental scale? *Global Ecology and Biogeography*, **20**, 343–53.

Letnic, M., Ritchie, E.G., and Dickman, C.R. (2012). Top predators as biodiversity regulators: the dingo *Canis lupus dingo* as a case study. *Biological Reviews*, **87**, 390–413.

Lord, A., Waas, J.R., Innes, J., and Whittingham, M.J. (2001). Effects of human approaches to nests of northern New Zealand dotterels. *Biological Conservation*, **98**, 233–40.

Loss, S.R., Will, T., and Marra, P.P. (2013). The impact of free-ranging domestic cats on wildlife of the United States. *Nature Communications*, **4**, 1396.

Lowe, S., Browne, M., Boudjelas, S., and De Poorter, M. (2004). *100 of the world's worst invasive alien species. A selection from the Global Invasive Species Database*. The Invasive Species Specialist Group, Species Survival Commission, World Conservation Union, Auckland, New Zealand.

Lowry, D.A. and McArthur, K.L. (1978). Domestic dogs as predators on deer. *Wildlife Society Bulletin*, **6**, 38–9.

Macdonald, D.W. and Carr, G.M. (1995). Variation in dog society: Between resource dispersion and social flux. In J. Serpell, Ed., *The Domestic Dog: Its Evolution, Behaviour, and Interactions with People*. Cambridge University Press, Cambridge, pp. 199–216.

Manor, R. and Saltz, D. (2004). The impact of free-roaming dogs on gazelle kid/female ratio in a fragmented area. *Biological Conservation*, **119**, 231–36.

Mech, D.L. (2012). Is science in danger of sanctifying the wolf? *Biological Conservation*, **150**, 143–9.

Mech, L.D. and Boitani, L. (Eds.) (2003). *Wolves: Behavior, Ecology and Conservation*. University of Chicago Press, Chicago.

Medina, F.M., Bonnaud, E., Vidal, E., Tershy, B.R., Zavaleta, E.S., Donlan, C.J., Keitt, B.S., Le Corre, M., Horwath, S.V., and Nogales, M. (2011). A global review of the impacts of invasive cats on island endangered vertebrates. *Global Change Biology*, **17**, 3503–10.

Meek, P. D. (1999). The movement, roaming behaviour and home range of free-roaming domestic dogs, *Canis lupus familiaris*, in coastal New South Wales. *Wildlife Research*, **26**, 847–55.

Nesbitt, W.H. (1975). Ecology of a feral dog pack on a wildlife refuge. In M.W. Fox, Ed., *The Wild Canids*. Van Nostrand Reinhold Company, New York, pp. 391–5.

Oskarsson, M.C.R., Klütsch, C.F.C., Boonyaprakob, U., Wilton, A., Tanabe, Y., and Savolainen, P. (2012). Mitochondrial DNA data indicate an introduction through Mainland Southeast Asia for Australian dingoes and Polynesian domestic dogs. *Proceedings of the Royal Society B: Biological Sciences*, **279**, 967–74.

Paschoal, A.M.O., Massara, R.L., Santos, J.L., and Chiarello, A.G. (2012). Is the domestic dog becoming an abundant species in the Atlantic forest? A study case in southeastern Brazil. *Mammalia*, **76**, 67–76.

Pople, A.R., Grigg, G.C., Cairns, S.C., Beard, L.A., and Alexander, P. (2000). Trends in the numbers of red kangaroos and emus on either side of the South Australian dingo fence: evidence for predator regulation? *Wildlife Research*, **27**, 269–76.

Prugh, L.R., Stoner, C.J., Epps, C.W., Bean, W.T., Ripple, W.J., Laliberte, A.S., and Brashares, J.S. (2009). The rise of the mesopredator. *BioScience*, **59**, 779–91.

Radford, C.G., Letnic, M., Fillios, M., and Crowther, M.S. (2012). An assessment of the taxonomic status of wild canids in south-eastern New South Wales: phenotypic variation in dingoes. *Australian Journal of Zoology*, **60**, 73–80.

Ritchie, E.G., Elmhagen, B., Glen, A.S., Letnic, M., Ludwig, G., and McDonald, R.A. (2012). Ecosystem restoration with teeth: What role for predators? *Trends in Ecology & Evolution*, **27**, 265–71.

Ritchie, E.G. and Johnson, C.N. (2009). Predator interactions, mesopredator release and biodiversity conservation. *Ecology Letters*, **12**, 982–98.

Robertshaw, J.D. and Harden, R.H. (1986). The ecology of the dingo in north eastern New South Wales. 4. Prey selection by dingoes, and its effect on the major prey species, the swamp wallaby, *Wallabia bicolor* (Desmarest). *Australian Wildlife Research*, **13**, 141–63.

Rueness, E.K., Asmyhr, M.G., Sillero-Zubiri, C., Macdonald, D.W., Bekele, A., Atickem, A., and Stenseth, N.C. (2011). The cryptic African wolf: *Canis aureus lupaster* is not a golden jackal and is not endemic to Egypt. *PLoS ONE*, **6**, e16385.

Savolainen, P., Leitner, T., Wilton, A.N., Matisoo-Smith, E., and Lundeberg, J. (2004). A detailed picture of the origin of the Australian dingo, obtained from the study of mitochondrial DNA. *Proceedings of the National Academy of Sciences, USA*, **101**, 12387–90.

Schaller, G. (1998). *Wildlife of the Tibetan Steppe*. University of Chicago Press, Chicago.

Scott, M.D. and Causey K. (1973). Ecology of feral dogs in Alabama. *Journal of Wildlife Management*, **37**, 263–5.

Shepherd, N.C. (1981). Predation of red kangaroos, *Macropus rufus*, by the dingo, *Canis familiaris dingo* (Blumenbach), in North-western New South Wales. *Australian Wildlife Research*, **8**, 255–62.

Short, J., Kinnear, J.E., and Robley, A. (2002). Surplus killing by introduced predators in Australia—evidence for ineffective anti-predator adaptations in native prey species? *Biological Conservation*, **103**, 283–301.

Sillero-Zubiri, C., Hoffmann, M., and Macdonald, D.W. (2004). *Canids: Foxes, Wolves, Jackals and Dogs. Status Survey and Conservation Action Plan*. IUCN/SSC Canid Specialist Group, Gland, Switzerland and Cambridge, UK.

Silva-Rodríguez, E.A. and Sieving, K.E. (2011). Influence of care of domestic carnivores on their predation on vertebrates. *Conservation Biology*, **25**, 808–15.

Silva-Rodríguez, E.A. and Sieving, K.E. (2012). Domestic dogs shape the landscape-scale distribution of a threatened forest ungulate. *Biological Conservation*, **150**, 103–10.

Silva-Rodríguez, E.A., Verdugo, C., Aleuy, O.A., Sanderson, J.G., Ortega-Solís, G.R., Osorio-Zúñiga, F. and González-Acuña, D. (2010). Evaluating mortality sources for the vulnerable pudu *Pudu puda* in Chile: implications for the conservation of a threatened deer. *Oryx*, **44**, 97–103.

Sludskii, A.A. (1962). The relationship between predator and prey. *Proceedings of the Zoological Institute of the Kazakhstan Academy of Science*, **17**, 24–143.

Smith, B.P. and Litchfield, C.A. (2009). A Review of the relationship between Indigenous Australians, dingoes (*Canis dingo*) and domestic dogs (*Canis familiaris*). *Anthrozoos*, **22**, 111–28.

Somaweera, R., Webb, J.K., and Shine, R. (2011). It's a dog-eat-croc world: Dingo predation on the nests of freshwater crocodiles in tropical Australia. *Ecological Research*, **26**, 957–67.

Stahler, D.R., Smith, D.W., and Guernsey, D.S. (2006). Foraging and feeding ecology of the gray wolf (*Canis lupus*): Lessons from Yellowstone National Park, Wyoming, USA. *Journal of Nutrition*, **136**. 1923–26.

Taborsky, M. (1988). Kiwis and dog predation: Observations in Waitangi state forest. *Notornis*, **35**, 197–202.

Torres, P.C. and Prado, P.I. (2010). Domestic dogs in a fragmented landscape in the Brazilian Atlantic Forest: Abundance, habitat use and caring by owners. *Brazilian Journal of Biology*, **70**, 987–94.

Van Bommel, L. and Johnson, C.N. (2012). Good dog! Using livestock guardian dogs to protect livestock from predators in Australia's extensive grazing systems. *Wildlife Research*, **39**, 220–9.

Vanak, A.T. (2008). Intraguild interactions between native and domestic carnivores in central India. PhD dissertation. University of Missouri.

Vanak, A.T. and Gompper, M.E. (2009a). Dogs *Canis familiaris* as carnivores: their role and function in intraguild competition. *Mammal Review*, **39**, 265–83.

Vanak, A.T. and Gompper, M.E. (2009b). Dietary niche separation between sympatric free-ranging domestic dogs and Indian foxes in central India. *Journal of Mammalogy*, **90**, 1058–65.

Wallach, A.D., Johnson, C.N., Ritchie, E.G., and O'Neill, A.J. (2010). Predator control promotes invasive dominated ecological states. *Ecology Letters*, **13**, 1008–18.

Wallach, A.D. and O'Neill, A.J. (2009). Threatened species indicate hot-spots of top-down regulation. *Animal Biodiversity and Conservation*, **32**, 127–33.

Wallach, A.D., Ritchie, E.G., Read, J., and O'Neill, A. (2009). More than mere numbers: The impact of lethal control on social stability of a top-order predator. *PLoS ONE*, **4**, e6861.

Woinarski, J.C.Z., Legge, S., Fitzsimons, J.A., Traill, B.J., Burbidge, A.A., Fisher, A., Firth, R.S.C., Gordon, I.J., Griffiths, A.D., Johnson, C.N., McKenzie, N.L., Palmer, C., Radford, I., Rankmore, B., Ritchie, E.G., Ward, S., and Ziembicki, M. (2011). The disappearing mammal fauna of northern Australia: Context, cause, and response. *Conservation Letters B*, 192–201.

Wroe, S., McHenry, C., and Thomason, J. (2005). Bite club: Comparative bite force in big biting mammals and the prediction of predatory behaviour in fossil taxa. *Proceedings of the Royal Society B-Biological Sciences*, **272**, 619–25.

Young, J.K., Olson, K.A., Reading, R.P., Amgalanbaatar, S., and Berger, J. (2011). Is wildlife going to the dogs? Impacts of feral and free-roaming dogs on wildlife populations. *BioScience*, **61**, 125–32.

CHAPTER 3

Top-dogs and under-dogs: competition between dogs and sympatric carnivores

Abi Tamim Vanak, Christopher R. Dickman, Eduardo A. Silva-Rodriguez, James R.A. Butler, and Euan G. Ritchie

3.1 Introduction

In many natural and semi-natural systems worldwide, dogs (*Canis familiaris*) are an integral part of the carnivore community (Gompper, Chapter 1). Until recently, the role of dogs as mid-sized members of the carnivore guild, and how they shape the structure of native carnivore communities, has garnered little attention. Of late, however, a suite of studies across the world has highlighted several key aspects of the competitive dynamics that occur between dogs and sympatric members of the carnivore community (Table 3.1). These studies have demonstrated that dogs often compete with native carnivores for food, and that smaller carnivores may react to dogs as they would to any other mid-sized predator: with increased vigilance, lowered food intake, avoidance of dog-dominated habitat types, or complete spatial separation. These studies have also shown that the types and intensity of interactions between dogs and subordinate predators are likely influenced by several factors, including, but not limited to:

1. Relative position within the native carnivore community: In several areas dogs are the largest mammalian carnivore and hence assume the role of the apex predator (e.g., Australasia), whereas in others they may be part of a guild of carnivores where they occupy a mesopredator position (e.g., India, Africa). The nature of the interactions of dogs with sympatric carnivores will depend on their position in the interference hierarchy, with dogs being either at the receiving or giving end of aggressive interactions, or both.
2. Dependence on human-derived food: Interference competition may be seen ultimately as a function of food acquisition. Although most dog populations are heavily dependent on human-derived food (HDF), this is variable depending on their degree of association with humans. Dogs in urban areas are almost completely dependent on HDF, whereas at the other extreme, wild dogs (such as the dingo) and feral dogs are fully dependent on wild-caught food (Vanak and Gompper, 2009b). Thus, it can be assumed that competition with wild carnivores is likely highest in areas where dependence on wild resources is highest.
3. Population density: The competitive effects of dogs can also be influenced by their density and their tendency to form packs. Even without human involvement, packs of dogs are capable of overpowering not just herbivores several times their size (Corbett, 1995), but also carnivores such as foxes, coyotes, and jackals (Vanak and Gompper, 2009b; Van Sittert, 1998). Furthermore, numerical superiority can improve competitiveness in obtaining resources such as carcasses. Thus, competitive ability may be a function of density at the local scale.

Free-Ranging Dogs and Wildlife Conservation. Edited by Matthew E. Gompper
© Oxford University Press 2014. Published 2014 by Oxford University Press.

Table 3.1 Summary of studies that examine or report the competitive effects of dogs on sympatric carnivores.

Country	Predator guild*, besides dog	Interactions studied	Effect of dogs	Reference
Australia	Large marsupials	Dog (dingo) → thylacine and Tasmanian devil	Direct killing and possibly more efficient resource exploitation, resulting in extinction of both marsupials on the Australian mainland§	Corbett 1995; Glen and Dickman 2005; Wroe et al. 2007; Fillios et al. 2012
Australia	Invasive red fox, feral cat, and small to mid-size native predators	Dog (dingo) → fox and cat	Killing, harassment, spatial interference (fox, cat) with corresponding benefit to native species via mesopredator release#	Corbett 1995; Glen and Dickman 2005; Mitchell and Banks 2005; Johnson and VanDerWal 2009; Letnic et al. 2009; 2011; 2012; Purcell 2010; Wallach et al. 2010; Brawata and Neeman 2011; Kennedy et al. 2012; Moseby et al. 2012
Brazil	Maned wolf, crab-eating fox, puma, South American coati, crab-eating raccoon, tayra	Dog → Maned wolf	Interference on maned wolf, no evidence of effects on crab-eating raccoon§	Lacerda et al. 2009
Brazil	Crab-eating fox, South American coati, crab-eating raccoon, tayra, lesser grison, oncilla, margay	Dog → Crab-eating fox and coati	Evidence of negative association with crab-eating fox and coati, no evidence of effects on tayra and small cats§	Espartosa 2009
Chile	Puma, chilla fox, guigna, Molina's hog-nosed skunk, American mink (invasive)	Dog → Chilla fox	Killing, harassment, spatial interference#	Silva-Rodriguez et al. 2010a
Ethiopia	Ethiopian wolf	Dog →← Ethiopian wolf	Outcome of interactions depended on numerical superiority#	Atickem et al. 2010
India	Wolf, golden jackal, jungle cat, Indian fox	Dog → Indian fox	Killing, harassment, spatial interference#	Vanak et al. 2009; Vanak and Gompper 2010
India	Striped hyena, wolf, jungle cat, Indian fox	Dog → golden jackal	Dominance at carcasses§	Aiyadurai and Jhala 2006
Madagascar	Fossa, small-toothed civet, Malagasy civet, ring-tailed mongoose, broad-striped mongoose, small Indian civet, domestic cat	Dog → Fossa and ring-tailed mongoose	Fossa nearly absent from sites with high dog occupancy. Capture rates of fossa decreased as capture rates of dogs increased. Ring-tailed mongoose changed activity patterns when dogs and the invasive small Indian civet were present. No effects reported for other three species analyzed§	Gerber et al. 2012; Barcala 2009
Poland	Red fox, domestic cat	Dog → Red fox, cat	Negative association between occurrence of dogs with cats and red foxes§	Krauze-Gryz et al. 2012
Spain	Red fox, badger, domestic cat, Egyptian mongoose,	Dog → Badger	Spatial exclusion§	Revilla et al. 2001
United Kingdom	Red fox	Dog → Red fox	Killing of adults and pups	Harris 1981
Zimbabwe	Lion, spotted hyena, leopard, black-backed jackal, side-striped jackal	Lion, spotted hyena, leopard → Dog → jackals	Dominance over jackals at carcasses on wildlife reserve peripheries; spatial exclusion of jackals (in communal lands and on commercial farmland peripheries)#	Butler 1998; Butler and Bingham 2000; Butler and du Toit 2002; Butler et al. 2004

* Arranged in order of decreasing body size, # Direct evidence, § Indirect or correlative evidence. Arrows indicate direction of dominant interactions.

4. Ranging behavior: The ability of dogs to range widely into natural habitats increases the potential for them to compete with sympatric carnivores not just at the periphery of human settlements but also in native habitats. Vanak and Gompper (2009b) proposed a conceptual model wherein the competitive effects of dogs were affected not just by their densities, but also by their ranging behavior (Figure 3.1). According to this model, the maximum impact on sympatric carnivores is expected from a high-density dog population that also exhibits wide-ranging behavior.

Other than in Australasia, few dog populations are truly independent of humans (Vanak and Gompper 2009b). Ultimately, their close relationship with human society strongly influences all the above factors. Whether dogs are herding dogs, guard dogs, village dogs, farm dogs, or simply household pets determines in large part their ranging behavior, population density, and feeding habits. Close association with human habitation may also provide refugia from top-down interference competition or predation in systems where larger predators exist (Butler et al., Chapter 5).

Here, we review the literature to examine competitive dynamics between dogs and sympatric carnivores as a function of their position in the carnivore community and the type of competition (see Box 3.1).

3.2 Dogs as interference competitors

Predators that share common resources will often show potential to compete with each other, but even if resources are abundant and overlap is minimal, interference competition may still occur via direct aggression and intraguild killing or via cues that drive spatial segregation (Box 3.1). Interference competition between mammalian carnivores is well documented, and it is expected that dogs as abundant mid-sized carnivores will be either recipients of top-down interference or will be the aggressors. However, the competitive dynamics that occur between dogs and sympatric carnivores can be expected to vary considerably across the world as a function of related human densities, land use types, and the place of dogs in the native carnivore community.

3.2.1 Where dogs are top-predators

In many regions of the world, dogs have assumed the role of top-predators for a variety of reasons: the absence of larger mammalian carnivores, the extermination of large carnivores either by humans or other factors, or the restriction of larger carnivores to non-human dominated landscapes (Gompper and Vanak, 2008). Some potential examples of the competitive ability of dogs as top-predators come

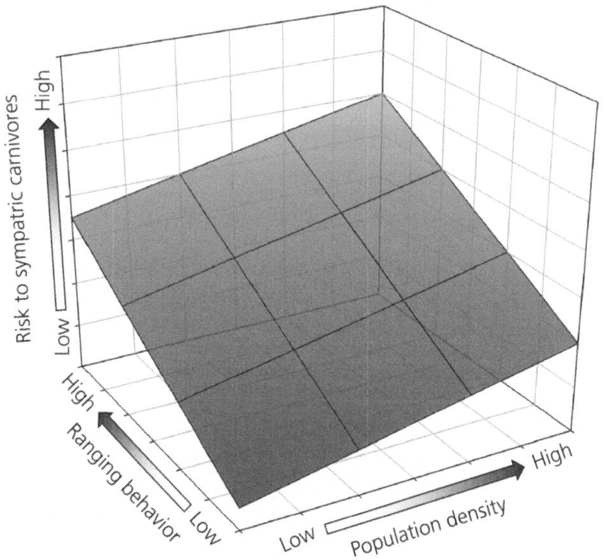

Figure 3.1 Conceptual model of the effect of dog population density and ranging behavior on the risk dogs may pose as interference competitors to sympatric native carnivores. Adapted from Vanak and Gompper (2009b). Reproduced with permission from John Wiley & Sons.

Box 3.1 Competition in its various forms

Within the carnivore guild, sympatric species compete fiercely with each other and this interaction manifests as either a form of: (1) exploitative competition, whereby differences in acquisition of limited resources determine the competitive outcome (Figure 3.2a); or (2) interference competition, whereby direct interactions such as spatial exclusion, harassment, or even intraguild killing determine the outcome (Figure 3.2b). Exploitative competition in carnivores has not been directly demonstrated but can be inferred, especially for species that kleptoparasitize the kills of others. Interference competition and intraguild killing, however, are common among carnivores and there are many well-documented examples of these interactions.

Interference competition

Mechanisms of interference competition usually follow an allometric relationship, with larger carnivores directly dominating smaller ones. Smaller species thus have a strong imperative to minimize interference interactions, especially when there are potentially lethal consequences (Creel et al., 2001; Palomares and Caro, 1999; Ritchie and Johnson, 2009). Studies that examine single or multiple features of interference competition among carnivores are numerous and have been reviewed extensively in the literature (Creel et al., 2001; Glen and Dickman, 2005; Ritchie and Johnson, 2009; Ritchie et al., 2012; Roemer et al., 2009). Here we highlight some of the patterns.

In general, interference competition can affect subordinate competitors in several ways: by limiting spatial distributions (Creel et al., 2001; Ritchie and Johnson, 2009), constraining habitat selection (Mitchell and Banks, 2005; Vanak and Gompper, 2010; Fisher et al., 2013), or reducing prey encounter rates (Atwood and Gese, 2008; Creel et al., 2001; Palomares et al., 1996; Vanak et al., 2009). In response, subordinate competitors use a suite of spatial and temporal behavioral strategies to minimize encounters with the dominant competitor(s). These include avoiding range overlap altogether, or modifying range use within overlapping home ranges (e.g., Berger and Gese, 2007; Crabtree and Sheldon, 1999; Elmhagen et al., 2002; Fedriani et al., 2000; Gosselink et al., 2003; Tannerfeldt et al., 2002). In these cases, interference competition results in non-overlapping territories, in favor of the dominant competitor.

There are also more subtle spatial and temporal behavioral choices that subordinate competitors make to avoid interference competition. African hunting dogs (*Lycaon pictus*) avoid prey-rich habitat due to the presence of lions (*Panthera leo*) (Creel et al., 2001), thereby incurring a reduction in hunting efficiency. After the reintroduction of wolves (*C. lupus*) into Yellowstone National Park, USA, coyotes (*C. latrans*) reduced foraging and increased vigilance when they were in wolf territories (Switalski, 2003). Kit foxes (*Vulpes macrotis*) are known to use several anti-predatory tactics, such as multiple diurnal den use and habitat partitioning, to avoid interference competition with coyotes (Nelson et al., 2007). These examples demonstrate that the mere presence of a dominant competitor results in risk aversion that is manifested by the reduced use of areas that would otherwise be selected for. Ultimately, interference competition can result in reduced individual fitness and population sizes (Ritchie and Johnson, 2009).

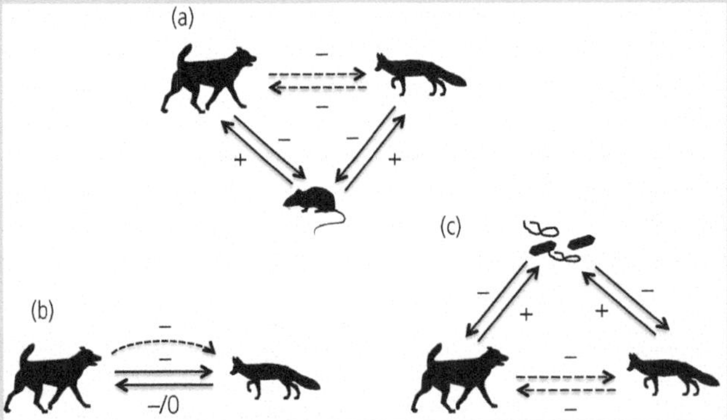

Figure 3.2 Illustrations of exploitative (a), interference (b), and apparent competition (c). Solid arrows denote direct effects, dashed arrows indicate indirect effects, and signs indicate effect on species. Adapted from Glen and Dickman (2005). Reproduced with permission from John Wiley & Sons.

continued

Box 3.1 *Continued*

Exploitative competition

Exploitative competition occurs when species share the same limited resources, and one species can potentially outcompete the other either through numerical or behavioral superiority in acquiring this shared resource (Petren and Case, 1996). Demonstrating the occurrence of exploitative competition between two species requires a number of factors to first be established: (1) reduced survivorship or reproduction as a function of the limited and shared resource; (2) reduced access to the shared resource; and (3) a lack of direct interference (Petren and Case, 1996). Because of this, exploitative competition among carnivores has rarely been demonstrated even though there is extensive evidence of dietary overlap between species. Indeed, exploitative competition has only been experimentally demonstrated in a few plants and animals (Bonaccorso et al., 2007; Dorchin, 2006; Smallegange et al., 2006). Nonetheless, the existence of exploitative competition is a common assumption among carnivore ecologists (Jhala and Giles, 1991; Johnson et al., 1996; Sillero-Zubiri and Gottelli, 1995).

Apparent competition

A third kind of interspecific interaction may also function to explain patterns of carnivore spatial distribution: apparent competition. While interference and exploitation competition are perceived as a function of resource limitation, the outcome of apparent competition results from a third factor indirectly mediating the distribution of apparently competing species (Figure 3.2c) (Holt, 1977; Holt and Lawton, 1994; Price et al., 1998). For example, populations of Sierra Nevada bighorn sheep (*Ovis canadensis californiana*) declined due to predation from mountain lion (*Puma concolor*), whose numbers were primarily sustained by the more abundant mule deer (*Odocoileus hemionus*). Hence, without factoring in predation by mountain lion, the decline in bighorn sheep appeared to be due to competition with mule deer (Johnson et al., 2013). Similarly, the critically endangered island fox (*Urocyon littoralis*) faced near extinction due to predation by golden eagles (*Aquila chrysaetos*), whose unnaturally large populations were in turn sustained by invasive feral pigs (*Sus scrofa*) (Roemer et al., 2001, 2002). Apparent competition among carnivores may result from one carnivore subsidizing the presence of another carnivore, and thus also putatively reducing the density of other sympatric carnivore species. However, apparent competition is more likely to result from shared parasites than from shared predators. Shared parasites may be an important driver in structuring assemblages of species (Rushton et al., 2000; Tompkins et al., 2000, 2003). This kind of interspecific interaction is a form of apparent competition because the parasite–host interactions result in the appearance of a competitive dynamic (Holt, 1977; Holt and Pickering, 1985; Price et al., 1998; Tompkins et al., 2000). Apparent competition has been documented in many taxa, ranging from bacteria and insects to birds and large mammalian predator–prey communities. In general, the rarer or more susceptible host species is reduced or eliminated as the parasite increases in the less susceptible or more abundant reservoir host (e.g., Morris et al., 2004; Power and Mitchell, 2004; Tompkins et al., 2000). Apparent competition via shared parasites may ultimately turn out to be a common and important component of community structure (Hatcher et al., 2006; Holt and Dobson, 2006) but has only rarely been explored as such in carnivore communities (Roemer et al., 2009).

from Australia. Prior to the arrival of the dingo in Australia around 4,000 years ago, the largest mammalian predators were the thylacine (*Thylacinus cynocephalus*) and the Tasmanian devil (*Sarcophilus harrisii*). Both of these marsupial predators disappeared from the Australian mainland after dingoes had become established, the thylacine around 3,500 years ago and the devil within the last millennium (Corbett, 1995). As both species persisted in the absence of dingoes in Tasmania, it has been commonly assumed that their demise on the mainland was caused by competition from the dingo (Corbett, 1995; Wroe et al., 2007). The evidence is ostensibly compelling. In Tasmania, thylacines (15–35 kg) reportedly hunted kangaroos and wallabies in open forest and woodland areas, pursuing their quarry singly or in pairs at night (Paddle, 2002). They may also have hunted a range of smaller prey (Attard et al., 2011; Wroe et al., 2007). Tasmanian devils (5–15 kg) occupy similar habitats and also pursue wallabies and smaller mammals, but include much scavenged material in their diet (Jones and Barmuta, 1998). Both species probably would have overlapped markedly in these aspects of their resource use with the dingo.

Although dingoes are smaller than the Tasmanian thylacine, at 12–22 kg, their ability to hunt in packs may have given them a competitive edge over the marsupial predators and suppressed their populations via either extreme interference competition or superior exploitation of shared prey (Letnic et al., 2012). Recent evidence further indicates that mainland thylacines were smaller than their counterparts in Tasmania and that mainland females were considerably smaller than dingoes (Fillios et al., 2012). This potentially would have placed mainland thylacines under great pressure, especially from direct interference interactions.

Despite the elegance of this interpretation, the arrival of the dingo coincided with several environmental changes that may have been more detrimental to the marsupials, making it difficult to resolve the competitive impact of the dingo on its own. The mid Holocene experienced intense climatic changes as the continent recovered from the extremes of the last glacial maximum (Brown, 2006). There was also an increase in the human population and in peoples' hunting efficiency due to the adoption of technologies, such as edge-ground and hafted stone tools, that improved considerably on previous weaponry (Johnson and Wroe, 2003; Letnic et al., 2012). Coincident with (and perhaps caused by) these changes, archeological records suggest that people exploited smaller-bodied prey progressively from the mid to late Holocene (Fillios et al., 2012). As this suite of changes affected the Australian mainland (but Tasmania to a much lesser extent), the arrival of the dingo would have had an additive impact on the mainland populations of the thylacine and devil and may perhaps have delivered the competitive *coup de grace*. The longer tenure of devils in the presence of dingoes on the Australian mainland may have arisen from their ability to exploit carrion and small prey that formed a more minor part of the diet of the dingo; that is, there was perhaps more dietary niche separation and hence less competition between devils and dingoes than between dingoes and thylacines (Corbett, 1995).

As the apex mammalian predator in Australia, dingoes could be expected to dominate in direct encounters with most other predators; moderate to large crocodiles would have ascendancy over them, although their eggs and young still may be vulnerable (Somaweera et al., 2011). No studies have yet addressed whether sympatric predators recognize and avoid cues of the presence of the dingo. There is some evidence that native prey species can distinguish the odor or other cues left by dingoes and domestic dogs and reduce their activity after exposure (Carthey and Banks, 2012; Parsons and Blumstein, 2010); native predators plausibly also would have experienced selection to reduce their risk of an encounter with dingoes.

In many parts of the world with high human population densities, native predators have either been

Box 3.2 Old dogs and new dogs: competition between dingoes and recent dogs

As a general rule, the form of competition that occurs between two species will be the same form of competition that is apparent between conspecifics. Thus, we might expect that interference will be the dominant form of competition between different groups of dogs, with exploitation playing no, or a very minor, role. Studies of wild animals generally bear out this expectation: pairs or packs of dingoes, for example, use howling and scent-marking to defend the boundaries of their territories, and chase intruders if they are detected (Purcell, 2010). There is no evidence for exploitative interactions between packs.

But what if dogs differ from each other in subtle ways? Dingoes and recently introduced domestic dogs co-occur throughout much of Australia and differ in aspects of their size, skull morphology, coloration, behavior, and seasonality of reproduction (Corbett, 1995). Dingoes occur in most parts of the continent except where they are heavily persecuted in the south-east and south-west, whereas wild domestic dogs tend to be most active near townships and settlements. There is some evidence that wild domestic dogs seldom successfully breach the territorial boundaries established by dingo packs, and fare poorly due to interference (Corbett, 1995). Conversely,

continued

Box 3.2 *Continued*

large breeds of domestic dog, such as the Maremma, are being used increasingly on rural properties to protect flocks of sheep from the ravages of dingoes. Termed livestock guardian dogs (LGDs), these large and aggressive animals are allowed to bond with flocks from an early age and are provided with all their food and shelter needs by their human owners. Once trained, they can be very effective in using aggressive interference behaviors to repel dingoes. In one survey of 150 livestock producers, two thirds of respondents indicated that predation from dingoes and other wild dogs ceased after LGDs were put in service; another 30% reported that predation on livestock decreased (van Bommel and Johnson, 2012). In these situations, LGDs are essentially the guardians of their own human-defined territories (usually fences and paddock boundaries) and effectively repel incursions by dingoes via direct interference. Without the continuous resource-subsidy from humans it is not clear whether LGDs would always prevail over dingoes, although interference would almost certainly be the mechanism of competition in any encounters.

Over the long term, any competitive superiority of dingoes over wild domestic dogs may not be sufficient for dingoes to survive in 'pure' form. Another kind of interaction—hybridization—may doom them. Although wild domestic dogs seldom gain access to dingo packs, they can do so more easily after packs have been fractured by misguided human attempts to control dingo numbers. Such attempts, using poisons, trapping, or shooting, often remove the dominant animals from a pack and may loosen the social and reproductive controls on remaining animals sufficiently that wild dogs then can gain access to the packs (Allen and Gonzalez, 1998; Purcell, 2010). If breeding is successful, hybridization ensues. Recent studies suggest that dingoes in remote regions of Australia have suffered little hybridization, with up to 90% of animals retaining the 'pure' dingo genotype (Newsome et al., 2013), whereas only 15% or less may be 'pure' in more heavily settled and rural areas where dingoes are controlled and wild domestic dogs are more prevalent (Elledge et al., 2009). Hybridization is not usually seen as a form of competition. However, the inexorable supplanting of dingo by dog genes in the present example gives the process the appearance of a strong and highly asymmetrical form of competition.

exterminated or occur at very low densities. Here, free-ranging dogs have become the *de facto* top-predator, and this is reflected in their interactions with the remaining native carnivores. One of the most comprehensive examples of competitive dynamics between dogs and sympatric carnivores comes from the dry grasslands of Maharashtra, India, where dogs interact with native Indian foxes (*V. bengalensis*). Here, dogs are not the largest predator, as the Indian wolf (*C. l. pallipes*) occupies that spot. However, wolves occur at very low densities (Habib, 2007), while dogs occur at very high densities (24 dogs/sq. km; Vanak, 2008; Vanak and Gompper, 2010). Thus, as the most abundant mid-sized carnivore in the landscape, dogs are dominant. Dogs and Indian foxes have low dietary overlap and thus do not appear to compete directly for food. This is due mainly to the fact that dogs subsist almost entirely on HDF such as garbage, human feces, direct feeding, and carrion, whereas foxes subsist almost entirely on rodents, fruit, and invertebrates (Vanak and Gompper, 2009a). Yet, Indian foxes react to dogs as they would to a dominant carnivore. When experimentally exposed to dogs, foxes reduced food consumption by as much as 70% (Figure 3.3a) and increased time spent in vigilance behavior 15-fold (Figure 3.3b; Vanak et al., 2009). Thus, at the fine scale, Indian foxes showed a strong foraging-vigilance trade-off, but risk-averse behaviors were also seen at larger spatial scales. The odds of foxes using an area decreased rapidly the closer that area was to a dog-dominated habitat. This happened even if these areas were associated with a higher abundance of their main food source (Vanak and Gompper, 2010).

Similar spatial avoidance in response to interference competition is also observed in other parts of the world. In the United Kingdom, the distribution of red foxes (*V. vulpes*) was negatively affected by the presence of high densities of street dogs. These dogs were not only the second most important cause of mortality of cubs, with as much as 15% of cub mortality attributed to dogs, but also killed adult foxes on occasion (Harris, 1981). In South America, three different studies suggest that dogs have negative effects on the occurrence of canids such as the maned wolf (*Chrysocyon brachyurus*), chilla fox (*Lycalopex*

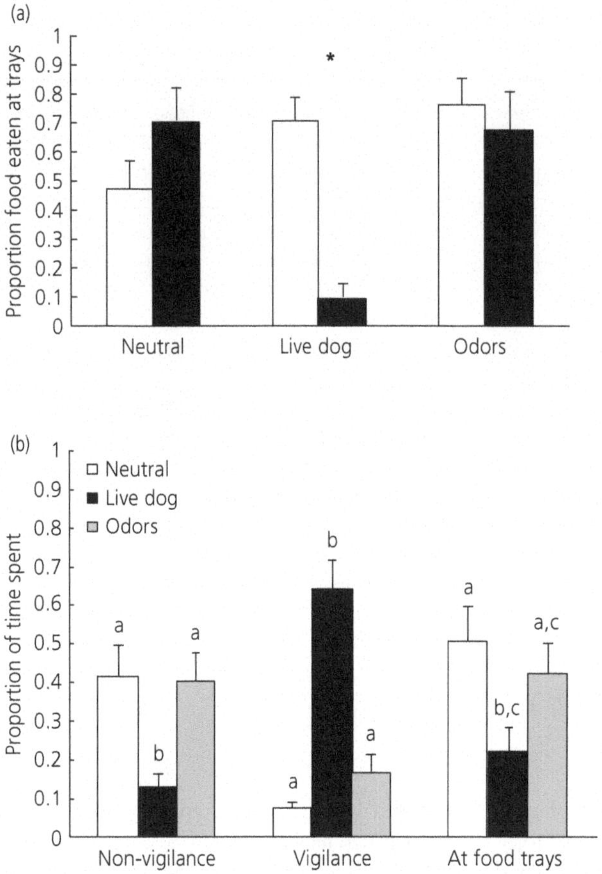

Figure 3.3 Indian foxes, when exposed to the presence of a live dog at experimentally placed food trays, (a) reduced the proportion of food eaten by ~70% compared to a control tray, and (b) increased time spent in vigilance behaviors by 15-fold when compared to trials when only dog odor or a control was placed. The asterisk and letters denote significant difference (Fisher's protected least squares difference $P<0.05$) among trials for each behavior and error bars represent mean ± SE, $n = 7$ dens. Reprinted from Vanak et al. (2009) with permission from Springer.

griseus), and crab-eating fox (*Cerdocyon thous*) (Espartosa, 2009; Lacerda et al., 2009; Silva-Rodríguez et al. 2010a). Two of these studies also addressed the effects of dogs on non-canid carnivores such as South American coati (*Nasua nasua*), crab-eating raccoon (*Procyon cancrivorus*), tayra (*Eira barbara*), and small cats (*Leopardus* spp.). Among these species, and with the sole exception of the coati (Espartosa, 2009), there was no evidence of potentially negative effects of dogs (Espartosa, 2009; Lacerda et al., 2009). The fact that canids seem to show stronger responses to dogs than other carnivores is not surprising as interference competition and intraguild killing should be more intense between members of the same family (Donadio and Buskirk, 2006).

Because of the scarcity of directed studies on competition between dogs and sympatric carnivores, the competitive effect of dogs can sometimes only be inferred without a direct link to causality (see Box 3.3). For example, in Spain the presence of dogs was inversely correlated with an indicator of Eurasian badger (*Meles meles*) presence (Revilla et al., 2001). In Madagascar, the fossa (*Cryptoprocta ferox*) is nearly absent from areas with high dog occupancy (Gerber et al., 2012) and trapping rates of this species decrease as dog trapping rates increase (Barcala, 2009; see Box 3.4). Similarly, the highest densities of the endangered narrow-striped mongoose (*Mungotictis decemlineata*) occur in areas where neither dogs nor people are detected and, in one area, sightings of this species increased after dog removal (Woolaver et al., 2006). These complementary distribution patterns are likely to be common throughout the world, especially in countries where dog densities are largely unregulated in rural areas (Boitani et al., 1995; Krauze-Gryz and Gryz, 2009; Slater et al., 2008; Vanak and Gompper, 2010).

Box 3.3 Competition between dingoes and introduced predators

In contrast to the relative paucity of studies invoking competition between dingoes and native predators, a substantial body of work has focused on interactions between dingoes and the more-recently introduced red fox and feral cat (*Felis catus*). Both predators show varying degrees of dietary overlap with the dingo, although their small size (fox 4–8 kg; cat 3–6 kg) means that they generally hunt smaller prey than their larger counterpart (Cupples et al., 2011; Glen et al., 2011). All three species are broadly sympatric over much of Australia and often co-occur in the same habitat, thus increasing the opportunity for competition. Evidence that competition does indeed occur is derived largely from inverse spatial patterns in the species' abundances. In the case of the fox, several studies have shown that fox numbers and/or activity are suppressed in the presence of the dingo at local, regional, and continental scales (Johnson and VanDerWal, 2009; Letnic et al., 2009, 2012; Mitchell and Banks, 2005; Wallach et al., 2010). In general these studies also show that where dingoes are abundant, foxes are rare, but where dingoes are scarce or absent, foxes can achieve much higher numbers (Figure 3.4). In the case of the feral cat, evidence for competition is more tenuous. Several studies show negative associations between the numbers of dingoes and cats (Brawata and Neeman, 2011; Kennedy et al., 2012; Purcell, 2010; Wallach et al., 2010), but others show weak patterns or no relationship (Letnic et al., 2009; Newsome, 2001). In contrast to the red fox, which is largely active on the ground, feral cats can climb trees and thus may be better able to avoid direct contact with dingoes. For example, in very barren areas of the Channel Country in western Queensland, the primary refuge sites are tree-lined riparian strips that allow cats to escape ground-active dingoes (Pettigrew, 1993).

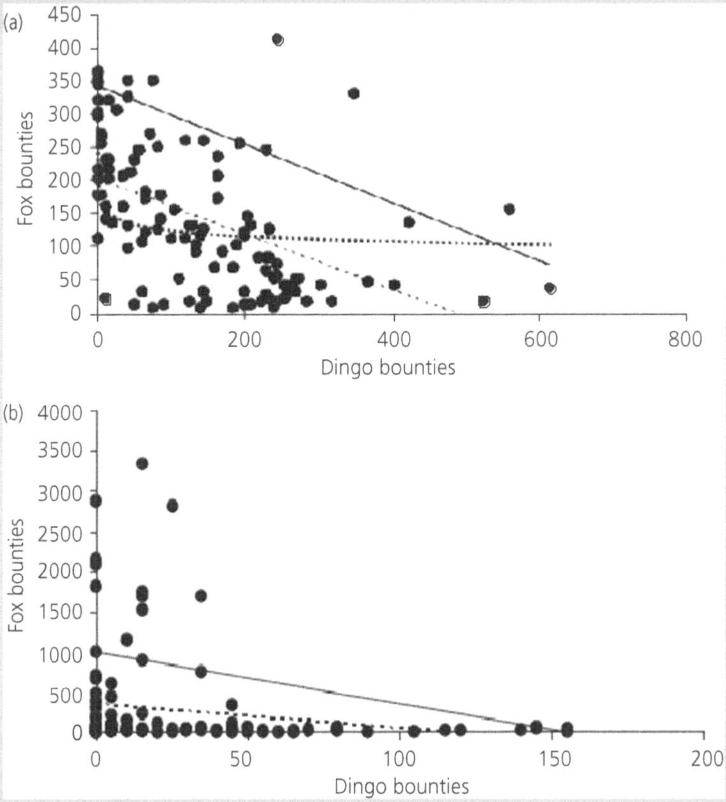

Figure 3.4 Numerical relationship between the dingo and red fox, based on bounty records collected (a) in Queensland in the financial year 1951–52 and (b) in Western Australia in the years 1947–52. Each dot represents the number of bounties paid within 100 × 100 km grid cells in each state over the bounty collection period. Solid lines show regressions for the 90% quantile of fox abundance against dingo abundance, light dotted lines show significant ($P<0.05$) least squares linear regressions, and the heavy dotted line in (a) indicates a significant ($P<0.05$) least squares logarithmic regression. Reprinted from Letnic et al. (2011) with permission from John Wiley & Sons.

continued

Box 3.3 *Continued*

The form of interaction between dingoes and the two smaller predators is almost certainly extreme interference competition. This is suggested by the fact that red fox and feral cat remains are found frequently but only at very low levels in scat or stomach samples of the dingo (Allen and Leung, 2012; Letnic et al., 2012), and observations that both the smaller predators will flee if an encounter with a dingo is imminent (Corbett, 1995; Purcell, 2010). However, the most compelling evidence for extreme interference comes from observations of dingoes killing but not eating both foxes and feral cats in a large (37 km^2) enclosure in treeless habitat in South Australia (Moseby et al. 2012). In this latter study, data derived from GPS collars confirmed that dingoes returned to the carcasses of foxes and cats that they had killed for several hours after the killing, with some dingoes returning again and again over the ensuing days. It is possible that these frequent return visits were to ensure that the smaller predators no longer posed a competitive threat.

Box 3.4 Landscapes of fear: spatial and temporal considerations

There is increasing recognition that predators, including dogs, may influence their prey in subtle ways that are not readily apparent by simply comparing measures of abundance of pairs of interacting species (Creel and Christianson, 2008; Ritchie and Johnson, 2009). This is important, because significant effects of dogs on other species may have been overlooked or dismissed previously due to a focus on abundance patterns only (Johnson and VanDerWal, 2009). Recent studies from two vastly separated regions (Australia and Madagascar) and with different predator communities serve to illustrate this point.

In Australia, dingoes are often killed to minimize their attacks on livestock, and the impact this has on their abundance and behavior, as well as that in turn on a sympatric mesopredator, the feral cat, was examined using remote camera surveys across nine pairs of large Australian rangeland (cattle) properties (Brook et al., 2012). Dingo abundance was generally reduced on properties where dingoes were controlled, but most interestingly dingo behavior varied between sites where dingoes were controlled and where they were not. Uncontrolled dingo populations were crepuscular, similar to their major prey (kangaroos). In populations subject to control, dingoes were less active around dusk, with activity concentrated in the period immediately before dawn. Shifts in feral cat abundance were inversely related to corresponding shifts in dingo abundance, but importantly there was also a negative relationship between predator visitation rates at individual camera stations, indicating that cats avoided areas where dingoes were most active, and that reduced activity by dingoes at dusk was associated with higher cat activity at dusk. This suggests that dingo control may not only lead to more feral cats (mesopredator release), but also allows cats to optimize their hunting when dingoes are less active (Brook et al., 2012).

In Madagascar, as part of an overall study examining activity patterns of carnivores, the authors found that the fossa selectively used crepuscular hours, but was cathemeral overall (Gerber et al., 2012). Fossas were conspicuously absent from sites where dogs were most abundant and active throughout the diel cycle, and the ring-tailed mongoose (*Galidia elegans*), another native, but smaller-bodied predator, shifted its activity away from its preferred periods in the presence of dogs. Together, these studies highlight that indicators other than abundance, such as spatial and temporal behavioral patterns, should also be considered when determining a predator's ability to affect trophic levels. However, abundance still remains an important consideration, as dogs have even been known to exclude larger-bodied predators, such as wolves, from accessing food; despite their smaller body-size as compared to wolves, dogs overcame this by being more aggressive and occurring in higher numbers than wolves (Boitani et al., 1995).

3.2.2 Dogs as mesocarnivores

In several ecosystems across the world, dogs are sympatric with larger carnivores such as the wolf, leopard (*Panthera pardus*), tiger (*P. tigris*), and bears. In natural systems, mesocarnivores are often suppressed by the presence of larger competitors (Crooks and Soulé, 1999). Thus, it would be expected that in areas where larger carnivores are present, dogs would be negatively affected. However, because of the lack of directed studies in such systems, we can only speculate on whether dogs are subject to top-down competitive effects from larger carnivores. What we do know is that in many systems dogs form an important part of the diet of larger carnivores (e.g., leopard; Butler et al., Chapter 5). Thus, it is likely that in such systems dogs will also display avoidance tactics as shown by other mesocarnivores. However, dogs may also be insulated from top-down effects if they can find refuge in human settlements or benefit from human association while ranging into native habitats.

Studies from southern Africa, where dogs have the potential to interact with possibly the greatest number, as well as diversity, of carnivores, well illustrate the complexity of the situations that are possible (Butler and du Toit, 2002; Butler et al., 2004). Detailed studies of interactions in Zimbabwe between dogs and large wild carnivores in Gokwe Communal Land (GCL) bordering the Sengwa Wildlife Research Area (SWRA) suggested that dogs were inefficient and infrequent predators. Their small group size (mean 1.7) and body mass (mean 15 kg) was reflected in their diet, which consisted of scavenged human refuse, feces and carrion, fruit, and some insects. The vast majority (87%) of this food was gained independently, with 55% of it derived from carrion, particularly mammalian carcasses. Leopards, lions, spotted hyenas (*Crocuta crocuta*), side-striped jackals (*Canis adustus*), and several smaller carnivore species travelled up to 3 km into GCL (Butler, 2000) while dogs were observed up to 3 km within the SWRA independently of people, and up to 6 km with people, and consequently were sympatric with these species in the SWRA–GCL interface. Dogs dominated the carnivore scavenging guild (see Section 3.3.1), but were preyed upon by lions, leopards, and spotted hyenas. Thus, inside wildlife reserves, and on their boundaries with communal lands, dogs were meso-level carnivores and had diverse relationships within a guild of larger and smaller native mammalian and avian species.

In most communal land areas and commercial farmlands where large carnivores are essentially absent, either due to high human population densities or persecution, dogs are the local apex predator. However, in commercial farmland typified by intensive cropping and more extensive cattle production, dogs occur at very low densities due to the lack of human settlement. As a consequence side-striped and black-backed jackals (*C. mesomelas*) have become the most numerous native carnivore through 'mesopredator release,' occurring at high densities relative to natural conditions (Cumming, 1982; Bingham et al. 1999). Hence within communal lands and where they border commercial farmland, relationships within the carnivore guild are limited to competitive interactions between dogs and jackals.

Both jackal species have omnivorous and opportunistic feeding habits in Africa (Skinner and Chimimba, 2005), and will efficiently scavenge carrion (Richardson, 1980) and human waste (Bothma et al., 1984), potentially overlapping the dietary niche of dogs. Consequently the ecological relationship between dogs and jackals is characterized by conflict and intolerance, with dogs dominating jackals in aggressive encounters due to their larger size (15 kg versus 8–10 kg for side-striped and 7–8 kg for black-backed jackals). Carcass experiments showed that side-striped jackals are attracted to carrion but do not feed, perhaps avoiding the scavenging dogs present (Butler and du Toit, 2002; see Section 3.3.1). At higher densities, dogs aggressively exclude jackals and jackals may become the subordinate species, avoiding contact with dogs and hence occurring at lower densities in communal lands. Commercial farmland therefore provides a refuge for jackals, which then 'spill over' into the peripheries of communal lands.

Competition between dogs and other carnivores may not always follow an allometric hierarchical relationship. In some cases, dogs can out-compete larger carnivores because of numerical superiority via frequency-dependent competition. The outcome of encounters between the endangered Ethiopian wolf (*C. simensis*) and dogs in the highlands of Ethiopia seems to depend on numbers. If wolves outnumber dogs, then the dogs are chased away, and vice-versa (Atickem et al., 2010). However, the authors concluded that interference competition with dogs was not a major concern for the conservation of Ethiopian wolves, but suggested that this could change in the future as the numbers of people and dogs grow (Atickem et al., 2010; Laurenson et al., 1998; Sillero-Zubiri and Macdonald, 1997).

3.2.3 Dogs as predators of other carnivores

The killing and eating of one carnivore by another is termed intraguild predation (Polis et al., 1989). Several reviews have highlighted the widespread occurrence of this interaction (Donadio and Buskirk, 2006; Glen and Dickman, 2005; Ritchie and Johnson, 2009). However, we draw the distinction between 'routine' predation, where the victim is killed mainly for consumption, from killing as a form of extreme interference competition, where the victim is not consumed or only partially consumed (see also Butler et al., Chapter 5). There are numerous examples of dogs killing other carnivores, both with and without consumption. For example, in Australia, dingoes have been recorded killing but not eating spotted-tailed quolls (*Dasyurus maculatus*) (Körtner, 2007; Körtner and Watson, 2006), red foxes, and feral cats (Moseby et al., 2012), but more generally they appear to eat or even cache their victims. Other than quolls, large varanids and snakes have been recorded in the diet of dingoes in several studies (e.g., Glen et al., 2011; Pascoe et al., 2012). However, the frequency of occurrence of other predators in dingo scats or stomach samples is usually very low (<1%), and it is seldom clear whether dingoes have killed or merely scavenged these items. One exceptional observation is that of Webb (1996) who saw three dingoes work together to kill and then consume a large (1.5–2 m long) lace monitor (*Varanus varius*).

Evidence of killing by dogs comes from almost everywhere they have been studied, and the list of mammalian carnivore species that are killed by dogs is long. The list includes taxa with a wide range of body sizes from dhole (*Cuon alpinus*) ~18 kg; (Williams, 1935), to civets ~1–6 kg (Ashraf et al., 1993; Dahmer, 2002), small cats ~5 kg (Barashkova and Smelansky, 2011) and mustelids ~1–2 kg (Butler and du Toit, 2002; Maran et al., 2009; Ross et al., 2008). In many cases intraguild killing (and not predation) can be deduced because the carcasses of even small carnivore species are recovered relatively intact. For example, dogs were the presumed cause of direct killing of nearly 29% of Indian foxes that were found dead (n = 13) (Vanak, 2008). Similarly, dogs killed 26% of spotted skunks (*Spilogale putorius*) known to have died (n = 77) during a study conducted in farmlands in southeast Iowa (Crabb, 1948). Dogs are also frequent causes of mortality for the Geoffroy's cat (*Leopardus geoffroyi*) and culpeo foxes (*Lycalopex culpaeus*) on certain Argentinean ranches (Novaro et al., 2005; Pereira et al., 2010). On the other hand, evidence of consumption comes from analyses of dog diet. One study showed consumption of carnivores such as lesser grison (*Galictis cuja*) and South American coati (Campos et al., 2007), but it was unclear whether dogs killed these carnivores or simply scavenged them. Despite these studies, the quality of the evidence for the importance of dogs as a source of mortality is mixed. In fact, most of the evidence available is constituted by anecdotal reports of single mortalities (see Section 3.5.1). Among these, the most attention-garnering report was the rediscovery of the black-footed ferret (*Mustela nigripes*), when a farm dog brought in a freshly killed individual (Clark, 1987; see Gompper, Preface).

3.3 Exploitative competition

Other than via direct interference competition and associated indirect effects, dogs can also compete with native carnivores exploitatively for food (Box 3.1). Because domestication has rendered dogs less efficient at hunting than wild carnivores (Clutton-Brock, 1995), they are less likely to be effective exploitative competitors (Petren and Case, 1996; Vanak and Gompper, 2009b). However, most populations of dogs are subsidized by humans and thus can attain numerical superiority, allowing them to potentially outcompete native carnivores in acquiring limited resources (as per the abundance-asymmetry hypothesis; Vázquez et al., 2007). Furthermore, the plasticity of dog behavior and the advantages of food security from human subsidies can potentially allow dogs to compete with a wide range of carnivore species, from top-predators such as wolves, lions, and leopards to mesopredators such as jackals and foxes as well as nonmammalian carnivores such as varanid lizards.

Although there are several examples worldwide of dogs killing often substantial numbers of prey and being an important cause of mortality for many species (Ritchie et al., Chapter 2), the evidence for dogs being effective exploitative competitors is not robust. This is mainly due to the fact that most dog populations are heavily dependent on HDF. Vanak and Gompper (2009b) found that of 21 studies of dog diet, only 5 (23%) reported subsistence on mainly wild-caught food whereas in the rest dogs were either fully or mainly dependent on HDF. Illustratively, all of those five studies were of truly feral dogs (i.e., dogs that were no longer tame or associated with human settlements and food subsidies). Even studies that have specifically examined the potential for exploitative competition between dogs and sympatric carnivores have found little evidence. Atickem et al. (2010) found that dogs in the Ethiopian highlands subsisted almost entirely on HDF, and that wild-caught prey such as rodents constituted a relatively minor part of their diet. The authors thus concluded that dogs are unlikely to be competing with endangered Ethiopian wolves. Similarly, in India, dogs and Indian foxes did not appear to compete directly for food. This was again due mainly to the fact that dogs subsisted almost entirely on HDF such as garbage, human feces, direct feeding, and carrion, whereas foxes subsisted almost entirely on rodents, fruit, and invertebrates (Vanak and Gompper, 2009a). Silva-Rodríguez et al. (2010a) found similar low levels of overlap in diet between dogs and chilla foxes in southern Chile. On the other hand, Campos et al. (2007) reported high rates of vertebrate consumption by dogs in suburban areas of Brazil. Although these authors suggested that dogs could compete with wild carnivores, there was some uncertainty in this regard since there was no assessment of dietary overlap.

3.3.1 Dogs competing with mammalian and non-mammalian scavengers

Almost all studies that examine the dietary habits of dogs conclude that an important part of their diet is obtained from scavenging human- and non-human generated carcasses of animals. This highly efficient facultative scavenging behavior brings dogs into direct competition with a suite of mammalian and non-mammalian species that are either specialist or opportunistic scavengers. Many of the same factors described previously, such as high population densities, proximity to humans, and 24-hour foraging ability, allow dogs to dominate carcasses, especially near human settlements. For instance, in western India, dogs dominated cattle carcasses at a village dump throughout the day, preventing scavengers such as golden jackal (*C. aucreus*), vultures, and striped hyena (*Hyaena hyaena*) from accessing these resources (Aiyadurai and Jhala, 2006).

A more detailed investigation into the potential scavenging interactions between dogs and wild carnivores was carried out in Zimbabwe (see Section 3.2.2). At experimental carcasses, Butler and du Toit (2002) found that dogs and vultures were the most successful species in the scavenging guild, which also consisted of eight mammalian and nine avian species. Dogs were the most efficient scavengers, consuming up to 60% of the total carcass biomass, compared to 15% for the vultures. However, the influence of dogs varied along transects. Dogs consumed the majority of carcass biomass 1 km within the communal land, whereas vultures, leopards, lions, and spotted hyenas were the principal scavengers at carcasses placed 1 km inside the wildlife reserve.

These patterns suggest that dogs primarily compete with vultures along the interface of communal lands and wildlife reserves. While most carnivores will scavenge given the opportunity, vultures have evolved specifically as scavengers of mammalian carrion, and the eagle-like vultures (hooded, *Necrosyrtes monachus*, white-headed, *Trigonoceps occipitalis*, and lappet-faced, *Torgos tracheliotus*) will also scavenge carrion and human refuse (Houston, 1979; Mundy, 1982). Dogs appear to out-compete vultures for four reasons. First, they physically dominate vultures at carcasses, aggressively excluding them from feeding due to their larger body mass (15 kg versus 6.2 kg for lappet-faced vultures, the largest species recorded scavenging; Houston, 1979). Second, because the experimental carcasses were small (<50 kg, as with the majority of carcasses found in GCL; Butler and du Toit, 2002), dogs were easily able to defend carcasses against vultures, and consume them quickly. Third, although vultures are attracted to carcasses in GCL, they are more sensitive to human disturbance and reluctant to settle at carcasses near human settlements. The majority of human activity occurs during the day, the only time when vultures forage. Fourth, dogs are active for most of the 24-hour cycle, enabling them to locate and then consume carrion more efficiently than vultures.

Similar evidence also points to dogs out-competing black vultures (*Coragyps atratus*) in South America. According to Pavés et al. (2008), dogs reduce food availability for vultures by preying and scavenging on South American sea lions (*Otaria flavescens*). Similarly, with the collapse of the vulture population in south Asia, several reports suggest that dog numbers have increased substantially and have replaced vultures as the main scavengers (Prakash et al., 2003).

By being effective scavengers, dogs can potentially out-compete much larger carnivores. For example, in Africa lions and leopards also scavenge opportunistically (Kruuk, 1972; Schaller, 1976; Smith, 1977) and considering that both species prey on dogs (Butler et al. 2004, Chapter 5), they could certainly outcompete dogs during direct interference confrontations. In spite of this, in Zimbabwe dogs were the more efficient scavengers at experimental carcasses owing to their higher numbers, enabling them to locate and consume larger quantities of carrion. Consequently dogs probably do successfully compete with lions and leopards in exploitative terms, but because both felid species are primarily predators (rather than scavengers), the impact of dogs is likely to be minimal. There is little evidence to show that dogs effectively kleptoparasitize recent kills of wild carnivores.

Spotted hyenas present a more complex issue. They are large (up to 86 kg) and highly efficient scavengers, and are able to locate and consume tissue and bones from carcasses of all but the largest species (Richardson, 1980). They often scavenge refuse around human settlements, for example in Ethiopia (Kruuk, 1968), Tanzania (Kruuk, 1972), and South Africa (Pienaar, 1969), but are also highly successful predators in their own right (Kruuk, 1972; Mills, 1990). Given that they prey on dogs in Zimbabwe (Butler et al., 2004), Tanzania (Kissui, 2008), Kenya (Kolowski and Holekamp, 2006) and Ethiopia (Atickem et al., 2010) (see also Butler et al., Chapter 5), they are also likely to dominate dogs in direct competitive encounters involving scavenging. Due to their higher densities, dogs may potentially reduce food supplies for spotted hyenas around human settlements, but this is unlikely to affect populations within wildlife reserves. The competitive impact of dogs on the smaller and more omnivorous brown (*H. brunnea*) and striped hyenas may potentially be more significant, but no research has been undertaken on this question.

3.3.2 Dogs competing as top-predators

The success of dogs as exploitative competitors seems to depend to a large extent on their association with humans. As mentioned earlier, in the review conducted by Vanak and Gompper (2009b), only 5 of 21 studies of dog food habits showed a primary dependence on wild-acquired food. It is notable that four of these dog populations were fully independent of humans and acted as wild predators. Thus, the true competitive ability of dogs with other carnivores can be assessed in situations where dogs are wild. Such a situation is best seen on a few islands (e.g., Kruuk and Snell, 1981) and on the continent of Australia.

Aside from the thylacine and Tasmanian devil, there is no evidence that any other predatory native species declined or became extinct in Australia following the introduction of the dingo, suggesting that competition between dingoes and extant native species may not be strong. Although contemporary studies generally support this interpretation, there is nonetheless considerable evidence that dingoes share food, habitat, or shelter resources with many native species. Hence, the potential for resource competition exists. Dingoes show some overlap in diet with spotted-tailed and western quolls (*D. geoffroii*); all species prey upon small to medium-sized mammals and birds, although dingoes are capable of hunting much larger prey than the marsupials, especially when in packs (Corbett, 1995; Glen and Dickman, 2008; Glen et al., 2011). It is also possible that the marsupial predators consume the remains of dingo-kills and gain an energetic benefit from doing so (Dickman, 1992). However, good evidence for this is scant; large mammals such as sheep have been identified in the diet of the spotted-tailed quoll, but it is not clear whether these prey had been killed initially by dingoes or had been scavenged after death from other causes (Glen et al., 2011).

Large predatory birds such as wedge-tailed eagles (*Aquila audax*) also share prey including rabbits

(*Oryctolagus cuniculus*) and small to medium-sized macropods with dingoes, as do large reptilian predators, such as varanid lizards (*Varanus* spp.) and saltwater crocodiles (*Crocodylus porosus*) (Glen et al., 2010; Parker et al., 2007; Pascoe et al., 2012; Taylor, 1979). During an irruption of native rodents in central Australia, Pavey et al., (2008) reported a dietary overlap of 88% between dingoes and letter-winged kites (*Elanus scriptus*), which provides considerable potential for food-based competition. Dingoes occupy the same broad range of habitats as native predators (Fleming et al., 2001) and overlap with some species on a fine spatial scale, although a degree of segregation via differential use of habitat components such as trees is usually still apparent (Glen et al., 2011). Dingoes also overlap broadly in their temporal activity with marsupial, some avian, and reptilian predators. However, until simultaneous monitoring of the dingo and native predators is attempted via remote tracking or camera trapping, it will not be possible to quantify patterns of temporal overlap or separation more precisely.

3.4 Apparent competition

The occurrence of apparent competition, wherein patterns of interaction that seem to be the result of competition are actually driven by the distribution and effects of a third species (Box 3.1), has not been robustly demonstrated among carnivores. However, apparent competition between carnivores as a result of shared parasites may be far more common. Several important microparasites of wild carnivores are also enzootic in dog populations and may be readily transmitted to native carnivores (Knobel et al., Chapter 6). For instance, almost all canid species are susceptible to infections from rabies virus, canine distemper virus (CDV), and canine parvovirus (CPV), and in some cases these three viruses are primary drivers of carnivore population dynamics (Cleaveland et al., 2007). This potential for disease-mediated changes in population densities of carnivores that are sympatric with dogs has been suggested as a form of apparent competition (Vanak and Gompper, 2009b). Prominent examples of the role of dogs as reservoirs of pathogens that significantly impact wild carnivore populations come from species in several carnivore families including canids, felids, hyaenids, phocids, mustelids, viverrids, and procyonids. The spillover of these pathogens from dogs to wild carnivores and the consequences of these events for the population dynamics of wild carnivores have been reviewed extensively (Cleaveland et al., 2007; Deem et al., 2000; Funk et al., 2001; Laurenson et al., 1998; Knobel et al., Chapter 6).

3.5 Conservation implications

3.5.1 Dog impacts on threatened carnivores

Most of the directed studies on the impacts of dogs on carnivores have used non-threatened species as study models (e.g., Silva-Rodríguez et al., 2010a; Vanak and Gompper, 2010). However, dogs have been reported to kill a large number of threatened carnivores (either for consumption or intraguild killing). To our knowledge there is (mostly anecdotal) evidence of dog-caused mortality for at least 14 (23%) of 61 species of extant threatened terrestrial carnivores (Silva, 2012). For example, in Nepal, dogs are thought to be one of the most frequent causes of mortality of the red panda (*Ailurus fulgens*) (Williams, 2004; Yonzon and Hunter Jr, 1991). In southern India, the death of 10 (of 22 dead) critically endangered Malabar civet (*Viverra civettina*) was attributable to dogs (Ashraf et al., 1993). In Estonia, dogs killed 4 of 54 European minks (*Mustela lutreola*) that had been released as part of a mink restoration program (Maran et al., 2009). In different areas of the world, local people inform that dogs have killed threatened carnivores such as the Andean cat (*Leopardus jacobita*), guigna (*L. guigna*), Darwin's fox (*Lycalopex fulvipes*), smooth-coated otter (*Lutrogale perspicillata*), and southern river otter (*Lontra provocax*) (Espinosa-Molina, 2011; Hon et al., 2010; Lucherini and Merino, 2008; Silva-Rodríguez and Sieving, 2011; Soler et al., 2004). Other examples of threatened carnivores with evidence of killing by dogs include the black-footed ferret (Clark, 1987), dhole (Williams, 1935), narrow-striped mongoose (Hawkins, 2008; Woolaver et al., 2006), marine otter (*L. felina*) (Pizarro-Neyra, 2008), marbled polecat (*Vormela peregusna*) (Dulamtseren et al., 2009), and Cozumel raccoon (*Procyon pygmaeus*) (McFadden et al., 2010). Considering the relative paucity of studies on free-ranging dogs, it is very likely that the proportion of species of conservation concern killed by dogs is much higher.

From the information available, the picture that emerges is that although the most compelling evidence supporting the impacts of dogs come from non-threatened carnivores such as the Indian fox (Vanak and Gompper, 2010), the same kind of problem may be affecting species that are critically endangered, such as the Malabar civet, the Cozumel raccoon, or Darwin's fox. Research directed at understanding the effects of dogs on both the distribution and population dynamics of endangered carnivores is urgently needed, but in the interim, the information available on common species could, under the precautionary principle, inform decision-making (Silva, 2012).

3.5.2 Dogs and human–carnivore conflict

Under certain conditions, the presence of dogs may be beneficial for conservation. Human–carnivore conflicts are one of the main threats for carnivore conservation worldwide (Inskip and Zimmermann, 2009; Treves and Karanth, 2003) because the economic losses from livestock depredation lead to intolerance of wild carnivores. Livestock-guarding dogs are most commonly used to mitigate this conflict (VerCauteren et al., Chapter 9). In such cases, the function of dogs is to keep native carnivores away from livestock. Hence, the negative correlation between the distribution of carnivores and dogs in many studies (Espartosa, 2009; Silva-Rodríguez et al., 2010a; Vanak and Gompper, 2010) may ultimately help in reducing carnivore mortality associated with human–wildlife conflict, and therefore benefit conservation in the long run (González et al., 2012).

In Australia, where the dingo is the 'top-dog,' much evidence has emerged in recent studies that this carnivore provides a net benefit to native wildlife by suppressing the impacts of the red fox and perhaps also the feral cat (Glen and Dickman, 2005; Letnic et al., 2012). However, as dingoes also cause livestock losses in rangeland areas, their management for conservation needs to be balanced against mitigating their effects in production landscapes. Intriguingly, one of the most promising means of doing this is to use domestic dogs as livestock guardians, to keep dingoes away from rangeland enterprises (Box 3.2; van Bommel and Johnson, 2012).

Although dogs can help reduce human–carnivore conflict, they can also intensify it. Just as for other carnivores, dogs prey on livestock, but distinguishing between losses caused by dogs and those by other carnivores is a difficult task. Despite the difficulty involved in the correct determination of cause of death, carnivores such as wolves are often blamed for losses that could have been caused by dogs (Cozza et al., 1996). For example, in northern Spain, wolves were blamed for 94% of the attacks on domestic animals (Cozza et al., 1996). Paradoxically, in the same area and during the same time period, dogs were more abundant than wolves. Furthermore, sheep represented 36% of prey items found in dog scats versus only 3% in wolf scats, thus suggesting that the actual impacts of wolves may have been overestimated (Echegaray and Vilà, 2010). These identification problems are a major issue for carnivore conservation, because they may lead to unjustified persecution of carnivores, and may increase the cost of compensation strategies to reduce human–carnivore conflicts (Cozza et al., 1996; Echegaray and Vilà, 2010; Sundqvist et al., 2008). Another situation where high densities of dogs can inadvertently trigger human–carnivore conflict is when they are prey for larger carnivores. As described by Butler et al. (Chapter 5), dogs are killed and often consumed by several species of large carnivores. In cases where the dog is a beloved pet or a prized animal, this causes resentment towards the presence of large carnivores and can be a major impetus for predator removal programs. Thus, it appears that dogs may be beneficial in mitigating human–carnivore conflict, but only in a narrow set of conditions where they are fully under human supervision.

3.5.3 Dogs as a large-scale edge effect

A body of evidence is starting to accumulate on the existence of large-scale edge effects as a function of human-driven activities (Laurance, 2000). Contrary to previous evidence, edge effects are not confined to just <150 m from a habitat edge, but can potentially extend up to 5 km from a disturbance edge (Laurance, 2000; Murcia, 1995). As shown above (Section 3.2.2), in Zimbabwe dogs were found 3 km within a wildlife reserve independently of their owners, and 6 km with people, and clearly modified the scavenging guild up to 1 km within the wildlife reserve. This study also demonstrated that dogs were the most common carnivore along the wildlife reserve boundary, occurring with a frequency six-fold greater than the most common wild carnivore, the leopard. Fragmentation and the loss of habitat, along with increasing human populations, have resulted in an ever-increasing proportion of natural areas subject to anthropogenic edge effects. Several recent studies have now proposed that domestic carnivores such as dogs and cats can constitute a large-scale edge effect (Torres and Prado, 2011).

Apart from these directed studies, a number of general faunal surveys across the world have noted that dogs are among the most commonly detected carnivore species (Table 3.2) in several natural areas with close proximity to human habitation. Dogs may be detected as far as 10 km from the nearest human habitation (and much further away in Australia), and thus the likely impacts they can have are not necessarily confined to short distances from human settlements. For most of the surveys and studies listed in Table 3.2, examination of the effects of dogs on other species was not their primary goal. Thus, the information gathered from across these studies is heterogeneous with regard to detail, and is intended to be purely illustrative. What emerges is that dogs are quite commonly detected in several surveys. Given the scope for negative interactions between dogs and native species discussed above and elsewhere in this book, there appears to be a legitimate cause for concern. There is thus not only a need for more directed research on the extent to which dogs can be viewed as an edge effect, but also for general surveys to explicitly report the presence of domestic carnivore species.

3.5.4 Dog management for conservation

The competitive dynamics between dogs and sympatric carnivores described above can clearly be problematic for conservation, particularly when dogs occur in and around protected areas (e.g., Lacerda et al., 2009; Srbek-Araujo and Chiarello, 2008; Torres et al., 1996; Vanak and Gompper, 2010). However, not all dogs are likely to be equal in their effects on carnivores. As postulated by Vanak and Gompper (2009b), the competitive ability of dogs is likely to be influenced by their population size and ranging behavior. The larger the population of dogs, and the wider ranging their behavior, the more likely they are to either directly or indirectly influence other carnivore species. Thus, to negate the possible deleterious effects of dogs on carnivores, it is necessary to not only reduce population size, but also restrain free-ranging behavior. Priority should be given to male dogs, since they have larger home ranges than females (Butler et al., Chapter 5) and travel greater distances when rabid (Knobel et al., Chapter 6), and therefore may be likely to interact more frequently with wild carnivores. Most of the emphasis on free-ranging dog populations worldwide has thus far been on reducing the potential for zoonotic diseases, in particular rabies. Even in cases where dogs have been implicated as the source of pathogens causing widespread mortality in carnivores, the emphasis has been on reducing the transmission potential via vaccination (e.g., Cleaveland et al., 2007), rather than on population control or restricting ranging-behavior.

Mitigating the risk of pathogen transmission does not, however, reduce the possibility of other kinds of dog–wildlife interactions (Vanak and Gompper, 2010). As discussed above, these depend on population size and ranging behavior. Therefore, controlling dog–wildlife interactions, particularly in the vicinity of conservation areas, must involve a multi-pronged approach. Pathogen transmission risk can be mitigated through vaccination, and the biotic potential of the population can be reduced via lethal control and sterilization. In areas of conservation concern, control measures must also include the removal of un-owned dogs, restriction of free-ranging activity (particularly among males), improvement of feeding, and a strong emphasis on responsible dog ownership (Vanak and Gompper, 2010; Silva-Rodríguez and Sieving, 2011). These management approaches must be implemented in a sustained and integrated manner for a long-term solution.

Table 3.2 Summary of a sample of surveys that recorded dogs in natural habitats as an example of a large-scale edge effect. We have excluded studies of dingoes from Australia as they are naturalized predators in the system and may not be considered as an edge effect.

Region	Habitat	Detection method	Frequency of detection relative to other carnivores	Distance from human habitation (km)	Reference
Europe					
Spain	Agricultural matrix	Transects (spotlight)	Red fox > dog > cat > others	Variable	Sobrino et al. 2009
Spain	Woodland	Camera traps	Cat > dog> fox > others	< 0.4	Fandos et al. 2012
Portugal	Mediterranean farmlands	Sign detection in transects	Red fox > dog > Egyptian mongoose > Eurasian badger > cat > Eurasian otter > weasel ~ European polecat ~ stone marten > genet	Variable	Pita et al. 2009
Asia					
Taiwan	Rainforest and secondary forest	Camera traps	Dogs detected, frequency not reported	2–10	Chen et al. 2010
Malaysia	Rainforest	Transects and camera traps	Dogs, jungle cats, and common palm civet frequently detected	<1	Sanei and Zakaria 2011
India	Grassland and plantation	Camera traps and track plots	Indian fox > dog > others	1–3	Vanak and Gompper 2007
North America					
USA	California sage scrub, annual grassland, chaparral, and oak woodland	Camera traps	Coyote > bobcat > dog > striped skunk > raccoon > gray fox > puma > others	Dogs associated with urban percentage but not with distance to edge	Ordeñana et al. 2010
USA	Deciduous/coniferous forests	Camera traps, Scent stations	Albany Pine Bush Preserve (APB): Cat > coyote ~ raccoon > dog > others. Adirondack sites: dogs not detected.	APB: Close to suburban and urban development	Gompper et al. 2006
South America					
Brazil	Atlantic forest	Camera traps	Dog > ocelot > crab-eating raccoon > puma > coati > crab-eating fox > others	0–1.1 from forest edge. A dog detected 0.9 from residence.	Srbek-Araujo and Chiarello 2008
Brazil	Cerrado	Track stations	Maned wolf > dog > crab-eating raccoon > crab-eating fox > coati ~ puma ~ tayra	Associated to edge of National Park (< 1)	Lacerda et al. 2009
Chile	Temperate rainforest	Camera traps and track stations	Dog > guigna > Darwin's fox > Southern river otter > others	Variable	Farías and Jaksic 2011
Chile	Temperate rainforest	Camera traps	Dog > skunk > guigna ~ puma ~ chilla fox	Variable	Silva-Rodriguez et al. 2010b
Africa					
Zimbabwe	Woodland savannah	Transects (tracks on sand, experimental carcass observation, wildlife monitoring transects)	Dog > leopard > side-striped jackal > lion > spotted hyena	3 (independent of humans), 6 (with humans)	Butler and du Toit, 2002; Butler et al. 2004

3.6 Further research

This review highlights the general lack of studies on the competitive dynamics between dogs and sympatric carnivores, both in terms of geographic representativeness as well as the range of species that are affected. Indeed, barely a handful of studies from Asia, Africa, and South America have directly examined these competitive dynamics. Even when direct aggression and intraguild predation can be detected among competing species, the subtler indirect effects of exploitation and apparent competition may sometimes escape notice. The various competitive roles of dogs are still largely unknown. As discussed in this review, dogs may have large-scale effects on native carnivores despite not competing strongly for food. Hence, it is clear that there are major gaps in directed research on the effects of dogs on sympatric carnivores worldwide, but particularly in Asia, Europe, and Africa, which together have the largest populations of dogs as well as a wide array of native carnivore species.

References

Aiyadurai, A. and Jhala, Y.V. (2006). Foraging and habitat use by golden jackals (*Canis aureus*) in the Bhal Region, Gujarat, India. *Journal of the Bombay Natural History Society*, **103**, 5–12.

Allen, L. and Gonzalez, T. (1998). *Baiting reduces dingo numbers, changes age structures yet often increase calf losses*. Proceedings of the Australian Vertebrate Pest Control Conference, Bunbury, WA.

Allen, B.L. and Leung, L.K.P. (2012). Assessing predation risk to threatened fauna from their prevalence in predator scats: dingoes and rodents in arid Australia. *PloS ONE*, **7**, e36426.

Ashraf, N.V.K., Kumar, A., and Johnsingh, A.J.T. (1993). Two endemic viverrids of the Western Ghats, India. *Oryx*, **27**, 109–14.

Atickem, A., Bekele, A., and Williams, S.D. (2010). Competition between domestic dogs and Ethiopian wolf (*Canis simensis*) in the Bale Mountains National Park, Ethiopia. *African Journal of Ecology*, **48**, 401–7.

Attard, M., Chamoli, U., Ferrara, T., Rogers, T., and Wroe, S. (2011). Skull mechanics and implications for feeding behaviour in a large marsupial carnivore guild: The thylacine, Tasmanian devil and spotted-tailed quoll. *Journal of Zoology*, **285**, 292–300.

Atwood, T.C. and Gese, E.M. (2008). Coyotes and recolonizing wolves: Social rank mediates risk-conditional behaviour at ungulate carcasses. *Animal Behaviour*, **75**, 753–62.

Barashkova, A. and Smelansky, I. (2011). Pallas's cat in the Altai Republic, Russia. *Cat News*, **54**, 4–7.

Barcala, O. (2009). Invasive stray and feral dogs limit fossa (*Cryptoprocta ferox*) populations in Ankarafantsika National Park, Madagascar. MS thesis, Duke University.

Berger, K.M. and Gese, E.M. (2007). Does interference competition with wolves limit the distribution and abundance of coyotes? *Journal of Animal Ecology*, **76**, 1075–85.

Bingham, J., Foggin, C.M., Wandeler, A.I. and Hill, F.W.G. (1999). The epidiemiology of rabies in Zimbabwe. 2. Rabies in jackals (*Canis adustus* and *Canis mesomelas*). Onderstepoort *Journal of Veterinary Research*, **66**, 11–23.

Boitani, L., Francisci, F., Ciucci, P. and Andreoli, G. (1995). Population biology and ecology of feral dogs in central Italy. In J. Serpell. Ed., *The Domestic Dog. Its Evolution, Behaviour, and Interactions with People*. Cambridge University Press, Cambridge, pp. 217–44.

Bonaccorso, F.J., Winkelmann, J.R., Shin, D., Agrawal, C.I., Aslami, N., Bonney, C., Hsu, A., Jekielek, P.E., Knox, A.K., Kopach, S.J., Jennings, T.D., Lasky, J.R., Menesale, S.A., Richards, J.H., Rutland, J.A., Sessa, A.K., Zhaurova, L., and Kunz, T.H. (2007). Evidence for exploitative competition: comparative foraging behavior and roosting ecology of short-tailed fruit bats (Phyllostomidae). *Biotropica*, **39**, 249–56.

Bothma, J.P., Nel, J., and Macdonald, A. (1984). Food niche separation between four sympatric Namib Desert carnivores. *Journal of Zoology*, **202**, 327–40.

Brawata, R.L. and Neeman, T. (2011). Is water the key? Dingo management, intraguild interactions and predator distribution around water points in arid Australia. *Wildlife Research*, **38**, 426–36.

Brook, L.A., Johnson, C.N., and Ritchie, E.G. (2012). Effects of predator control on behaviour of an apex predator and indirect consequences for mesopredator suppression. *Journal of Applied Ecology*, **49**, 1278–86.

Brown, O.J.F. (2006). Tasmanian devil (*Sarcophilus harrisii*) extinction on the Australian mainland in the mid-Holocene: Multicausality and ENSO intensification. *Alcheringa*, **30**, 49–57.

Butler, J.R.A. (1998). The ecology of domestic dogs *Canis familiaris* in the communal lands of Zimbabwe. PhD dissertation, University of Zimbabwe.

Butler, J. (2000). The economic costs of wildlife predation on livestock in Gokwe communal land, Zimbabwe. *African Journal of Ecology*, **38**, 23–30.

Butler, J.R.A., and Bingham, J. (2000). Demography and dog-human relationships of the dog population in Zimbabwean communal lands. *Veterinary Record*, **147**, 442–6.

Butler, J.R.A. and du Toit, J.T. (2002). Diet of free-ranging domestic dogs (*Canis familiaris*) in rural Zimbabwe:

Implications for wild scavengers on the periphery of wildlife reserves. *Animal Conservation*, **5**, 29–37.

Butler, J.R.A., du Toit, J.T., and Bingham, J. (2004). Free-ranging domestic dogs (*Canis familiaris*) as predators and prey in rural Zimbabwe: Threats of competition and disease to large wild carnivores. *Biological Conservation*, **115**, 369–78.

Campos, C.B., Esteves, C.F., Ferraz, K., Crawshaw, P.G., and Verdade, L.M. (2007). Diet of free-ranging cats and dogs in a suburban and rural environment, south-eastern Brazil. *Journal of Zoology*, **273**, 14–20.

Carthey, A.J.R. and Banks, P.B. (2012). When does an alien become a native species? A vulnerable native mammal recognizes and responds to its long-term alien predator. *PLoS ONE*, **7**, e31804.

Chen, M.T., Tewes, M.E., Pei, K.J., and Grassman, L.I. (2010). Activity patterns and habitat use of sympatric small carnivores in southern Taiwan. *Mammalia*, **73**, 20.

Clark, T.W. (1987). Black-footed ferret recovery: a progress report. *Conservation Biology*, **1**, 8–13.

Cleaveland, S., Mlengeya, T., Kaare, M., Haydon, D.A.N., Lembo, T., Laurenson, M.K., and Packer, C. (2007). The conservation relevance of epidemiological research into carnivore viral diseases in the Serengeti. *Conservation Biology*, **21**, 612–22.

Clutton-Brock, J. (1995). Origins of the dog: Domestication and early history. In J. Serpell, Ed., *The Domestic Dog. Its Evolution, Behaviour, and Interactions with People*. Cambridge University Press, Cambridge, pp. 7–20.

Corbett, L.K. (1995). *The Dingo in Australia and Asia*. University of New South Wales Press, Sydney.

Cozza, K., Fico, R., Battistini, M.L., and Rogers, E. (1996). The damage-conservation interface illustrated by predation on domestic livestock in central Italy. *Biological Conservation*, **78**, 329–36.

Crabb, W.D. (1948). The ecology and management of the prairie spotted skunk in Iowa. *Ecological Monographs*, **18**, 201–32.

Crabtree, R.L. and Sheldon, J.W. (1999). Coyotes and canid coexistence in Yellowstone. In T. Clark, P. Curlee, P. Kareiva and S. Minta, Eds., *Carnivores in Ecosystems: The Yellowstone Experience*. Yale University Press, New Haven, pp. 127–63.

Creel, S. and Christianson, D. (2008). Relationships between direct predation and risk effects. *Trends in Ecology and Evolution*, **23**, 194–201.

Creel, S., Spong, G., and Creel, N. (2001). Interspecific competition and the population biology of extinction prone carnivores. In J.L. Gittleman, S.M. Funk, D.W. Macdonald, and R.K. Wayne, Eds., *Carnivore Conservation*. Cambridge University Press, Cambridge, pp. 35–60.

Crooks, K.R. and Soulé, M.E. (1999). Mesopredator release and avifaunal extinctions in a fragmented system. *Nature*, **400**, 563–66.

Cumming, D.H.M. (1982). A case history of the spread of rabies in an African country. *South African Journal of Science*, **78**, 443–7.

Cupples, J.B., Crowther, M.S., Story, G., and Letnic, M. (2011). Dietary overlap and prey selectivity among sympatric carnivores: could dingoes suppress foxes through competition for prey? *Journal of Mammalogy*, **92**, 590–600.

Dahmer, T.D. (2002). Feral/stray dogs and civet mortality on Kau Sai Chau, 2001–2002. *Porcupine!*, **27**, 7–9.

Deem, S.L., Spelman, L.H., Yates, R.A., and Montali, R.J. (2000). Canine distemper in terrestrial carnivores: a review. *Journal of Zoo and Wildlife Medicine*, **31**, 441–51.

Dickman, C.R. (1992). Commensal and mutualistic interactions among terrestrial vertebrates. *Trends in Ecology and Evolution*, **7**, 194–7.

Donadio, E. and Buskirk, S.W. (2006). Diet, morphology, and interspecific killing in Carnivora. *American Naturalist*, **167**, 524–36.

Dorchin, N. (2006). Resource partitioning and exploitative competition in stem-galling midges mediated by the host plant. *Entomologia Experimentalis et Applicata*, **118**, 111–9.

Dulamtseren, S., Shar, S., Murdoch, J.D., Reading, R.P., Gantulga, J., Usukhjargal, D., and Buyandelger, S. (2009). Notes on the distribution of marbled polecat *Vormela peregusna* in Mongolia. *Small Carnivore Conservation*, **40**, 29–32.

Echegaray, J. and Vilà, C. (2010). Noninvasive monitoring of wolves at the edge of their distribution and the cost of their conservation. *Animal Conservation*, **13**, 157–61.

Elledge, A.E., Allen, L.R., Carlsson, B.L., Wilton, A.N., and Leung, L.K.P. (2009). An evaluation of genetic analyses, skull morphology and visual appearance for assessing dingo purity: implications for dingo conservation. *Wildlife Research*, **35**, 812–20.

Elmhagen, B., Tannerfeldt, M., and Angerbjorn, A. (2002). Food-niche overlap between arctic and red foxes. *Canadian Journal of Zoology*, **80**, 1274–85.

Espartosa, K.D. (2009). Mamíferos terrestres de maior porte e a invasão de cães domésticos em remanescentes de uma paisagem fragmentada de Mata Atlântica: avaliação da eficiência de métodos de amostragem e da importância de múltiplos fatores sobre a distribuição das espécies. MS thesis, Universidad de Sao Paulo.

Espinosa-Molina, M.I. (2011). Dieta y uso de hábitat del huillín (*Lontra provocax*) en ambientes de agua dulce y su relación con comunidades locales en el bosque templado lluvioso, Isla Grande de Chiloé, Chile. DVM thesis, Universidad Mayor, Santiago.

Fandos, G., Fernández-López, J., and Tellería, J.L. (2012). Incursion of domestic carnivores around urban areas: A test in central Spain. *Mammalia*, **76**, 223–5.

Farias, A.A. and Jaksic, F.M. (2011). Low functional richness and redundancy of a predator assemblage in native forest fragments of Chiloe Island, Chile. *Journal of Animal Ecology*, **80**, 809–17.

Fedriani, J.M., Fuller, T.K., Sauvajot, R.M., and York, E.C. (2000). Competition and intraguild predation among three sympatric carnivores. *Oecologia*, **125**, 258–70.

Fillios, M., Crowther, M.S., and Letnic, M. (2012). The impact of the dingo on the thylacine in Holocene Australia. *World Archaeology*, **44**, 118–34.

Fisher, J.T., Anholt, B., Bradbury, S., Wheatley, M., and Volpe, J.P. (2013). Spatial segregation of sympatric marten and fishers: the influence of landscapes and species-scapes. *Ecography*, **36**, 240–248.

Fleming, P., Corbett, L., Harden, R., and Thomson, P. (2001). *Managing the Impacts of Dingoes and Other Wild Dogs*. Bureau of Rural Sciences, Canberra.

Funk, S.M., Fiorello, C.V., Cleaveland, S., and Gompper, M.E. (2001). The role of disease in carnivore ecology and conservation. In J.L. Gittleman, S.M. Funk, D.W. MacDonald, and R.K. Wayne, Eds., *Carnivore Conservation*. Cambridge University Press, Cambridge, pp. 443–66

Gerber, B.D., Karpanty S.M., and Randrianantenaina, J. (2012). Activity patterns of carnivores in the rain forests of Madagascar: Implications for species coexistence. *Journal of Mammalogy*, **93**, 667–76.

Glen, A.S., de Tores, P.J., Sutherland, D.R., and Morris, K.D. (2010). Interactions between chuditch (*Dasyurus geoffroii*) and introduced predators: A review. *Australian Journal of Zoology*, **57**, 347–56.

Glen, A.S. and Dickman, C.R. (2005). Complex interactions among mammalian carnivores in Australia, and their implications for wildlife management. *Biological Reviews*, **80**, 387–401.

Glen, A.S. and Dickman, C.R. (2008). Niche overlap between marsupial and eutherian carnivores: Does competition threaten the endangered spotted-tailed quoll? *Journal of Applied Ecology*, **45**, 700–7.

Glen, A.S., Pennay, M., Dickman, C.R., Wintle, B.A., and Firestone, K.B. (2011). Diets of sympatric native and introduced carnivores in the Barrington Tops, eastern Australia. *Austral Ecology*, **36**, 290–6.

Gompper, M.E., Kays, R.W., Ray, J.C., Lapoint, S.D., Bogan, D.A., and Cryan, J.R. (2006). A Comparison of noninvasive techniques to survey carnivore communities in northeastern North America. *Wildlife Society Bulletin*, **34**, 1142–51.

Gompper, M.E. and A.T. Vanak. (2008). Subsidized predators, landscapes of fear and disarticulated carnivore communities. *Animal Conservation*, **11**, 13–14.

González, A., Novaro, A., Funes, M., Pailacura, O., Bolgeri, M.J., and Walker, S. (2012). Mixed-breed guarding dogs reduce conflict between goat herders and native carnivores in Patagonia. *Human-Wildlife Interactions*, **6**, 327–34.

Gosselink, T.E., Van Deelen, T.R., Joselyn, M.G., and Warner, R.E. (2003). Temporal habitat partitioning and spatial use of coyotes and red foxes in east-central Illinois. *Journal of Wildlife Management*, **67**, 90–103.

Habib, B. (2007). Ecology of Indian wolf (*Canis lupus pallipes* Sykes, 1831), and modeling its potential habitat in the Great Indian Bustard Sanctuary, Maharashtra, India. PhD dissertation, Aligarh Muslim University.

Harris, S. (1981). An estimation of the number of foxes (*Vulpes vulpes*) in the city of Bristol, and some possible factors affecting their distribution. *Journal of Applied Ecology*, **18**, 455–65.

Hatcher, M.J., Dick, J.T.A., and Dunn, A.M. (2006). How parasites affect interactions between competitors and predators. *Ecology Letters*, **9**, 1253–71.

Hawkins, A.F.A. (2008). *Mungotictis decemlineata. IUCN Red List of Threatened Species. Version 2012.1.* <http://www.iucnredlist.org>. Retrieved July 04, 2012.

Holt, R.D. (1977). Predation, apparent competition, and the structure of prey communities. *Theoretical Population Biology*, **70**, 197–229.

Holt, R.D. and Dobson, A.P. (2006). Extending the principles of community ecology to address the epidemiology of host-pathogen systems. In S.K. Collinge and C. Ray, Eds., *Disease Ecology—Community Structure and Pathogen Dynamics*. Oxford University Press, Oxford, pp. 6–27.

Holt, R.D. and Lawton, J.H. (1994). The ecological consequences of shared natural enemies. *Annual Review of Ecology and Systematics*, **70**, 495–520.

Holt, R.D. and Pickering, J. (1985). Infectious disease and species coexistence: a model of Lotka-Volterra form. *American Naturalist*, **126**, 196–211.

Hon, N., Neak, P., Khov, V., and Cheat, V. (2010). Food and habitat of Asian small-clawed otter in northeastern Cambodia. *IUCN Otter Specialist Group Bulletin*, **21**, 12–23.

Houston, D.C. (1979). The adaptations of scavengers. In A.R.E. Sinclair and M. Norton-Griffiths, Eds., *Serengeti: Dynamics of an Ecosystem*. Chicago University Press, Chicago, pp. 281–308.

Inskip, C. and Zimmermann, A. (2009). Human-felid conflict: a review of patterns and priorities worldwide. *Oryx*, **43**, 18–34.

Jhala, Y.V. and Giles, R.H.J. (1991). The status and conservation of the wolf in Gujarat and Rajasthan, India. *Conservation Biology*, **5**, 476–83.

Johnson, W.E., Fuller, T.K., and Franklin, W.L. (1996). Sympatry in canids: a review and assessment. In J.L.

Gittleman, Ed., *Carnivore Behavior, Ecology, and Evolution. Volume 2*. Cornell University, Ithaca, pp. 189–218.

Johnson, H.E., Hebblewhite, M., Stephenson, T.R., German, D.W., Pierce, B.M. and Bleich, V.C. (2013). Evaluating apparent competition in limiting the recovery of an endangered ungulate. *Oecologia*, **171**, 295–307.

Johnson, C.N. and VanDerWal, J. (2009). Evidence that dingoes limit abundance of a mesopredator in eastern Australian forests. *Journal of Applied Ecology*, **46**, 641–6.

Johnson, C.N. and Wroe, S. (2003). Causes of extinction of vertebrates during the Holocene of mainland Australia: Arrival of the dingo, or human impact? *The Holocene*, **13**, 941–8.

Jones, M.E. and Barmuta, L.A. (1998). Diet overlap and relative abundance of sympatric dasyurid carnivores: A hypothesis of competition. *Journal of Animal Ecology*, **67**, 410–21.

Kennedy, M., Phillips, B.L., Legge, S., Murphy, S.A., and Faulkner, R.A. (2012). Do dingoes suppress the activity of feral cats in northern Australia? *Austral Ecology*, **37**, 134–9.

Kissui, B.M. (2008). Livestock predation by lions, leopards, spotted hyenas, and their vulnerability to retaliatory killing in the Maasai steppe, Tanzania. *Animal Conservation*, **11**, 422–32.

Kolowski, J.M. and Holekamp, K.E. (2006). Spatial, temporal, and physical characteristics of livestock depredations by large carnivores along a Kenyan reserve border. *Biological Conservation*, **128**, 529–41.

Körtner, G. (2007). 1080 aerial baiting for the control of wild dogs and its impact on spotted-tailed quoll (*Dasyurus maculatus*) populations in eastern Australia. *Wildlife Research*, **34**, 48–53.

Körtner, G. and Watson, P. (2006). The immediate impact of 1080 aerial baiting to control wild dogs on a spotted-tailed quoll population. *Wildlife Research*, **32**, 673–80.

Krauze-Gryz, D. and Gryz, J. (2009). *Biocenotical role of dogs and cats in rural areas of Central Poland*. Paper presented at the 2nd European Congress of Conservation Biology.

Krauze-Gryz, D., Gryz, J.B., Goszczyńskia, J., Chylareckic, P., and Śmihorski, M. (2012). The good, the bad and the ugly: space use and intraguild interactions among three opportunistic predators—cat *Felis catus*, dog *Canis familiaris* and fox *Vulpes vulpes*—under human pressure. *Canadian Journal of Zoology*, **90**, 1402–13.

Kruuk, H. (1968). Hyaenas: the hunters nobody knows. *National Geographic*, **134**, 44–57.

Kruuk, H. (1972). *The Spotted Hyena: A Study of Predation and Social Behavior*. University of Chicago Press, Chicago.

Kruuk, H. and Snell, H. (1981). Prey selection by feral dogs from a population of marine iguanas (*Amblyrhynchus cristatus*). *Journal of Applied Ecology*, **18**, 197–204.

Lacerda, A., Tomas, W., and Marinho-Filho, J. (2009). Domestic dogs as an edge effect in the Brasília National Park, Brazil: Interactions with native mammals. *Animal Conservation*, **12**, 477–87.

Laurance, W. (2000). Do edge effects occur over large spatial scales? *Trends in Ecology and Evolution*, **15**, 134–5.

Laurenson, K., Sillero-Zubiri, C., Thompson, H., Shiferaw, F., Thirgood, S., and Malcolm, J. (1998). Disease as a threat to endangered species: Ethiopian wolves, domestic dogs and canine pathogens. *Animal Conservation*, **1**, 273–80.

Letnic, M., Greenville, A., Denny, E., Dickman, C.R., Tischler, M., Gordon, C., and Koch, F. (2011). Does a top predator suppress the abundance of an invasive mesopredator at a continental scale? *Global Ecology and Biogeography*, **20**, 343–53.

Letnic, M., Koch, F., Gordon, C., Crowther, M.S., and Dickman, C.R. (2009). Keystone effects of an alien top-predator stem extinctions of native mammals. *Proceedings of the Royal Society of London Series B: Biological Sciences*, **276**, 3249–56.

Letnic, M., Ritchie, E.G., and Dickman, C.R. (2012). Top predators as biodiversity regulators: the dingo *Canis lupus dingo* as a case study. *Biological Reviews*, **87**, 390–413.

Lucherini, M. and Merino, M.J. (2008). Perceptions of human-carnivore conflicts in the High Andes of Argentina. *Mountain Research and Development*, **28**, 81–5.

Maran, T., Põpra, M., Põlma, M., and Macdonald, D.W. (2009). The survival of captive-born animals in restoration programmes: case study of the endangered European mink *Mustela lutreola*. *Biological Conservation*, **142**, 1685–92.

McFadden, K.W., García-Vasco, D., Cuarón, A.D., Valenzuela-Galván, D., Medellín, R.A., and Gompper, M.E. (2010). Vulnerable island carnivores: the endangered endemic dwarf procyonids from Cozumel Island. *Biodiversity and Conservation*, **19**, 491–502.

Mills, M. (1990). *Kalahari Hyaenas: Comparative Behavioural Ecology of Two Species*. Unwin Hyman, London.

Mitchell, B.D. and Banks, P.B. (2005). Do wild dogs exclude foxes? Evidence for competition from dietary and spatial overlaps. *Austral Ecology*, **30**, 581–91.

Morris, R.J., Lewis, O.T., and Godfray, H.C.J. (2004). Experimental evidence for apparent competition in a tropical forest food web. *Nature*, **428**, 310–13.

Moseby, K.E., Neilly, H., Read, J.L., and Crisp, H.A. (2012). Interactions between a top order predator and exotic mesopredators in the Australian rangelands. *International Journal of Ecology*, **2012**, 250–352.

Mundy, P.J. (1982). *The Comparative Biology of Southern African Vultures*. Vulture Study Group, Johannesburg.

Murcia, C. (1995). Edge effects in fragmented forests: implications for conservation. *Trends in Ecology and Evolution*, **10**, 58–62.

Nelson, J.L., Cypher, B.L., Bjurlin, C.D., and Creel, S. (2007). Effects of habitat on competition between kit foxes and coyotes. *Journal of Wildlife Management*, **71**, 1467–75.

Newsome, A.E. (2001). The biology and ecology of the dingo. In D. Lunney and C.R. Dickman, Eds., *A Symposium on the Dingo*. Royal Zoological Society of New South Wales, Mosman, pp. 20–33.

Newsome, T.M., Stephens, D., Ballard, G.-A., Dickman, C.R., and Fleming P.J.S. (2013). Genetic profile of dingoes (*Canis lupus dingo*) and free-roaming domestic dogs (*C. l. familiaris*) in the Tanami Desert, Australia. *Wildlife Research*, **40**, 196–206.

Novaro, A.J., Funes, M.C., and Walker R.S. (2005). An empirical test of source-sink dynamics induced by hunting. *Journal of Applied Ecology*, **42**, 910–20.

Ordeñana, M.A., Crooks, K.R., Boydston, E.E., Fisher, R.N., Lyren, L.M., Siudyla, S., Haas, C.D., Harris, S., Hathaway, S.A., Turschak, G.M., Miles, A.K., and Van Vuren, D.H. (2010). Effects of urbanization on carnivore species distribution and richness. *Journal of Mammalogy*, **91**, 1322–31.

Paddle, R. (2002). *The Last Tasmanian Tiger: The History and Extinction of the Thylacine*. Cambridge University Press, Cambridge.

Palomares, F. and Caro, T.M. (1999). Interspecific killing among mammalian carnivores. *American Naturalist*, **153**, 492–508.

Palomares, F., Ferreras, P., Fedriani, J.M., and Delibes, M. (1996). Spatial relationships between Iberian lynx and other carnivores in an area of south-western Spain. *Journal of Applied Ecology*, **33**, 5–13.

Parker, B.D., Hume, I.D., and Boles W.E. (2007). Diet of breeding Wedge-tailed Eagles *Aquila audax* in south-central Queensland. *Corella*, **31**, 50–62.

Parsons, M.H. and Blumstein, D.T. (2010). Familiarity breeds contempt: kangaroos persistently avoid areas with experimentally deployed dingo scents. *PLoS ONE*, *5*, e10403.

Pascoe, J.H., Mulley, R.C., Spencer, R., and Chapple, R. (2012). Diet analysis of mammals, raptors and reptiles in a complex predator assemblage in the Blue Mountains, eastern Australia. *Australian Journal of Zoology*, **595**, 295–301.

Pavés, H.J., Schlatter, R.P., and Espinoza, C.I. (2008). Scavenging and predation by black vultures *Coragyps atratus* at a South American sea lion breeding colony. *Vulture News*, **58**, 4–15.

Pavey, C.R., Eldridge, S.R., and Heywood, M. (2008). Population dynamics and prey selection of native and introduced predators during a rodent outbreak in arid Australia. *Journal of Mammalogy*, **89**, 674–83.

Pereira, J.A., Fracassi, N.G., Rago, V., Ferreyra, H., Marull, C.A., McAloose, D., and Uhart, M.M. (2010). Causes of mortality in a Geoffroy's cat population—a long-term survey using diverse recording methods. *European Journal of Wildlife Research*, **56**, 939–42.

Petren, K. and Case, T.J. (1996). An experimental demonstration of exploitation competition in an ongoing invasion. *Ecology*, **77**, 118–32.

Pettigrew, J.D. (1993). *A burst of feral cats in the Diamantina–a lesson for the management of pest species*. Paper presented at the Cat Management Workshop Proceedings.

Pienaar, U.V. (1969). Predator-prey relationships amongst the larger mammals of the Kruger National Park. *Koedoe*, **12**, 108–76.

Pita, R., Mira, A., Moreira, F., Morgado, R., and Beja, P. (2009). Influence of landscape characteristics on carnivore diversity and abundance in Mediterranean farmland. *Agriculture, Ecosystems and Environment*, **132**, 57–65.

Pizarro-Neyra, J. (2008). Mortality of the marine otter (*Lontra felina*) in Southern Peru. *IUCN Otter Specialist Group Bulletin*, **25**, 94–9.

Polis, G.A., Myers, C.A., and Holt, R.D. (1989). The ecology and evolution of intraguild predation: potential competitors that eat each other. *Annual Review of Ecology and Systematics*, **20**, 297–330.

Power, A.G., and Mitchell, C.E. (2004). Pathogen spillover in disease epidemics. *American Naturalist*, **164**, 79–89.

Prakash, V., Pain, D.J., Cunningham, A.A., Donald, P.F., Prakash, N., Verma, A., Gargi, R., Sivakumar, S., and Rahmani, A.R. (2003). Catastrophic collapse of Indian white-backed *Gyps bengalensis* and long-billed *Gyps indicus* vulture populations. *Biological Conservation*, **109**, 381–390.

Price, P.W., Westoby, M., and Rice, B. (1998). Parasite mediated competition: some predictions and tests. *American Naturalist*, **131**, 544–55.

Purcell, B. (2010). *Dingo*. Csiro Publishing, Collingwood, Australia.

Revilla, E., Palomares, F., and Delibes, M. (2001). Edge-core effects and the effectiveness of traditional reserves in conservation: Eurasian badgers in Doñana National Park. *Conservation Biology*, **15**, 148–58.

Richardson, P.R.K. (1980). The natural removal of ungulate carcasses, and the adaptive features of the scavengers involved. MSc thesis, University of Pretoria.

Ritchie, E.G., Elmhagen, B., Glen, A.S., Letnic, M., Ludwig, G., and McDonald, R.A. (2012). Ecosystem restoration with teeth: what role for predators? *Trends in Ecology and Evolution*, **27**, 265–71.

Ritchie, E.G. and Johnson, C.N. (2009). Predator interactions, mesopredator release and biodiversity conservation. *Ecology Letters*, **12**, 982–98.

Roemer, G.W., Coonan, T.J., Garcelon, D.K., Bascompte, J., and Laughrin L. (2001). Feral pigs facilitate hyperpredation by golden eagles and indirectly cause the decline of the island fox. *Animal Conservation*, **4**, 307–18.

Roemer, G.W., Donlan, C.J., and Courchamp, F. (2002). Golden eagles, feral pigs, and insular carnivores: how exotic species turn native predators into prey. *Proceedings of the National Academy of Sciences*, **99**, 791–96.

Roemer, G.W., Gompper, M.E., and VanValkenburgh, B. (2009). The ecological role of the mammalian mesocarnivore. *BioScience*, **52**, 165–73.

Ross, S., Murdoch, J., Mallon, D., Sanderson, J., and Barashkova, A. (2008). *Otocolobus manul. IUCN Red List of Threatened Species. Version 2012.2.* <http://www.iucnredlist.org>. Retrieved November 3, 2012.

Rushton, S.P., Lurz, P.W.W., Gurnell, J., and Fuller, R. (2000). Modelling the spatial dynamics of parapoxvirus disease in red and grey squirrels: a possible cause of the decline in the red squirrel in the UK? *Journal of Applied Ecology*, **37**, 997–1012.

Sanei, A. and Zakaria, M. (2011). Occupancy status of Malayan leopard prey species in a fragmented forest in Selangor, Malaysia. *Asia Life Sciences, Supplement* **7**, 41–55.

Schaller, G.B. (1976). *The Serengeti Lion: A Study of Predator-Prey Relations*. University of Chicago Press, Chicago.

Sillero-Zubiri, C. and Gottelli, D. (1995). Diet and feeding behavior of Ethiopian wolves (*Canis simensis*). *Journal of Mammalogy*, **76**, 531–41.

Sillero-Zubiri, C. and Macdonald, D.W. (1997). *The Ethiopian Wolf: Status Survey and Conservation Action Plan*. IUCN Canid Specialist Group, Gland, Switzerland and Cambridge, UK.

Silva, E.A. (2012). Domestic dogs as invasive species: from local to global impacts. PhD dissertation, University of Florida, Gainesville.

Silva-Rodríguez, E.A., Ortega-Solís, G.R., and Jimenez, J.E. (2010a). Conservation and ecological implications of the use of space by chilla foxes and free-ranging dogs in a human-dominated landscape in southern Chile. *Austral Ecology*, **35**, 765–77.

Silva-Rodríguez, E.A., Verdugo, C., Aleuy, O.A., Sanderson, J.G., Ortega-Solís, G.R., Osorio-Zúñiga, F., and González-Acuña, D. (2010b). Evaluating mortality sources for the Vulnerable pudu *Pudu puda* in Chile: implications for the conservation of a threatened deer. *Oryx*, **44**, 97–103.

Silva-Rodríguez, E.A. and Sieving, K.E. (2011). Influence of care of domestic carnivores on their predation on vertebrates. *Conservation Biology*, **25**, 808–15.

Skinner, J.D. and Chimimba, C.T. (2005). *The Mammals of the Southern African Subregion*. Cambridge University Press, Cambridge.

Slater, M.R., Di Nardo, A., Pediconi, O., Villa, P.D., Candeloro, L., Alessandrini, B., and Del Papa, S. (2008). Cat and dog ownership and management patterns in central Italy. *Preventive Veterinary Medicine*, **85**, 267–94.

Smallegange, I.M., van der Meer, J., and Kurvers, R. (2006). Disentangling interference competition from exploitative competition in a crab-bivalve system using a novel experimental approach. *Oikos*, **113**, 157–67.

Smith, R.M. (1977). Movement Patterns and Feeding Behaviour of Leopard in the Rhodes Matopos National Park, Rhodesia National Museums and Monuments of Rhodesia, Salisbury, *Arnoldia*, **8**, 13.

Sobrino, R., Acevedo, P., Escudero, M.A., Marco, J., and Gortázar, C. (2009). Carnivore population trends in Spanish agrosystems after the reduction in food availability due to rabbit decline by rabbit haemorrhagic disease and improved waste management. *European Journal of Wildlife Research*, **55**, 161–5.

Soler, L., Carenton, J.M., Cáceres, C., Pérez, P., Fleita, A., Taboada, A., Miller, G., and Gonzalez, R. (2004). La conservación de los carnívoros silvestres del Nordeste Argentino a través de la participación de los pobladores locales. Imagen y respuestas de los lugareños rurales hacia los predadores. In R. Bodmer, Ed., *Memoria VI Congreso sobre Manejo de Fauna Silvestre en la Amazonía y Latinoamérica*. Universidad Nacional de la Amazonía Peruana, Iquitos, pp. 634–43.

Somaweera, R., Webb, J.K., and Shine, R. (2011). It's a dog-eat-croc world: Dingo predation on the nests of freshwater crocodiles in tropical Australia. *Ecological Research*, **26**, 957–67.

Srbek-Araujo, A. and Chiarello, A. (2008). Domestic dogs in Atlantic forest preserves of south-eastern Brazil: A camera-trapping study on patterns of entrance and site occupancy rates. *Brazilian Journal of Biology*, **68**, 771–9.

Sundqvist, A.K., EllegrenH., and Vilà, C. (2008). Wolf or dog? Genetic identification of predators from saliva collected around bite wounds on prey. *Conservation Genetics*, **9**, 1275–9.

Switalski, T.A. (2003). Coyote foraging ecology and vigilance in response to gray wolf reintroduction in Yellowstone National Park. *Canadian Journal of Zoology*, **81**, 985–93.

Tannerfeldt, M., Elmhagen, B., and Angerbjorn, A. (2002). Exclusion by interference competition? The relationship between red and arctic foxes. *Oecologia*, **132**, 213–20.

Taylor, J.A. (1979). Foods and feeding habits of sub-adult *Crocodylus porosus* Schneider in northern Australia. *Australian Wildlife Research*, **6**, 347–59.

Tompkins, D.M., Draycott, R.A.H., and Hudson, P.J. (2000). Field evidence for apparent competition mediated via the shared parasites of two gamebird species. *Ecology Letters*, **3**, 10–14.

Tompkins, D.M., White, A.R., and Boots, M. (2003). Ecological replacement of native red squirrels by invasive greys driven by disease. *Ecology Letters*, **6**, 189–96.

Torres, P. and Prado, P. (2010). Domestic dogs in a fragmented landscape in the Brazilian Atlantic Forest: abundance, habitat use and caring by owners. *Brazilian Journal of Biology*, **70**, 987–994.

Treves, A. and Karanth, K.U. (2003). Human-carnivore conflict and perspectives on carnivore management worldwide. *Conservation Biology*, **17**, 1491–9.

van Bommel, L. and Johnson, C.N. (2012). Good dog! Using livestock guardian dogs to protect livestock from predators in Australia. *Wildlife Research*, **39**, 220–9.

Van Sittert, L. (1998). "Keeping the enemy at bay": the extermination of wild Carnivora in the Cape colony, 1889–1910. *Environmental History*, **3**, 333–56.

Vanak, A.T. (2008). Intraguild interactions between native and domestic carnivores in central India. PhD dissertation, University of Missouri.

Vanak, A.T. and Gompper, M.E. (2007). Effectiveness of non-invasive techniques for surveying activity and habitat use of the Indian fox *Vulpes bengalensis* in southern India. *Wildlife Biology*, **13**, 219–24.

Vanak, A.T. and Gompper, M.E. (2009a). Dietary niche separation between sympatric free-ranging domestic dogs and Indian foxes in central India. *Journal of Mammalogy*, **90**, 1058–65.

Vanak, A.T. and Gompper, M.E. (2009b). Dogs *Canis familiaris* as carnivores: Their role and function in intraguild competition. *Mammal Review*, **39**, 265–83.

Vanak, A.T. and Gompper, M.E. (2010). Interference competition at the landscape level: the effect of free-ranging dogs on a native mesocarnivore. *Journal of Applied Ecology*, **47**, 1225–32.

Vanak, A.T., Thaker, M., and Gompper, M.E. (2009). Experimental examination of behavioural interactions between free-ranging wild and domestic canids. *Behavioral Ecology and Sociobiology*, **64**, 279–87.

Vázquez, D.P., Melián, C.J., Williams, N.M., Blüthgen, N., Krasnov, B.R., and Poulin, R. (2007). Species abundance and asymmetric interaction strength in ecological networks. *Oikos*, **116**, 1120–1127.

Wallach, A.D., Johnson, C.N., Ritchie, E.G., and O'Neill, A.J. (2010). Predator control promotes invasive dominated ecological states. *Ecology Letters*, **13**, 1008–18.

Webb, J.K. (1996). Observation of three dingoes killing a large lace monitor (*Varanus varius*). *Australian Mammalogy*, **19**, 55–6.

Williams, J. (1935). Wild dog killed by domestic dogs. *Journal of the Bombay Natural History Society*, **37**, 949–50.

Williams, B.H. (2004). The status of the red panda in Jamuna and Mabu villages of eastern Nepal. MS thesis, San José State University, California.

Woolaver, L., Nichols, R., Rakotombololona, W.F., Volahy, A.T., and Durbin, J. (2006). Population status, distribution and conservation needs of the narrow-striped mongoose *Mungotictis decemlineata* of Madagascar. *Oryx*, **40**, 67–75.

Wroe, S., Clausen, P., McHenry, C., Moreno, K., and Cunningham, E. (2007). Computer simulation of feeding behaviour in the thylacine and dingo as a novel test for convergence and niche overlap. *Proceedings of the Royal Society B: Biological Sciences*, **274**, 2819–28.

Yonzon, P.B. and Hunter Jr M.L. (1991). Conservation of the red panda *Ailurus fulgens*. *Biological Conservation*, **57**, 1–11.

CHAPTER 4

Dogs as agents of disturbance

Michael A. Weston and Theodore Stankowich

4.1 Introduction

The mere presence of a predator in an environment can affect prey in subtle, sublethal, indirect, yet apparently deleterious ways (Preisser et al., 2005; Zanette et al., 2011). The response of wildlife to the presence of a threatening stimulus, such as a dog, is referred to as 'disturbance,' and these responses involve the disruption of normal activities or states, and often evoke antipredator behaviors, commonly vigilance, flight, retreat to refuge, freezing behavior, or hiding (Hockin et al., 1992). Behavioral changes in the presence of a threatening stimulus have been widely documented and often involve ceasing normal activities (e.g., foraging, parental care, resting, display). A growing body of literature also points to physiological changes, such as hormone release or altered heart rates (e.g., MacArthur et al., 1982). Studies (e.g., Gill et al., 1996) demonstrate population-level effects of disturbance, essentially because disturbance effectively lowers habitat quality and thus reduces carrying capacities. A diverse range of stimuli can disturb wildlife. Dogs, often as companions to humans, are increasingly recognized as prevalent, wide-ranging stimuli that often evoke particularly strong and typically deleterious responses among wildlife (Williams et al., 2009). This may be especially true where wildlife and dogs co-occur at high densities in constrained areas, such as coasts and recreational parks. Increasingly, management solutions are being sought to mitigate the problem of dog disturbance to wildlife (Williams et al., 2009). One key information gap in relation to disturbance of wildlife, however, is differentiating the extent to which disturbance is a welfare issue, primarily impacting individual animals, and the extent to which it is a conservation issue, reducing viability of wildlife populations (Hockin et al., 1992). This, and the way humans value and protect wildlife and their dogs, means managing dog disturbance to wildlife is a controversial topic (Williams et al., 2009).

While most studies focus on 'pet' dogs accompanying their owners, a few have described wildlife being disturbed by herding or hunting dogs (e.g., Sastre et al., 2009), and only a handful have considered free-ranging (a.k.a., 'free-running' or 'free-roaming'), mostly unaccompanied dogs (Berger et al., 2007). Very little is known about the disturbance caused by dogs not accompanied by humans (but see Miller et al., 2001; Sastre et al., 2009), though an expanding literature examines the interaction between wildlife and dogs accompanied by people (Box 4.1). Thus, this chapter necessarily emphasizes the latter, and we acknowledge that more information is needed on disturbance caused by unaccompanied dogs. We also acknowledge a bias in available literature, which tends to focus on dog–wildlife conflicts in urban, coastal, forest, and heathland recreational areas (i.e., those areas where humans engage in leisure time activities; Box 4.1). Additionally, while there are many critical information gaps in relation to disturbance to wildlife caused specifically by dogs, some general principles of wildlife responses to threats are used here to discuss likely factors influencing disturbance to wildlife by dogs. This chapter emphasizes wild birds and mammals; while dog disturbance is also likely to occur to many reptilian and amphibian species (see Holderness-Roddam, 2011), publications are limited, and less text is devoted to these groups.

Free-Ranging Dogs and Wildlife Conservation. Edited by Matthew E. Gompper
© Oxford University Press 2014. Published 2014 by Oxford University Press.

Box 4.1 The limited information base for dog disturbance of wildlife

Not only is little information evident, but clear biases exist in the available information pertaining to dog disturbance of wildlife, and we acknowledge these biases will necessarily be reflected in the emphases of this chapter. Of those biases evident, we note that most available studies are from the developed world, mostly from temperate regions, and most deal with accompanied dogs (thus not representative of most of the world's dogs). Table 4.1 presents and characterizes selected studies which deal with the issue of dog disturbance in some substantive way ($n = 35$). These studies have increased exponentially over time (Figure 4.1); 40 and 34% originate from North America and Europe respectively, the remainder from Australasia and the Middle East. Most studies (60%) deal with a single species of wildlife. Most focus on open habitats, notably coasts (49%), and 40% involved some form of experimental delivery of stimuli to wildlife, the remainder used observational techniques.

Although literature specific to dog disturbance of wildlife is limited, there has been a growing body of literature concerning the broader topic of anthropogenic disturbance of wildlife over the past 40 years, and this has been frequently and adequately reviewed (e.g., Hockin et al., 1992; Weston et al., 2012). The vast majority of this literature involves wildlife responses to humans on foot, with less attention paid to motorized transport (vehicles, boats, aircraft, etc.) or dogs (see Weston et al., 2012). This literature describes great variation in response (extent, type, etc.) and consequences of disturbance to wildlife, and documents disturbance regimes (type, extent, and frequency of occurrence of stimuli, and the rate at which they evoke responses). The literature also elucidates some general principles that undoubtedly apply to the response of wildlife to dogs. These include: an inverse relationship between distance of a stimulus and probability and extent of the response to it, that unpredictable and 'non-benign' stimuli are associated with enhanced response, and that several attributes of wildlife (most notably increasing body mass) are associated with increased response distances (Hockin et al., 1992; Weston et al., 2012).

Despite this substantial body of work, key information gaps remain. These center around the need to explore the higher-order consequences of individual responses to disturbance (Weston et al., 2012). In particular, to date only a few studies document the population-level impacts of disturbance (Mallord et al., 2007). Few examine the consequences of disturbance-mediated declines in habitat quality. If disturbance represents an influential process

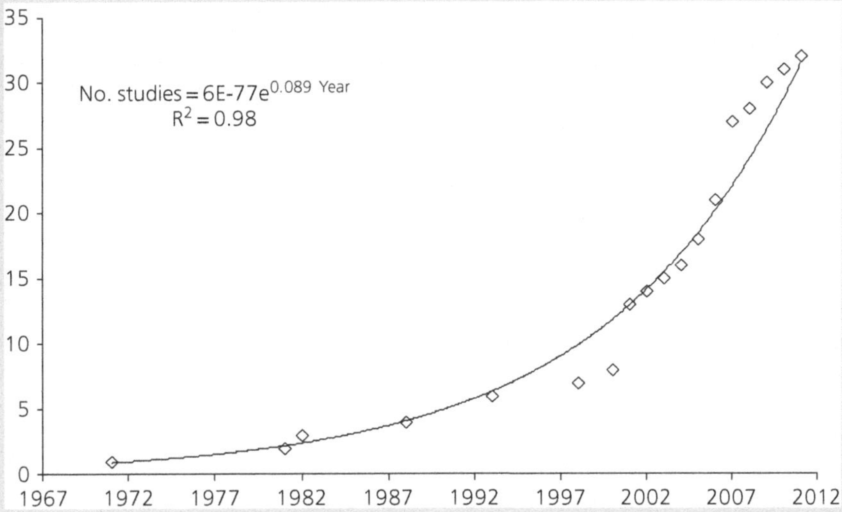

Figure 4.1 The cumulative number of 'major' studies of disturbance to birds and mammals that include dogs as a substantive stimulus, over time.

continued

Box 4.1 *Continued*

Table 4.1 Selected 'major' studies of dog disturbance to wildlife presented in alphabetical order within group (bird or mammal). Criteria for inclusion are that studies involve dogs as a stimulus (directly, not indirectly via dog management zonation or incidental occurrence) and focus on the response and its impact on wildlife, present data, and are published in peer-reviewed journals. Similar articles are grouped.

Group	Study	Approach	Wildlife species	Predominant type of dog	Habitat	Country
Birds	Banks and Bryant 2007	Experimental	Bird assemblages	Leashed	Woodland	Australia
	Burger 1981; Burger et al. 2007; Lafferty 2001a, b	Observational	Coastal birds	Accompanied dogs	Marine shores	USA
	Dunbrack and Dunbrack 2010	Experimental	Glacous-winged gulls and north-western crows	Closely accompanied	Open fields	USA
	Fernández-Juricic and Tellería 2000	Observational	Common blackbird	Accompanied	Urban parks	Spain
	Dowling and Weston 1999; Weston and Elgar 2005, 2007	Observational	Hooded plover (breeding)	Accompanied	Beaches	Australia
	Fitzpatrick, and Bouchez 1998; Kirby et al. 1993; Robinson and Pollitt 2002	Observational	Shorebirds	Accompanied	Coasts, estuaries, wetlands	UK
	Glover et al. 2011	Experimental	Shorebirds	Leashed	Coasts and wetlands	Australia
	Lafferty et al. 2006	Experimental	Coastal birds	Accompanied	Coasts	USA
	Lord et al. 2001	Experimental	New Zealand dotterel (breeding)	Leashed	Beaches	New Zealand
	Mallord et al. 2007	Observational and modeling	Woodlark	Accompanied	Heathlands	UK
	Miller et al. 2001	Experimental	Grassland and forest birds, plus one deer species	Accompanied (but one treatment where some separation from people occurred)	Forest	USA
	Murison et al. 2007	Observational	Dartford warbler	Accompanied	Heathlands	UK
	Randler 2006	Experimental	Eurasian Coot	Recording of barks	Wetlands	Germany
	Sastre et al. 2009	Observational	Great bustard	Accompanied and unaccompanied	Dry agriculture	Spain
	Taylor et al. 2007	Observational	Stone curlew	Accompanied	Fields (open access)	UK
Mammals	Cooper et al. 2008	Experimental	Eastern gray squirrel	Leashed	University campus	USA
	Gingold et al. 2009	Observational	Mountain gazelle	Guard dogs	Grassland	Jordan

continued

Box 4.1 *Continued*

Table 4.1 *Continued*

Group	Study	Approach	Wildlife species	Predominant type of dog	Habitat	Country
	Hamr 1988	Experimental and observational	Alpine chamois	Leashed and unleashed	Alpine forests and pastures	Austria
	Kloppers et al. 2005	Experimental	Elk	Unleashed herding dogs	Urban, wetlands, forest	Canada
	MacArthur et al. 1982	Observational and experimental	Bighorn sheep	Leashed	Forest	Canada
	Mainini et al. 1993	Experimental	Marmot	Leashed	Alpine areas	Switzerland
	Manor and Saltz 2003, 2004	Observational	Mountain gazelle	Feral dogs	Coastal plain	Israel
	Martinetto and Cugnasse 2001	Experimental	Mouflon	Leashed and unleashed	Woodland and rocky	France
	Pelletier 2006	Observational	Bighorn sheep	Leashed and unleashed	Grassy slopes	Canada
	Sweeney et al. 1971	Experimental	White-tailed deer	Unleashed hunting dogs	Forests, pastures, cropland	USA

that degrades habitat quality, then it might be one of several ecological processes which effectively create 'ecological traps'; insidious situations whereby animals select habitat based on cues that no longer reflect actual habitat quality (Schlaepfer et al., 2002). For example, animals may select habitat on the basis of the presence of resources for foraging and breeding (e.g., the presence of suitable hollows or prey), but may be unable to breed successfully because of disturbance while breeding. Such populations may represent attractive 'population sinks' (Schlaepfer et al., 2002). The episodic and extreme variation in human (and dog) presence in many areas means animals might settle in highly disturbed habitats during undisturbed periods.

4.2 Dogs as stimuli

The depth of evolutionary history and extent of wildlife interactions with wild canids have presumably shaped how wildlife perceive dogs, and the way dogs and wildlife behave during encounters (instances when wildlife and dogs interact). Canids may instinctively hunt wildlife and therefore dogs may be perceived as particularly threatening by wildlife (Gabrielsen and Smith, 1995). Among the diverse array of stimuli encountered by wildlife (e.g., humans, vehicles, predators, etc.), dogs as stimuli are associated with a specific set of features. First, like other predators, dogs evoke some of the most dramatic responses among wildlife and are therefore apparently perceived as especially threatening (Weston and Elgar, 2007). Unlike benign stimuli, in which wildlife responses can be considered unnecessary (e.g., to recreationists on beaches), dogs are frequently 'non-benign' stimuli that often actively pursue wildlife during encounters, for example, by chasing (9% of dogs chased birds on a Californian beach, Lafferty, 2001b; 11.1% of disturbance to shorebirds on beaches around Mackay, Queensland, involved dogs chasing birds, Bloor, 2005). The lack of an adequate response by wildlife may result in injury or death during such encounters. Dogs are associated with unique visual, auditory, and olfactory cues (e.g., shape, barking or howling, and scent marking) as well as more holistic cues such as posture, gait, and behavior. These cues may evoke responses among wildlife separately or in combination, however the specific canine cues (stimulus attributes) that are detected by wildlife and used to judge risk and inform response are unknown. Apparently, few studies test these cues separately, but it is unlikely that the visual cues are the only ones used to detect and assess risk by wildlife. Barking increases vigilance among preening Eurasian coots (*Fulica atra*; Randler, 2006), but apparently no information exists on disturbance by scent marking. Other native carnivores may respond to the scent of dogs on trails or in areas where dogs are allowed to roam freely, resulting in changes in activity in these areas (Lenth et al., 2008; Vanak et al., 2009).

4.2.1 Dogs as agents of disturbance

Several pieces of evidence suggest that dogs are prominent agents of wildlife disturbance, and that their role as agents of disturbance is often underestimated. First, experimental studies may underestimate dog disturbance. Most experimental studies of dog disturbance to wildlife mimic the most common types of stimuli because they seek management solutions and do not unravel specific aspects of a stimulus that cause disturbance. Studies of humans or other stimuli behaving as dogs do (e.g., by roaming) could enlighten as to whether it is dog behavior or dogs per se that contribute to the observed intense responses by wildlife (see Box 4.2). One key limitation of the current data available for dogs as an agent of disturbance to wildlife is the reliance on restrained dogs as experimental stimuli (Banks and Bryant, 2007; Faillace, 2010; Glover et al., 2011; Lord et al., 2001; Vanak et al., 2009). Thus, the most extreme wildlife responses may go unreported by experimental studies, while observational studies may better reflect the wildlife responses during more realistic encounters.

Second, dogs are among the most commonly encountered predator stimulus in at least some areas and circumstances (e.g., urban and recreational parks and coasts; Antos et al., 2007; Butler et al., 2004; Scott, 1989; Underhill-Day and Liley, 2007). In some natural areas, walkers and dogs are the most common source of disturbance (response) recorded, such as in wetlands in the UK (Robinson and Pollitt, 2002). The frequency of dog encounters with wildlife probably stems from three main factors: (1) dogs are common companions of humans; (2) they are capable of roaming over large areas of habitat; and (3) they tend to be year-round residents. When unrestrained they can occupy larger parts of the wildlife habitats in which they occur compared to humans (Figure 4.2). The higher area of occupancy of owned dogs results from both the mobility of their owners and from dog 'roaming' (i.e., the distance they move from their owners). Dog walkers can be rather mobile; on the Thames Basin heaths, UK, the mean length of route for dog walkers was 2,500 m, more than walkers and picnickers (2,300 and 1,200 m, respectively), but less than joggers, cyclists, and horse riders (3,900, 4,900, and 3,200 m, respectively) (Underhill-Day and Liley, 2007). Unrestrained dogs roam within coastal habitats perhaps more than any other stimulus type (Coombes et al., 2008) except possibly predatory birds. They also roam in non-coastal habitats (e.g., Sastre et al., 2009), although in at least some areas their roaming has been regarded as more modest, perhaps because of low penetrability of thick trail-side vegetation (Bekoff and Meaney, 1997; Forrest and St. Clair, 2006; Mallord et al., 2007). Owned dogs, especially in urbanized societies, enjoy regular walks; for example, of 380 coastal residents in south-eastern Australia, 36.8% owned a dog of which 93.6% took their dog to the beach (Maguire et al., 2011a). This means dogs tend to be present in wildlife habitat year-round (Figure 4.3 provides an example of complete temporal overlap between unleashed companion dogs and a vulnerable life history stage of a sensitive wildlife species). For owned dogs, human social factors such as weekends and holidays influence their occurrence in many areas (Sastre et al., 2009), and presumably climate also dictates seasonality of occurrence, perhaps especially in higher latitudes. Resident village dogs are also often present year round, and presumably so too are free-ranging dogs. Despite the already high densities of accompanied dogs in many parts of the

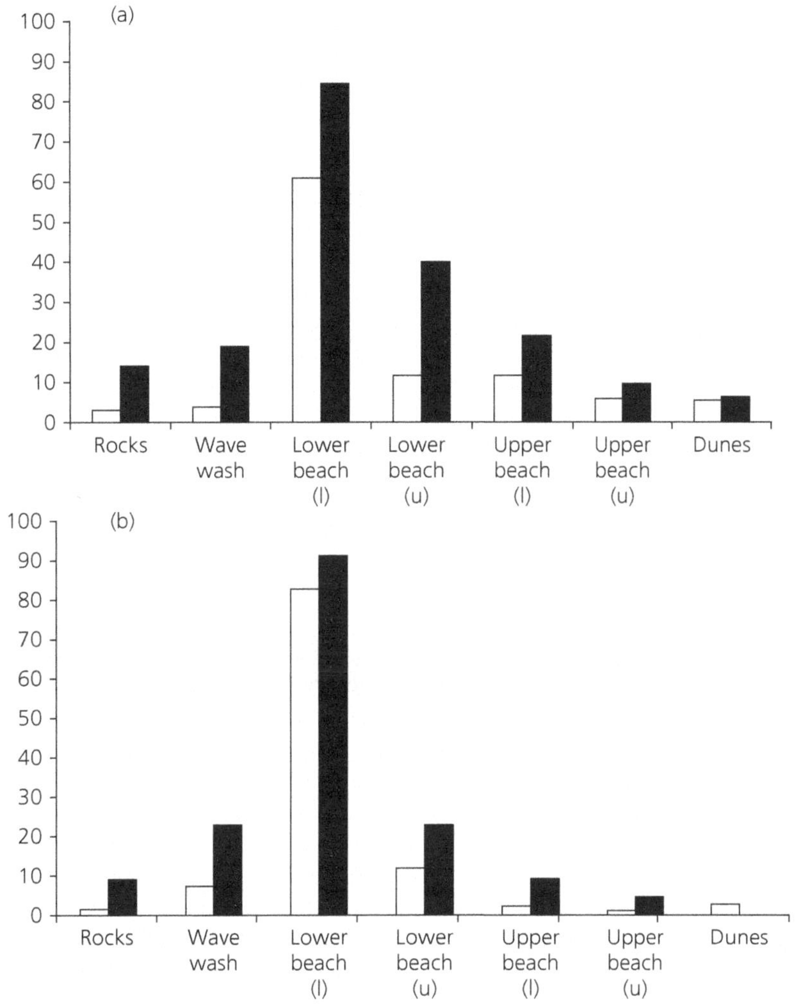

Figure 4.2 The habitat use of walkers (a, top panel) and joggers (b, lower panel) as they passed hooded plover (*Thinornis rubricollis*) nests on beaches in Victoria, Australia (see Weston and Elgar, 2007 for details; 'l' is the lower half, 'u' is the upper half) (unpublished data). The percentage of recreationists in each beach zone (averaged across nests) is shown. Open bars indicate that no dogs accompanied recreationists and closed bars indicate recreationists were accompanied by dogs off the leash. Walkers and joggers with dogs on leash were omitted due to small sample sizes.

world, projections suggest this activity will increase in future (Brickner, 2000); in some parts of the UK, 3.8–7.3% increases in dog walkers by 2080 are predicted (Coombes et al., 2008).

Finally, unrestrained dogs often move 'unpredictably' (i.e., their direction and speed varies frequently) and sometimes harass wildlife, traits that do not promote 'habituation,' the process whereby wildlife learn to reduce response intensities or frequencies with increasing exposure to a stimulus (Lafferty, 2001b; Sastre et al., 2009). Rather, these attributes promote 'sensitization,' or enhanced response frequencies or intensities with increasing exposure to stimuli (Glover et al., 2011). Roaming (usually erratic central place movements around an owner) influences three factors used by many wildlife species to judge degree of threat: predictability (in behavior and to some extent occurrence), proximity, and speed (Glover et al., 2011). While some dogs roam without accompanying humans, many others are kept indoors or in yards, and roam during 'walks.' During walks, some highly trained

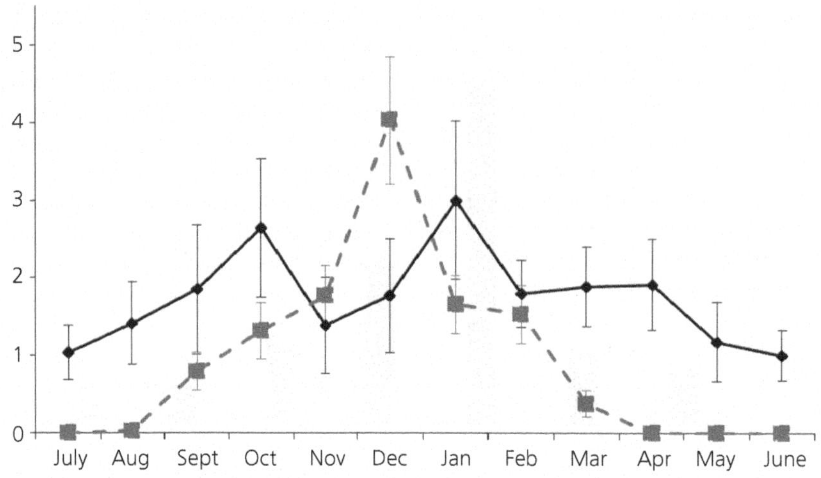

Figure 4.3 The average number of walkers with dogs off the leash (1,571 surveys of 69 beaches in Victoria, Australia, 1995–98; unpublished data; black lines) in relation to the average number of nests of hooded plover on those beaches (gray dotted line). Means and one standard error are shown; 6.1% of 743 dogs were leashed and are excluded from the graph.

dogs are effectively controlled by voice commands, but restraint in the form of a leash is by far the most common method of effectively managing dog roaming during walks. However, leashing rates are often low, with unleashed dogs apparently occupying more habitat than leashed dogs (Box 4.2).

4.2.2 Birds

A prerequisite for disturbance of wildlife is the temporal and/or spatial co-occurrence of stimuli and wildlife. The extent of overlap with wildlife populations and the frequency with which encounters occur is critical when judging possible impacts of disturbance. Birds probably frequently encounter dogs, though information on encounter rates (i.e., where an interaction is possible) is limited mostly to parks and beaches (see Section 4.2.1), where dogs are reported as the most, or among the most, frequently occurring stimuli (Antos et al., 2007; Mallord et al., 2007). While these studies often focus on recreational disturbance, and thus presumably present a biased 'heavily disturbed' sample, the potential for conflict between dogs and birds is high. Some information is available on the occurrence of dogs in or near bird habitat, and that information, while restricted to urban and wetland areas, suggests dogs are common sometimes even in 'off-limits' nature reserves dedicated to bird conservation (e.g., 8.5 times per weekend day; Antos et al., 2007). The hooded plover, a threatened beach-nesting shorebird, experiences many natural and anthropogenic stimuli on Victorian beaches, Australia, where 18–19% of encounters with nests or broods involved dogs, at a rate of 0.47 encounters per hour (Weston and Elgar, 2005, 2007). Off-leash dogs and dogs chasing birds were the third and fourth most common causes of disturbance to shorebirds around Mackay, Queensland (Bloor, 2005). Dogs accompanying people were the third most common stimulus causing flushing among blackbirds *Turdus merula* in urban parks in Madrid, Spain (Fernández-Juricic and Tellería, 2000). Of all stimuli encountered by roosting shorebirds on the Dee Estuary, UK, 1986–91, 26–41% involved dogs (Kirby et al., 1993). Little information is available on encounter rates between unattended dogs and birds; in Madrid unattended dogs represented 1.3% of potentially disturbing activities for great bustards (*Otis tarda*; Sastre et al., 2009), and on Victorian beaches, unattended but apparently owned dogs represented 0.9% of stimuli passing hooded plover nests (Weston and Elgar, 2007).

Not all birds are threatened by dogs in the same way or to the same extent, so the perception of dogs as threatening probably varies taxonomically (see

Box 4.2 Leashing as a tool to reduce roaming

'Leashing' refers to the attachment of a tether to a collar or harness on a dog to control the dog's movements in relation to a mobile human. Leashing is considered the most effective way of reducing harmful dog–wildlife interactions in areas where pet dogs are exercised in areas where they may encounter wildlife. It is often considered more effective and enforceable than 'effective control' (e.g., through voice commands), which are difficult to measure and enforce.

Leashing presumably reduces wildlife responses (e.g., Weston and Elgar, 2005) by reducing dog roaming rather than the presence of a leash per se. Thus, there is presumably a maximum leash length that effectively reduces the roaming of dogs to the point where most wildlife decrease their responses (this is likely to vary between species; Glover et al., 2011). A review of commercially available leashes on the Internet reveals a significant difference in length of different types of leashes ($n = 58$, Kruskal Wallis = 38.83, $p < 0.001$). Traditional, commercially available, leashes constrain dogs to within 1–2 m of their owners, though retractable leashes are longer (typically 5–8 m in length) and some non-retractable leashes are up to 15 m in length (Figure 4.4). As leashing laws have become more prevalent, and with the advent of retractable (longer) leashes, it seems likely that, on average, leash length has increased over time.

Despite a great variety of leashing options, in at least some areas, the majority or at least a substantial proportion of dogs are unrestrained. For example, on Australian (90% unleashed, Weston and Elgar, 2005; Williams et al., 2009) or US beaches (93%, Lafferty, 2001b), including areas where dogs are not permitted off-leash or at all, such as national parks (88%, 1991–98, Dowling and Weston, 1999; Arnberger et al., 2005), recreation reserves (22%, Austria, Arnberger and Eder, 2008), wetland reserves (100%, Antos et al., 2007) and buffers (68%, Weston et al., 2009). In heathland sites (UK), generally most or all dogs were unleashed (92%, Mallord et al., 2007; Underhill-Day and Liley, 2007). Thus, in at least many parts of the world, wildlife most frequently encounter free-ranging dogs regardless of prevailing local regulations (Lafferty et al., 2006). Miller et al. (Chapter 12) discuss the decisions made by owners in relation to leashing.

Although there have been virtually no studies, leashing appears to constrain dog roaming, at least in habitats where dog roaming is not constrained by vegetation or other impediments to movement. For example, on beaches in Victoria, Australia, where active hooded plover nests occurred, walkers and joggers accompanied by unleashed dogs occupied more levels of the beach than walkers or joggers without dogs (Figure 4.2). Walkers and joggers without dogs occupied fewer beach zones compared with when their recreational group (people and dogs) included unleashed dogs (walkers, 1.0 versus 2.0 beach zones occupied respectively (medians), $n = 1081$, $U = 21.69$, $p < 0.001$; joggers, 1.0 versus 1.5 zones, $n = 161$, Kruskal Wallis = 28.25, $p < 0.001$).

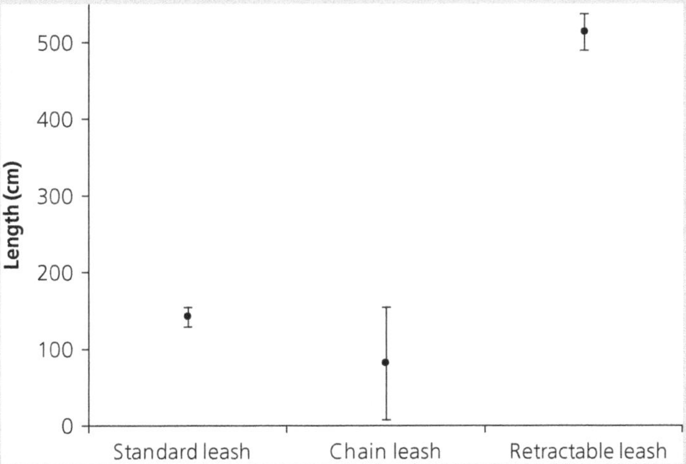

Figure 4.4 The length of commercially available leashes for dogs (not puppies; in cm), as revealed by an Internet search of several major pet supply stores ($n = 58$ products). Standard leashes are made of nylon or leather. Means and 95% confidence intervals are shown. Two 'recall' leashes (9 and 15 m) and leash extenders (up to 1 m) are excluded.

also Box 4.3). Many birds use trees, cliffs, open water, or spend much of their lives on the wing; such species are presumably immune or buffered from any negative impacts of dogs, and it might be predicted that dogs are not perceived as especially threatening by these species (but see Banks and Bryant, 2007). Dogs sometimes enter waterbodies where they can disturb waterbirds (Kramer, 1986). However, flightless species or those that become flightless when molting, ground-dwelling, and especially ground-nesting birds are most likely to interact with dogs, and have been the subject of most research on dog–bird interactions. In this way, there may be a bias in existing literature, whereby the species most vulnerable to negative interactions with dogs may have been documented most. Nevertheless, bird–dog "conflicts" are reported from around the world.

4.2.3 Mammals

Unlike birds, where for some species researchers can monitor all disturbances for a group of birds over an extended period of time, encounter rates between wild mammals and dogs are less evident. In regions where dogs primarily travel with owners, the rate at which mammals may encounter dogs should be positively related to their distance from restricted trails where leashed dogs may occur and the prevalence of dogs off-leash in the area. Accompanied dogs, however, are typically found in recreational areas during daylight hours, while many mammals are nocturnal, reducing the potential for direct encounters. Free-ranging dogs, however, tend to be nocturnal and show great variation in their home range sizes (from 1 ha up to about 2,500 ha) with potentially much greater disturbance effects on wild mammals (Meek, 1999). Unlike birds, which typically have the option to fly away, most mammals (excluding bats, aquatic, semi-aquatic, and arboreal mammals) are limited to terrestrial escape and are, therefore, more greatly affected by the tendency for dogs to roam widely across a landscape. While most types of dogs are better chasers than they are efficient hunters (cf., dingoes), they are able to capture and kill a variety of mammals (see Ritchie et al., Chapter 2) and generations of selective breeding have provided subsets of dogs with specialized traits favorable for hunting, locating and capturing prey, fighting, and racing (Serpell, 1995). Trained hunting dogs can induce prolonged disturbance bouts. Sweeney et al. (1971) found that hunting dogs ($n = 65$) chased white-tailed deer (*Odocoileus virginianus*) for an average of 33 min (up to 155 min) and an average of 3.9 km (up to 21.6 km). While no mortality occurred during these chases, 78% of the experimental chases resulted in the deer leaving their home range, with most returning within a day. While these were controlled experimental cases, untrained free-ranging dogs have much greater home range sizes (Meek, 1999) and should, therefore, be more likely to initiate long chases and move wildlife from their own home ranges. Unleashed dogs on beaches may be particularly dangerous for marine mammals. Several studies report harassment and killing of harbor seals (*Phoca vitulina*; Allen et al., 1984) and Hawaiian monk seals (*Monachus scbauinslandi*; Gerrodette and Gilmartin, 1990; Kenyon, 1972) by dogs, suggesting that uncontrolled dogs could be particularly detrimental to populations of pinnipeds, which move slowly and awkwardly on land. Clearly, the impact of free-ranging dogs is a function of breed, training, past experiences with wildlife, and the prey encountered. Ultimately, dogs can prey upon a wide variety of wild mammals, including terrestrial (e.g., hedgehogs *Erinaceus europaeus*; Doncaster, 1994), arboreal (Lumholtz's tree-kangaroos *Dendrolagus lumholtzi*; Newell, 1999), and marine mammals (Allen et al., 1984; Barnett and Rudd, 1983), thus mammalian responses to dogs are expected to be substantial.

Wild mammals (perhaps except very large herbivores or carnivores) may alter their spatial distribution in areas where dogs are permitted to roam to reduce the likelihood of encountering potentially dangerous canines. Lenth et al., (2008) studied signs of mammal activity nearby and far from trails in parks where leashed dogs were permitted on-trail and in parks where dogs were prohibited. They found lower signs of activity of mule deer (*O. hemionus*), rabbits (*Sylvilagus* spp.), prairie dogs (*Cynomys ludovicianus*), and other small mammals

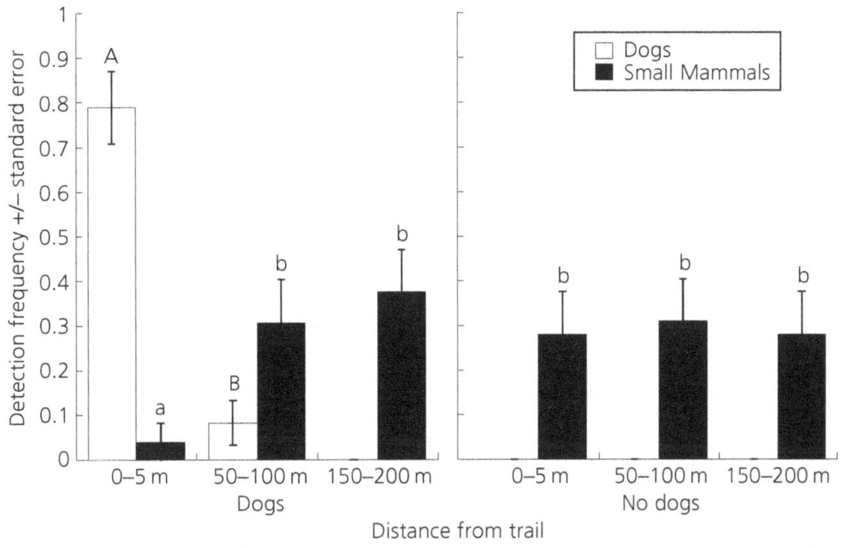

Figure 4.5 Detection frequencies of dogs and small mammals on track plates at different distances from trails where dogs are permitted or prohibited in a natural area. Different uppercase letters above columns indicate significant differences ($p < 0.1$) in dog detection frequencies, and different lowercase letters indicate significant differences ($p < 0.1$) in small mammal detection frequencies. Redrawn and reprinted with permission from the Natural Areas Association from Lenth et al., (2008).

on trails where dogs were permitted, compared to dog-free trails (Figure 4.5). Native carnivore activity, however, was higher near the end of trails where dogs were permitted but lower near trail heads, suggesting that carnivores avoided dog cues when abundant (i.e., trail heads) but may be attracted to them as novelty when they are present but rare (i.e., ends of trails). Bobcats (*Lynx rufus*) avoided dog trails altogether, perhaps because of the similarity between dog cues and those of coyotes (*Canis latrans*), a natural potential predator (which showed no difference in activity between sites). Similarly, mesocarnivores tend to avoid areas of high dog activity; bobcats (George and Crooks, 2006) and Indian foxes (*Vulpes bengalensis*) (Vanak and Gompper, 2010) showed reduced activity in areas where dogs were most active.

4.2.4 Other vertebrates

Comparatively little information is available on the role dogs play as stimuli for non-avian or non-mammalian vertebrates (henceforth 'other vertebrates'). In addition to visual, auditory, and olfactory cues, some reptiles (e.g., snakes) also perceive vibrations in the ground associated with the approach of a threat (Young, 1983).

In many terrestrial areas, there is probably substantial temporal and spatial overlap between other vertebrates and dogs, though the extent of any interaction is virtually undocumented. Despite this, dog disturbance and 'harassment' is considered a conservation problem for amphibians and reptiles (e.g., British Columbia Government, 2004). While dogs have been present in many habitats for millennia, in some habitats they are relatively new arrivals, and have quickly established themselves as predators, and agents of disturbance, of vertebrates apart from birds and mammals. One example is on the Galapagos Islands, where marine iguanas (*Amblyrhynchus cristatus*), isolated from terrestrial predators for 5–15 million years, apparently first encountered dogs on some islands only *ca.* 150 years (Berger et al., 2007). The increase in dogs (and cats) has coincided with human settlement in the islands and now causes disturbance and mortality among iguanas.

Box 4.3 Persecutor to protector; dog disturbance protecting wildlife

Not all dog–wildlife disturbance results in negative outcomes for wildlife. Dogs also disturb pest and nuisance species, and in circumstances where wildlife are more tolerant or less vulnerable than pests, or where dogs are trained specifically to defend wildlife, then dogs essentially become their protectors. Additionally, disturbance can be used as a non-lethal alternative to achieve management objectives that reduce human–wildlife conflict, and so effectively benefit wildlife.

In urban backyards of some Australian cities, the presence of pet dogs decreases the probability of denning by the introduced pest species, the red fox (*V. vulpes*; Marks and Bloomfield, 2006). This presumably permits a variety of wildlife to persist which otherwise could not. Carefully trained guard dogs defend a handful of threatened species or significant wildlife colonies (e.g., burrow and surface nesting seabirds) against introduced or problematic predators (van Bommel, 2010); in the same way they can defend stock against predators and so reduce farmer–wildlife conflict (Coppinger et al., 1987; VerCauteren et al., Chapter 9). Some airports use dogs to reduce bird hazards to aircraft as a non-lethal management alternative (Froneman and van Rooyan, 2003). Dogs have even been used as aversive conditioning stimuli to dishabituate elk *Cervus canadensis* and other ungulates in areas where they are heavily encroaching on human settlements (Kloppers et al., 2005; VerCauteren et al., 2008; Walter et al., 2010). In these cases, dogs of different breeds have been effective hazing tools for wildlife managers. Livestock and crop protection dogs have also been used to chase away wild ungulates from cattle ranches (thereby limiting the spread of zoonoses like brucellosis to cattle), golf courses, orchards, and forest plantations (VerCauteren et al., 2005, 2008; Walter et al., 2010); and breeds that are territorial and patrol open spaces (e.g., Siberian Husky, Alaskan Malamute) have been most effective (VerCauteren et al., 2005).

Finally, the ability of dogs to detect wildlife that would be otherwise undetectable, often by honing in on their scent and evoking a flight reaction, has supported the conservation effort of many cryptic species such as kiwi (*Apteryx australis*) in New Zealand (Taborsky, 1988), black grouse (*Tetrao tetrix*) in England (Baines and Richardson, 2007), or Mojave desert tortoises (*Gopherus agassizii*) in the USA (Heaton et al., 2008). This ability to locate cryptic wildlife can be harnessed to capture individuals for their use in threatened species programs or to survey sites to assess their suitability for human development, or general wildlife surveys (Gutzwiller, 1990; Woollett et al., Chapter 10). Additionally, dogs have played a critical role in pest eradication aimed at ecological restoration, such as in the attempts to eradicate European rabbits (*Oryctolagus cuniculus*) from the sub-Antarctic Macquarie Island (Australian Government, 2012). Such efforts cause short-term disturbance but can result in long-term beneficial conservation outcomes.

4.3 The response of wildlife

Disturbance responses among wildlife are generally accepted to have evolved as anti-predator responses, and are adaptive among populations exposed to predators. Wildlife responses to dogs range from vigilance and crypsis to active defense such as aggression or flight, and are adjusted in relation to a range of internal and external factors (Glover et al., 2011). Optimal escape theory posits that responses can be considered in a cost–benefit framework, whereby escape entails costs (often energetic costs) but delivers benefits (notably enhanced survival) (Ydenberg and Dill, 1986), and in an optimality framework, whereby escape permits gain of fitness after the interaction whereby death leads to loss of all future fitness (Cooper and Frederick, 2007, 2010).

Based on differential responses, wildlife have the capacity to discriminate between stimuli, including discriminating between dogs and other stimuli (Glover et al., 2011; Lord et al., 2001). Given all wildlife have to respond to threatening stimuli in some way, a key question is how response rates or intensities in relation to dogs compare with those caused by other stimuli, natural or anthropogenic.

Many studies of disturbance report wildlife responding frequently and substantially to dogs, despite at least many decades, if not centuries or millennia, of exposure. Wildlife responses are likely to be shaped at two distinct time-scales: within lifetimes and over evolutionary time. Changes to responses within lifetimes are driven by learning, that is altered responses on the basis of individual experience. The persistence of responses in circumstances where the stimuli do not apparently represent substantive threats (i.e., to human recreationist) suggests that habituation, if it occurs, is often modest (Glover et al., 2011). Responses will also have been shaped by evolution. Continuing (apparently costly) responses of wildlife to dogs, despite long-term exposure, presumably mean that such responses are adaptive, in other words, the avoidance of dog predation despite the cost of responses has presumably conferred fitness benefits. In many places (such as Europe and North America) dogs may be considered as functional analogs, albeit at elevated densities, of now-extinct or rare predators such as wolves *C. lupus*; in other areas (such as New Zealand and many other islands) mammalian terrestrial predators are evolutionarily novel. This creates an interesting paradox (Figure 4.6); among wildlife which is evolutionarily 'better-prepared,' dog-related disturbance may sometimes be higher (and yet levels of predation lower) than in areas with largely naïve native faunas, where disturbance may be relatively modest but predation levels high (e.g., Berger et al., 2007).

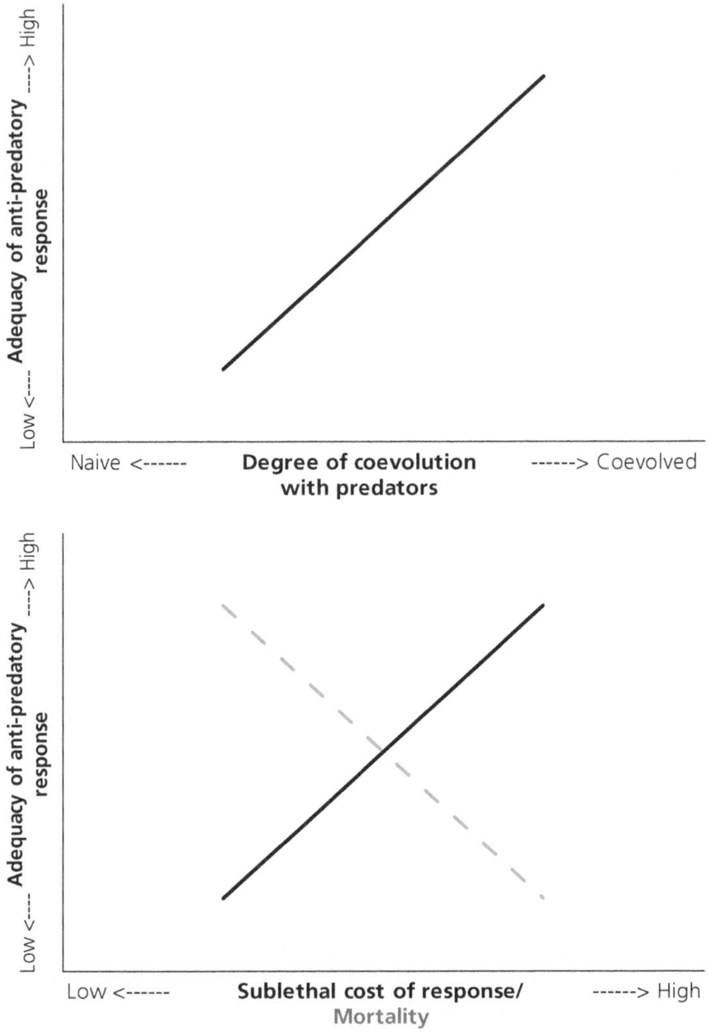

Figure 4.6 Diagrammatic representation of a conceptual model of the 'Disturbance–Predator Paradox.' Evolutionarily naïve species may not exhibit strong anti-predator responses (top panel) which means they may not incur substantial sublethal costs from maladaptive responses to benign stimuli. However, they may experience high mortality when stimuli are not benign (bottom panel). Black solid lines indicate sublethal effects; dashed gray lines indicate lethal effects.

4.3.1 Birds

Birds flee dogs—on foot, on the wing, by climbing, swimming, or diving. One measure of response intensity is Flight Initiation Distance (FID), the distance at which an animal flees from an approaching stimulus. Apart from flight (fleeing) a variety of responses are given to dogs. These include vigilance, absences from nests or young, distraction, and reductions in, or cessation of, foraging (Colwell and Sundeen, 2000; Weston and Elgar, 2005, 2007). Aggression is also reported, with aggressive ground- (e.g., lapwings, *Vanellus vanellus*) and tree-nesting species (e.g., Australian magpies, *Gynmorhina tibicens*) swooping some dogs (M.A. Weston, unpublished data). These responses are all associated with energetic and other consequences, which are generally poorly known and require more study.

Observational studies report higher response rates or intensities of ground-nesting birds to unleashed dogs over other anthropogenic stimuli, although naturally occurring stimuli may still evoke more frequent or longer lasting responses (Burger, 1981; Taylor et al., 2007; Weston and Elgar, 2005; 2007). Walkers accompanied by dogs often evoke greater responses in ground-dwelling birds than humans alone (Lord et al., 2001; Sastre et al., 2009; Sime, 1999). For example, stone curlews (*Burhinus oedicnemus*) show FIDs to dog walkers that sometimes exceed 500 m (Taylor et al., 2007). Dogs caused higher rates of flushing at prairie chicken (*Tympanuchus cupido*) leks compared with visits by foxes (Hamerstrom et al., 1965). Few studies of the response of birds to unattended dogs are available, but those results that are available suggest that the degree to which dogs unaccompanied by people disturb birds is context specific. Unattended dogs in steppes around Madrid represented 1.3% of potentially disturbing stimuli for great bustards, but caused 2.9% of responses; the probability of causing a disturbance was higher for dogs than for any anthropogenic activity recorded, including hunting, vehicles, and aircraft (Sastre et al., 2009). On the other hand, two grassland birds in the USA (vesper sparrows, *Pooecetes gramineus*, and western meadowlarks, *Sturnella neglecta*), were disturbed least by unattended lone dogs, and more so by walkers and walkers with dogs (Miller et al., 2001). Perhaps the height of an approaching stimulus alters the distance at which stimuli are detected, especially where lower strata, such as grass, limit the field of view.

Observational studies of disturbance to birds have focused on attended dogs, usually on coasts, wetlands, or in parks, and have demonstrated that dog–bird interactions are common worldwide, and that birds respond frequently and substantially to dogs. Walkers and dogs were the most common source of disturbance recorded at coastal and inland wetlands in the UK, although they were more likely to cause disturbance to waterbirds at coastal sites, probably because of the nature of the habitats and the degree of spatial overlap between human activities and birds (Robinson and Pollitt, 2002). On the Dee Estuary, UK, 1986–91, dogs caused high rates of disturbance to shorebirds (27–72% of all disturbance events) compared with walkers (20–34%), even though walkers were more commonly encountered (Kirby et al., 1993). On the other hand, dogs did not influence vigilance rates of shorebirds foraging on a rocky beach at Belfast Lough, UK (Fitzpatrick and Bouchez, 1998). On a beach in Santa Barbara, California, 10% of humans and 39% of dogs disturbed birds (Lafferty, 2001a). On beaches around Mackay, 14.8% of all disturbances to shorebirds (involving 24.6% of shorebirds present) were caused by dogs (birds flew up to at least 250 m; Bloor, 2005). Dogs were the most common cause of disturbance to breeding Dartford warblers (*Sylvia undata*) in heathland sites in Dorset, UK (Murison et al., 2007). At Esquimalt Lagoon, Canada, dog walking caused 18% of observed anthropogenic disturbances to waterbirds (Clowater, 2008). Observational studies such as these have underpinned the perception among managers that dog disturbance of wildlife is a high priority for management (Le Corre et al., 2009). While observational studies map the occurrence of stimuli and response in space and time, they don't unravel the specific stimulus–response–consequence mechanisms that permit deeper analysis of the issue. The consequences of disturbance responses are not obvious at the individual level, let alone the population level. For example, it is generally unknown how the rate or intensity of response is associated with fitness, and whether the relationship is linear or nonlinear, with or without 'thresholds.'

Several experimental studies of wildlife disturbance explicitly investigate the influence of stimulus type on response (reviewed in Weston et al., 2012) but very few use dogs as one of the stimuli tested. Glover et al. (2011) showed that of eight shorebirds tested, stimulus type (walker, jogger, walker with leashed dog) significantly influenced FID of three species. Excluding joggers, all three species had highest FID when approached by a person with a leashed dog, rather than by a person alone. Snowy plovers (*Charadrius alexandrinus*) react at twice the distance to dogs than to pedestrians (Fahy and Woodhouse, 1995; Lafferty, 2001b), and disruptions to incubation caused by investigator approaches to northern New Zealand dotterel (*C. obscurus aquilonius*) nests were longer when a leashed dog was present (Lord et al., 2001). North-western crows (*Corvus caurinus*) and glaucous-winged gulls, (*Larus glaucescens*) took bread at greater distances from a human and dog than from a human alone, reversing the outcome of food competition between these species (Dunbrack and Dunbrack, 2010).

Other measures of response include the amount of time before the resumption of normal activities. Similar to the New Zealand dotterel, snowy plovers remained away from their nests for longer durations when a walker with a leashed dog approached, compared with a walker alone, on one Florida island (although not on another) (Faillace, 2010). A variety of internal and external factors influence response rates to stimuli, and these presumably also apply to responses of birds to dogs. Body mass, wing shape, diet, age, sex, group size, experience including geographical isolation from predators, personality, site attributes including distance from cover and the presence of barriers such as fences or canals, whether stimuli occur on- or off-trail, and weather, among other things, may influence responses (see Stankowich and Blumstein, 2005; Weston et al., 2012). There will also doubtless be attributes of dogs that alter response, potentially including size and personality (reflecting breeds in some cases), group size, vocalizations, propensity to roam, age, and so on. Habitat mediates the responses of some birds, perhaps because some habitats are impenetrable to dogs (Mallord et al., 2007; Robinson and Pollitt, 2002). In forests, the American robin (*Turdus migratorius*) responded similarly to walkers alone and those accompanied by a dog, while two grassland species responded more strongly when a walker with a dog approached (Miller et al., 2001). Thus, it is possible that habitat mediates the responses of birds to dogs.

4.3.2 Mammals

Many studies have measured the direct disturbance effects of human recreation on wild mammals (see Knight and Gutzwiller, 1995; Stankowich, 2008; Stankowich and Blumstein, 2005). Mammals typically respond to dogs by fleeing to a burrow or tree, but larger mammals that lack these options often escape over long distances and are more likely to be displaced from their home range, stressed physiologically, and experience the negative effects of escape for longer durations. As a consequence, ungulates have been a common focus of disturbance studies, especially in response to dogs. Generally, the presence of a human has a greater impact on wildlife behavior in areas with relatively low human density, and one explanation of this is the possibility that animals habituate to humans in a non-threatening context (Stankowich, 2008). The effect of dogs relative to humans is less clear. While some studies find wild mammals to be less fearful of humans alone compared to humans accompanied by dogs (e.g., Hone, 1934), others have observed that mammals responded in the same manner irrespective of the presence of dogs (e.g., Hamr, 1988). While the most common response to the presence of a dog is to become alert and flee, the presence of dogs may also alter physiological responses. Bighorn sheep (*Ovis canadensis*) had greater heart rates when approached by a human with a dog compared to a human alone (MacArthur et al., 1982) and domestic sheep (*O. aries*) showed greater fear and avoidance of a dog compared to a human or a goat (Beausoleil et al., 2005). The presence of dogs during captures by wildlife managers may exacerbate these physiological effects (Sime, 1999). Marmots (*Marmota marmota*) were more likely to flee to burrows and to emit warning whistles, and they took longer to re-emerge from burrows after escape when dogs were present with humans, compared to humans alone (Figure 4.7). Interestingly, in some cases, the presence of a dog may actually reduce the degree of response altogether (where 'degree of response' refers to the level of behavioral response in a typical hierarchy of responses that escalate with increasing risk). Eastern gray squirrels (*Sciurus carolinensis*), in the presence of a dog compared to a human alone, tended to run *less* often and were more likely to freeze, erect, or flick their tails (Cooper et al., 2008). In addition to these immediate reactions to the presence of dogs, some species may increase their group size to gain protection. Mountain gazelles (*Gazella gazella*) had larger group sizes in areas with more feral dogs (Manor and Saltz, 2003), which prey on gazelle neonates, and the culling of feral dogs significantly increased the kid:female ratio in subsequent years (Manor and Saltz, 2004).

Wild mammals may be more attuned to dogs as potential predators, and in their presence may have a greater zone of awareness (the bounds of which are the maximum distance at which an animal will become alert and monitor a potential predator; Stankowich and Coss, 2006). Marmots became alert to trail hikers led by dogs and hikers with free-ranging dogs at significantly greater distances than to off-trail hikers or trail hikers without dogs (Mainini et al., 1993), but there was no difference between leashed and free-ranging dogs. Mule deer were more likely to become alert, and became alert at greater distances, when hikers were accompanied by leashed dogs off-trail compared to when hikers were alone (Miller et al., 2001). While the presence of a dog with a hiker elicited greater alert distances from eastern gray squirrels in areas with high human activity, interestingly, there was no corresponding effect in areas of low human activity, where alert distances were much greater for both types of stimuli (Cooper et al., 2008), suggesting that wild mammals may have an easier time habituating to humans when they are common, but dogs will always be perceived as dangerous regardless of their ubiquity.

The decision to flee is based on a substantial array of factors, including past experience with the stimulus type (i.e., learning effects), the type of stimulus (i.e., predator identity), and threatening behavior of the stimulus (e.g., approach speed, directness) (Stankowich, 2008; Stankowich and Blumstein, 2005). Therefore, it is predicted that, similar to alert distances, animals should flee from more threatening stimuli at greater distances than less threatening stimuli. Indeed, marmots fled from hikers with leashed or free-ranging dogs at greater distances than hikers without dogs, but, again there was no difference between leashed and unleashed dogs, although the minimum FIDs in response to off-leash dogs were much greater than for leashed dogs (Figure 4.7). Identical results were found for mouflon (*O. musimon*; Martinetto and Cugnasse, 2001) and mule deer also had a greater probability of fleeing and a greater FID in response to hikers with a

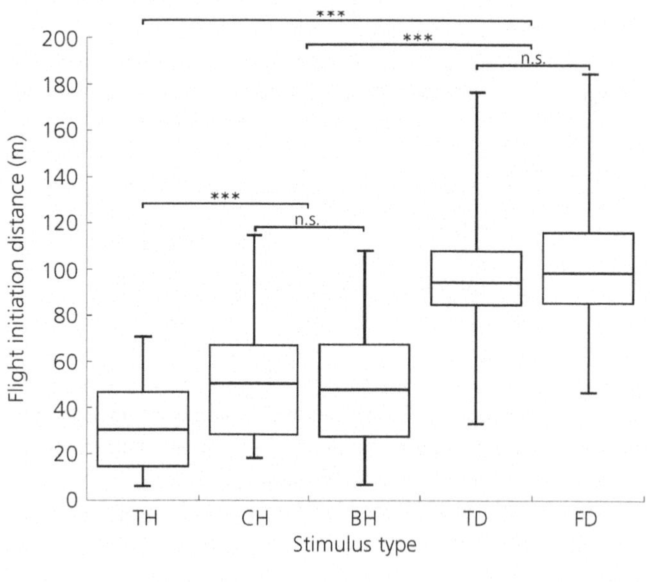

Figure 4.7 Flight initiation distance (median, IQR, min., max.) of marmots in response to people hiking on marked trails (TH), hikers walking cross-country off trails (CH), people hiking off trails across burrows (BH), people hiking on trails with leashed dogs (TD), and people hiking off trails with dogs on a 10 m leash to simulate free-ranging dogs (FD). $n = 20$ for each stimulus; *** $p < 0.001$. Redrawn and reprinted with permission from Elsevier from Mainini et al., (1993).

leashed dog, compared with hikers without dogs (Miller et al., 2001). Interestingly, elk fled at similar distances to humans alone and humans accompanied by dogs, both before aversive conditioning (experimental harassment by humans and dogs) and afterwards (Kloppers et al., 2005). This population, however, was highly habituated to humans and was encroaching on a settlement prior to conditioning, so the generality of this result is questionable. Finally, due to thousands of years of co-evolution with other wild canids, wildlife escape responses to dogs may be influenced by subtle behavioral cues shared by hunting canids: caribou (*Rangifer tarandus*) herds allow wolves to approach closely until they recognize behaviors indicating the wolves' intentions (Bergerud, 1974).

Once an animal has fled, the distance they move from the source of disturbance may also be an indicator of the perceived threat of the disturbance, however, results are conflicting. Similar to the findings for FID and alert distance noted above, bighorn sheep fled similar distances in response to humans with leashed dogs in comparison with free-ranging dogs (Pelletier, 2006). While mule deer tended to flee greater distances when humans approached with dogs off-trail compared to humans without dogs, they found no such difference on-trail (Miller et al., 2001). In contrast, alpine chamois (*Rupicapra rupicapra*) that haven't been exposed to wolves for many generations, were more curious of trained stationary dogs; and when unaccompanied-but-trained dogs were allowed to pursue, chamois fled shorter distances compared to humans alone but defended themselves with horns if overtaken and cornered (Hamr, 1988). Given these results, it appears that the distance that wild mammals move in response to dogs may depend more upon their past experiences with them and the landscape in which the encounter occurs.

Being on or off trail influences many wildlife responses to dogs (Mainini et al., 1993; Miller et al., 2001), but many other factors may mediate fright responses. Larger group sizes may ameliorate physiological effects on mammals of dog presence and increase perceptions of safety; MacArthur et al., (1982) found a negative association between group size and heart rate in mountain sheep when humans approached with a dog but not when humans approached without a dog. Many seasonal differences influence the response of wildlife to dogs, including difficulties of escape in snow, over rugged terrain, or when accompanied by offspring (Sime, 1999; Stankowich, 2008). As the effects of the presence of dogs during human–mammal encounters have received little attention, other interacting effects have yet to be studied, although they likely include past experiences of wildlife with dogs or other wild canids, human density in the area, size and defensive ability of the wildlife species, and the availability of, and distance to, refuge.

4.3.3 Other vertebrates

Reptiles and amphibians are preyed on by dogs (e.g., Koenig et al., 2002), so it is unsurprising that they respond to the presence of dogs. They respond to disturbance by fleeing on land or in water, climbing, and often use refuges, for example in crevices or vegetation. Snakes may defend themselves against dogs by rearing and striking, resulting in a much publicized dog–wildlife interaction, which is often characterized in the media as 'snake attacks' (e.g., Levy, 2011), but which are more likely to be an aggressive defensive response to the approach of a dog. Snake bites of dogs occur worldwide; 44% of domestic animals in Australia suffering from snake bites, and which were presented to veterinarians, were dogs ($n = 1590$; Mirtschin et al., 2008).

While escape behavior in reptiles, usually running, has been used as a general model of developing and testing theoretical frameworks for flight (e.g., Cooper and Wilson, 2007), little information is available of the response of reptiles to the presence of dogs. Mojave Desert tortoises (*Gopherus agassizii*) did not alter their movement patterns when detected by dogs as opposed to by a person without a dog (Heaton et al., 2008). Marine iguanas on islands with free-ranging dogs (and cats) exhibited higher FIDs in response to human approaches and human chasing and had higher corticosterone levels than those on islands without dogs (Berger et al., 2007). The species is capable of habituating to human disturbance, but dogs are an actual predator (i.e., a non-benign stimulus) so responses are likely to be adaptive, although currently they are not effective at avoiding predation (Berger et al., 2007; Rödl et al., 2007). Blue-tongued lizards (*Tiliqua scincoides*) may persist in suburbia, partly because their use of hard cover refuges in response to threats may avoid predation by dogs (Koenig et al., 2001).

We were unable to locate any literature on dog disturbance of amphibians, though fleeing, cypsis including the cessation of calling, and other responses are expected.

4.4 The impacts of dog disturbance on wildlife

The question as to whether disturbance is a welfare and/or a conservation issue is critical to the way dogs should be managed in natural areas. Management priorities may not include mitigating disturbance unless it is perceived to be a conservation risk. Once again, the most studied impacts of disturbance by dogs on wildlife involve coasts and parks, and owned dogs.

4.4.1 Birds

Disturbance represents a conservation threat if it negatively influences wildlife population viability. Population parameters that influence viability include average reproductive success, recruitment, survival, and habitat use. Most evidence of deleterious impacts of dog disturbance derives from readily measured behavioral responses, which involve the disruption of critical behaviors such as compromised parental care (Weston and Elgar, 2005, 2007). A few findings infer that such disruption can affect key population parameters, such as through depressed reproductive success. For example, effective dog management zones are associated with higher reproductive success among hooded plovers (Dowling and Weston, 1999).

Only a handful of studies link population or community level impacts of disturbance, and even fewer focus specifically on dogs as agents of disturbance. An increase in recreational disturbance, which mostly constituted dog walkers and unleashed dogs in heathland sites at Dorset, UK, resulted in a 17% decrease in breeding productivity of the ground-nesting woodlark (*Lullula arborea*). A range of access scenarios indicated that a doubling of current recreational levels does not apparently influence the woodlark breeding population size, but a more evenly distributed occurrence of recreationists would impact populations substantially (Mallord et al., 2007). Little direct evidence is available to link dogs with altered usage of habitat. In an Australian woodland park, there was a 35% reduction in bird diversity and 41% reduction in abundance due to the presence of leashed dogs, both in areas where dog walking is common and where dogs are prohibited (Banks and Bryant, 2007). The critical issue of whether dog disturbance is a conservation issue for birds remains virtually undocumented, partly because studies are complex and require substantial quantities of data.

4.4.2 Mammals

Like birds, there is a lack of empirical work on the broader impacts of dog disturbance on wild mammals. The scant evidence available implies broad and potentially significant effects: (1) in areas where dogs on leashes are common; (2) where feral or free-ranging dogs are abundant; and (3) for small mammals. In addition to the potential effects on group sizes discussed in Section 4.3.2 above, free-ranging dogs also have a direct negative influence on the kid:female ratio of mountain gazelles (Manor and Saltz, 2004), suggesting the potential for significant impact on population growth and viability (Gaillard et al., 1998). When some of the dogs were removed from a particularly abundant population near a garbage dump, the kid:female ratio increased significantly. Gingold et al. (2009) found similar results for mountain gazelle responses to guard dogs, where increased vigilance and time spent running came at the expense of time resting and walking, which led to decreased numbers of fawns per female. The burrows of small mammals may become damaged by dogs (Sime, 1999), and even dogs walking over top of burrows may cause a disturbance (Mainini et al., 1993). Finally, Gerrodette and Gilmartin (1990) attribute the recovery of Hawaiian monk seals at Kure Atoll, after the species was listed as endangered in 1976, to US Coast Guard efforts aimed at reducing disturbance by dogs and automobiles.

Free-ranging dogs can act as predators of a variety of mammals (Ritchie et al., Chapter 2), and this may have significant population- and community-level effects. These effects, however, may be most relevant for feral or free-ranging dogs as there is little evidence that leashed or controlled dogs in recreational areas have any effect on species richness or abundance. Forrest and St. Clair (2006) found minimal effects of off-leash dogs on the diversity and abundance of small mammals (and birds) in urban parks. While native carnivore species richness was lower in public areas where dogs were permitted, compared to non-public protected sites (especially for coyotes and bobcats), overall carnivore species richness and abundance was unaffected (Reed and Merenlender, 2011). Therefore, while controlled dogs may influence the activity and movement patterns of wild mammals, there is currently little evidence to suggest they have broad community-level effects.

4.4.3 Other vertebrates

Virtually nothing is known about the impacts of disturbance by dogs on other vertebrates. However, Section 4.3.4 clearly documents a range of responses to dogs, which are likely to have consequences at least at the individual level. While not specific to dogs, human disturbance can decrease the habitat occupancy of amphibians (Rodríguez-Prieto and Fernández-Juricic, 2005).

4.5 Managing dog disturbance

The high usage of natural areas by dog walkers, their high numbers and mobility, and their high potential to cause disturbance means that in some areas they may represent a high management priority for mitigating disturbance to wildlife (Le Corre et al., 2009; Underhill-Day and Liley, 2007). This section focuses on owned dogs. Managing disturbance by dogs will involve either constraining their occurrence, or altering the way they are perceived by wildlife by reducing threatening aspects of dog behavior or by mitigating the deleterious effects of wildlife responses.

4.5.1 Constraining the occurrence of dogs

'Off limit' areas, or restrictions on seasons or periods of the day when dogs are permitted, are commonplace (e.g., banning dogs from islands with monk seal colonies; Gilmartin, 1983), but available data indicate that compliance is rather low (see Box 4.2). Variants of spatial restrictions include buffers (separation distances between natural values and incompatible uses) but recreationists, including dog walkers, are allowed in some 'buffers' (Weston et al., 2009). In particular, natural parks and reserves in many parts of the world prohibit owned dogs, though exceptions occur in some coastal parks (e.g., in Australia). Such restrictions can work (Lafferty et al., 2006). Key to the success of restriction is achieving adequate compliance, which can be promoted through the provision of 'dog-areas' that allow off-leash exercise for dogs and educational initiatives (Williams et al., 2009).

4.5.2 Altering the stimulus

Leashing reduces the speed, degree of roaming, and chasing by dogs and generally decreases response rates and distances among wildlife (Bloor, 2005; Hudson, 1982; Lafferty, 2001b; Weston and Elgar, 2007). For example, unleashed dogs disturbed seven times more red grouse (*Lagopus lagopus*) than leashed dogs (Hudson, 1982). Leashing not only alters problematic aspects of dog behavior, but could also underpin habituation on the part of the wildlife. On one Californian beach, once a protection area for birds became established, leashed dogs no longer caused any bird disturbance, while all disturbance in the protected area from dogs was from unleashed dogs (Lafferty et al., 2006). However, leashing does not prevent barking or other potentially disturbing cues (Randler, 2006).

The designation of leashing laws has not proven to be a major barrier to effective coexistence between dog walking and wildlife, and in many countries 'leash-only' zonation is common in public lands. Rather, the main barrier appears to be the low compliance rates with these regulations (see Box 4.2). Low compliance with leashing laws may explain the failure of some 'gradient' studies to correlate avian or mammal species diversity with prevailing dog regulations (Forrest and St. Clair, 2006). The requirement for human social change with respect to complying with leash laws is undeniable, though better information and signage is often required to more clearly demarcate different dog zones. Managers may be reluctant to tackle the unpopular and apparently insurmountable problem of low compliance with leashing laws, but over the medium to long term, compliance may be improved. Even where compliance with leashing regulations is low, sustained efforts by managers can increase leashing rates (Dowling and Weston, 1999). The presence of a strong social norm among dog walkers suggests that if leashing becomes frequent enough, and therefore expected, many dog owners would leash their dogs on beaches (Williams et al., 2009). Many codes of conduct are available, such as advice to take particular care with dogs around seals or beach-nesting birds (e.g., New Zealand Department of Conservation, 2007).

4.5.3 Mitigating deleterious responses

Responses potentially compromise energy balances, reduce survival, or compromise parental care. Theoretically, if responses cannot be prevented, then management that mitigates the processes that lead to deleterious effects can reduce the impact of disturbance. Examples may include the use of nest cages or shelters for flightless young, which provide thermal insulation and protection of unattended young from predators including dogs (Maguire et al., 2011b).

4.6 Research needs

Compared with other sources of disturbance to wildlife, such as walkers and aircraft, relatively few studies consider disturbance by dogs (e.g., only 2.4% of 211 articles on disturbance to waterfowl mention dogs; Dahlgren and Korschgen, 1992). This chapter has demonstrated a series of critical information gaps regarding disturbance of wildlife by dogs.

First, in terms of dogs as a stimulus, there is a poor understanding of dog occurrence in space and time, in relation to the occurrence of wildlife. An important aspect of space use by dogs is their roaming in natural habitats. How far owned dogs stray from their owners and which types of habitat are penetrated and to what extent, represent research questions that, if addressed, would map the extent of any problem and offer management solutions. While relatively inexpensive, commercially available GPS loggers exist for dogs, these do not appear to have been used to study dog movements. Virtually nothing is known about disturbance by un-owned dogs or unmonitored owned dogs. Additionally, few studies have attempted to separate the visual, auditory, and olfactory cues associated with dogs, and the extent to which these disturb wildlife. Barking, in particular, may be detectable at greater distances than sight or smell of dogs and therefore warrants investigation.

Second, two aspects of wildlife response to stimuli represent research priorities:

1. With few exceptions, only behavioral studies of the response of wildlife to dogs are available, so physiological responses remain largely unknown (but see, for example, Berger et al., 2007; MacArthur et al., 1982). The available evidence suggests they may occur in the absence of behavioral responses, and so be subtle and underestimated. Additionally, physiological responses may occur at greater distances than behavioral responses (i.e., may be precursors to behavioral responses) and so may occur more frequently. They may also be costly, at the individual level manifesting potentially themselves as poorer condition or lower 'health' (e.g., disease resistance), and at the population level potentially manifesting themselves as lower average survival or longevity.
2. The consequences of responses for individuals have rarely been investigated, in general or specifically for dogs, yet these will underpin population responses to disturbance. Scaling up, perhaps the most critical information gap is the link between wildlife population viability and disturbance by dogs, in particular the specific role of dogs in systems where dog disturbance is one of a variety of forms of disturbance. Tolerable disturbance thresholds for populations, if they exist, remain unknown. The influence of disturbance on population viability is likely to be highly context-specific, for both sites and species. Beyond populations, further investigation of the evolutionary costs and benefits of disturbance responses in relation to predator environments might aid predator and species management programs.

Finally, very few instances exist of successful management of disturbance by dogs (possibly some remain undocumented), and this hampers management. A critical element of this will involve social research. The lack of uptake of adaptive dog–wildlife management is regrettable, because this could engage behavioral ecologists with managers to tackle, and hopefully help resolve, the controversial issue of dogs as agents of disturbance.

Acknowledgments

We thank Grainne Maguire (BirdLife Australia's beach-nesting bird project) and two anonymous reviewers. Australian data were collected under RP97208, RP96113, RP95067, NP978/095, NP945/156, NP67/011, and AEEC6540720921120. Funding involved an Australian Postgraduate Research Award, the Holsworth Wildlife Research Fund, Australian Bird Environment Foundation, Stuart Leslie Bird Research Fund, Barwon Heads Committee of Management, and the M.A. Ingram Trust.

References

Allen, S.G., Ainley, D.G., Page, G.W., and Ribic, C. (1984). The effect of disturbance on harbor seal haul out patterns at Bolinas Lagoon, California. *Fishery Bulletin*, **82**, 493–500.

Antos, M.J., Ehmke, G.C., Tzaros, C.L., and Weston, M.A. (2007). Unauthorised human use of an urban coastal wetland sanctuary: current and future patterns. *Landscape and Urban Planning*, **80**, 173–83.

Arnberger, A. and Eder, R. (2008). Assessing user interactions on shared recreational trails by long-term video monitoring. *Managing Leisure*, **13**, 36–51.

Arnberger, A., Brandenburg, C., and Eder, R. (2005). Dog walkers as management challenge in urban forests. In P. Neuhöferová, Ed., *Management of Urban Forests Around Large Cities*. Czech University of Agriculture, Prague, pp. 3–5.

Australian Government (2012). No let up in the hunt for rabbits. *Macquarie Island Pest Eradication Project Newsletter*, **11**, 1–2.

Baines, D. and Richardson, M. (2007). An experimental assessment of the potential effects of human disturbance on black grouse *Tetrao tetrix* in the North Pennines, England. *Ibis*, **149**, 56–64.

Banks, P.B. and Bryant, J.V. (2007). Four-legged friend or foe? Dog walking displaces native birds from natural areas. *Biology Letters*, **22**, 611–3.

Barnett, B.D. and Rudd, R.L. (1983). Feral dogs of the Galápagos Islands: impact and control. *International Journal for the Study of Animal Problems*, **4**, 44–58.

Beausoleil, N.J., Stafford, K.J., and Mellor, D.J. (2005). Sheep show more aversion to a dog than to a human in an arena test. *Applied Animal Behaviour Science*, **91**, 219–32.

Bekoff, M. and Meaney, C.A. (1997). Interactions among dogs, people, and the environment in Boulder, Colorado: A case study. *Anthrozoos*, **10**, 23–31.

Berger, S., Wikelski, M., Romero, L.M., Kalko, E.K.V., and Rödl, T. (2007). Behavioral and physiological adjustments to new predators in an endemic island species, the Galápagos marine iguana. *Hormones and Behavior*, **52**, 653–63.

Bergerud, A.T. (1974). The role of the environment in the aggregation, movement and disturbance behaviour of caribou. In V. Geist and F. Walther, Eds., *The Behavior of Ungulates and its Relation to Management*. International Union for Conservation of Nature and Natural Resources (IUCN), Morges, Switzerland, pp. 552–584.

Bloor, M. (2005). *Mackay Shorebird Conservation Project Action Plan*. WWF-Australia, Sydney.

Brickner, I. (2000). The impact of domestic dogs (*Canis familiaris*) on wildlife welfare and conservation: a literature review. With a situation summary from Israel. MS thesis, Tel Aviv University.

British Columbia Government (2004). *Best Management Practices for Amphibians and Reptiles in Urban and Rural Environments in British Columbia*. Ministry of Water, Land and Air Protection, Nanaimo, BC.

Burger, J. (1981). The effect of human activity on birds at a coastal bay. *Biological Conservation*, **21**, 231–41.

Burger, J., Carlucci, D., Jeitner, S.A., and Niles, C.W. (2007). Habitat choice, disturbance, and management of foraging shorebirds and gulls at a migratory stopover. *Journal of Coastal Research*, **23**, 1159–66.

Butler, J.R.A., du Toit, J.T., and Bingham, J. (2004). Free-ranging domestic dogs (*Canis familiaris*) as predators and prey in rural Zimbabwe: threats of competition and disease to large wild carnivores. *Biological Conservation*, **115**, 369–78.

Clowater, J.S. (2008). Disturbance to Birds on Esquimalt Lagoon Migratory Bird Sanctuary. Consultancy Report.

Colwell, M.A. and Sundeen, K.D. (2000). Shorebird distributions on ocean beaches of northern California. *Journal of Field Ornithology*, **71**, 1–15.

Coombes, E.J., Jones, A.P., and Sutherland, W.J. (2008). The biodiversity implications of changes in coastal tourism due to climate change. *Environmental Conservation*, **35**, 319–30.

Cooper, W.E. and Frederick, W.G. (2007). Optimal flight initiation distance. *Journal of Theoretical Biology*, **244**, 59–67.

Cooper, W.E. and Frederick, W.G. (2010). Predator lethality, optimal escape behavior, and autotomy. *Behavioral Ecology*, **21**, 91–6.

Cooper, W. and Wilson, D. (2007). Beyond optimal escape theory: Microhabitats as well as predation risk affect escape and refuge use by the phrynosomatid lizard *Sceloporus virgatu*. *Behaviour*, **144**, 1235–54.

Cooper, C.A., Neff, A.J., Poon, D.P., and Smith, G.R. (2008). Behavioral responses of eastern gray squirrels in suburban habitats differing in human activity levels. *Northeastern Naturalist*, **15**, 619–25.

Coppinger, R., Lorenz, J., and Coppinger, L. (1987). New uses of livestock guarding dogs to reduce agriculture/wildlife conflicts. *Third Eastern Wildlife Damage Control Conference*. Paper 12.

Dahlgren, R.B. and Korschgen, C.E. (1992). *Human Disturbances of Waterfowl: An Annotated Bibliography*. U.S. Fish and Wildlife Service Resource Publication 188, Northern Prairie Wildlife Research Center, Jamestown, North Dakota. <http://www.npwrc.usgs.gov/resource/literatr/disturb/index.htm [Version 16JUL1997]>.

Doncaster, C.P. (1994). Factors regulating local variations in abundance: Field tests on hedgehogs, *Erinaceus europaeus*. *Oikos*, **69**, 182–92.

Dowling, B. and Weston, M.A. (1999). Managing the hooded plover in a high-use recreational environment. *Bird Conservation International*, **9**, 255–70.

Dunbrack, T.R.E. and Dunbrack, R.L. (2010). Why take your dog on a picnic: presence of a potential predator (*Canis lupus familiaris*) reverses the outcome of food competition between northwestern crows (*Corvus caurinus*) and glaucous-winged gulls (*Larus glaucescens*). *Northwestern Naturalist*, **91**, 94–8.

Fahy, K.A. and Woodhouse, C.D. (1995). *Snowy Plover Linear Restriction Monitoring Project*. Vandenberg Air Force Base, Project No. 0S005097.

Faillace, C.A. (2010). Breeding snowy plovers (*Charadrius alexandrinus*) exhibit variable response to human disturbance on two islands in southwest Florida. MS. thesis, Rutgers University.

Fernández-Juricic, E. and Tellería, J.L. (2000). Effects of human disturbance on spatial and temporal feeding patterns of blackbird *Turdus merula* in urban parks in Madrid, Spain. *Bird Study*, **47**, 13–21.

Fitzpatrick, S. and Bouchez, B. (1998). Effects of recreational disturbance on the foraging behaviour of waders on a rocky beach. *Bird Study*, **45**, 157–71.

Forrest, A. and St. Clair, C. (2006). Effects of dog leash laws and habitat type on avian and small mammal communities in urban parks. *Urban Ecosystems*, **9**, 51–66.

Froneman, A. and van Rooyan, M. (2003). The successful implementation of a border collie bird scaring program at Durban International Airport, South Africa. International Bird Strike Committee IBSC26/WP-AE1.

Gabrielsen, G.W. and Smith, E.N. (1995). Physiological responses of wildlife to disturbance. In R.L. Knight and K.J. Gutzwiller, Eds., *Wildlife and Recreationists. Coexistence Through Management and Research*. Island Press, Washington D.C., pp. 95–106.

Gaillard, J.-M., Festa-Bianchet, M., and Yoccoz, N.G. (1998). Population dynamics of large herbivores: Variable recruitment with constant adult survival. *Trends in Ecology and Evolution*, **13**, 58–63.

George, S.L. and Crooks, K.R. (2006). Recreation and large mammal activity in an urban nature reserve. *Biological Conservation*, **133**, 107–17.

Gerrodette, T.I.M., and Gilmartin, W.G. (1990). Demographic consequences of changed pupping and hauling sites of the Hawaiian monk seal. *Conservation Biology*, **4**, 423–30.

Gill, J.A., Sutherland, W.J., and Watkinson, A.R. (1996). A method to quantify the effects of human disturbance on animal populations. *Journal of Applied Ecology*, **33**, 786–92.

Gilmartin, W.G. (1983). Recovery plan for the Hawaiian monk seal, *Monachus schauinslandi*. National Oceanic and Atmospheric Administration, USA.

Gingold, G., Yom-Tov, Y., Kronfeld-Schor, N., and Geffen, E. (2009). Effect of guard dogs on the behavior and reproduction of gazelles in cattle enclosures on the Golan Heights. *Animal Conservation*, **12**, 155–62.

Glover, H.K., Weston, M.A., Maguire, G.S., Miller, K.K., and Christie, B.A. (2011). Towards ecologically meaningful and socially acceptable buffers: Response distances of shorebirds in Victoria, Australia, to human disturbance. *Landscape and Urban Planning*, **103**, 326–34.

Gutzwiller, K.J. (1990). Minimizing dog-induced biases in game bird research. *Wildlife Society Bulletin*, **18**, 351–6.

Hamerstrom, F., Berger, D.D., and Hamerstrom, F.N. (1965). The effect of mammals on prairie chickens on booming grounds. *Journal Wildlife Management*, **29**, 536–42.

Hamr, J. (1988). Disturbance behavior of chamois in an alpine tourist area of Austria. *Mountain Research and Development*, **8**, 65–73.

Heaton, J.S., Cablk, M.E., Nussear, K.E., Esque, T.C., Medica, P.A., Sagebiel, J.C., and Francis, S.S. (2008). Comparison of effects of humans versus wildlife-detector dogs. *Southwestern Naturalist*, **53**, 472–9.

Hockin, D., Ounsted, M., Gorman, M., Hill, D., Keller, V., and Barker, M.A. (1992). Examination of the effects of disturbance on birds with reference to its importance in ecological assessments. *Journal of Environmental Management*, **36**, 253–86.

Holderness-Roddam, B. (2011). The effects of domestic dogs (*Canis familiaris*) as a disturbance agent on the natural environment. MEM thesis, University of Tasmania.

Hone, E. (1934). The present status of the muskox in Arctic North America and Greenland. *Special Publications of the American Committee for International Wildlife Protection*, **5**, 1–87.

Hudson, P.J. (1982). Red grouse production and management in relation to tourism. *Moorland Wildlife Conservation Amenity Recreation Research Group*, **8**, 45–54.

Kenyon, K.W. (1972). Man versus the monk seal. *Journal of Mammalogy*, **53**, 687–96.

Kirby, J.S., Clee, C., and Seager, V. (1993). Impact and extent of recreational disturbance to wader roosts on the

Dee estuary: some preliminary results. *Wader Study Group Bulletin*, **68**, 53–58.

Kloppers, E.L., St. Clair, C.C., and Hurd, T.E. (2005). Predator-resembling aversive conditioning for managing habituated wildlife. *Ecology and Society*, **10**(1), 31.

Knight, R.L. and Gutzwiller, K.J. (1995). *Wildlife and Recreationists: Coexistence Through Management and Research*. Island Press, Washington, DC.

Koenig, J., Shine, R., and Shea, G. (2001). The ecology of an Australian reptile icon: how do blue-tongued lizards (*Tiliqua scincoides*) survive in suburbia? *Wildlife Research*, **28**, 215–27.

Koenig, J., Shine, R., and Shea, G. (2002). The dangers of life in the city: patterns of activity, injury and mortality in suburban lizards (*Tiliqua scincoides*). *Journal of Herpetology*, **36**, 62–68.

Kramer, D. (1986). The effects of recreational activity on wintering wildfowl populations at Priory Park Lake, Bedford. *Bedfordshire Naturalist*, **41**, 21–6.

Lafferty, K.D. (2001a). Birds at a southern California beach: Seasonality, habitat use and disturbance by human activity. *Biodiversity and Conservation*, **10**, 1949–62.

Lafferty, K.D. (2001b). Disturbance to wintering western snowy plovers. *Biological Conservation*, **101**, 315–25.

Lafferty, K.D., Goodman, D., and Sandoval, C.P. (2006). Restoration of breeding by snowy plovers following protection from disturbance. *Biodiversity and Conservation*, **15**, 2217–30.

Le Corre, N., Gélinaud, G., and Brigand, L. (2009). Bird disturbance on conservation sites in Brittany (France): The standpoint of geographers. *Journal Coastal Conservation*, **13**, 109–18.

Lenth, B.E., Knight, R.L., and Brennan, M.E. (2008). The effects of dogs on wildlife communities. *Natural Areas Journal*, **28**, 218–27.

Levy, A. (2011). Adders kill two pet dogs: Vets blame warmer weather for surge in deadly snake attacks. *The Daily Mail* (UK). July 6, 2011. <http://www.dailymail.co.uk/news/article-2011568/Two-dogs-killed-poisonous-adder-brought-hot-weather.html>.

Lord, A., Waas, J.R., Innes, J., and Whittingham, M.J. (2001). Effects of human approaches to nests of northern New Zealand dotterels. *Biological Conservation*, **98**, 233–40.

MacArthur, R.A., Geist, V., and Johnston, R.H. (1982). Cardiac and behavioral responses of mountain sheep to human disturbance. *Journal of Wildlife Management*, **46**, 351–8.

Maguire, G.S., Miller, K.K., Weston, M.A., and Young, K.A. (2011a). Being beside the seaside: Beach use and preferences among coastal residents of south-eastern Australia. *Ocean & Coastal Management*, **54**, 781–88.

Maguire, G.S., Duivenvoorden, A.K., Weston, M.A., and Adams, R. (2011b). Provision of artificial shelter on beaches is associated with improved shorebird fledging success. *Bird Conservation International*, **21**, 172–85.

Mainini, B., Neuhaus, P., and Ingold, P. (1993). Behaviour of marmots *Marmota marmota* under the influence of different hiking activities. *Biological Conservation*, **64**, 161–4.

Mallord, J.W., Dolman, P.M., Brown, A.F., and Sutherland, W.J. (2007). Linking recreational disturbance to population size in a ground-nesting passerine. *Journal of Applied Ecology*, **44**, 185–95.

Manor, R., and Slatz, D. (2003). Impact of human nuisance disturbance on vigilance and group size of a social ungulate. *Ecological Applications*, **13**, 1830–34.

Manor, R. and Saltz, D. (2004). The impact of free-roaming dogs on gazelle kid/female ratio in a fragmented area. *Biological Conservation*, **119**, 231–6.

Marks, C.A. and Bloomfield, T.E. (2006). Home-range size and selection of natal den and diurnal shelter sites by urban red foxes (*Vulpes vulpes*) in Melbourne. *Wildlife Research*, **33**, 339–347.

Martinetto, K. and Cugnasse, J.M. (2001). Reaction distance in Mediterranean mouflon (*Ovis gmelini musimon* x *Ovis* sp.) in the presence of hikers with a dog on the Caroux Plateau (Hérault, France). *Revue d'Ecologie-La Terre et la Vie*, **56**, 231–42.

Meek, P.D. (1999). The movement, roaming behaviour and home range of free-roaming domestic dogs, *Canis lupus familiaris*, in coastal New South Wales. *Wildlife Research*, **26**, 847–55.

Miller, S.G., Knight, R.L., and Miller, C.K. (2001). Wildlife responses to pedestrians and dogs. *Wildlife Society Bulletin*, **29**, 124–32.

Mirtschin, P.J., Masci, P., Paton, D.C., and Kuchel, T. (2008). Snake bites recorded by veterinary practices in Australia. *Australian Veterinary Journal*, **76**, 195–98.

Murison, G., Bullock, J.M., Underhill-Day, J., Langston, R., Brown, A.F., and Sutherland, W.J. (2007). Habitat type determines the effects of disturbance on the breeding productivity of the Dartford warbler *Sylvia undata*. *Ibis*, **149 (Suppl. 1)**, 16–26.

Newell, G.R. (1999). Responses of Lumholtz's tree-kangaroo (*Dendrolagus lumholtzi*) to loss of habitat within a tropical rainforest fragment. *Biological Conservation*, **91**, 181–9.

New Zealand Department of Conservation (2007). *The Seal Deal: Caring for Kekeno Together*. Department of Conservation, Wellington, New Zealand.

Pelletier, F. (2006). Effects of tourist activities on ungulate behavior in a mountain protected area. *Journal of Mountain Ecology*, **8**, 15–19.

Preisser, E.L., Bolnick, D.I., and Benard, M.F. (2005). Scared to death? The effects of intimidation and consumption in predator-prey interactions. *Ecology*, **86**, 501–9.

Randler, C. (2006). Disturbances by dog barking increase vigilance in coots *Fulica atra*. *European Journal of Wildlife Research*, **52**, 265–70.

Reed, S.E., and Merenlender, A.M. (2011). Effects of management of domestic dogs and recreation on carnivores in protected areas in Northern California. *Conservation Biology*, **25**, 504–13.

Robinson, J.A., and Pollitt, M.S. (2002). Sources and extent of human disturbance to waterbirds in the UK: An analysis of Wetland Bird Survey data, 1995/96 to 1998/99. *Bird Study*, **49**, 205–11.

Rödl, T., Berger, S., Romero, L.M., and Wikelski, M. (2007). Tameness and stress physiology in a predator-naive island species confronted with novel predation threat. *Proceedings of the Royal Society B*, **274**, 577–82.

Rodríguez-Prieto, I., and Fernández-Juricic, E. (2005). Effects of direct human disturbance on the endemic Iberian frog *Rana iberica* at individual and population levels. *Biological Conservation*, **123**, 1–9.

Sastre, P., Ponce, C., Palacín, C., Martín, C.A., and Alonso, J.C. (2009). Disturbances to great bustards (*Otis tarda*) in central Spain: Human activities, bird responses and management implications. *European Journal of Wildlife Research*, **55**, 425–32.

Schlaepfer, M.A., Runge, M.C., and Sherman, P.W. (2002). Ecological and evolutionary traps. *Trends in Ecology and Evolution*, **17**, 474–80.

Scott, F.E. (1989). Human disturbance of wading birds on the Ythan estuary. Unpublished BSc Thesis, Department of Zoology, University of Aberdeen.

Serpell, J. (Ed.). (1995). *The Domestic Dog: Its Evolution, Behaviour, and Interactions with People*. Cambridge University Press, Cambridge.

Sime, C.A. (1999). Domestic dogs in wildlife habitats. In G. Joslin and H. Youmans, Eds., *Effects of recreation on Rocky Mountain wildlife: A review for Montana*. Committee on Effects of Recreation on Wildlife, Montana Chapter of the Wildlife Society, Helena, Montana, pp. 8.1–8.17.

Stankowich, T. (2008). Ungulate flight responses to human disturbance: A review and meta-analysis. *Biological Conservation*, **141**, 2159–73.

Stankowich, T. and Blumstein, D.T. (2005). Fear in animals: a meta-analysis and review of risk assessment. *Proceedings of the Royal Society B*, **272**, 2627–34.

Stankowich, T. and Coss, R.G. (2006). Effects of predator behavior and proximity on risk assessment by Columbian black-tailed deer. *Behavioral Ecology*, **17**, 246–54.

Sweeney, J.R., Marchinton, R.L., and Sweeney, J.M. (1971). Responses of radio-monitored white-tailed deer chased by hunting dogs. *Journal of Wildlife Management*, **35**, 707–16.

Taborsky, M. (1988). Kiwis and dog predation: Observations in Waitangi State Forest. *Notornis*, **35**, 197–202.

Taylor, E.C., Green, R.E., and Perrins, J. (2007). Stonecurlews *Burhinus oedicnemus* and recreational disturbance: Developing a management tool for access. *Ibis*, **149 (Suppl.1)**, 37–44.

Underhill-Day, J.C. and Liley, D. (2007). Visitor patterns on southern heaths: A review of visitor access patterns to heathlands in the UK and the relevance to Annex I bird species. *Ibis*, **149 (Suppl. 1)**, 112–9.

Vanak, A.T. and Gompper, M.E. (2010). Interference competition at the landscape level: the effect of free-ranging dogs on a native mesocarnivore. *Journal of Applied Ecology*, **47**, 1225–32.

Vanak, A.T., Thaker, M., and Gompper, M.E. (2009). Experimental examination of behavioural interactions between free-ranging wild and domestic canids. *Behavioural Ecology and Sociobiology*, **64**, 279–87.

van Bommel, L. (2010) *Guardian Dogs: Best Practice Manual for the use of Livestock Guardian Dogs*. Invasive Animals CRC, Canberra.

VerCauteren, K., Seward, N., Hirchert, D., Jones, M., and Beckerman, S. (2005). Dogs for reducing wildlife damage to organic crops: a case study. *Proceedings of the Wildlife Damage Management Conference*, **11**, 286–93.

Vercauteren, K.C., Lavelle, M.J., and Phillips, G.E. (2008). Livestock protection dogs for deterring deer from cattle and feed. *Journal of Wildlife Management*, **72**, 1443–48.

Walter, W.D., Lavelle, M.J., Fischer, J.W., Johnson, T.L., Hygnstrom, S.E., and VerCauteren, K.C. (2010). Management of damage by elk (*Cervus elaphus*) in North America: A review. *Wildlife Research*, **37**, 630–46.

Weston, M.A. and Elgar, M.A. (2005). Disturbance to brood-rearing hooded plover *Thinornis rubricollis*: Responses and consequences. *Bird Conservation International*, **15**, 193–209.

Weston, M.A. and Elgar, M.A. (2007). Responses of incubating hooded plovers (*Thinornis rubricollis*) to disturbance. *Journal of Coastal Research*, **23**, 569–76.

Weston, M., Antos, M.J., and Glover, H.K (2009). Birds, buffers and bicycles: A review and case study of wetland buffers, *Victorian Naturalist*, **126**, 79–86.

Weston, M.A. McLeod, E.M. Blumstein, D.T., and Guay, P.-J. (2012). A review of flight initiation distances and their application to managing disturbance to Australian birds. *Emu*, **112**, 269–86.

Williams, K.J., Weston, M.A., Henry, S., and Maguire, G.S. (2009). Birds and beaches, dogs and leashes: Dog owner's sense of obligation to leash dogs on beaches in Victoria, Australia. *Human Dimensions of Wildlife*, **14**, 89–101.

Ydenberg, R.C. and Dill, L.M. (1986). The economics of fleeing from predators. *Advances in the Study of Behavior*, **16**, 229–49.

Young, B.A. (1983). Snake bioacoustics: Toward a richer understanding of the behavioral ecology of snakes. *Quarterly Review of Biology*, **78**, 303–25.

Zanette, L.Y., White, A.F., Allen, M.C., and Clinchy, M. (2011). Perceived predation risk reduces the number of offspring songbirds produce per year. *Science*, **334**, 1398–1401.

CHAPTER 5

Dog eat dog, cat eat dog: social-ecological dimensions of dog predation by wild carnivores

James R.A. Butler, John D.C. Linnell, Damian Morrant, Vidya Athreya, Nicolas Lescureux, and Adam McKeown

5.1 Introduction

Due to their close relationship with humans, and their ability to adapt to a wide range of social-ecological systems, dogs (*Canis familiaris*) are probably the most numerous carnivores in the world today. As the global human population continues to grow and expand, dogs are more abundant and widely distributed than ever before. This is particularly evident in developing nations, where the majority of the world's human population exists and growth rates are highest (Gompper, Chapter 1). Consequently, although dogs have long been a part of the ecology of many landscapes, they are becoming an even more influential agent of anthropogenic impact on biodiversity, interacting with native wildlife and hence potentially modifying ecosystems to an unprecedented degree (Hansen et al., 2005; Young et al., 2011).

One rarely studied form of interaction is predation on dogs by wild carnivores. As a preliminary step for this chapter, we undertook a review of the scientific literature using Google, Google Scholar, and ISI Web of Knowledge to assess the range of carnivore species recorded as being responsible for killing dogs around the world. We also examined predation studies of potential dog-killing species to find records of dogs being killed or consumed. Most of this literature was in English, but we also accessed some publications from the Spanish and Russian literature. In total 13 species have been documented to kill dogs in 83 studies (Table 5.1, Figure 5.1). The gray wolf (*C. lupus*) occurred most frequently (28 records), largely in Europe but also in North America and Asia, followed by the leopard (*Panthera pardus*; 18 records) in Africa and Asia. Other felids documented were pumas (*Puma concolor*) in North and South America, jaguars (*Panthera onca*) in South America, Amur tigers (*P. tigris altaica*) in Asia (Siberia), and lions (*P. leo*) in Africa. Other canids were coyotes (*C. latrans*) in North America, dingoes (*C. f. dingo*) in Australia, and black-backed jackals (*C. mesomelas*) in Africa. Spotted hyenas (*Crocuta crocuta*) also accounted for a relatively large number of records in Africa. Striped hyenas (*Hyaena hyaena*) regularly scavenge dog carcasses in their African and Asian range and were assumed to prey on dogs by three studies. There was one published record of polar bears (*Ursus maritimus*) killing dogs in Arctic Canada, and another of Asiatic black bears (*U. thibetanus*) killing young dogs in Nepal.

In addition, there were records of other, non-carnivorous species killing dogs. In Africa, adult male baboons (*Papio ursinus*) can kill dogs when villagers use them to repel baboon troops raiding their crops and livestock (Butler et al., 2004), and in Australia southern cassowaries (*Casuarius casuarius*) can

Free-Ranging Dogs and Wildlife Conservation. Edited by Matthew E. Gompper
© Oxford University Press 2014. Published 2014 by Oxford University Press.

Table 5.1 Details of the 83 records of wild carnivore species documented in published scientific literature as having preyed upon dogs in different regions and countries of the world. Species are listed in descending order of total records.

Species	Region	Country	Record
Gray wolf	Asia	India	Jethva and Jhala 2004
		Mongolia	Hovens and Tungaslaktuja 2005
		Tajikistan	Bibikov 1988
	Europe	Albania and Macedonia	Keci et al. 2008
		Belarus	Sidorovich et al. 2003
		Estonia and Latvia	Valdmann et al. 2005; Zunna et al. 2009
		Finland	Kojola and Kuittinen 2002; Kojola et al. 2004
		Italy	Boitani 1982; Ciucci et al. 1996
		Poland	Nowak et al. 2005, 2011; Gula 2008
		Portugal	Vos 2000
		Russia	Bibikov 1988; Pozio et al. 2001
		Spain	Salvador and Abad 1987; Cuesta et al. 1991; Llaneza et al. 2000; Cortés 2001; Barja 2009
		Ukraine	Bibikov 1988
	North America	USA	Fritts and Paul 1989; Bangs et al. 2004, 2005; Ruid et al. 2009; Edge et al. 2011
Leopard	Africa	Côte d'Ivoire	Bodendorfer et al. 2006
		Ethiopia	Yirga et al. 2011; 2012
		Kenya	Kock et al. 1998; Kolowski and Holekamp 2006
		Namibia	Dabe 1997
		Tanzania	Kissui 2008
		Zimbabwe	Butler et al. 2004
	Asia	Bhutan	Wang and Macdonald 2009
		India	Johnsingh 1983; Edgaonkar and Chellam 1998; Ramakrishnan et al. 1999; Mukherjee and Mishra 2001; Daniels 2009; Shah et al. 2009; Mondal et al. 2011
		Iran	Sanei et al. 2011
		Nepal	Bhattarai and Kindlmann 2012
		Pakistan	Dar et al. 2009
Puma	North America	USA	Robinette et al. 1959; Aune 1991; Davies 1991; Mansfield 1991; Russ 1995; Sanders and Halfpenny 1991; Torres et al. 1996; Leberg et al. 2004
	South America	Brazil	Mazzolli 2009
		Venezuela	Farrell et al. 2000
Coyote	North America	Canada	Alexander and Quinn 2011
		USA	Howell 1982; Timm et al. 2004; Farrar 2007; Timm and Baker 2007; White and Gehrt 2009
Spotted hyena	Africa	Ethiopia	Atickem et al. 2010; Yirga et al. 2012
		Kenya	Kolowski and Holekamp 2006
		Tanzania	Holmern et al. 2007; Kissui 2008
		Zimbabwe	Butler, du Toit, and Bingham 2004

continued

Table 5.1 *Continued*

Species	Region	Country	Record
Tiger	Asia	Russia	Miquelle et al. 1996; Goodrich and Miquelle 2005; Goodrich et al. 2011; Tkachenko 2012
Lion	Africa	Tanzania	Kissui 2008
		Zimbabwe	Butler et al. 2004
Dingo	Australasia	Australia	Burger and Knowles 1976; Woodall et al. 1996
Striped hyena	Asia	India	Gajera et al. 2009
		Iran	Monchot and Mashkour 2010
Jaguar	South America	Belize	Foster et al. 2010
Black-backed jackal	Africa	Tanzania	Holmern et al. 2007
Polar bear	North America	Canada	Dyck 2006
Asiatic black bear	Asia	Nepal	Stubblefield and Shrestha 2007

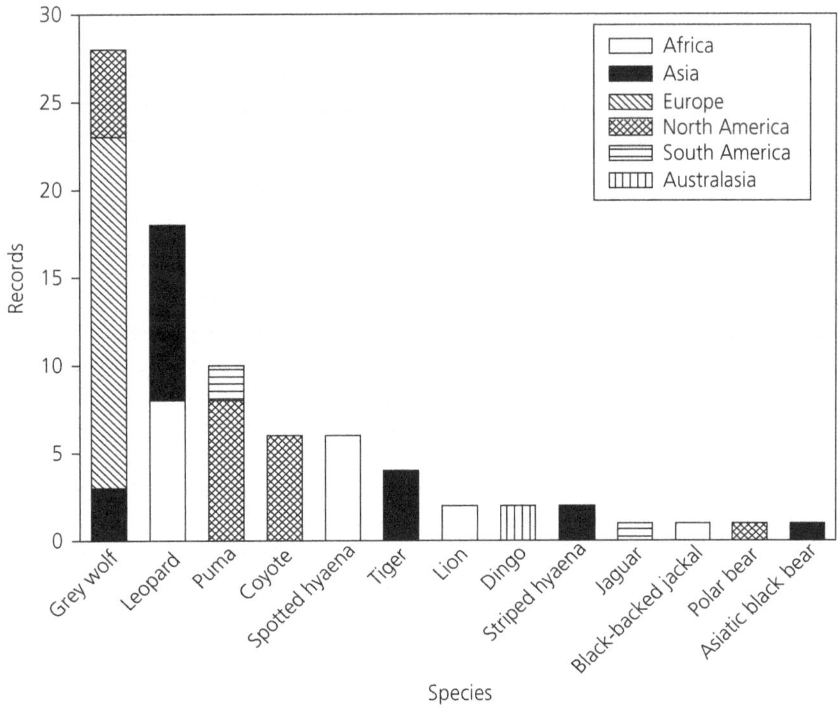

Figure 5.1 Summary of the 83 records of wild carnivore species documented in published scientific literature as having preyed upon dogs in different regions of the world. See Table 5.1 for details.

kill dogs (Kofron, 1999). Also, reptilian carnivores attack and eat dogs, for example saltwater (*Crocodylus porosus*) and freshwater crocodiles (*C. johnstoni*) in Australia (Mawson, 2004).

To augment this review we used email to survey our own contacts, the authors of recent papers or reports on potential dog-killing species, and various electronic mailing lists for records of dogs being killed by predators.

This survey elicited replies from 55 researchers and wildlife managers in Europe, North America, South America, Australia, and Asia. In addition, we used online search engines to review e-newspapers. Responses showed that brown or grizzly bears (*U. arctos*), black bears (*U. americanus*), Eurasian lynx (*Lynx lynx*), and golden eagles (*Aquilla chrysaetos*) have also been occasionally documented killing dogs in Europe and North America. Similarly, there are media reports of amethystine pythons (*Morelia amethistina*) and wedge-tailed eagles (*A. audax*) killing and consuming dogs in Australia, and Burmese pythons (*Python molurus bivittatus*) in Asia. In fact, predation or killing of dogs by a broader range of species is evident from numerous researchers' experiences, management agency records, and media reports, but is rarely mentioned in scientific papers.

These anecdotes imply that records in the scientific literature are unlikely to be an accurate representation of the extent of predation on dogs by the listed carnivores. Numbers of records will be a function of the geographic range of the carnivore concerned, contact rates related to dog and carnivore population densities within the species' overlapping ranges, variable investment in research among countries, and scientists' publication rates and interests. Furthermore, most studies mentioned an isolated incident anecdotally, while only a minority specifically investigated predatory interactions.

However, the literature review and survey results do illustrate the diversity of carnivores that dogs may interact with, which raises questions about the differing characteristics of these ecological relationships. Considering that dogs and native carnivores are potentially members of the same trophic guild, predatory interactions may be considered as intraguild predation, defined as the killing (and sometimes eating) of potential competitors because both species utilize the same prey resources and also benefit nutritionally from preying upon one another (Linnell and Strand, 2000; Palomares and Caro, 1999; Polis and Myers, 1989). Within this broad definition there is a continuum from asymmetrical predation, in which one species kills and eats the other, to symmetrical predation, in which both species may kill each other. There also appears to be a separate, non-predatory dimension where dogs are killed by sympatric species in self-defense (e.g., baboons and cassowaries).

The results also suggest an anthropogenic perspective. Because dog–human relationships vary so greatly between and within social contexts (Serpell, 1995), the loss of dogs to wild carnivore predation may have differing emotional and economic impacts on dog owners and their livelihoods. These impacts may result in diverse responses by humans to dog predation and their perceptions of the carnivores responsible. Unlike other interactions between dogs and wildlife, such as competition for food resources (Vanak et al., Chapter 3) and pathogen transmission (Knobel et al., Chapter 6), predation on dogs may be a more direct driver of conflict between humans and wild carnivores, mediated by complex social and institutional factors.

This chapter explores the nature of dog predation from the published scientific literature collated above, augmented by our own field experience, unpublished data, and wildlife researchers' and managers' anecdotal information. These data illustrate the range of potential predator–prey interactions between wild carnivores and dogs, and their influence on human–carnivore conflict. To synthesize these patterns we present a typology of dog predation based on the theory of intraguild competition. By integrating anthropogenic perspectives we then develop a human impact gradient, and present a hypothetical 'heat map' of ensuing human–carnivore interactions for different social-ecological contexts. We conclude that dog predation is a widespread but poorly understood phenomenon, which has varying impacts for both humans and wild carnivores. Furthermore, because the majority of dogs exist in developing nations, interactions with wild carnivores are likely to result in a new and unique set of challenges as these societies continue their rapid socio-economic development.

5.2 Focal examples of wild carnivore predation on dogs

5.2.1 Wolves in Asia, Europe, and North America

Gray wolves are the most widespread wild carnivore species in the world, and across much of their range they occur in areas with a substantial presence of humans and dogs. Wolves occupy a diversity of habitats, exploiting an equally large range of prey, and coexisting with human cultures with varied land use and socio-economic status. It is well known that wolves kill dogs throughout most of their area of overlap, although only a few studies have specifically investigated this phenomenon (e.g., Edge et al., 2011; Fritts and Paul, 1989; Kojola and Kuittinen, 2002; Kojola et al., 2004). In some countries, detailed records are kept of domestic animals killed by wolves as part of conflict management protocols, often associated with the existence of compensation programs. Such data are available from several states in the USA (Minnesota, Michigan, Wisconsin, Idaho, Montana, Wyoming) and countries in Europe (Norway, Sweden, Finland, Estonia, Latvia, Poland, Croatia).

There is great variation in the average number of dogs killed each year, both within and between sites. Typical rates are <20 dogs/yr killed within each of these states or countries, with the exception of Croatia where >90 dogs/yr are reported killed. Furthermore, there are a range of anecdotal accounts of wolf–human relationships from across the wolf's range that also document wolf predation. These include Albania, Macedonia, Bulgaria, Slovakia, Germany, Spain, Italy, Mongolia, and Kyrgyzstan, although in some of these areas the extent of killing may be very low. Dog remains have also been recorded in wolf diet studies from Estonia, Latvia, Lithuania, Belarus, Russia, Poland, Italy, Spain, Portugal, Romania, Ukraine, Tajikistan and India (see Table 5.1 and multiple personal communications).

The overall indication is that dog-killing is widespread, occurring wherever wolves and dogs are sympatric. However, the extent of the killing is usually infrequent and irregular. There is little evidence that dogs constitute a major part of wolf diet, although this could often be due to the tendency of researchers to focus on remote study sites with relatively low human (and dog) densities. In studies where dogs do occur in wolf diet, they tend to be represented in <5% of scats or stomachs. In only a handful of cases can dogs be described as a major source of wolf nutrition. Most of these are in areas or periods where natural prey occurs at very low density, such as parts of western Russia, Spain, and Croatia (Bibikov, 1988; Cuesta et al., 1991; Pozio et al., 2001).

There have been several attempts to explain variation in dog-killing by wolves. Three factors are commonly cited. First, there is a broad positive correlation between dog-killing and increasing wolf population size, for example in cases of recovering or reintroduced wolf populations (Ruid et al., 2009). A second correlation is for dog killing to be associated with areas (Cuesta et al., 1991; Kojola and Kuittinen, 2002; Pozio et al., 2001) and periods (Pozio et al., 2001; Sidorovich et al., 2003) of low prey density, implying that dogs can be targeted if natural prey is extremely rare. A third trend is for specific packs to become habitual dog-killers (Kojola et al., 2004). These patterns are also clearly modulated by the local availability of dogs, which depends on how the local human population use and keep them. Throughout Europe, Siberia, and North America dogs are commonly used by recreational hunters. In most cases hunting dogs, either singly or in groups, are released to drive or locate game. Many of the attacks are on free-ranging hunting dogs during the process of training or hunting. Of all lethal attacks on dogs, the percentage involving hunting dogs varies from 30% in Finland (Kojola and Kuittinen, 2002), 47% in Belarus (Sidorovich et al., 2003), 59% in Michigan (Edge et al., 2011), 80% in Sweden (Swedish Wildlife Damage Center unpublished data), to 87% in Wisconsin (Ruid et al., 2009). Most of the remaining cases were dogs killed in yards, and a few while guarding livestock.

This implies that dogs are killed in three contexts. The first is when dogs such as hunting dogs are running free in wolf habitat. This requires no active effort of the wolf to find the dog and the attack may even be provoked by the dog seeking the wolves. The second is where dogs are killed in villages or yards, often when chained to a building, suggesting that the wolf actively sought out the dog and killed it without provocation. The third is when livestock-guarding or herding dogs are killed during a wolf attack on livestock. An additional context concerns wolves killing feral dogs. Although it is widely believed that wolves control feral dog numbers in Spain, possibly through predation (Blanco et al., 1992), there have been no formal studies to confirm this. The extent to which dogs are then consumed also varies enormously, but occurs in around half the cases.

Hence dog-killing by wolves is a complex ecological phenomenon, involving aspects of predation, defense, or dominance (Karlsson and Jaxgård, 2004). The relative extent to which one of these mechanisms is responsible is likely to vary over space and time, and the mechanisms are not mutually exclusive. Studying wolf predation on dogs is also challenging because it is so poorly reported and is such a rare event when viewed from either the perspective of an individual dog at risk or from an individual wolf. For example, in Scandinavia the Scandinavian Wolf Project (SKANDULV) has been conducting intensive telemetry-based studies of wolf predation since the late 1990s. This project has found more than 800 wolf-killed prey items, but dogs are not among them. However, during the same period the region's wildlife management system has documented 293 dogs killed by wolves (Swedish Wildlife Damage Center; Norwegian Directorate for Nature Management, unpublished data).

Although the number of dogs killed may be statistically insignificant relative to other livestock predation, it can have a dramatic impact on conservation discourses. In some cultures humans and dogs have strong social and emotional links, and dogs are treated as family or team members (Haraway, 2003; Sanders, 1993, 2003; Serpell, 1995). Good hunting and livestock guarding dogs are valuable and cannot be replaced quickly (Lescureux and Linnell, 2010). The loss of such an animal to a wolf triggers strong emotional responses of grief. Finally, the fact that wolves often enter villages and farmyards to take dogs close to houses may induce fear because of the threat that they also pose to human life. All of these mechanisms increase animosity towards wolves and weaken community and political support for their conservation (Bisi et al., 2007; Sjölander-Lindqvist, 2010; Skogen and Krange, 2003; Skogen et al., 2006).

5.2.2 Leopards, lions, spotted hyenas, and jackals in Africa

Most records of dog predation in Africa are incidental within studies focusing on livestock predation by large carnivores (e.g., Atickem et al., 2010; Holmern et al., 2007; Kissui, 2008; Kolowski and Holekamp, 2006; Yirga et al., 2011, 2012). In rural regions where traditional agro-pastoralism co-exists with large carnivores, lions, leopards, and spotted hyenas are the main predators of livestock and dogs. Holmern et al. (2007) recorded an instance of a black-backed jackal killing a dog in Tanzania, and similar records exist in Zimbabwe (where dogs also kill jackals: Butler, 1998; Vanak et al., Chapter 3), but we are not aware of records of side-striped (*Canis adustus*) or golden jackals (*C. aureus*) killing dogs. African wild dogs (*Lycaon pictus*) and cheetahs (*Acinonyx jubatus*) are also predators of livestock but, again, we are unaware of records of either killing dogs. Due to their similar omnivorous ecology brown hyenas (*Hyaena brunnea*) may compete with dogs where their ranges overlap (Vanak et al., Chapter 3), but there are no records of predatory interactions between them.

Throughout African agro-pastoralist systems, dogs are kept to deter wildlife from raiding livestock and crops (Atickem et al., 2010; Kissui, 2008; Kolowski and Holekamp, 2006) and also for livestock herding, hunting, and protecting homes (Butler and Bingham, 2000; Knobel et al., 2008). Homesteads are usually open or only partially fenced, allowing dogs to roam and breed freely, while also providing little protection from predators (Atickem et al., 2010; Butler and Bingham, 2000; Kolowski and Holekamp, 2006). The relationship between dogs and humans observed in most rural areas of sub-Saharan Africa has remained largely unchanged since dog immigration with the Bantu 2,000–4,000 years ago (Gallant, 2002). Modern day dog densities have been recorded as ranging between 6 and 21 per km^2 in Kenya (Kitala et al., 1993) and 8 and 53 per km^2 in rural Zimbabwe, and are likely to be increasing rapidly (Butler and Bingham, 2000).

Only one detailed analysis of the predator–prey relationships between wild carnivores and dogs has been undertaken, in a 33 km^2 section of Gokwe Communal Land (GCL) bordering the Sengwa Wildlife Research Area (SWRA), Zimbabwe (Butler and du Toit, 2002; Butler et al., 2004). The study area contained 130 households, 937 people, and 236 dogs, plus 537 cattle (*Bos indicus*), 819 goats (*Capra hircus*), 157 donkeys (*Equus africanus asinus*), and small numbers of sheep (*Ovis aries*) and pigs (*Sus scrofa*). Cattle were the most valuable livestock (US$100 per head in 1996 values), followed by donkeys ($40), pigs ($25), sheep ($15), and goats ($10) (Butler, 2000). By comparison, dogs were valued at $10. Leopards, lions, and spotted hyenas traveled up to 3 km into GCL at night, while dogs were sighted up to 6 km within the SWRA, and these species were therefore effectively sympatric.

Household surveys revealed that 23 dogs were preyed upon during the study, representing approximately 10% of the population. By comparison, 5% of livestock holdings were taken by wild carnivores (Butler, 2000). For the 19 cases of dog predation where the predator could be identified, leopards were responsible for the most (53%), followed by lions (42%) and spotted hyenas (5%). Eighteen (78%) were killed away from the homestead, and five within the homestead perimeter. Leopards were responsible for taking all dogs killed within homesteads. All incidents took place at night, and all dogs were wholly or partially consumed. There was a seasonality to dog predation, with the monthly rate of kills for all three predators being at least twice as high in the dry season as that for the wet season. In total, the rate of dry season kills (0.75 per month) was almost four times greater than for the wet season (0.19 per month). These patterns were also reflected in livestock predation, with 80% of losses occurring in dry season months (Butler, 2000). Radio-tracking of dogs indicated that they were particularly vulnerable to predation due to their solitary scavenging away from homesteads and human protection (Box 5.1).

Drought potentially escalated predatory interactions between wild carnivores and dogs. The failed wet season of 1994–95 was followed by a sharp increase in dog sightings within the SWRA, partially due to increased illegal hunting within the reserve by GCL inhabitants related to food shortages. Dogs also entered the SWRA independently, perhaps due to a dwindling availability of waste human food in GCL (Butler and du Toit, 2002). This is corroborated by household surveys in seven communal lands that recorded greater incidences of stray or feral dogs during droughts when food became scarce (Butler, 1998). Leopards and lions rely upon vegetative cover to hunt wild prey successfully (Schaller, 1972; van Orsdol, 1984), and therefore may resort to killing livestock more regularly in drier conditions, as also indicated by the higher incidence of livestock predation in dry season months. Hence dogs may become more regular incidental prey during drought years, both within communal lands and adjoining protected areas. However, due to the low economic value of dogs, losses to predators had an insignificant economic effect on GCL households relative to the impact of cattle, goat, and donkey predation.

5.2.3 Leopards in Asia

Leopards are known for being highly adaptable in terms of their habitat and diet requirements, and this is demonstrated clearly in India where they inhabit intact tropical forests, farmlands, and suburban environments. The tendency of leopards to kill and consume dogs has been widely noted in the Indian colonial era hunting literature, with many records of leopards taking dogs from close to, and even inside, human habitation (Daniels, 2009). However, there has been little formal study of dog predation. Most wildlife research in India occurs in the relatively intact ecosystems of protected areas where dogs are not normally present. In such areas it is not surprising that dogs do not register frequently in leopard diet, and predation on dogs by leopards is often only documented in anecdotal footnotes (Johnsingh, 1983), or via occasional occurrence in scats (<5% frequency; Ramakrishnan et al., 1999). In intact ecosystems, leopards show a clear preference for small- to medium-sized ungulates (Hayward et al., 2006). However, the few studies that have occurred in landscapes where natural ungulate prey are rare or absent have frequently found that dogs can be the major prey item. For example, dogs occurred in 25% of scats in Kashmir (Shah et al., 2009), 64% of scats in Sanjay Gandhi National Park (Edgaonkar and Chellam, 1998), and 39% of scats in rural Maharashtra (V. Athreya, unpublished data). No other prey species was more common than dog in these three studies. Farmers and migratory shepherds in Maharashtra have also identified leopards as major predators of dogs (V. Athreya, unpublished data).

Box 5.1 The ranging behavior and vulnerability of dogs to predation in rural Zimbabwe

Radio-tracking of 14 adult dogs from households adjacent to the SWRA enabled an analysis of dog behavior, and hence their potential vulnerability to predation. Dogs had a mean independent home range area (i.e., where they moved freely without their owners) of 97.2 ha (range 0.3–316.0), and males had larger home ranges (mean 145.0 ha) than females (mean 33.5 ha). Seven of these home ranges included areas up to 1 km inside the SWRA. The outer limits of home ranges were often determined by the locations of wild or domestic mammal carcasses which the dogs scavenged, and for males the locations of females in estrus (Figure 5.2). Dogs spent the majority of their time with their owners (76%) and in the immediate vicinity of human habitation (63%). However, they scavenged in their home ranges independently of people and usually alone: 87 and 84% of dogs recorded at experimental carcasses arrived and fed alone, respectively (Figure 5.3).

Of the radio-tracked dogs, one was killed and eaten by a lion within its home range, at night, 400 m from its owner's homestead (Figure 5.2), and another disappeared possibly due to predation. These detailed data suggest that dogs were largely taken opportunistically as prey when ranging independently of people away from their homestead, usually at night. However, leopards appeared to target dogs within homestead perimeters. In contrast to Atickem et al. (2010), who recorded dogs being killed while defending homesteads and livestock against spotted hyena attacks in Ethiopia, none of the cases in GCL involved similar agonistic interactions. The

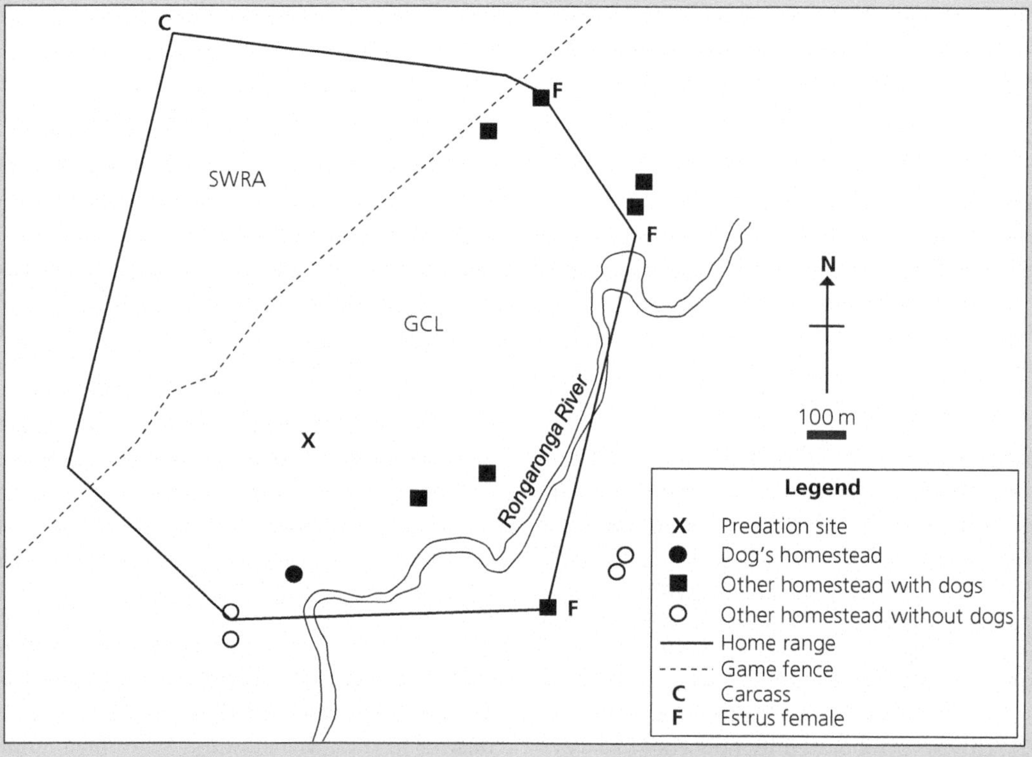

Figure 5.2 Convex polygon independent home range (316 ha) for a 16 kg adult male dog in the Gokwe Communal Land (GCL), relative to the neighboring Sengwa Wildlife Research Area (SWRA), derived from 94 fixes over 188 consecutive days. The dog was killed and consumed by a lion on the night of June 8, 1996 in a field 400 m from the dog's homestead.

continued

Box 5.1 *Continued*

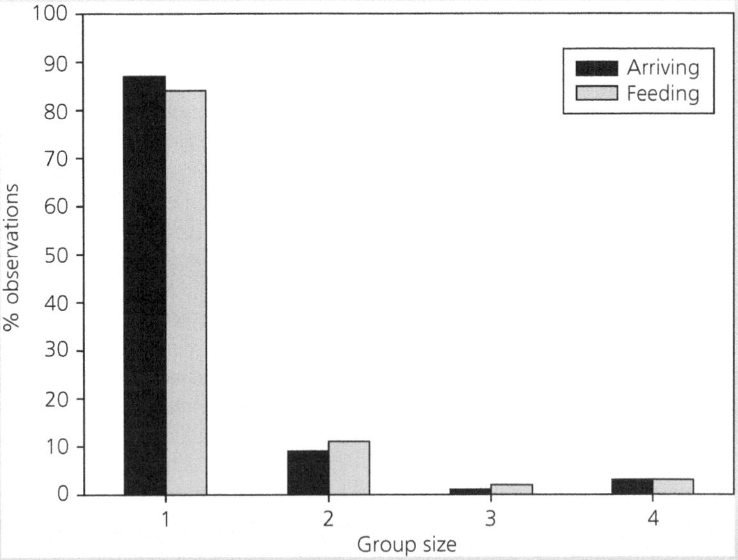

Figure 5.3 Group sizes of dogs arriving and feeding at 12 experimental carcasses in the Gokwe Communal Land and Sengwa Wildlife Research Area, determined from 229 separate observations of dog meals.

tendency of dogs to forage and feed alone may render them particularly vulnerable to predation. This is most likely to occur at carcasses, which also attracted leopards, lions, and spotted hyenas on the GCL–SWRA boundary at night. Human activity is known to be at its lowest between 18:00 and 04:00, providing little interference to nocturnal wild carnivores or protection to dogs during this period (Butler and du Toit, 2002).

The implication is that in the presence of preferred natural prey, dog-killing remains a widespread but uncommon activity, but that in the absence of wild prey, leopards can sustain themselves on a diet of dogs, supplemented with livestock. In India, dogs weigh approximately 15 kg, are usually owned but unrestricted and therefore largely unprotected, and are found throughout the landscape at high densities (30–50 per km^2; V. Athreya, unpublished data; Punjabi et al., 2012). The deliberate targeting of dogs by leopards is illustrated by the frequent cases of leopards pursuing owned dogs into houses (V. Athreya, unpublished data). In some areas, feral dogs are also targeted by leopards. Similar situations probably exist elsewhere in the region, as indicated by the presence of dog remains in leopard diet in Bhutan (Wang and Macdonald, 2009), Nepal (Bhattarai and Kindlmann, 2012), Iran (Sanei et al., 2011), and Pakistan (Dar et al., 2009).

5.2.4 Pumas and jaguars in the Americas

There are only a few published records in the scientific literature of pumas killing dogs, and dogs are rarely reported in scats or among kills found (Aune, 1991; Davies, 1991; Dettmann, 1991; Leberg et al., 2004; Mansfield, 1991; Russ, 1995). However, our survey of online sources, including newspapers

Box 5.2 The challenges of documenting rare events: the case of the puma in the Americas

Reviewing the extent of large carnivore predation on dogs is a challenging task for a number of reasons. First, although our review indicates that it is a widespread phenomenon, it does not appear to be frequent. This implies that within any project researching wild carnivore diet it is highly unlikely it will document many, if any, kills of dogs or scats containing dog remains. Second, the study of pet predation falls between an ecologically relevant focus on predation of wild prey, and the economically relevant issues related to predation on livestock. This leads to a reporting bias, such that even if data exist they do not find their way into the published literature.

Our search for records of predation on dogs by the puma illustrates this challenge. Initial searches using Google Scholar and ISI Web of Knowledge, and keywords associated with 'puma,' 'dog,' 'pet,' and 'predation' led us to only a few articles that either mentioned in passing that pets are occasionally killed by pumas (Torres et al., 1996), record the occurrence of dog hair in a single puma scat (Leberg et al., 2004 (in Louisiana, USA); Farrell et al., 2000 (in Venezuela)), or report anecdotally a single event (Mazzolli, 2009 (in Brazil)). We then searched manually through all relevant publications in our library collections, including data-based scientific articles, technical reports, books, and conference proceedings. We searched for records of dogs as prey species, which led to one record (Robinette et al., 1959), and a few anecdotes in papers dealing with human–wildlife conflict (Aune, 1991; Dettman 1991; Davies, 1991; Mansfield, 1991; Russ, 1995; Sanders and Halfpenny, 1991), with almost all coming from a single conference proceedings. We then sent emails to researchers and wildlife managers who had published on pumas in either scientific articles or conference proceedings during recent years. This elicited many replies that reported records of pumas killing dogs in eight US states and two Central American countries. The information was based on results from research projects, interviews with local people, and public records concerning wildlife damage and depredation reports. As a final line of enquiry we used Google to search English language newspapers and newsfeeds in the USA and Canada. This revealed 38 individual cases of dogs being attacked and killed in 10 US states and 2 Canadian provinces between 2000 and 2012 (Table 5.2).

Table 5.2 The regional coverage of records of pumas killing dogs in North and South America as reflected by sourcing different types of information: scientific papers data-based; manual search of articles and proceedings; email survey of researchers and managers; Internet search of online newspapers and newsfeeds.

State, province, or country	Scientific papers[1]	Articles and proceedings	Email surveys[2]	Online media[3]
United States and Canada				
Alberta	0	0	0	2
Arizona	0	0	0	1
British Columbia	0	0	0	6
California	0	2	1	8
Colorado	0	3	0	6
Florida	0	0	1	0
Idaho	0	0	0	3
Louisiana	1	0	0	0
Montana	0	1	0	2
Nevada	0	0	1	0
New Mexico	0	0	0	1
Oregon	0	0	1	2

continued

Box 5.2 *Continued*

Table 5.2 *Continued*

State, province, or country	Scientific papers[1]	Articles and proceedings	Email surveys[2]	Online media[3]
South Dakota	0	0	1	0
Texas	0	1	0	1
Utah	0	0	1	2
Washington	0	0	1	4
Wyoming	0	0	1	0
Latin America				
Brazil	1	0	0	
Guatemala	0	0	1	
Mexico	0	0	1	
Venezuela	1	0	0	
TOTAL	3	7	10	38

[1] Number of articles.
[2] Number of respondents giving a positive reply.
[3] Number of unique cases mentioned by diverse media. Only North American media were searched.

Hence the peer-reviewed scientific literature only gives limited insight into the extent and frequency of pumas killing dogs, especially when searches are confined to the use of keywords in database searches. Secondary information from publications, the gray literature, media comment, and local experts reveal a different picture. Pumas appear to kill dogs at low frequency in most places where pumas occur (Table 5.2). While there is nothing to indicate that dogs are of dietary importance for pumas anywhere, the occasional killing of a pet has potentially large repercussions for public tolerance of puma presence, an issue that emerges clearly in the media coverage.

This example has fundamental implications for the way that we review information related to wild carnivore conservation. As scientists we like to insist on the use of peer-reviewed material because it is easy to access and has been quality-controlled. However, as shown above, this insistence can lead to a bias and underestimation of the importance of certain issues. While the recent focus on evidence-based conservation is laudable, we must not ignore other forms of knowledge, including unpublished data and media reports. Furthermore, the mutual recognition and integration of local knowledge and expert opinion is particularly important because conflict over large wild carnivores must be resolved through multi-stakeholder co-management processes (Butler, 2011; Young et al., 2012).

and news reports from wildlife management agencies, indicated that predation on dogs is widespread, but occurs at low frequencies (Box 5.2). We found reports of pumas killing dogs in 15 US states and 2 Canadian provinces during recent years, with California, Colorado, and British Columbia contributing the most records. The dogs were often killed in close proximity to houses, with pumas frequently taking dogs from the yard or veranda, and in one case even entering the house in pursuit of the dogs. The houses tended to be on the suburban–forest interface. Dogs varied in size from Miniature Poodles to German Shepherds, and many were consumed. In a few cases the pumas were apparently injured and emaciated, although this did not seem to be a general pattern (Box 5.2). Pumas also kill and occasionally eat coyotes, implying that dog killing is likely to have both nutritional and intraguild competitive motivations. In South America there is only one published report of a puma killing a dog (Brazil; Mazzolli, 2009), and one other of dog hair occurring in a scat (Venezuela; Farrell et al., 2000).

Similarly, there is limited published scientific information on dog predation by jaguars in South America. In Belize, jaguars preyed on dogs and other livestock on the periphery of a protected area where wild prey was scarce, while sympatric pumas avoided these areas and were not recorded preying on dogs, perhaps due to competitive interactions with jaguars (Foster et al., 2010). Our online search and survey among researchers for anecdotal information also revealed occasional incidents of jaguar predation on dogs in many parts of Central and South America.

5.2.5 Coyotes in North America

The last few decades have seen a shift in North American research on conflict between coyotes and humans. From early concern about the role of coyotes as predators of livestock, there is an emerging focus on direct conflict between coyotes and humans. These interactions occur frequently in suburban areas, especially at the suburban–wildland interface. In many parts of North America, coyotes have shown an ability to occupy these suburban habitats and exploit human food sources. This includes attacking and killing dogs and cats (*F. catus*), and coyotes are often killed by dogs (Kamler et al., 2003). The fact that coyotes kill and occasionally consume pet dogs has been reported in many studies (Alexander and Quinn, 2011; Farrar, 2007; Lukasik and Alexander, 2011; Timm and Baker, 2007; White and Gehrt, 2009), although little quantitative data has been published because pet attacks are overshadowed by the controversy caused by coyote attacks on humans (Timm and Baker, 2007). The development of problem behavior in suburban coyotes has been hypothesized to follow a habituation gradient, whereby coyotes begin to lose their fear of people, then begin to exploit human foods, and ultimately try to attack pets or people (Schmidt and Timm, 2007). The focus of these studies has been to use pet attacks as an early warning for situations of potential danger to humans. Dog-killing by suburban coyotes therefore appears to only begin once coyotes have become sufficiently habituated to human presence. However, once they have overcome this fear, it seems that dog-killing may be motivated by both predation and territorial defense.

5.2.6 Tigers in Asia

Dietary studies have not identified dogs as prey of tigers in most regions of Asia, largely because tiger research there tends to occur in protected areas with low human densities. However, Amur tigers in Siberia exist in multi-use landscapes and dog-killing and their consumption as prey is well documented, both as remains in scats and among recorded kills (Miquelle et al., 1996; Tkachenko, 2012). The main data source is reports of tiger–human conflicts recorded by researchers and wildlife management authorities. Dogs are by far the most commonly killed domestic animal. For example, in one study they constituted 63% of the 254 documented domestic animal kills (Goodrich et al., 2011). The dogs were often killed in the middle of villages and when chained outside houses, implying that the tigers were deliberately targeting the dogs rather than killing them following chance encounters. However, there were also cases of predation on free-ranging hunting dogs in the forest. Because these dogs are often highly valued, their death triggers considerable negative feelings and retaliatory killing by their owners (Goodrich et al., 2011). Ironically, Tkachenko (2012) found that in some cases the tigers responsible for attacks on dogs were injured and/or emaciated as a direct result of such human persecution. The tiger–dog relationship is especially interesting considering the evidence for competitive exclusion of wolves by tigers. Existing data indicate that Amur tigers have a dramatic effect on wolf distribution (Miquelle et al., 2005); hence it is possible that dogs are also viewed as intraguild competitors and killed.

5.2.7 Dingoes in Australia

In Australia, dingoes and feral 'wild dogs' (dingo x domestic dog hybrids) are a cause of major economic losses for the sheep and cattle industry, resulting in specific legislation in some states requiring landholders to control themon their property using lethal measures, such as trapping with cage or padded-jaw traps, shooting, and/or poisoning with 1080 (sodium monofluoroacetate) or strychnine baits (Fleming et al., 2001). Dogs are kept by Australians primarily as pets, but in rural areas they are highly valued as working dogs, and for hunting feral pigs (ACAC, 2010). Consequently dogs, dingoes, and wild dogs come into regular contact, particularly in rural areas (Allen and Fleming, 2004), but there are surprisingly few scientific records of dog predation. Burger and Knowles (1976) reported dogs being killed by dingoes on Fraser Island, and Woodall et al. (1996) noted dogs being 'lured' away from human habitation by a female dingo in estrus, and then being attacked by the other members of the dingo pack. Box 5.3 highlights one such study in the Wet Tropics bioregion of Queensland that may be representative of the growing conflict between dingoes, domestic dogs, and people in suburban and peri-urban regions of Australia.

Box 5.3 Conflict between dingoes, domestic dogs, and people in the peri-urban Wet Tropics of Australia

The Wet Tropics bioregion of north-east Queensland is characterized by native rainforest covering the coastal escarpment, and much of the region is protected by the Wet Tropics World Heritage Area (WHA). Following European settlement in the late nineteenth century, the coastal floodplains and inland tablelands were cleared for timber and agriculture. Today this landscape is a mosaic of diverse habitat types, including remnant coastal rainforest and mangroves, interspersed with sugar cane, horticultural production, and cattle grazing (Pert et al., 2012). The Wet Tropics has a rapidly-growing human population which has been forecast to increase from 216,000 in 2004 to 300,000 in 2024 (McDonald and Weston, 2004). This growth is driven by immigrants from large southern Australian cities seeking an idyllic tropical lifestyle (Bohnet and Pert, 2010). The resulting suburban expansion is encroaching on surrounding agricultural land and protected areas, which brings owned dogs into increasing contact with dingoes and wild dogs.

Dingoes in the Wet Tropics maintain home ranges of up to 120 km^2, moving between peri-urban areas and WHA forest (D. Morrant, unpublished data). When roaming dingoes move into human-populated areas, their territorial and predatory behaviors bring them into direct conflict with humans. While they primarily prey on small to medium-sized wild animals (50 g–19 kg), they also occasionally attack and kill domestic animals. In 2011, a questionnaire survey of 3000 households across all land use types was undertaken to assess the frequency and characteristics of such attacks on owned dogs. Records of attacks were also obtained from Pest Management Officers from Cairns Regional Council, the most populous council in the Wet Tropics, and AgForce Queensland, the representative body for the state's cattle industry.

Twelve respondents (6%) reported attacks on their dogs by dingoes. An additional nine incidents were reported by the Pest Management Officers, and three during discussions with farmers. Of the 24 incidents, 10 (42%) occurred on suburban fringes and 13 in rural agricultural areas (Figure 5.4). Five of the seven interactions that were witnessed by respondents involved packs of up to six dingoes. Domestic dogs were injured in twenty incidents, and killed in four, but not consumed. Witness descriptions often emphasized the determination of the dingoes to attack the dogs concerned. One cattle grazier fought off three dingoes with a stock whip when they attacked his dog within 50 m of his house. Another farmer reported that a "pack of wild dogs" returned on three consecutive nights to attack her dog. Another respondent claimed that a male domestic dog was "lured from [the] yard by a female in heat," and then mauled by six other dingoes. This owner physically separated the animals to protect his dog.

The responses revealed a strong emotional reaction and a willingness on the part of dog-owners to protect their pets. Owners of attacked dogs were angry and upset, and were also concerned for the safety of local children and other pets, prompting dingo and wild dog population control. Legal

continued

Box 5.3 *Continued*

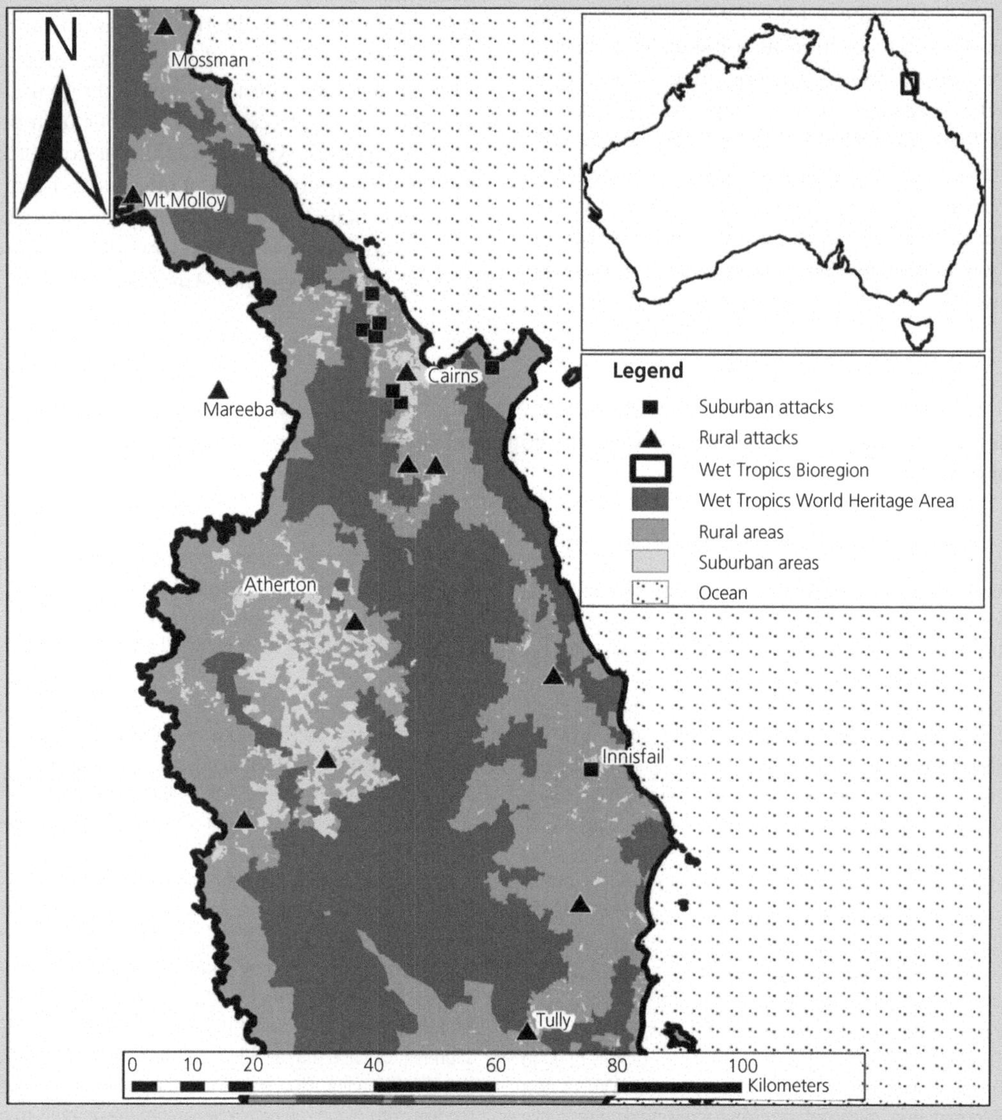

Figure 5.4 The location of 20 dingo attacks on domestic dogs in the Wet Tropics of Queensland, Australia, relative to land use. Four records are not shown because locations were not provided by respondents.

retaliatory control measures followed six of the incidents. Approximately 10% of the landholders surveyed had engaged in lethal control of dingoes and wild dogs in the past 12 months. However, Pest Management Officers admitted that they intentionally avoided killing suspected 'pure' dingoes, in the belief that dingoes with a stable pack structure pose less of a threat to pets and livestock than wild dogs, and exclude wild dogs from their home ranges.

5.2.8 Striped hyenas in Africa and Asia

Striped hyenas occur from the Horn of Africa through the Middle East and into Asia, and throughout their range are noted as being omnivorous and generalist carnivores with a predilection for scavenging, particularly human waste around settlements (Kruuk, 1976; Kuhn, 2005; Monchot and Mashkour, 2010). They also regularly prey upon small livestock such as goats and sheep (Gajera et al., 2009; Leakey et al., 1999). Monchot and Mashkour (2010) have suggested that their close association with humans may even qualify as a commensal relationship. Consequently striped hyenas are likely to come into regular contact with dogs and compete with them, particularly since in many of the regions within their range dogs are likely to be free-ranging and also scavenge human-derived food (Vanak et al., Chapter 3).

In spite of the likelihood of interactions there are relatively few records of hyenas preying upon dogs. There is evidence of hyenas feeding on dogs in India (Gajera et al., 2009), Kenya (Leakey et al., 1999), and Jordan (Kuhn, 2005), and Monchot and Mashkour (2010) state that in Iran sick dogs are likely to be preyed upon, and hyenas interact aggressively with dogs over carcasses. Hence the relationship between dogs and hyenas appears to be driven by direct interference competition for shared food resources, although hyenas may opportunistically prey upon and consume young or sick dogs. However, we could find no evidence of dogs killing striped hyenas.

5.3 Synthesis

5.3.1 Ecological dimensions

The examples of dog predation reviewed above reveal a wide range of ecological relationships between dogs and wild mammalian carnivores along the continuum of intraguild predation. At the asymmetrical extreme dogs are killed and eaten by lions, leopards, tigers, jaguars, and pumas as food. Records suggest that in many cases dogs are specifically targeted as prey and eaten, for example by leopards in India and Zimbabwe and tigers in Siberia. In other cases predation may be more opportunistic, for example with lions and spotted hyenas in Zimbabwe. At the symmetrical extreme dogs are attacked and killed as competitors but rarely eaten, for example by dingoes in Australia. At this extreme, dogs are equally likely to attack but not necessarily consume competitors (see Vanak et al., Chapter 3).

In addition, there are intermediate relationships which involve elements of both. This is most evident for wolves, which exhibit a complex range of behaviors towards dogs, both targeting them as competitors, preying upon them for food, and also killing them when challenged by hunting or livestock guard dogs. Given some evidence that they kill wild canids, predation of dogs by tigers and pumas may also contain elements of competitive exclusion. Spotted hyenas may also compete with dogs in Africa for carrion and human refuse (Butler and du Toit, 2002; Vanak et al., Chapter 3), and hence a similar mix of predatory and competitive elements could exist. Striped hyenas in Africa and Asia may have a similar relationship, but with a greater emphasis on competition for human-derived food.

By considering the adult body mass of the wild carnivores concerned relative to dogs and then mapping these specific relationships along the continuum of intraguild predation, a typology emerges (Figure 5.5). This indicates that predators that have a body mass at least twice that of dogs, such as lions, leopards, pumas, jaguars, and tigers, have a predominantly asymmetrical, predatory relationship with dogs. This is to be expected since dogs fall within the prey range of all of these carnivores, and for leopards in particular, which are known to favor small- to medium-sized mammals of 5–20 kg (e.g., Bodendorfer et al., 2006; Hayward et al., 2006). Those with a smaller or equivalent body mass to dogs may have exclusively symmetrical relationships involving competitive killing, for example among black-backed jackals, coyotes, and dingoes. Given their similar size, dogs are capable of killing these species in agonistic interactions (Vanak et al., Chapter 3). Wolves, spotted and striped hyenas occupy an intermediate position, having a range of predatory and competitive killing interactions with dogs. In the case of wolves and striped hyenas, this is perhaps because their body sizes overlap and also exceed those of dogs.

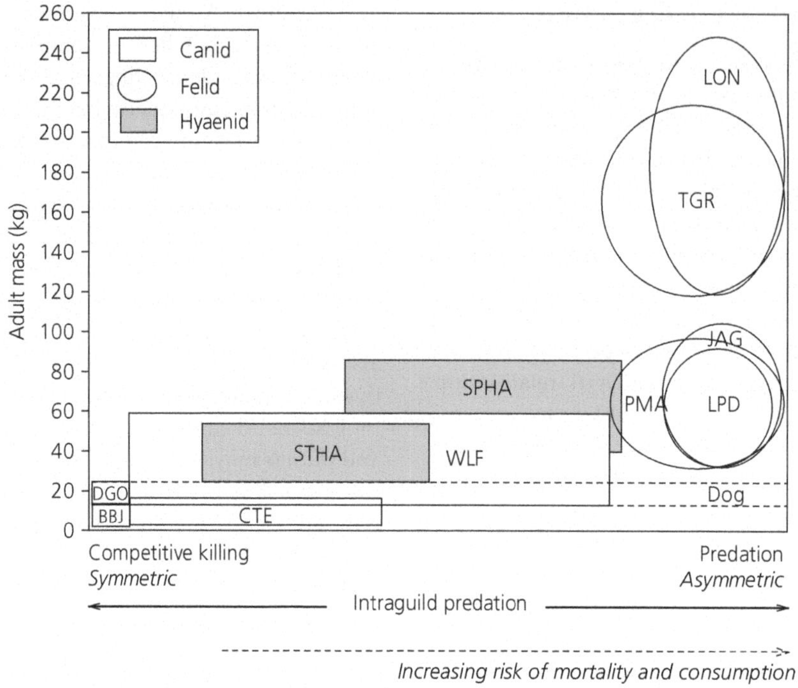

Figure 5.5 A typology of intraguild relationships between wild mammalian carnivores and dogs, assuming an adult dog body mass of 15–25 kg. The wider the symbol along the intraguild predation axis, the broader the range of intraguild relationships possible. The greater the body mass relative to dogs, the more likely the relationship will involve predation, and the risk of mortality for dogs in interactions will increase. Abbreviations (and sources for adult body mass data) are: BBJ black-backed jackal (Loveridge and Nel, 2004); DGO dingo (Letnic et al., 2011); CTE coyote (Kennedy-Stoskopf, 2003); WLF grey wolf (Kennedy-Stoskopf, 2003); SPHA spotted hyena (Ramsay, 2003); TGR tiger (Slaght et al., 2005); PMA puma (Wack, 2003); LPD leopard (Wack, 2003); JAG jaguar (Wack, 2003); LON lion (Wack, 2003); STHA striped hyena (Monchot and Mashkour, 2010).

Polar, brown/grizzly, and black bears in North America are among the largest carnivores in the world and would therefore be expected to have an extreme predatory relationship with dogs. However, reports of dog killing by bears are rare, which is consistent with brown/grizzly and black bears' more omnivorous ecology and the specialization of polar bears as predators of marine mammals, plus their very low levels of sympatry with dogs. Consequently we have omitted these species, plus the Asiatic black bear, from our typology.

This analysis is highly simplified, since all wild carnivores considered have a wide range of body masses that may influence relationships with dogs under different circumstances. These body masses also ignore their immature life stages. Equally, dogs vary in size depending on their breed and age: adult Chihuahuas may only weigh 1–2 kg, while Bull Mastiffs can weigh 50–60 kg. Also, the social behavior and feeding strategy of wild carnivores may vary between and within species, influencing their ability to prey upon or dominate dogs. Furthermore, we have not included reptilian and avian carnivores, which could distort the influence of body mass. In spite of these shortcomings, the typology highlights two fundamental principles. First, carnivores that prey on dogs for food are most likely to be large felids, while those involved in competitive killing are most likely to be canids of a similar size to dogs. Second, a dog's risk of mortality during an intraguild interaction increases with the body mass of the carnivore, because it is more likely to be taken as prey (Figure 5.5).

5.3.2 Social dimensions

Since their domestication from wolves, dogs have become part of human society and culture, providing benefits including transport, companionship, livestock protection and herding, hunting aides, and a source of food. Dogs can be perceived to have crossed the barrier between animality and humanity, but their status varies widely between and within cultures (Serpell, 1995). In Western occidental cultures, dogs are often referred to as 'man's best friend.' They can be given human names, which contribute to their individuality and personality in anthropomorphic terms (Haraway, 2003; Sanders, 2003). In addition, they contribute significantly to human wellbeing (Hart, 1995; Wells, 2007, 2009).

Beyond cultural differences, it appears in several countries that the number of dogs owned is correlated with socio-economic status and livelihood profile. Hence, in Tanzania, wealthier and better educated households tended to own more dogs, and rural households keeping livestock were also most likely to own dogs (Knobel et al., 2008). A similar correlation is evident in more developed countries such as the United Kingdom (Westgarth et al., 2007). Furthermore, an owners' investment in dog husbandry and health may increase with socio-economic status. In Zimbabwe, dog condition score, frequency of rabies vaccination, and rates of neutering were correlated with indicators of rural households' affluence (Butler, 1995, 1998).

Generalizing these dog–human relationships assists in anticipating the impacts that predation on dogs by different wild carnivores could have on humans, and thus the potential for conflict. The Human Development Index (HDI) is commonly used to measure the standard of living amongst different nations, based on indicators of education, health, and per capita income (UNDP, 2011). Assuming that there is a correlation between HDI and the nature of a society's relationship with dogs, a logical parallel may be a notional 'dog development index' (DDI). Dogs with a high DDI typically have clear ownership, high levels of husbandry and selective breeding, and contribute significantly to their owners' wellbeing. These dogs have higher or similar economic value relative to other domestic animals, and their owners have disposable income to invest in their health, resulting in high life expectancy (Table 5.3). Examples of such animals include urban pet dogs or rural hunting, herding, and livestock guarding dogs in nations with a very high HDI, such as Norway, Australia, the United States, and Canada.

By contrast, dogs of low DDI have less clear ownership, limited or no selective breeding, and hence less direct benefit for their owners' wellbeing. They are of generally low economic value relative to other livestock, and consequently receive little investment in their health or husbandry, exacerbated by dog owners' limited financial resources—often resulting in unrestricted movements and breeding and low life expectancy—but retaining a high dependence on human-derived resources. Such dogs are typified by 'village' or 'neighborhood' dogs ubiquitous to many rural and suburban areas of countries with low HDIs such as Zimbabwe, Nepal, and Pakistan (Table 5.3). Feral dogs are by definition un-owned and therefore receive no investment in their husbandry, and may also have negative impacts as pests. These animals have been recorded in varying numbers in many different development contexts, such as rural Italy (Boitani and Ciucci, 1995; Boitani et al., 1995), the United States (Causey and Cude, 1980), Australia (Fleming et al., 2001), and suburban India (Oppenheimer and Oppenheimer, 1975). This emphasizes an important point that in any one nation a range of dog–human relationships may exist along a DDI continuum (Table 5.3), which is determined by local variations in socio-economic status and levels of human development.

Table 5.3 A continuum of dog–human relationships and relevant indicators, termed the Dog Development Index.

Indicator	Dog Development Index		
	High	**Low**	**Feral**
1. Owner's Human Development Index	High	Intermediate	None
2. Clarity of dog ownership	High	Low	None
3. Dog's selective breeding	High	Intermediate	None
4. Dog's contribution to owner's wellbeing	High	Low	None
5. Dog's economic value relative to other livestock	High	Low	None
6. Owner investment in dog's health and husbandry	High	Low	None
7. Dog's life expectancy	High	Intermediate	Low

5.3.3 Human impact

Combining the ecological and social dimensions presented above allows the characterization of the potential impacts of dog predation on human wellbeing and livelihoods (Figure 5.6). This suggests an impact gradient ranging from neutral for feral dogs, to highly negative for dogs with high DDIs. Predation on 'village' dogs with low DDIs is of intermediate impact. However, there is a subtle distinction in impact along the intraguild predation continuum, with an increasing risk of mortality for dogs towards the predation extreme, which will potentially result in a greater human impact due to the death of the dog. Although competitive attacks can also result in lethal outcomes, there is a higher probability of dog survival from these interactions. Hence the greatest human impact is likely to occur where dogs with high DDIs are killed by felids and the largest canids as prey.

From the data reviewed and presented in this chapter it is also possible to identify the location of specific wild carnivores on the impact gradient, and the regions of the world where such interactions are most likely to occur. The zone of highest impact is North America, where pumas and wolves prey on hunting or working dogs in rural areas, or pet dogs in the suburban–wildlands interface. Tigers occupy a similar zone when they kill hunting and other dogs in rural Siberia. Wolves also have a widespread impact across Europe due to their broad range of ecological interactions with dogs, ranging from competitive killing to predation. Dingoes in Australia and coyotes in North America have a marginally less acute impact, often in the interface between suburban and rural or protected areas. By comparison, predation on dogs by lions, leopards, spotted and striped hyenas, and black-backed jackals in Africa is likely to generate lesser human impact, largely due to the low economic value of dogs relative to cattle and other livestock. The same is probable for rural or peri-urban areas of India and other parts of Asia (Figure 5.6).

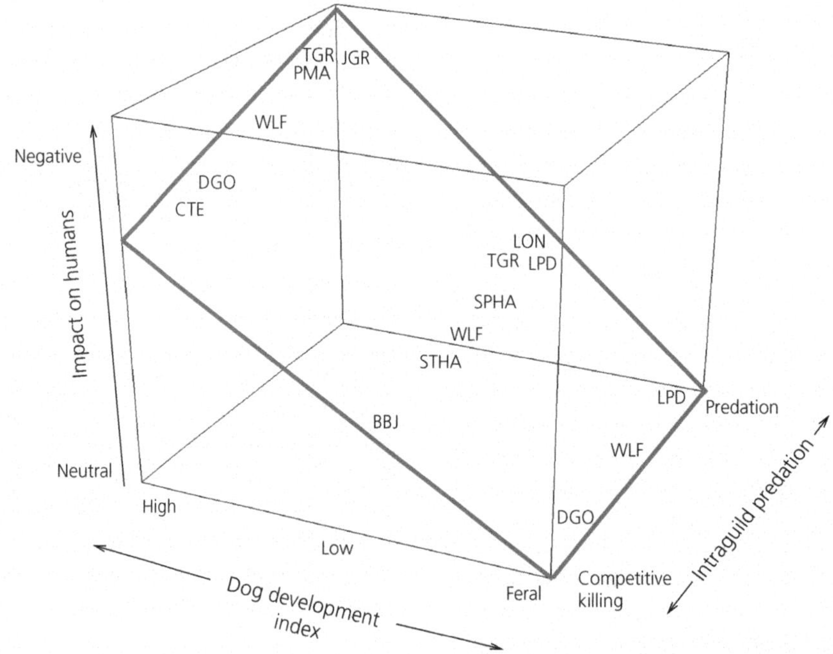

Figure 5.6 A hypothetical human impact gradient resulting from predation on dogs by wild mammalian carnivores in different ecological and social contexts. This combines the dimensions of intraguild predation (see Figure 5.5) and dog–human relationships (see Table 5.3). Note that negative impacts are greatest for predation on high dog development index dogs because the risk of mortality for dogs is higher than in competitive killing. Abbreviations are: BBJ black-backed jackal; DGO dingo; CTE coyote; WLF grey wolf; SPHA spotted hyena; TGR tiger; PMA puma; LPD leopard; JAG jaguar; LON lion; STHA striped hyena.

It is also possible to infer the positions on the impact gradient of these carnivores for other regions where they occur. For example, wolves and tigers in less economically developed regions of Asia (e.g., rural India) are likely to have an intermediate impact due to their predation on village dogs with low DDIs. Also, it is possible that wolf and leopard predation on feral dogs in Europe and Asia, respectively, and the killing or competitive exclusion of feral wild dogs by dingoes in Australia, may have a minor positive impact on humans given the pest status of these animals.

There is a third potentially important variable that may intensify the degree of negative human impact in any given context. Native habitat modification, and related to this the availability of natural prey, appears to drive an increased frequency of dog predation. For example, wolves in Europe are known to kill and consume more dogs when natural food availability is low, as do leopards in India. Related to this is the conflict that occurs along urban interfaces with surrounding forest or modified habitats, for example for pumas and coyotes in North America and dingoes in Australia. Seasonal or climatic factors may also influence the availability of natural prey where dogs and wild carnivores are sympatric. In Zimbabwe, dog predation by leopards, lions, and spotted hyenas escalated in the dry season in parallel with higher levels of livestock predation, perhaps due to a lack of vegetative cover that aided hunting wild prey. Drought also resulted in more extensive ranging by dogs away from homesteads and into the neighboring protected area, increasing their vulnerability to predation by wild carnivores. The availability of wild prey linked to seasonal migrations also influences temporal variations in livestock killing by large carnivores in Kenya (Kolowski and Holekamp, 2006), although this was less evident in Tanzania (Holmern et al., 2007).

5.4 Implications for wild carnivore conservation

5.4.1 Impacts on wild carnivores

To assess the implications of dog predation for wild carnivores and their conservation, it is important to analyze the potential impacts of dog predation on the carnivores themselves. While all mammalian carnivores considered in the intraguild typology (Figure 5.5) must benefit to some extent from the removal of potential competitors, levels of predation on dogs appear to be so low and infrequent that dogs are probably not a primary source of nutrition. However, an exception may be leopards in rural landscapes of India, where dogs clearly form a significant part of their diet, probably due to high dog densities and the relative scarcity of natural prey.

Potential negative impacts can therefore be considered to outweigh these limited benefits. The first route of impact is through direct retaliatory killing by local people. This may occur legally, for example the targeted poisoning of dingoes in Australia under legislation, or illegally through the indiscriminate hunting, snaring, or poisoning of large carnivores, for example the killing of spotted hyenas, leopards, and jackals in the Bale Mountains of Ethiopia (Atickem et al., 2010). Some carnivores may be targeted specifically for killing dogs (e.g., Amur tigers in Siberia; Goodrich et al., 2011), but in many cases retaliation is a reaction to livestock predation more generally (e.g., in Kenya; Kolowski and Holekamp, 2006), which in the case of the Maasai in Tanzania is enshrined within traditional lion-hunting ceremonies (Kissui, 2008). In a more benign form of retaliation, some 'problem animals' may be captured and translocated (e.g., Amur tigers; Goodrich, 2010; Goodrich and Miquelle, 2005), but this is more feasible in economically developed contexts where wildlife management is well resourced. Dog predation may also be such an emotive issue that it becomes a powerful argument against large carnivore conservation, particularly when there is also a threat of attacks on humans, for example in the case of wolf recovery programs in Finland (Bisi et al., 2007).

The second more indirect route is through the transmission of pathogens. Predation or agonistic interactions with dogs offer an ideal route of transmission for canid pathogens, such as rabies and canine distemper viruses, due to the opportunity for close contact or consumption of infective tissue (Butler et al., 2004; Knobel et al., Chapter 6). Dogs have been implicated as the reservoir hosts underpinning a canine distemper epidemic in 1994, which caused widespread mortality of lions in the Serengeti, Africa (Roelke-Parker et al., 1996; Kock et al., 1998), and also underpinning canine distemper and rabies epizootics amongst the highly endangered African wild dog (Cleaveland et al., 2000). Spotted hyenas may play an important role as an intermediate host facilitating the spillover of pathogens from dogs to other wild carnivores in African ecosystems (Butler et al., 2004; Harrison et al., 2004). Similarly, Amur tigers are known to have been exposed to canine distemper virus, probably due to close contact with dogs (Goodrich et al., 2012; Quigley et al., 2010). However, the risk of mortality for carnivores preying upon dogs will vary according to the susceptibility of the species, and the health status of the dog population concerned. In general it can be assumed from our typology of dog–human relationships (Table 5.3) that dogs with a high DDI are likely to have higher levels of investment in their health and vaccination than 'village' dogs in contexts with lower HDIs, and feral dogs will have none. Hence, on a per capita basis, dogs with a low DDI and feral dogs pose the greatest risk of pathogen transmission to wild carnivores, but this will be determined by dog densities and hence contact rates with carnivores.

5.4.2 Potential human–carnivore conflict

By combining the human impact gradient (Figure 5.6) with these impacts on carnivores, it is possible to develop a 'heat map' of potential conflict and the nature of that conflict (Figure 5.7). Based on the evidence for species reviewed here, this suggests that there may be three broad types of conflict (Table 5.4). The first and most intense is 'Pet Predation,' where dogs of high DDI are taken as prey or through intraguild killing, and human retaliation results in the killing or translocation of problem animals and increased opposition to large carnivore conservation. However, the risk of pathogen transmission is relatively low given the generally healthy status of

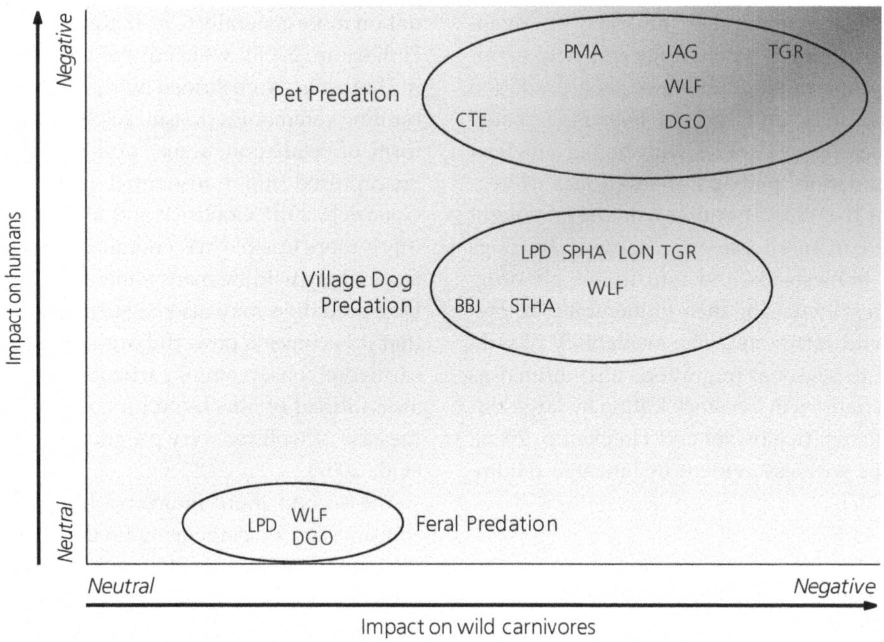

Figure 5.7 A 'heat map' of potential human–wild carnivore conflict as a result of predation on dogs by wild mammalian carnivores. The darker shading indicates greater potential for conflict. Based on the data reviewed, three types are identified: Pet Predation, Village Dog Predation, and Feral Predation. Abbreviations are: BBJ black-backed jackal; DGO dingo; CTE coyote; WLF grey wolf; SPHA spotted hyena; TGR tiger; PMA puma; LPD leopard; JAG jaguar; LON lion; STHA striped hyena.

Table 5.4 Features of the three types of human–wild carnivore conflict generated by predation on dogs (see Figure 5.7).

Predation type	Impact on humans	Impact on wild carnivores	Example and region
Pet	Loss or injury to few valuable pets and working dogs; high impacts on wellbeing and livelihoods; intensified by risk of carnivore attack on people and pets in peri-urban and modified landscapes.	Targeted retaliation, lethal and non-lethal, legal and illegal; decreased support for wild carnivore conservation; some pathogen transmission risk from dogs with lower dog development index.	Amur tiger: Asia (Siberia) Gray wolf: North America Dingo: Australia Puma: North America Coyote: North America
Village Dog	Dogs of low value relative to other livestock; largely in rural areas and on protected area boundaries.	Indiscriminate illegal retaliation against all large carnivores driven by livestock predation; high risk of patho gen transmission from all dogs.	Lion, leopard, spotted hyena: Africa Leopard, tiger, gray wolf: Asia Striped hyena: Africa and Asia
Feral	Dogs of no value or pest status; predation or competitive exclusion provides neutral or positive impact; limited by low dog densities.	No retaliation; high risk of pathogen transmission, mitigated by low densities of feral dogs.	Dingo: Australia Leopard: Asia (India) Gray wolf: Europe and Asia

dogs, although this will be dependent on the densities of dogs with poorer health status. The most extreme example of this is for Amur tigers in Siberia, where hunting dogs are preyed upon regularly and represent a large proportion of livestock kills, resulting in targeted retaliatory killing by local communities and some disease risk from canine distemper. Slightly less intense are situations involving wolves in Europe and North America and dingoes in Australia, which may kill dogs but with less frequency than do Amur tigers. Conflict is also generated by unease amongst communities about the risk of predation on people, combined with their impacts on other livestock, resulting in targeted killing. Puma and coyote predation may have similar human impacts to those of tigers, dingoes, and wolves, but threats of retaliation may be less.

By comparison 'Village Dog Predation' may result in similar levels of carnivore impact, but a lesser degree of human impact due to the generally lower value of dogs, which mitigates the loss of animals. In this type, the per capita risk of pathogen transmission is high, and human retaliation often stems from the cumulative impact of livestock predation, rather than on dogs specifically, and tends to be indiscriminate resulting in a more diffuse impact. Examples include interactions between communities and their dogs with lions, spotted and striped hyenas and jackals in Africa, and leopards and striped hyenas in Asia. 'Feral Predation' occurs where feral dogs of minimal value or pest status are taken as prey or through intraguild killing, and there are neutral or even positive human impacts. Given the limited abundance of feral dogs relative to dogs of higher DDI, the potential for conflict is much reduced (Table 5.4).

Clearly this model is highly generalized, and limited by the paucity of data and case studies of interactions between wild carnivores and dogs. Furthermore, the response of local communities to wild carnivore predation on livestock is likely to vary even between individuals depending on their education, age, and ethnicity (e.g., Marchini and Macdonald, 2012; Thorn et al., 2012). In addition, local responses will be countered or modified by the wider institutional and stakeholder setting, resulting in different outcomes for the same wild carnivore in different locations within its geographical distribution. The specific conservation status of the carnivore concerned will also influence this context. There may also be other locally-specific issues that will influence the nature of the conflict, such as the presence of wolves that specialize in attacking dogs (e.g., Kojola et al., 2004), or reintroduction or recovery programs that instigate wider social conflict (e.g., Skogen et al., 2006).

5.5 Conclusions

Due to the unique relationship between dogs and humans, the study of dog predation requires analysis of both ecological and social dimensions. Our review shows that, in spite of the importance of understanding this relationship for human wellbeing, livelihoods, and wild carnivore conservation, there is only limited scientific data available. Because of the rarity of the events there are likely to be many challenges to studying them in a systematic manner. Yet there appears to be ample anecdotal information and knowledge amongst researchers, managers, and local experts which show that predation of dogs by wild carnivores is widespread, and the impacts for humans and wild carnivores are varied but often considerable. Given the rapid growth of global human and dog populations, and hence the increasing fragmentation of natural habitats and 'hard edges' between modified and wild environments, the potential for conflict between dogs and wild carnivores can only escalate.

Therefore our review represents a useful starting point for characterizing the nature of these conflicts, and the underlying ecological and social determinants. Utilizing this framework it is possible to design and prioritize appropriate responses, although we acknowledge that there will be context-specific issues that our general models have overlooked. For example, it is clear that within the 'Pet Predation' typology the high value of dogs significantly intensifies the potential for conflict. Consequently, in some cases the only option for reconciling conflict may be to ban dogs altogether, as has been proposed in the Wet Tropics of Australia (D. Morrant, unpublished data). Under the Village Dog Predation typology dogs are of lesser importance relative to other livestock, but pose a greater direct threat to carnivores because of pathogen transmission risk, facilitated by their vulnerability to predation due to their free-ranging, solitary behavior. Clearly, linked improvements in both dog and livestock husbandry are a key to mitigating these risks.

Finally, with socio-economic development, human–dog relationships typical of countries or regions with lower HDIs (and therefore DDIs) are likely to shift towards those that are characteristic of higher HDIs. Hence, in developing and urbanizing regions of the world, the nature of wild carnivore conflict resulting from dog predation may further intensify, amplified by the likelihood that with increased socio-economic status people will own more dogs per capita, and livestock-owning households in particular, and exacerbated by the contraction of carnivore habitat and related availability of wild prey. Our review provides a hypothetical framework that can project the likely characteristics of such future conflict. However, it requires further development, testing, and improvement through more targeted research into the social-ecological systems and typologies presented here.

Acknowledgments

We thank the 55 wildlife researchers and managers surveyed who provided much of the anecdotal and unpublished material presented in this review. Two anonymous reviewers also provided useful comments that improved earlier drafts.

References

ACAC. (2010). *Contribution of the Pet Care Industry to the Australian Economy*, 7th edn. Australian Companion Animal Council, Mornington, Australia.

Alexander, S.M. and Quinn, M.S. (2011). Coyote (*Canis latrans*) interactions with humans and pets reported in the Canadian print media (1995-2010). *Human Dimensions of Wildlife*, **16**, 345–59.

Allen, L.R. and Fleming, P.J.S. (2004). Review of canid management in Australia for the protection of livestock and wildlife: potential application to coyote management. *Sheep and Goat Research Journal*, **19**, 97–104.

Atickem, A., Williams, S., Bekele, A., and Thirgood, S. (2010). Livestock predation in the Bale Mountains, Ethiopia. *African Journal of Ecology*, **48**, 1076–82.

Aune, K.E. (1991). Increasing mountain lion populations and human-mountain lion interactions in Montana. In C.S. Braun, Ed., *Mountain Lion-Human Interaction. Symposium and Workshop. Colorado Division of Wildlife*, Denver, Colorado, pp. 86–94.

Bangs, E.E., Fontaine, J.A., Meier, T., Niemeyer, C., Jimenez, M.D., Smith, D.W., Mack, C.M., Asher, V., Handegard, L., Collinge, M., Krischke, R., Sime, C., Nadeau, S., and Moody, D. (2004). Restoration and conflict management of the gray wolf in Montana, Idaho and Wyoming. *Transactions of the North American Wildlife and Natural Resources Conference*, **69**, 89–105.

Bangs, E.E., Fontaine, J.A., Jimenez, M.D., Meier, T.J., Bradley, E.H., Niemeyer, C., Smith, D.W., Mack, C.M., Asher, V., and Oakleaf, J.K. (2005). Managing wolf-human conflict in the northwestern United States. In R. Woodroffe,

S. Thirgood and A. Rabinowitz, Eds., *People and Wildlife: Conflict or Co-Existence?* Cambridge University Press, Cambridge, pp. 340–56.

Barja, I. (2009). Prey and prey-age preference by the Iberian wolf *Canis lupus signatus* in a multiple-prey ecosystem. *Wildlife Biology*, **15**, 1–8.

Bhattarai, B.P. and Kindlmann, P. (2012). Interactions between Bengal tiger (*Panthera tigris*) and leopard (*Panthera pardus*): Implications for their conservation. *Biodiversity Conservation*, **21**, 2075–94.

Bibikov, D.I. (1988). *Der Wolf*. A. Ziemsen Verlag, Wittenberg Lutherstadt.

Bisi, J., Kurki, S., Svensberg, M., and Liukkonen, T. (2007). Human dimensions of wolf (*Canis lupus*) conflicts in Finland. *European Journal of Wildlife Research*, **53**, 304–14.

Blanco, J.C., Reig, S., and Cuesta, L. (1992). Distribution, status and conservation problems of the wolf *Canis lupus* in Spain. *Biological Conservation*, **60**, 73–80.

Bodendorfer, T., Hoppe-Dominik, B. Fischer, F., and Eduard Linsenmair, K. (2006). Prey of the leopard (*Panthera pardus*) and the lion (*Panthera leo*) in the Comoé and Marahoué National Parks, Côte d'Ivoire, West Africa. *Mammalia*, **70**, 231–46.

Bohnet, I.C. and Pert, P.L. (2010). Patterns, drivers and impacts of urban growth—a study from Cairns, Queensland, Australia from 1952-2031. *Landscape and Urban Planning*, **97**, 239–48.

Boitani, L. (1982). Wolf management in intensively used areas of Italy. In F.H. Harrington and P.C. Paquet, Eds., *Wolves of the World: Perspectives of Behaviour, Ecology, and Conservation*. Noyes, New Jersey, pp. 158–72.

Boitani, L., and Ciucci, P. (1995). Comparative social ecology of feral dogs and wolves. *Ethology Ecology and Evolution*, **7**, 49–72.

Boitani, L., Francisci, F., Ciucci, P., and Andreoli, G. (1995). Population biology and ecology of feral dogs in central Italy. In J. Serpell, Ed., *The Domestic Dog. Its Evolution, Behaviour, and Interactions with People*. Cambridge University Press, Cambridge, pp. 218–44.

Burger, A. and Knowles, P. (1976). *Fraser Island*. Rigby, Adelaide.

Butler, J.R.A. (1995). A survey of the communal land dog population in Zimbabwe and its rabies vaccination status. In *Proceedings of the Third International Conference of the Southern and Eastern African Rabies Group, 7-March 9, 1995, Harare, Zimbabwe*. Fondation Marcel Merieux, Lyon, France, pp. 139–42.

Butler, J.R.A. (1998). The ecology of domestic dogs *Canis familiaris* in the communal lands of Zimbabwe. PhD dissertation, University of Zimbabwe.

Butler, J.R.A. (2000). The economic costs of wildlife predation on livestock in Gokwe communal land, Zimbabwe. *African Journal of Ecology*, **38**, 23–30.

Butler, J.R.A. (2011). The challenge of knowledge integration in the adaptive co-management of conflicting ecosystem services provided by seals and salmon. *Animal Conservation*, **14**, 599–601.

Butler, J.R.A. and Bingham, J. (2000). Demography and dog-human relationships of the dog population in Zimbabwean communal lands. *Veterinary Record*, **147**, 442–46.

Butler, J.R.A. and du Toit, J.T. (2002). Diet of free-ranging domestic dogs (*Canis familiaris*) in rural Zimbabwe: implications for wild scavengers on the periphery of wildlife reserves. *Animal Conservation*, **5**, 29–37.

Butler, J.R.A., du Toit, J.T., and Bingham, J. (2004). Free-ranging domestic dogs *Canis familiaris* as predators and prey in rural Zimbabwe: threats of competition and disease to large wild carnivores. *Biological Conservation*, **115**, 369–78.

Causey, M.K. and Cude, C.A. (1980). Feral dog and white-tailed deer interactions in Alabama. *Journal of Wildlife Management*, **44**, 481–84.

Ciucci, P., Boitani, L., Pelliccioni, E.R., Rocco, M., and Guy, I. (1996). A comparison of scat-analysis methods to assess the diet of the wolf *Canis lupus*. *Wildlife Biology*, **2**, 37–48.

Cleaveland, S., Appel, M.J.G., Chalmers, W.S.K., Chillingworth, C., Kaare, M., and Dye, C. (2000). Serological and demographic evidence for domestic dogs as a source of canine distemper virus infection for Serengeti wildlife. *Veterinary Microbiology*, **72**, 217–27.

Cortés, Y. (2001). Ecología y conservación del lobo (*Canis lupus*) en medios agrícolas. PhD dissertation, Universidad Complutense de Madrid.

Cuesta, L., Barcena, F., Palacios, F., and Reig, S. (1991). The trophic ecology of the Iberian wolf (*Canis lupus signatus* Cabrera 1907). A new analysis of stomach data. *Mammalia*, **55**, 240–54.

Dabe, D. (1997). Non-consumptive utilisation of leopards: community conservation and ecotourism in practice. In *Proceedings of a Conference On Lions and Leopards as Game Ranch Animals*. Onderstepoort, South Africa, pp. 50–57.

Daniels, J.C. (2009). *The Leopard in India: A Natural History*. Natraj Publishers, Delhi.

Dar, N.I, Minhas, R.A., Zaman, Q., and Linkie, M. (2009). Predicting the patterns, perceptions and causes of human–carnivore conflict in and around Machiara National Park, Pakistan. *Biological Conservation*, **142**, 2076–82.

Davies, R.B. (1991). Lion damage to pets in urban Colorado Springs, Colorado. In C.S. Braun, Ed., *Mountain Lion—Human Interactions. Symposium and Workshop*. Colorado Division of Wildlife, Denver, Colorado, pp. 79–80.

Dettmann, R. (1991). The role of local, county, state, and federal agencies in the prevention of conflict with large predators. In C.S. Braun, Ed., *Mountain Lion—Human*

Interactions. Symposium and Workshop. Colorado Division of Wildlife, Denver, Colorado, p. 24.

Dyck, M.G. (2006). Characteristics of polar bears killed in defense of life and property in Nunavut, Canada, 1970-2000. *Ursus*, **17**, 52–62.

Edgaonkar, J.J. and Chellam, R. (1998). A preliminary study on the ecology of the leopard, *Panthera pardus fusca*, in the Sanjay Gandhi National Park, Maharashtra. Wildlife Institute of India, Dehradun.

Edge, J.L., Beyer, D.E., Belant, J.L., Jordan, M.J., and Roell, B.J. (2011). Livestock and domestic dog predations by wolves in Michigan. *Human-Wildlife Interactions*, **5**, 66–78.

Farrar, R.O. (2007). Assessing the impact of urban coyote on people and pets in Austin, Travis County, Texas. *Proceedings of the Wildlife Damage Management Conference*, **12**, 334–41.

Farrell, L.E., Roman, J., and Sunquist, M.E. (2000). Dietary separation of sympatric carnivores identified by molecular analysis of scats. *Molecular Ecology*, **9**, 1583–90.

Fleming, P., Corbett, L.K., Harden, R., and Thomson, P. (2001). *Managing the Impacts of Dingoes and other Wild Dogs*. Bureau of Rural Sciences, Canberra.

Foster, R.J., Harmsen, B.J., Valdes, B., Pomilla, C., and Doncaster, C.P. (2010). Food habits of sympatric jaguars and pumas across a gradient of human disturbance. *Journal of Zoology*, **280**, 309–18.

Fritts, S.H. and Paul, W.J. (1989). Interactions of wolves and dogs in Minnesota. *Wildlife Society Bulletin*, **17**, 121–23.

Gajera, N., Dave, S.M., and Nishith, D. (2009). Feeding patterns and den ecology of striped hyena (*Hyaena hyaena*) in North Gujarat, India. Tiger Paper, **36**, 13–17.

Gallant, J. (2002). *The Story of the African Dog*. University of Natal Press, Pietermaritzburg.

Goodrich, J.M. (2010). Human-tiger conflict: A review and call for comprehensive plans. *Integrative Zoology*, **5**, 300–12.

Goodrich, J.M. and Miquelle, D.G. (2005). Translocation of problem Amur tigers *Panthera tigris altaica* to alleviate tiger-human conflicts. *Oryx*, **39**, 1–4.

Goodrich, J.M., Seryodkin, I., Miquelle, D., and Bereznuk, S.L. (2011). Conflicts between Amur (Siberian) tigers and humans in the Russian Far East. *Biological Conservation*, **144**, 584–92.

Goodrich, J., Quigley, K.S., Lewis, J.C.M., Astafiev, A.A., Slabi, E.V., Miquelle, D., Smirnov, E.N., Kerley, L.L., Armstrong, D.L., Quigley, H.B., and Hornocker, M.G. (2012). Serosurvey of free-ranging Amur tigers in the Russian Far East. *Journal of Wildlife Diseases*, **48**, 186–89.

Gula, R. (2008). Wolf depredation on domestic animals in the Polish Carpathian Mountains. *Journal of Wildlife Management*, **72**, 283–89.

Hansen, A.J., Knight, R.L., Marzluff, J.M., Powell, S., Brown, K., Gude, P.H., and Jones. K. (2005). Effects of exurban development on biodiversity: Patterns, mechanisms, and research needs. *Ecological Applications*, **15**, 1893–1905.

Harrison, T.M., Mazet, J.K., Holekamp, K.E., Dubovi, E., Engh, A.L., Nelson, K., Van Horn, R.C., and Munson, L. (2004). Antibodies to canine and feline viruses in spotted hyena (*Crocuta crocuta*) in the Masai Mara National Reserve. *Journal of Wildlife Diseases*, 40, 1–10.

Haraway, D. (2003). *The Companion Species Manifesto: Dogs, People, and Significant Otherness*. Prickly Paradigm Press, Chicago.

Hart, L.A. (1995). Dogs as human companions: a review of the relationship. In J. Serpell, Ed., *The Domestic Dog. Its Evolution, Behaviour, and Interactions with People*. Cambridge University Press, Cambridge, pp. 161–78.

Hayward, M.W., Henschel, P., O'Brien, J., Hofmeyr, M., Balme, G., and Kerley, G.I.H. (2006). Prey preferences of the leopard (*Panthera pardus*). *Journal of Zoology*, **270**, 298–313.

Holmern, T., Nyahongo, J., and Roskaft, E. (2007). Livestock loss caused by predators outside the Serengeti National Park, Tanzania. *Biological Conservation*, **135**, 518–26.

Hovens, J.P.M. and Tungaslaktuja, K. (2005). Seasonal fluctuations of the wolf diet in the Hustali National Park (Mongolia). *Mammalian Biology—Zeitschrift für Säugetierkunde*, **70**, 210–17.

Howell, R.G. (1982). The urban coyote problem in Los Angeles County. *Proceedings of the Vertebrate Pest Conference*, **10**, 21–23.

Jethva, B.D. and Jhala, Y. (2004). Foraging ecology, economics and conservation of Indian wolves in the Bhal region of Gujarat, western India. *Biological Conservation*, **115**, 351–57.

Johnsingh, A.J.T. (1983). Large mammalian prey—predators in Bandipur. *Journal of the Bombay Natural History Society*, **80**, 1–57.

Kamler, J. F., Keeler, K., Wiens, G., Richardson, C., and Gipson, P.S. (2003). Feral dogs, *Canis familiaris*, kill coyote, *Canis latrans*. *Canadian Field-Naturalist*, **117**, 123–4.

Karlsson, J. and Jaxgård, P. (2004). Wolf attacks on dogs [Vargangrepp på hundar]. In G. Jansson, C. Seiler, and H., andrén, Eds., Skogsvilt III: vilt och landskap i förändring. Grimsö Wildlife Research Station, Swedish University of Agricultural Science, pp. 243–8.

Keci, E., Trajce, A., Mersini, K., Bego, F., Ivanov, G., Melovski, D., Stojanov, A., Breitenmoser, U., Von Arx, M., Schwaderer, G., Spangenberg, A., and Linnell, J.D.C. (2008). Conflicts between lynx, other large carnivores, and humans in Macedonia and Albania. *Proceedings of the Congress of Ecologists of Macedonia*, **3**, 257–64.

Kennedy-Stoskopf, S. (2003). Canidae. In M.E. Fowler, and E.R. Miller, Eds., *Zoo and Wild Animal Medicine*. Saunders, St. Louis, pp. 482–91.

Kissui, B. (2008). Livestock predation by lions, leopards, spotted hyenas, and their vulnerability to retaliatory killing in the Maasai steppe, Tanzania. *Animal Conservation*, **11**, 422–32.

Kitala, P.M., McDermott, J.J., Kyule, M.N., and Cathuma, J.M. (1993). Features of dog ecology relevant to rabies spread in Machakos District, Kenya. *Onderstepoort Journal of Veterinary Research*, **60**, 445–9.

Knobel, D.L., Laurenson, M.K., Kazwala, R.R., Boden, L.A., and Cleaveland, S. (2008). A cross-sectional study of factors associated with dog ownership in Tanzania. *BMC Veterinary Research*, **4**, 5.

Kock, R., Chalmers, W., Mwanzia, J., Chillingworth C., Wambua, J., Coleman P., and Baxendale, W. (1998). Canine distemper antibodies in lions of the Masai Mara. *Veterinary Record*, **142**, 662–5.

Kofron, C.P. (1999). Attacks to humans and domestic animals by the southern cassowary (*Casuarius casuarius johnsonii*) in Queensland, Australia. *Journal of Zoology*, **249**, 375–81.

Kojola, I. and Kuittinen, J. (2002). Wolf attacks on dogs in Finland. *Wildlife Society Bulletin*, **30**, 498–501.

Kojola, I., Ronkainen, S., Hakala, A., Heikkinen, S., and Kokko, S. (2004). Interactions between wolves *Canis lupus* and dogs *C. familiaris* in Finland. *Wildlife Biology*, **10**, 101–6.

Kolowski, J. and Holekamp, K. (2006). Spatial, temporal, and physical characteristics of livestock depredations by large carnivores along a Kenyan reserve border. *Biological Conservation*, **128**, 528–41.

Kruuk, H. (1976). Feeding and social behavior of the striped hyaena (*Hyaena vulgaris* Desmaret). *East African Wildlife Journal*, **14**, 91–111.

Kuhn, B. (2005). The faunal assemblages and taphonomic signatures of five striped hyaena (*Hyaena hyaena syriaca*) dens in the desert of Eastern Jordan. *Levant*, **37**, 221–34.

Leakey, L.N., Milledge, S.A.H., Leakey, S.M., Edung, J., Haynes, P., Kiptoo, D.K., and McGeorge, A. (1999). Diet of striped hyaena in northern Kenya. *African Journal of Ecology*, **37**, 314–26.

Leberg, P.L., Carloss, M.R., Dugas, L.J., Pilgrim, K.L., Mills, L.S., Green, M.C., and Scognamillo, D. (2004). Recent record of a cougar (*Puma concolor*) in Louisiana, with notes on diet, based on analysis of fecal materials. *Southeastern Naturalist*, **3**, 653–8.

Lescureux, N. and Linnell, J.D.C. (2010). Knowledge and perceptions of Macedonian hunters and herders: the influence of species specific ecology of bears, wolves, and lynx. *Human Ecology*, **38**, 389–99.

Letnic, M., Ritchie, E.G., and Dickman, C.R. (2011). Top predators as biodiversity regulators: The dingo *Canis lupus dingo* as a case study. *Biological Reviews*, **87**, 390–413.

Linnell, J.D.C. and Strand, O. (2000). Conservation implications of aggressive intra-guild interactions among mammalian carnivores. *Diversity and Distributions*, **6**, 169–76.

Llaneza, L., Iglesias, J., and Rico, M. (2000). Hábitos alimenticios del lobo ibérico en el antiguo Parque Nacional de la Montaña de Covadonga. *Galemys*, **12**, 93–102.

Loveridge, A.J. and Nel, J.A.J. (2004). Black-backed jackal. In C. Sillero-Zubiri, M. Hoffman, and D.W. Macdonald, Eds., *Canids: Foxes, Wolves, Jackals and Dogs. Status Survey and Conservation Action Plan*. IUCN/SSC Canid Specialist Group, Oxford, pp. 151–155.

Lukasik, V.M. and Alexander, S.M. (2011). Human-coyote interactions in Calgary, Alberta. *Human Dimensions of Wildlife*, **16**, 114–27.

Mansfield, T.M. (1991). Mountain lion damage to property in California. In C.S. Braun, Ed., *Mountain Lion–Human Interactions. Symposium and Workshop*. Colorado Division of Wildlife, Denver, Colorado, pp. 75–78.

Marchini, S. and Macdonald, D.W. (2012). Predicting ranchers' intention to kill jaguars: Case studies in Amazonia and Pantanal. *Biological Conservation*, **147**, 213–21.

Mawson, P. (2004). Crocodile management in Western Australia. In *Crocodiles: Proceedings of the 17th Working Meeting of the IUCN-SSC Crocodile Specialist Group*. IUCN, Gland, Switzerland, pp. 28–37.

Mazzolli, M. (2009). Mountain lion *Puma concolor* attacks on a maned wolf *Chrysocyon brachyurus* and a domestic dog in a forestry system. *Mastozoologia Neotropical*, **16**, 465–70.

McDonald, G. and Weston, N. (2004), *Sustaining the Wet Tropics: A Regional Plan for Natural Resource Management, Volume 1: Background to the Plan*. Rainforest CRC and FNQ NRM Ltd., Cairns.

Miquelle, D.G., Smirnov, E.N., Quigley, H., Hornocker, M.G., Nikolaev, I.G., and Matushkin, E. (1996). Food habits of Amur tigers in Sikhote-Alin Zapovednik and the Russian Far East. *Journal of Wildlife Research*, **1**, 138–47.

Miquelle, D.G., Stephens, P.A., Smirnov, E.N., Goodrich, J.M., Zaumyslova, O.J., and Myslenkov, A.E. (2005). Tigers and wolves in the Russian Far East: Competitive exclusion, functional redundancy, and conservation implications. In J.C. Ray, K.H. Redford, R.S. Steneck, and J. Berger, Eds., *Large Carnivores and the Conservation Of Biodiversity*. Island Press, Washington, DC, pp. 179–207.

Monchot, H. and Mashkour, M. (2010). Hyenas around the city (Kashan, Iran). *Journal of Taphonomy*, **8**, 17–32.

Mondal, K., Gupta, S., Qureshiand, Q., and Sankar, K. (2011). Prey selection and food habits of leopard (*Panthera pardus fusca*) in Sariska Tiger Reserve, Rajasthan, India. *Mammalia*, **75**, 201–5.

Mukherjee, S. and Mishra, C. (2001). Predation by leopard in Majhatal Wildlife Sanctuary, western Himalaya. Journal of the Bombay Natural History Society, **98**, 2467–8.

Nowak, S., Myslajek, R., Klosinska, A., and Gabrys, G. (2011). Diet and prey selection of wolves (*Canis lupus*) re-colonising western and central Poland. *Mammalian Biology*, **76**, 709–15.

Nowak, S., Myslajek, R.W., and Jedrzejewska, B. (2005). Patterns of wolf *Canis lupus* predation on wild and domestic ungulates in the western Carpathian Mountains (S Poland). *Acta Theriologica*, **50**, 263–76.

Oppenheimer, E.C. and Oppenheimer, J.R. (1975). Certain behavioral features in the pariah dog (*Canis familiaris*) in West Bengal. *Applied Animal Ethology*, **2**, 81–92.

Palomares, F. and Caro, T.M. (1999). Interspecific killing among mammalian carnivores. *American Naturalist*, **153**, 492–508.

Pert, P.L., Butler, J.R.A., Bruce, C.M., and Metcalfe, D.J. (2012). A composite threat indicator approach to monitoring vegetation condition in the Wet Tropics, Queensland, Australia. *Ecological Indicators*, **18**, 191–9.

Polis, G.A. and Myers, C.A. (1989). The ecology and evolution of intraguild predation: Potential competitors that eat each other. *Annual Review of Ecology and Systematics*, **20**, 297–330.

Pozio, E., Casulli, a., Bologov, V., Marucci, G., and La Rosa, G. (2001). Hunting practices increase the prevalence of Trichinella infection in wolves from European Russia. *Journal of Parasitology*, **87**, 1498–1501.

Punjabi, G.A., Athreya, V., and Linnell, J.D.C. (2012). Using natural marks to estimate free-ranging dog *Canis familiaris* abundance in a MARK-RESIGHT framework in suburban Mumbai, India. *Tropical Conservation Science*, **5**, 510–20.

Quigley, K.S., Evermann, J.F., Leathers, C.W., Armstrong, D.L., Goodrich, J., Duncan, N.M., and Miquelle, D. (2010). Morbillivirus in a wild Siberian tiger in the Russian Far East. *Journal of Wildlife Diseases*, **46**, 1252–6.

Ramakrishnan, U., Coss, R.G., and Pelkey, N.W. (1999). Tiger decline caused by the reduction of large ungulate prey: Evidence from a study of leopard diets in southern India. *Biological Conservation*, **89**, 113–20.

Ramsay, E.C. (2003). Ursidae and Hyaenidae. In M.E. Fowler, and E.R. Miller, Eds., *Zoo and Wild Animal Medicine*. Saunders, St. Louis, pp. 523–38.

Robinette, W.L., Gashwiler, J.S., and Morris, O.W. (1959). Food habits of the cougar in Utah and Nevada. *Journal of Wildlife Management*, **23**, 261–73.

Roelke-Parker, M.E., Munson, L., Packer, C., Kock, R., Cleaveland, S., Carpenter, M., O'Brien, S.J., Pospischil, A., Hofmann-Lehmann, R., Lutz, H., Mwamengele, G., Mgasa, M.N., Machange, G.A., Summers, B.A., and Appel, M.J.G. (1996). A canine distemper virus epidemic in Serengeti lions (*Panthera leo*). Nature, **379**, 441–445.

Ruid, D.B., Paul, W.J., Roell, B.J., Wydeven, A.P., Willging, R.C., Jurewicz, R.L., and Lonsway, D.H. (2009). Wolf-human conflicts and management in Minnesota, Wisconsin, and Michigan. In A.P. Wydeven, T.R. Van Deelen, and E.J. Heske, Eds., *Recovery of Gray Wolves in the Great Lakes Region of the United States: An Endangered Species Success Story*. Springer, New York, pp. 279–98.

Russ, W.B. (1995). The status of mountain lions in Texas. *Proceedings of the Annual Conference of the Southeastern Association of Fish and Wildlife Agencies*, **49**, 545–51.

Salvador, A. and Abad, P.L. (1987). Food habits of a wolf population (*Canis lupus*) in Léon Province, Spain. *Mammalia*, **51**, 45–52.

Sanders, C.R. (1993). Understanding dogs: Caretakers' attributions of mindedness in canine-human relationships. *Journal of Contemporary Ethnography*, **22**, 205–26.

Sanders, C.R. (2003). Actions speak louder than words: Close relationships between humans and nonhuman animals. *Symbolic Interaction*, **26**, 405–426.

Sanders, M.R. and Halfpenny, J.C. (1991). Human—lion interactions in Boulder County, Colorado: behavioral patterns. In C.S.Braun, Ed., *Mountain Lion - Human Interaction. Symposium and Workshop*. Colorado Division of Wildlife, Denver, Colorado, p. 17.

Sanei, A., Zakaria, M., and Hermidas, S. (2011). Prey composition in the Persian leopard distribution range in Iran. *Asia Life Sciences Supplement*, **7**, 19–30.

Schaller, G.B. (1972). *The Serengeti Lion: A Study of Predator-Prey Relations*. Chicago University Press, Chicago.

Schmidt, R.H. and Timm, R.M. (2007). Bad dogs: why do coyotes and other canids become unruly? *Proceedings of the Wildlife Damage Management Conference*, **12**, 287–302.

Serpell, J. (1995). From paragon to pariah: some reflections on human attitudes to dogs. In J. Serpell, Ed., *The Domestic Dog. Its Evolution, Behaviour, and Interactions with People*. Cambridge University Press, Cambridge, pp. 217–44.

Shah, G.M., Jan, U., Bhat, B.A., Ahmad, F., and Ahmad, J. (2009). Food habits of the leopard *Panthera pardus* in Dachigam National Park, Kashmir, India. *Journal of Threatened Taxa*, **1**, 184–5.

Sidorovich, V.E., Tikhomirova, L.L., and Jedrzejewska, B. (2003). Wolf *Canis lupus* numbers, diet and damage to livestock in relation to hunting and ungulate abundance in northeastern Belarus during 1990-2000. *Wildlife Biology*, **9**, 103–12.

Sjölander-Lindqvist, A. (2010). Social-natural landscape re-organised: Swedish forest edge farmers and wolf recovery. *Conservation and Society*, **7**, 130–40.

Skogen, K. and Krange, O. (2003). A wolf at the gate: The anti-carnivore alliance and the symbolic construction of community. *Sociologia Ruralis*, **43**, 309–25.

Skogen, K., Mauz, I., and Krange, O. (2006). Wolves and eco-power. A French-Norwegian analysis of the narratives of the return of large carnivores. *Journal of Alpine Research*, **94**, 78–87.

Slaght, J.C., Miquelle, D.G., Nikolaev, I.G., Goodrich, J.M., Smirnov, E.N., Traylor-Holzer, K., Christie, S., Arjanova, T., Smith, J.L.D., and Karanth. K.U. (2005). Who's king of the beasts? Historical and recent body weights of wild and captive Amur tigers, with comparisons to other subspecies. In D.G. Miquelle, E.N. Smirnov, and J.M. Goodrich, Eds., *Tigers in Sikhote-Alin Zapovednik: Ecology And Conservation*. PSP, Vladivostok, pp. 25–35.

Stubblefield, C.H. and Shrestha, M. (2007). Status of Asiatic black bears in protected areas of Nepal and the effects of political turmoil. *Ursus*, **18**, 101–8.

Thorn, M., Green, M., Dalerum, F., Bateman, P.W., and Scott, D.M. (2012). What drives human–carnivore conflict in the North West Province of South Africa? *Biological Conservation*, **150**, 23–32.

Timm, R.M. and Baker, R.O. (2007). A history of urban coyote problems. *Proceedings of the Wildlife Damage Management Conference*, **12**, 272–86.

Timm, R.M., Baker, R.O., Bennett, J.R., and Coolahan, C.C. (2004). Coyote attacks: An increasing suburban problem. *Proceedings of the Vertebrate Pest Conference*, **21**, 47–57.

Tkachenko, K.N. (2012). Specific features of feeding of the Amur tiger *Panthera tigris altaica* (Carnivora, Felidae) in a densely populated locality (with reference to Bol'shekhekhtsirskii Reserve and its environs). *Biology Bulletin*, **39**, 279–87.

Torres, S.G., Mansfield, T.M., Foley, J.E., Lupo, T., and Brinkhaus, A. (1996). Mountain lion and human activity in California: testing speculations. *Wildlife Society Bulletin*, **24**, 451–60.

UNDP (2011). *Human Development Report 2011: Sustainability and Equity, A Better Future for all*. United Nations Development Program, New York.

Valdmann, H., Andersone-Lilley, Z., Koppa, O., Ozolins, J., and Bagrade, G. (2005). Winter diets of wolf *Canis lupus* and lynx *Lynx lynx* in Estonia and Latvia. *Acta Theriologica*, **50**, 521–7.

Van Orsdol, K.G. (1984). Foraging and hunting success of lions in Queen Elizabeth National Park, Uganda. *African Journal of Ecology*, **22**, 79–99.

Vos, J. (2000). Food habits and livestock depredation of two Iberian wolf packs (*Canis lupus signatus*) in the north of Portugal. *Journal of Zoology*, **251**, 457–62.

Wack, R.F. (2003). Felidae. In M.E. Fowler, and E.R. Miller, Eds., *Zoo and Wild Animal Medicine*. Saunders, St. Louis, pp. 491–501.

Wang, S. and Macdonald, D.W. (2009). Feeding habits and niche partitioning in a predator guild composed of tigers, leopards and dholes in a temperate ecosystem in central Bhutan. *Journal of Zoology*, **277**, 275–83.

Wells, D.L. (2007). Domestic dogs and human health: an overview. *British Journal of Health Psychology*, **12**, 145–56.

Wells, D.L. (2009). The effects of animals on human health and well-being. *Journal of Social Issues*, **65**, 523–43.

Westgarth, C., Pinchbeck, G., Bradshaw, J., Dawson, S., Gaskell, R., and Christley, R. (2007). Factors associated with dog ownership and contact with dogs in a UK community. *BMC Veterinary Research*, **3**, 5.

White, L.A. and Gehrt, S.D. (2009). Coyote attacks on humans in the United States and Canada. *Human Dimensions of Wildlife*, **14**, 419–32.

Woodall, P.F., Pavlov, P., and Twyford. K.L. (1996). Dingoes in Queensland, Australia: Skull dimensions and the identity of wild canids. *Wildlife Research*, **23**, 581–8.

Yirga, G., Bauer, H., Worasi, Y., and Asmelash, S. (2011). Farmers' perception of leopard (*Panthera pardus*) conservation in a human dominated landscape in northern Ethiopian highlands. *International Journal of Biodiversity and Conservation*, **3**, 160–6.

Yirga, G., De Iongh, H.H., Leirs, H., Gebrehiwot, K., Berhe, G., Asmelash, T., Gebrehiwot, H., and Bauer, H. (2012). The ecology of large carnivores in the highlands of northern Ethiopia. *African Journal of Ecology*, **51**, 78–86.

Young, J.K., Olson, K.A., Reading, R.P., Amgalanbaatar, S., and Berger, J. (2011). Is wildlife going to the dogs? Impacts of feral and free-roaming dogs on wildlife populations. *BioScience*, **61**, 125–32.

Young J.C., Butler J.R.A., Jordan A., and Watt, A.D. (2012). Less government intervention in biodiversity management: Risks and opportunities. *Biodiversity and Conservation*, **21**, 1095–1100.

Zunna, A., Ozolins, J., and Pupila, A. (2009). Food habits of the wolf *Canis lupus* in Latvia based on stomach analyses. *Estonian Journal of Ecology*, **58**, 141–52.

CHAPTER 6

Dogs, disease, and wildlife

Darryn L. Knobel, James R.A. Butler, Tiziana Lembo, Rob Critchlow, and Matthew E. Gompper

6.1 Introduction

Populations are limited by three factors: resource availability, predation, and disease. Ecologists know a great deal about how the first two factors work, but our understanding of the third factor is more limited. While studies of humans and domestic animals have revealed tremendous insights into the proximate mechanisms by which parasites[1] that cause diseases influence individual hosts, an ecological perspective that involves understanding how parasites might alter the broader dynamics of populations and the structure of communities is only now emerging. Part of the difficulty is that we know very little about the basic natural history of the parasites that are harbored by most host species. Dogs (*Canis familiaris*) are a partial exception to this generalization, as their close associations with humans, human interest in dog welfare, as well as a greater accessibility for handling and sampling, have resulted in a good understanding of their interactions with many parasitic species. Indeed, along with economically important livestock, we arguably know more about the parasites of these animals than those of any other non-human host species.

This chapter examines the ecological parasitology of systems that are influenced by dogs. While there has long been recognition that parasites of dogs (most notably, rabies virus) may influence other species, consideration of the role of dogs as vectors of agents pathogenic to wildlife increased considerably in the 1990s when a series of well publicized and large-scale die-offs of wild carnivore populations occurred following the introduction of pathogenic viruses (lions, *Panthera leo*: Roelke-Parker et al., 1996; Ethiopian wolves, *C. simensis*: Laurenson et al., 1998; Sillero-Zubiri et al., 1996; African wild dogs *Lycaon pictus*: Gascoyne et al., 1993; Hofmeyr et al., 2000; Arctic foxes, *Vulpes lagopus*: Goltsman et al., 1996; Caspian seals, *Pusa caspica*: Kennedy et al., 2000). Efforts to trace these events to their origin concluded that the viruses derived from dogs and that dogs acted as sources of infection for wildlife, causing considerable changes to rates of fecundity, morbidity, and mortality. With this conclusion came the recognition that dog populations could be fundamentally important drivers of wildlife population dynamics and that a multidisciplinary approach, linking veterinary sciences, public policy, ecology, and wildlife conservation was needed to better address this issue.

The dynamics of parasite populations, their likelihood of causing disease in individuals and thereby influencing populations, and the ensuing effects that these disease-induced fluctuations in populations have on communities are complex. In an effort to address these issues, here we focus on three main themes: the primary pathogens of concern, the ecological processes by which these pathogens move within and between dog and wildlife populations, and the prospects and strategies for managing pathogens that are considered particularly problematic when transmitted from dogs to wildlife.

[1] The focus of this chapter is on parasitic organisms, which include microparasites such as viruses, fungi, and bacteria, as well as macroparasites and ectoparasites such as helminths, mites, and ticks.

Free-Ranging Dogs and Wildlife Conservation. Edited by Matthew E. Gompper
© Oxford University Press 2014. Published 2014 by Oxford University Press.

6.2 The pathogen community

In a review of disease-causing organisms of humans and domestic animals, Cleaveland et al. (2001) identified 358 pathogens that are reported to infect dogs. We compared these 358 pathogens to the Global Mammal Parasite Database (GMPD; Nunn and Altizer, 2005). The GMPD is comprised of records from published literature of pathogens in free-living wild mammal populations in the orders Artiodactyla and Perissodactyla (together forming the wild ungulates), and Carnivora. In our analysis, we excluded those records from the GMPD where either the host or parasite was only identified to a genus level. Of the 358 dog pathogens identified by Cleaveland et al. (2001), 168 (47%) are reported in the GMPD: in other words, are shared between dogs and wild mammals. By taxonomic group, 71.4% of viruses that infect dogs are shared with wild mammals, as are 56.1% of helminths, 50% of protozoa, 46.7% of bacteria, and 15.4% of fungi.

Thus a significant proportion of the parasites associated with dogs are also shared with wildlife. For instance, Figure 6.1 shows the proportion of these shared pathogens that occur in wild carnivores, wild ungulates, or in both of these host groups. From this, we conclude that the majority of those shared parasites are principally parasites of carnivores. Yet two points deserve consideration. First, the observation that dogs share parasites with wildlife does not necessarily mean that these parasites are maintained in dog populations such that in the absence of dogs the parasites would not persist. Second, even if dogs have an important influence on the prevalence of these parasites in wildlife host species, many of these shared parasites may have little if any effect on host health and host population trajectories relative to factors such as resource availability and predation. Only a handful of these parasites are known or suspected to be a concern for wildlife populations and, in the context of transmission from dogs to wildlife, just three of these parasites receive the bulk of attention from wildlife conservationists and managers. These 'big three' are rabies virus, canine distemper virus, and canine parvovirus.

6.2.1 Rabies virus

Rabies is an acute, typically fatal, encephalomyelitis caused by infection with any of the virus species in the *Lyssavirus* genus in the Rhabdoviridae family. Cases of the disease in dogs are, with few exceptions, caused by infections with rabies virus (RABV, genotype/species 1), the type member of the genus. Rabies virus has a worldwide distribution, with the exception of Antarctica and some isolated landmasses. Although all mammals are susceptible

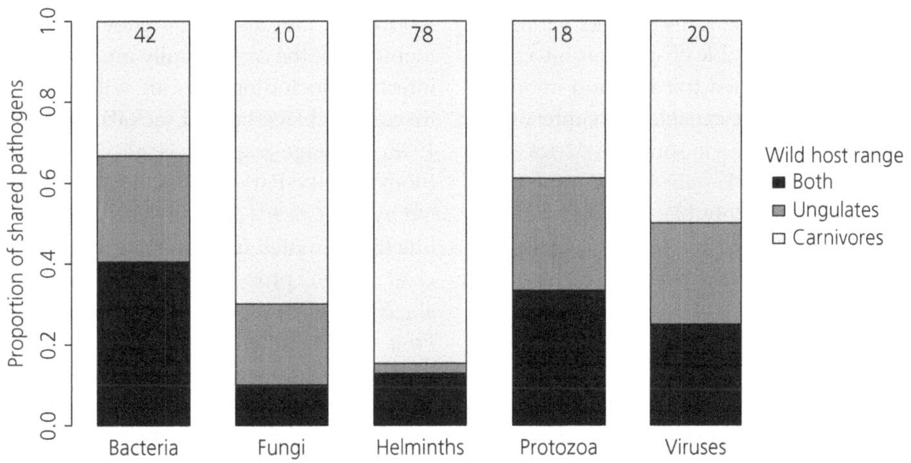

Figure 6.1 The proportion of dog pathogens that are reported to infect wild carnivores, wild ungulates, or both wild host groups according to the Global Mammal Parasite Database. Values for each taxon represent the total number of shared pathogens per group.

to infection with RABV, certain species are capable of sustained intraspecies maintenance of particular viral variants that seem to be well adapted to those species. Such reservoir species are found among members of the order Carnivora (in the families Canidae, Herpestidae, Procyonidae, and Mephitidae) and Chiroptera. Molecular epidemiology studies of RABV isolates (Nadin-Davis, 2007) reveal several distinct lineages. The most widely distributed is a cosmopolitan lineage thought to have originated in Europe and spread across the globe with the movement of dogs during colonial times (Smith et al., 1992). Within this lineage, which includes viruses from all over the world, variants may cluster by geographic region into particular clades, such as the Africa 1a (north-eastern Africa) and Africa 1b (south-eastern Africa) clades within the Africa 1 lineage. Distinct clades are also represented by virus variants circulating in particular host species, such as those in mongooses in the central plateau of southern Africa. Typing RABVs through antigenic or genetic methods may therefore assist in identifying the source of the virus in outbreaks in wildlife (see Section 6.3.1).

Certain wildlife species are capable of maintaining RABV in the absence of a dog reservoir population. In the USA, dog rabies has been eliminated by vaccination, but still persists in wild terrestrial hosts, including, striped skunks (*Mephitis mephitis*), raccoons (*Procyon lotor*), and gray foxes (*Urocyon cinereoargenteus*) (Blanton et al., 2008; CDC, 2007). These species maintain phylogenetically distinct variants, which are nevertheless capable of spillover infection and of establishing sustained transmission in other species. Other wild species capable of maintenance of RABV include mongooses in southern Africa and the Caribbean, red foxes (*V. vulpes*) in Europe and North America, and possibly black-backed jackals (*C. mesomelas*) and bat-eared foxes (*Otocyon megalotis*) in southern Africa, and raccoon dogs (*Nyctereutes procyonoides*) in north-eastern Europe (Kuzmin and Tordo, 2012). Unusually, in that it is a herbivorous species, kudu (*Tragelaphus strepsiceros*) in southern Africa also appear to be capable of sustained maintenance of RABV under certain conditions, although virus isolates were all associated with regional canine rabies variants (Mansfield et al., 2006) and the epidemiology of these cycles is poorly understood.

Given the broad host range of RABV, essentially all mammalian species are at risk of spillover infection from these reservoir host species. Spillover into small host populations can have dramatic consequences when followed by even short chains of intraspecies transmission, as has been the case in endangered Ethiopian wolf (Randall et al., 2004) and African wild dog populations (Gascoyne et al., 1993). In both cases, strong evidence pointed to dog populations as the sources of the virus spillover, although in another outbreak of rabies in African wild dogs, black-backed jackals were strongly implicated as the vectors and possible reservoirs of the virus, with no evidence of domestic dog involvement (Hofmeyr et al., 2000). When populations of unvaccinated dogs are present in a given host community, they likely act as maintenance hosts of canid variants of RABV and sources of infection for other species (see Box 6.1).

6.2.2 Canine distemper virus

Canine distemper virus (CDV) is a member of the genus *Morbillivirus* of the family Paramyxoviridae. It causes a range of clinical signs, including fever, ocular and nasal discharge, cough, vomiting, diarrhea, as well as progressive neurologic signs (ataxia, muscle twitching, tremors, and seizures). Severe generalized distemper, and severe neurologic manifestations, have a high fatality rate. Canine distemper virus is widespread in dog populations worldwide and also has a wide host range. Other members of the canid family are also susceptible to infection, including African wild dogs, bat-eared foxes, and black-backed jackals (Carpenter et al., 1998; Gowtage-Sequeira et al., 2009), as are felids (lions; Roelke-Parker et al., 1996), hyaenids (spotted hyena, *Crocuta crocuta*: Haas et al., 1996), mustelids (black-footed ferrets, *Mustela nigripes*: Williams et al., 1988), procyonids (raccoons: Lemberger et al., 2005), and phocids (Caspian seal; Baikal seal, *Pusa sibirica*: Kennedy et al., 2000; Mamaev et al., 1995). The virus has caused declines in populations of threatened wildlife, including African wild dogs (Alexander and Appel, 1994) and lions (Roelke-Parker et al., 1996), and the extirpation of what was at the time the last wild population of black-footed ferrets (Williams et al., 1988).

Box 6.1 Which dogs drive rabies transmission to wild carnivores?

The spillover of rabies from dogs to wild carnivores is a high profile issue. However, the epidemiology of rabies in multi-host communities and its transmission from dogs to other carnivores, often across land use boundaries, has only been analyzed at macro-scales. For example, in Zimbabwe outbreaks of rabies among black-backed and side-striped jackals (*C. adustus*) in commercial farmland, where they are the predominant wild carnivore, have been linked to cases among dogs living in neighboring agro-pastoralist communal lands (Bingham, 2005; Bingham et al. 1999a, b; Foggin, 1988). The analysis of similar spillover from dogs to carnivores in protected areas is, however, limited by a paucity of surveillance data in remote regions, and by the naturally low densities of the carnivores concerned (Cumming, 1982). Data on the movements of individual rabid dogs is therefore helpful to fully understand the individual-level mechanisms of transmission between species and across land use boundaries.

Data of this kind are available from a study conducted in a 33 km^2 area of Gokwe Communal Land (GCL) bordering the Sengwa Wildlife Research Area (SWRA) in Zimbabwe. From January 1995 to June 1996 14 adult dogs were radio-collared at homesteads and tracked for ≥119 consecutive days. The 'independent' (i.e., where they moved freely without their owners) and 'accompanied' home ranges (i.e., where they followed their owners' movements) of dogs were calculated. The area was simultaneously monitored regularly for reports of possibly rabid or otherwise diseased dogs. Where direct observation was not possible, case histories for dogs were constructed based on interviews with witnesses. Samples of brain stem were taken from the base of the skull from any dog carcasses reported, and were analyzed using the fluorescent antibody test (FAT). During the study the coverage of the government rabies vaccination program was approximately 25% of dogs (Butler, 1998).

Table 6.1 Demographic and movement data on: (a) $n = 15$ rabid dogs that left their homes and (b) $n = 9$ dogs that stayed and died at home in a study area of Gokwe Communal Land and neighboring Sengwa Wildlife Research Area, Zimbabwe, in 1995–96 (Butler, 1998). Information on the form of rabies and the diagnostic method is also provided.

Case	Age	Sex	Distance traveled (km)	Form	Diagnosis
(a)					
1	Adult	Male	11.1	Furious	FAT +
2	Adult	Male	≥8.7*	Furious	Clinical
3	11 months	Male	≥4.6*	Dumb	FAT +
4	3 months	Female	4.4	Furious	Clinical
5	Adult	Male	≥3.7*	Furious	Clinical
6	Adult	Male	≥2.8	Furious	Clinical
7	Adult	Female	≥1.9*	Furious	Clinical
8	10 months	Male	≥1.8*	Furious	Clinical
9	9 months	Female	1.5	Furious	FAT +
10	Adult	Female	0.9	Dumb	Clinical
11	Adult	Male	≥0.8*	Dumb	Clinical
12	Adult	Male	≥0.8*	Furious	Clinical
13	Adult	Male	≥0.7*	Unknown	FAT +
14	Adult	Female	≥0.5*	Furious	Clinical
15	8 months	Female	0.3	Furious	Clinical

continued

Box 6.1 *Continued*

Table 6.1 *Continued*

Case	Age	Sex	Distance traveled (km)	Form	Diagnosis
(b)					
16	6 months	Male	0	Dumb	Clinical
17	6 months	Male	0	Dumb	Clinical
18	6 months	Male	0	Dumb	FAT +
19	Adult	Female	0	Furious	FAT +
20	1 month	Male	0	Unknown	Clinical
21	1 month	Female	0	Unknown	Clinical
22	6 months	Male	0	Dumb	FAT +
23	Adult	Male	0	Dumb	Clinical
24	9 months	Female	0	Furious	Clinical

*Unknown dogs that entered the study area.

Twenty-four suspected dog rabies cases were recorded, of which seven were tested by FAT and all were found to be positive. Sufficient clinical and circumstantial information was gathered for the remaining 17 to be confidently diagnosed as rabid, having been bitten or in close contact with the FAT-diagnosed animals. Fifteen cases (63%) wandered from their own home with the onset of clinical signs (Table 6.1). Among those that wandered, most exhibited furious rabies (11 of 14). Most dogs were also adults (10 of 15), and of these most were males (7 of 10). Of six dogs that traveled ≥2 km, five were males. However, 10 of the 15 cases were unknown dogs from outside the study area and therefore could have traveled considerably greater distances than those recorded. Two of these (cases 3 and 5) traveled through the study area and into the SWRA, and both were males. Nine cases (37%) stayed and died at home (Table 6.1). A greater proportion (five of seven) exhibited dumb rabies (Fishers Exact Test, $p = 0.056$) and were juveniles (seven of nine; Fishers Exact Test, $p = 0.089$) relative to dogs that wandered.

Assuming that those dogs that left their homesteads were more likely to encounter wild carnivores, either within the GCL or SWRA, and that the probability of contact increased with distance traveled away from their home, then only a minority of these cases were of significance for interspecies and transboundary transmission. The greatest confirmed distance traveled was by a radio-collared adult male that traveled 11.1 km in one night, leaving its independent range but remaining within its accompanied home range, visiting a store and school frequented by its owners in GCL, before returning home and dying (Figure 6.2). However many of the dogs entered the study area from unknown locales, indicating potentially extensive movements by a subset of infectious animals. Foggin (1988) also recorded cases of two rabid dogs traveling 25 km and 35 km in Zimbabwean communal lands, supporting the contention that a minority of animals are responsible for the transmission of rabies over large distances, and hence across land use boundaries. The limited sample size from the GCL–SWRA study area suggests that these are most likely to be adult dogs exhibiting furious rabies, and adult males in particular, perhaps related to their larger independent home ranges (Butler et al., Chapter 5), and hence they may pose the greatest risk of transmission to wild carnivores on wildlife reserve boundaries in Africa.

continued

Box 6.1 *Continued*

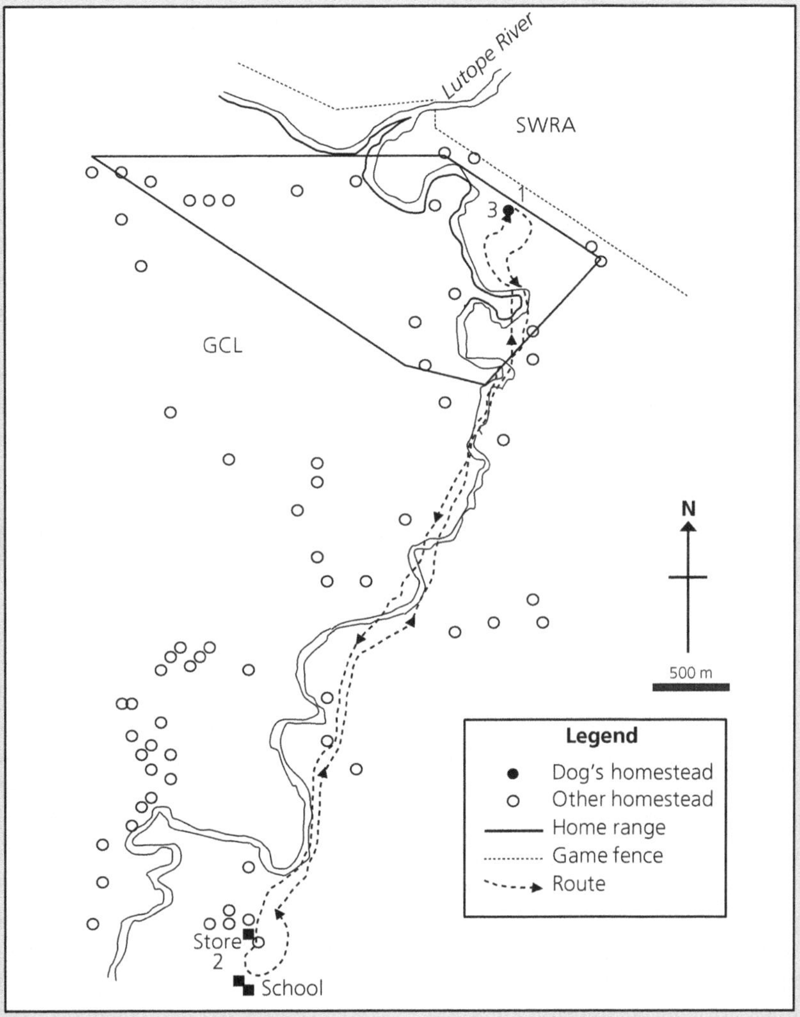

Figure 6.2 Movements of a radio-collared rabid dog (Case 1 in Table 6.1) over three days in the Gokwe Communal Land (GCL) and Sengwa Wildlife Research Area (SWRA), Zimbabwe. (1) June 13, 1996: Starts behaving aggressively, barking and chasing livestock. Bites a sheep and a 3-month old dog in the homestead. Stops eating. (2) June 14, 1996: Stays in home during day, barking but not eating. Leaves homestead at 18:00, wandering out of independent home range along the Lutope River, avoiding other houses. Reaches a store and school at 21:00 and is stoned and chased by people; (3) June 15, 1996: Injured and bleeding, the dog returns home along the same route, arriving at 5:00. Lies in house for the rest of the day. Dies at 17:00.

Based on phylogenetic analyses of the haemagglutinin gene, CDV strains cluster in at least nine lineages within different geographic areas (America-1, America-2, Asia-1, Asia-2, Europe-1, European wildlife, Arctic-like, South America, and Southern Africa; Martella et al., 2006; McCarthy et al., 2007). Several studies have shown that distinct genotypes of CDV, belonging to different lineages, can circulate in dog and wildlife populations in the same geographic area during a relatively short time period (Benetka et al., 2011; Nikolin et al., 2012). Different variants may have different phenotypic effects, which may affect morbidity and mortality rates in the same host population (Lednicky et al., 2004). By contrast, studies of other outbreaks show that the same CDV variant may circulate among susceptible animals of several host species in a given geographic area, with evidence of frequent interspecies transmission (Carpenter et al., 1998). Concurrent infections have also been shown to contribute to higher rates of mortality in CDV epidemics (Munson et al., 2008). This potential variation in factors influencing the host–pathogen dynamics and expression of disease makes it difficult to elucidate the role of a single species in the maintenance and transmission of CDV in multi-host communities, through either epidemiological or phylogenetic analyses.

6.2.3 Canine parvovirus

Canine parvovirus (CPV) was identified in the late 1970s as the cause of a new disease in dogs, characterized by hemorrhagic enteritis and a high mortality rate. Phylogenetic studies and analysis of historical samples suggested that the virus emerged a few years prior to this, in Europe or Eurasia, after which it spread rapidly to dog populations around the world (Hoelzer and Parrish, 2010). Canine parvovirus is one of three closely-related parvoviruses that infect members of the order Carnivora, the others being feline panleukopenia virus (FPV) and mink enteritis virus; however, as there is no evidence that dogs are able to transmit infection with either of these two viruses (Hoelzer and Parrish, 2010), we focus here on infections in wildlife with CPV.

Canine parvovirus is extremely stable in the environment, able to survive for several months. Large amounts of virus are shed in the feces of infected animals, but for only relatively short periods (7–10 days) before the immune system rids the host of the infection (McCaw and Hoskins, 2006). Indirect transmission through environmental contamination therefore likely plays an important role in the maintenance and transmission of this virus. The virus requires actively dividing cells to replicate, resulting in more severe clinical illness in young, rapidly-growing animals. In populations in which the virus is endemic, disease is usually seen in young animals after 2–4 months of age, when maternally-derived antibodies have declined to low levels.

Among free-ranging wild canids, CPV infection has been confirmed only in coyotes (*C. latrans*; Evermann et al., 1980) and gray wolves (*C. lupus*; Mech et al., 1997), where the virus has been associated with mortality and poor pup survival in particular (Gese et al., 1997; Mech and Goyal, 1993, 1995). High prevalence of antibodies to parvovirus has also been found in these species, as well as in black-backed, golden (*C. aureus*), and side-striped jackals (Alexander et al., 1994), gray foxes (Davidson et al., 1992), island foxes (*U. littoralis*; Garcelon et al., 1992), kit foxes (*V. macrotis*; McCue and O'Farrell, 1988), swift foxes (*V. velox*; Miller et al., 2000), and African wild dogs (Alexander et al., 1993; Creel et al., 1997). However, serological tests are unable to distinguish between exposure to CPV or other related parvoviruses. Other canids considered susceptible to CPV include bush dogs (*Speothos venaticus*) and maned wolves (*Chrysocyon brachyurus*), but infection has not been confirmed in either species, whereas FPV has been isolated from both. Although the original CPV virus was unable to replicate in domestic cats (*Felis catus*), the later variants that soon replaced it (CPV-2a, CPV-2b, CPV-2c) are able to infect cats and related wild carnivores. These CPV variants have been detected in free-ranging leopard cats (*Prionailurus bengalensis*; Ikeda et al., 1999, 2000), captive cheetahs (*Acinonyx jubatus*; Steinel et al., 2000), a captive tiger (*Panthera tigris*; Steinel et al., 2000), a captive raccoon (Kapil et al., 2010), and a captive red panda (*Ailurus fulgens*; Qin et al., 2007). Infection with CPV in cats generally results in a milder disease than that seen in dogs or that which is caused by FPV infection in cats (Hoelzer and Parrish, 2010).

Among wild carnivores, CPV has been isolated from animals with acute enteric illness, as well as from animals displaying no apparent clinical signs. However, despite the apparent ubiquity of exposure of wild carnivores to parvoviruses, few studies have demonstrated an impact of CPV on wild populations, and in no cases has the role of dogs in the maintenance and transmission of the virus been explored in detail.

6.2.4 Other microparasites

Although CPV, CDV, and rabies virus might be considered the 'big three' when dog–wildlife disease risks are considered, there are a multitude of other pathogens worthy of concern. For instance, surveys of dogs have revealed high levels of exposure to, or active infection by, viruses (such as canine hepatitis virus, canine coronavirus, canine herpes virus, canine parainfluenza virus, and canine adenovirus), bacteria (such as *Leptospira interrogans* and *Ehrlichia canis*), and protozoa (such as *Neospora caninum* and *Babesia* spp.), often with these taxa co-occurring, in localities where these parasites also occur in wildlife (Alexander et al., 2010; Curi et al., 2010; Fiorello et al., 2004; Woodroffe et al., 2012). These parasites can have significant health impacts on individual dogs, and as such it is parsimonious to assume similar effects on wild carnivore taxa. Such effects are, however, almost entirely unstudied.

6.3 Reservoirs of infection

Of the 374 species of parasites that cause diseases in domestic carnivores (dogs and cats), 91% are multi-host pathogens (Cleaveland et al., 2001), meaning that infection is not limited to a single species, but rather may occur in multiple host species, including humans, domestic livestock, and wildlife. Thus, the vast majority of pathogens of dogs are also capable of causing infections in other species, including wildlife. But do dogs serve as 'infection reservoirs' within these multi-host systems? Answering this question may be important for the design of effective disease control programs, and requires a firm understanding of the concept of 'infection reservoirs'. Haydon et al. (2002) propose that reservoirs can only be understood with reference to defined target populations, for example a threatened population of a wildlife species. A reservoir can then be defined as "one or more epidemiologically connected populations or environments in which the pathogen can be permanently maintained and from which infection is transmitted to the defined target population" (Haydon et al., 2002).

A key conceptual threshold in this respect is that of critical community size (CCS), which defines the persistence of infections as a function of population size (Bartlett, 1960). Pathogens can persist in populations larger than a certain threshold size (maintenance populations), whereas smaller populations cannot maintain a pathogen independently (non-maintenance populations). Nonetheless, smaller populations can, together with other maintenance or non-maintenance populations, constitute part of a reservoir. Following these definitions, for a pathogen that can infect dogs, a given population of dogs that is epidemiologically linked to the target wildlife population may then serve as: (1) a maintenance population capable of maintaining the pathogen indefinitely in the absence of transmission from other host populations, or a maintenance population within a maintenance community, with other populations contributing to the reservoir as maintenance or non-maintenance populations (see Section 6.3.1); (2) a non-maintenance population that forms part of a maintenance community, in which the pathogen is maintained through transmission between several non-maintenance populations (that may include the target wildlife population); or (3) neither a maintenance population nor part of a maintenance community (i.e., not required to maintain the infection), but a source that nonetheless transmits infection from such a maintenance population/community directly to the target population. The particular contribution of dogs in a given host–pathogen community may have important implications when considering management options for a target wildlife population.

6.3.1 Dogs as reservoirs: rabies in the Serengeti ecosystem

A detailed study of rabies in the Serengeti ecosystem, Tanzania, illustrates the first scenario well, while also highlighting the difficulties of identifying all or even just essential components of pathogen reservoirs in complex systems (Lembo et al., 2008). Several 'spillover' hosts,[2] including endangered

[2] The term 'spillover' is often used in a generic sense when referring to a parasite that has been transmitted from one taxon (generally humans or domestic animals) to another (wildlife), regardless of the potential for the receiving taxon to sustain the parasites (the reverse pattern, transmission for wildlife to domestic animals, is sometimes termed 'spillback'). More precisely, however, not all interspecific transmission events represent spillover, as some pathogens exist in a multi-taxon host community (Fenton and Pedersen, 2005; see Section 6.4.1). Nor are all such events, even when dealing with similarly pathogenic agents, likely to have similar outcomes.

wildlife, can be considered target populations of concern in the context of rabies transmission. Despite the predominance of reported dog rabies cases in African ecosystems, wildlife hosts are considered capable of maintaining distinct rabies virus variants, such as mongoose rabies, or independent cycles of transmission of the canid rabies variant, such as in black-backed and side-striped jackals, and bat-eared foxes (Davis et al., 2007). Lembo et al. (2008) attempted to tackle the issue of rabies reservoirs in the species-rich community of the Serengeti that is comprised of a wide range of potential maintenance hosts. A combination of 'practical indicators' and quantitative analyses were employed to determine that dogs to the west of the Serengeti occur at high densities and are the only maintenance population of the rabies reservoir in the ecosystem, with wild carnivore populations contributing to the reservoir as non-essential components. This conclusion was supported by several lines of evidence including: (1) the identification of a single southern Africa canid-associated rabies virus variant (Africa 1b) from a range of species and a lack of species-specific virus–host associations; and (2) genetic and epidemiological associations between rabies cases in dogs and other species suggestive of cross-species transmission leading to short-lived chains of transmission with no evidence for long-term persistence in wildlife. In other parts of Africa, wild Canidae such as jackals can maintain rabies cycles independently of dogs, but cycles have not been sustained longer that a few years without the contribution of dogs, due to the limited geographical extent and inter-connectedness of their populations (Bingham et al. 1999b; Bingham, 2005; Zulu et al. 2009). However, the absence of independent wildlife maintenance hosts in a species-rich area such as the Serengeti indicates that removing the contribution of dogs from the reservoir system through canine rabies control programs would have the benefit of protecting populations of concern, as well as of resolving outstanding questions related to the dynamics of complex reservoir systems (Fitzpatrick et al., 2012).

6.3.2 Dogs as reservoirs in other host–pathogen systems

The role of dogs as reservoirs of pathogens that infect wildlife in other systems is less clear-cut. Woodroffe et al. (2012) found that increasing contact with domestic dogs was associated with increased exposure to CPV, *E. canis* and *N. caninum* in African wild dogs, but not with exposure to CDV or canine coronavirus. This is consistent with the hypothesis that domestic dogs serve as reservoir hosts of the first three pathogens; however, as the authors point out, the evidence is insufficient to confirm this. Indeed, increased pathogen exposure in a target wildlife population following increased contact with a domestic dog population is consistent with any of the scenarios described above: the dog population may be the sole maintenance population, a maintenance or a non-maintenance population within a maintenance community, or merely a transmission vector of the pathogen from a maintenance population/community to the target population. By contrast, the absence of an association between pathogen exposure and domestic dog contact, as was the case with CDV in the study, provides some evidence that domestic dogs are *not* the sole reservoir for this pathogen, and that other wildlife hosts may play an important role in maintaining this pathogen. This has been suggested for CDV by other studies (Craft et al., 2008; Prager et al., 2012), and may have important implications for determining management approaches to CDV in endangered wildlife populations.

Understanding the respective roles of dogs and wildlife populations in maintaining pathogens may also be important when considering other target populations, such as humans or domestic livestock. Zoonotic visceral leishmaniasis is an important vector-borne disease of humans. The principal reservoir hosts are dogs and sandflies, with the latter also acting as the primary vector. However, a number of other animal species may potentially act as reservoirs. In Latin America, crab-eating foxes (*Cerdocyon thous*) show a high prevalence of infection with *Leishmania infantum*, and have long been considered a potential source of human infection. Courtney et al. (2002) investigated the role of crab-eating foxes in maintaining and transmitting this pathogen. They demonstrated that, although foxes became infected following exposure to infected sandflies, they were unable to pass the infection back to naïve sandflies. Foxes most likely acquire the infections in the peri-domestic environment from dogs (via sandflies).

In terms of Haydon et al.'s (2002) framework, when considering either humans or the foxes themselves as the target population, foxes fall outside of the reservoir system, which is made up of two non-maintenance populations (dogs and sandflies). Control of zoonotic visceral leishmaniasis (in humans or wildlife) should therefore focus on either or both of these species.

The importance of correctly identifying the reservoir host before implementation of control measures is illustrated by Chaves et al. (2007), using the example of a closely-related disease, American cutaneous leishmaniasis (ACL). They used a mathematical model developed with parameters from an outbreak of ACL in Venezuela to explore the prevalence dynamics of the disease in humans (the target population) following culling of one of two epidemiologically linked populations: dogs (which are incidental hosts in the case of ACL, Reithinger and Davies, 1999), and donkeys (*Equus africanus asinus*; a true reservoir host). Modeling the culling of donkeys reduced the prevalence of the disease in humans, as expected. However, the model showed that culling dogs would in fact *increase* the prevalence in humans. This is an example of the dilution effect seen in some vector-borne diseases, in which an increase in the relative abundance of a competent vertebrate host (i.e., one able to efficiently transmit the infection back to the vector—in this example, donkeys) at the expense of less competent hosts leads to increased pathogen transmission to all remaining hosts (Ostfeld and Keesing, 2000). Such examples show that, although it may be difficult to establish the role of dogs in the reservoir system of pathogens of wildlife, such a step is advisable before implementing any control measures aimed at dogs as putative reservoir hosts.

Box 6.2 Are dogs an important source of macroparasites for wildlife?

While microparasite transmission to wildlife is the primary disease-associated concern where dogs and wild carnivores interact, spillover of macroparasites (both endoparasites and ectoparasites) may also warrant scrutiny. Dogs, and free-ranging dogs in particular, may harbor a diverse community of macroparasites (e.g., Bryan et al., 2011; Oliveira-Sequeira et al., 2002; Papazahariadou et al., 2007; Sowemimo, 2009). Do these parasites represent a significant risk to wildlife? On the one hand, many of these parasitic species may require intermediate hosts to complete their life cycles, thereby muting the potential for spillover events. Furthermore, the receipt of food and healthcare subsidies from people may reduce the prevalence and abundance of parasitic species for any given dog and across dog populations. While this has not been studied in dogs, work on humans, livestock, and wildlife that received increased food identified a decrease in parasitism, especially by indirectly transmitted parasites, most likely due to a reduction in dietary diversity, associated reduced exposure to infective stages of parasites, enhanced nutrition, and enhanced host immune response to nematode infections (Ezenwa, 2004; Koski and Scott, 2001; Monello and Gompper, 2011). On the other hand, the high densities of dogs may enhance the prevalence of parasitic species in these intermediate hosts and thereby increase opportunities for transmission to native wildlife species who feed on these intermediate hosts. Currently, such scenarios are conjectural as relatively little attention has been paid to the role of dogs in the community ecology of macroparasites in wildlife host communities. This is unfortunate, as multi-host species of macroparasites may have strong impacts on less common species when a sympatric reservoir species occurs at high densities and results in a sustained and strong force of infection (e.g., Tompkins et al., 2001), a process sometimes termed apparent competition because the host dynamics are suggestive of interference or resource competition (Vanak et al., Chapter 3).

For directly transmitted macroparasites and ectoparasites, however, several likely cases of spillover have been documented. For example, the dog biting louse *Trichodectes canis* was introduced from dogs to coyotes and wolves in south-central Alaska, USA, resulting in moderate-to-severe pediculosis and a virtual 100% prevalence rate (Woldstad, 2010;

continued

Box 6.2 *Continued*

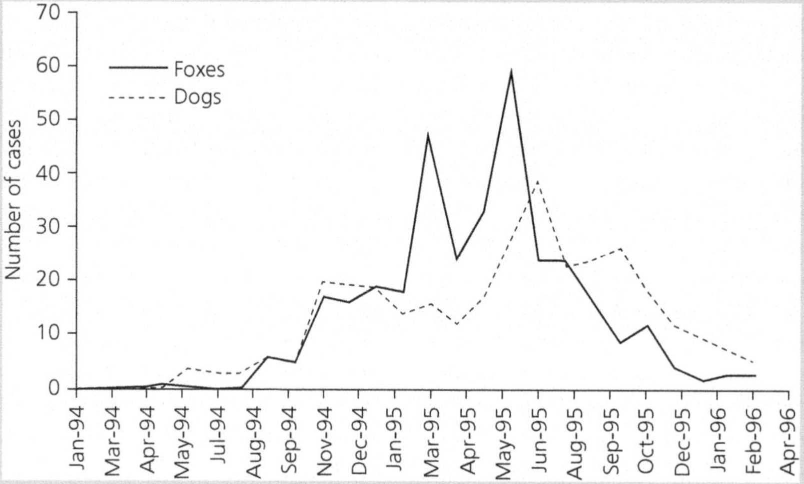

Figure 6.3 Temporal correlation in the occurrence of cases of sarcoptic mange in foxes and dogs in Bristol, UK, from 1994 t o 1996. Figure from Soulsbury et al. (2007). Reproduced with permission from John Wiley & Sons.

Gardner et al., 2013). To combat this infection, within a 13,000 km^2 zone of infection, oral ivermectin-injected baits were aerially distributed at den and rendezvous sites. Treated wolf packs were lice-free the winter following treatment whereas control packs remained infested. Thus the louse can be eliminated from the wolf population in some circumstances, although reinfection from either dogs or dispersing wolves from other infected regions remains possible.

Perhaps the most important ectoparasite from a dog–wildlife spillover perspective is *Sarcoptes scabiei*, a mite species that is the causative agent of sarcoptic mange. The introduction of infested domestic animals and the success of the mite in adapting to new host species can lead to epidemics in previously mange-free wildlife populations (Pence and Ueckermann, 2002). Thus endemic and epidemic mange in dogs could result in spillover into wildlife. This interpretation is, however, complicated by the observation that experimental cross-contamination of *Sarcoptes* mites between hosts of different species often fails, and that mites from sympatric host species show a lack of gene flow or recent admixture between carnivore-, herbivore-, and omnivore-derived *Sarcoptes* populations (Arlian, 1989; Rasero et al., 2010). On the other hand, the mite variety found on dogs can sometimes be transmitted to other carnivores: Soulsbury et al. (2007) documented co-occurrence of mange in dogs and red foxes in Bristol, UK (Figure 6.3), although they suggest that the direction of transmission was likely from foxes to dogs. Furthermore, Gakuya et al. (2011) show that African predators from the Masai Mara, Kenya, may become infested through consumption of their primary prey species. Thus, while much remains unclear regarding mange ecology and epidemiology, endemic mange in dogs may represent a risk to other carnivores, and perhaps also to large carnivores that prey on dogs (Butler et al., 2004, Chapter 5).

Many other macroparasites are likely worthy of a closer look, especially those parasites that have high prevalence rates in dogs and are also able to infect other host species. Examples include *Dirofilaria immitis* and *Echinococcus multilocularis*. The former is globally distributed and mosquito-transmitted, and is the causative agent of heartworm, which can affect many other species, especially native species of canids. *Echinococcus multilocularis* is the cause of human alveolar echinococcosis. A global increase in the range of this parasite has been attributed in part to dog population growth and the translocation of dogs by people (Davidson et al., 2012).

6.4 How species boundaries are crossed

Transmission of a parasite from an infectious host to a susceptible host, whether of the same or a different taxon, requires contact of the two hosts. The extent of contact necessary to allow transmission, however, depends on the parasite. In some cases (e.g., rabies virus, CDV) direct interaction of the animals is necessary as transmission primarily involves oronasal exposure to aerosolized respiratory secretions or bite wound exposure to saliva. In other cases (e.g., CPV, canine coronavirus) exposure occurs via ingestion or inhalation of infected fecal materials. In still other cases infection requires intermediate hosts or arthropod vectors. Thus at a population scale, the contact rate between dogs and wildlife needs to be considered both in the context of a specific parasite and in light of the capacity of the various taxa that comprise a community to act as competent hosts (that is, to not only become infected but also to allow pathogen replication within the host and to further transmit the pathogen). It is therefore necessary to recognize that parasites often occur in a multi-host community and that the composition of the community may mediate the impact of the pathogen on any given community member.

6.4.1 The ecology of multi-host pathogens

Haydon et al.'s (2002) model for identifying reservoirs of infection is limited in that the only epidemiological parameter considered is host population size. Fenton and Pedersen (2005) provide an extended conceptual framework (Figure 6.4) for the transmission of multi-host pathogens, which takes into account variance in rates of interspecific and intraspecific transmission. This approach is useful for considering management options for the control of transmission of pathogens from dogs to wildlife (see Section 6.5).

In this framework, a dog population (H_1) represents a maintenance or reservoir host population. The pathogen is transmitted at some rate (f_{12}) to the wildlife population. Once within the wildlife population (H_2), the pathogen infects new individuals at some rate (f_{22}). Depending on the transmission rates f_{12} and f_{22}, one can categorize the likely transmission dynamic into four scenarios: spillover

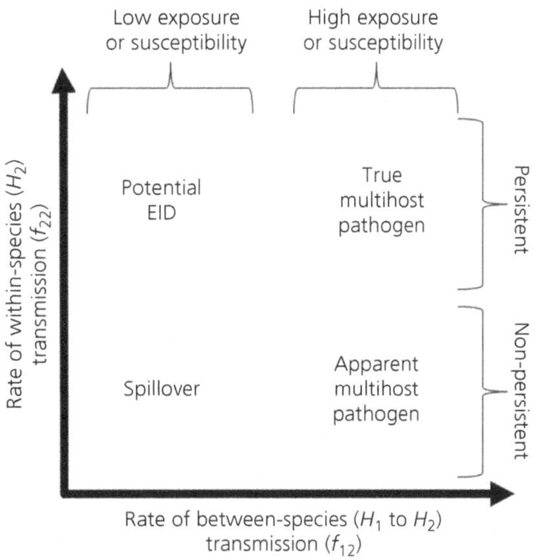

Figure 6.4 A framework for visualizing how the combined rates of parasite transmission (f) among individuals within a wildlife species (H_2) and between dogs (H_1) and wildlife will result in the emergence and persistence of different categories of pathogens, including potential EIDs (Emerging Infectious Diseases) in the wildlife species. Figure modified from Fenton and Pedersen (2005).

(low between species transmission and low rates of transmission within the wildlife species), potential emerging infectious disease (low between species transmission but high rates of transmission within the wildlife species), and two types of multi-host pathogen scenarios. In the latter cases, these can reflect that the pathogen is a true multi-host pathogen with high rates of between species transmission as well as high rates of within species transmission, or it may only seem this way, with high between species transmission rates masking low within species transmission rates (Fenton and Pedersen, 2005).

One of the take home messages of this framework is that if f_{22} is low, then the pathogen will not persist in the wildlife species, but rather will 'fade-out' because R_0 is <1, although spillover impacts can be variable.[3] Therefore, in many cases,

[3] R_0, the basic reproductive number, is an epidemiological measure of the growth rate of a parasite population. For microparasites, R_0 is defined as the average number of infected hosts that each infectious host generates. When R_0 is <1, the pathogen will fade out and be lost from the population. When R_0 is >1, the pathogen will spread in the host population.

spillover should not be of significant concern. For example, Fiorello (2004) modeled the likelihood of CDV epidemics occurring in jaguar (*Panthera onca*) in Bolivia. While jaguars might become infected by contact with dogs, the likelihood of the infected jaguars transmitting the pathogen to other members of the population was low given the low density of jaguars in the region, and thus there was a low likelihood of an epidemic. Similarly, in Kenyan rangelands direct transmission of rabies from domestic dogs to African wild dogs might occasionally occur, but opportunities for onward transmission to other wild dog packs are limited because pack interactions are relatively rare (Woodroffe and Donnelly, 2011). This finding helps explain how the African wild dog population in this region of Kenya has grown, despite occasional rabies outbreaks and high domestic dog populations near human settlements.

In other circumstances, however, the impact of spillover events can be devastating because intraspecific transmission rates (f_{22}) are sufficient to allow the pathogen to persist for a short period, with associated high rates of mortality, prior to fade out. Perhaps the best documented example of this is the CDV epidemic that occurred among lions in the Serengeti. Between late 1993 and 1994, CDV killed approximately 30% of the Serengeti lion population (Roelke-Parker et al., 1996) and also affected other Serengeti wildlife and the Mara lion population (Kock et al., 1998). Viruses recovered from Serengeti wildlife were indistinguishable from an isolate from a dog adjacent to the Serengeti National Park (Carpenter et al., 1998; Haas et al., 1996; Harder et al., 1995; Roelke-Parker et al., 1996), and serological, demographic, and case-surveillance data pointed to the large population of dogs to the west of the Serengeti as the most likely source of infection for wildlife (Cleaveland et al., 2000). Although Serengeti lion prides are sufficiently well-connected to sustain lion-to-lion transmission for a brief period, multiple-host transmission likely fuelled the outbreak (Craft et al., 2009). While mechanisms of long-term CDV maintenance and transmission have yet to be resolved, this particular outbreak demonstrated that spillover of CDV from dogs to wildlife can impact population viability considerably.

6.4.2 Persistent spillover and emerging infectious diseases

The other scenario for which spillover might be particularly problematic is the case wherein the rates of dog–wildlife transmission (f_{12}) are so high that, despite low intraspecies transmission rates (f_{22}), the pathogen repeatedly is introduced and drives the mortality rate higher than the birth rate. In a sense, such scenarios might be thought of as a variant of an emerging infectious disease in the wildlife population in that, while the pathogen may not be new to the host, its consistent prevalence and incidence are far above what could be supported in the absence of dogs.

An example of this scenario is the transmission dynamics of rabies in the Ethiopian wolf, one of the world's rarest Carnivora. The longer-term decline of this species can be traced to altered landscapes, which have resulted in a global population of only several hundred adult animals divided into about seven isolated subpopulations. A more proximate and insidious concern, however, is rabies transmission from sympatric dogs (Haydon et al., 2006; Laurenson et al., 1997; Randall et al., 2006). Notably, the rabies epidemics are recurrent, having occurred in 1991–92, 2003–04, and 2008–09, and resulting in high mortality in affected subpopulations. Vaccination campaigns for local dogs were implemented, but this was insufficient to curtail the repeated interspecific transmission events (Laurenson et al., 1997), and the focus now is on vaccination of the Ethiopian wolves themselves (Haydon et al., 2006; Johnson et al., 2010; Knobel et al., 2008; Randall et al., 2006). Note that the assumption here is that rabies would not be able to persist within the Ethiopian wolf population: while the pack structure facilitates transmission among geographically associated individuals, the population as a whole is too small to sustain rabies without periodic input, and geographic isolation of subpopulations would hinder the spread of rabies.

The emerging infectious disease (EID) is an eco-epidemiological construct that describes those diseases caused by pathogens that are new to a species or a region (Daszak et al., 2000; Williams et al., 2002). While occasionally these pathogens are truly 'new' to the taxon of interest, or indeed even new to science, in most cases the idea of novelty is broadly defined based on a strong increase in prevalence within the past several years or decades. An important nuance for differentiating EIDs is that, once in the new host population, the intraspecific transmission rate is high (that is, R_0 >1 and f_{22} is thus sufficient to maintain the pathogen in the host to which it was introduced) such that the pathogen persists (although in situations involving small initial population size, R_0 could only be >1 for a brief period before either extinction or fade-out, and so persistence in such scenarios is a likely function of periodic reintroduction). Most EIDs are caused by new variants of previously known microparasites (e.g., influenza, CDV, parvoviruses) or by previously known variants colonizing a new host or new geographic region. Macroparasites and ectoparasites may, however, also generate EIDs (see Box 6.2). The global size and density of dog populations and the veterinary attention to these populations facilitates the identification of new EIDs within dogs; recent examples include CPV and canine influenza (Crawford et al., 2005; Parrish and Kawaoka, 2005). The close contact of dogs and humans also means that dogs are an important source of EIDs and re-emerging diseases in humans (Salb et al., 2008).

The role of dogs as a source of EIDs of wildlife is unclear, perhaps in part because it is rare for wildlife populations to be closely monitored to the point that EIDs that do not cause severe impacts are recognized. Furthermore, the genetic and evolutionary mechanisms that determine how pathogens traverse species boundaries and adapt to new taxa are only partially understood (Parrish et al., 2008), and well-documented examples of virus emergence, in which both the source and recipient species are known with certainty, are rare. For example, canine parvovirus type 2 (CPV-2) emerged in the 1970s, likely from FPV or an FPV-like variant, and then was replaced worldwide by CPV type 2a (CPV-2a). This strain in turn mutated into variants CPV-2b and CPV-2c, and local variants are increasingly common (Hoelzer and Parrish, 2010; Hoelzer et al., 2008). While dogs have been implicated in the emergence of pathogenic strains of CPV in wild canids, other common wildlife taxa may also have been important (Allison et al., 2012), indicating that, even for EIDs it is necessary to consider such microparasites in a broader, multi-host framework. Similarly, the epidemics of CDV in felids (Roelke-Parker et al., 1996), pinnipeds (Kennedy et al., 2000), and non-Carnivora mammals (Kameo et al., 2012; Sun et al., 2010) may also represent EIDs but, as with CPV, the role of dogs in relation to other carnivore taxa is often unclear. Wildlife may mediate the transmission of CDV from dogs to these taxa, or may be the reservoir source with respect to these particular outbreaks.

6.4.3 True multi-host parasites—how pathogens persist without dogs

In a landscape where dogs are abundant and where dogs often closely interact with both wildlife and humans, two broad generalizations seem to persist. The first is that the high density of (often unvaccinated) dogs suggests that dogs are an underlying cause of disease outbreaks in wild carnivores. For example, the occurrence of antibodies to four microparasites in giant pandas (*Ailuropoda melanoleuca*) and dogs in Wolong Reserve, China was reported as suggestive of the risk of pathogen transmission from dogs to pandas (Mainka et al., 1994). Potentially spurious conclusions that dogs were the source of CPV infections in maned wolves have also been drawn, on the basis of a temporal association between the emergence of the disease in dogs and the observation of clinical disease and microscopic lesions in maned wolves, despite the fact that CPV infection was not confirmed. It is known that maned wolves are susceptible to FPV, and that outbreaks due to this virus occurred prior to the emergence of CPV (Barker and Parrish, 2001). The second generalization is that wild carnivores are often the cause of disease outbreaks in dogs. For example, in the United States coyotes and dogs are both parasitized by *Dirofilaria immitis*, the causative agent of heartworm, and coyotes are sometimes purported to be the reservoir (Miller and Crosbie, 2011; Paras et al., 2012) despite population densities that are likely far below that of dogs. In these examples the authors may be correct; pathogens maintained by dogs may be a risk to giant pandas and maned wolves, and coyotes may be the reservoir of *D. immitis*. The point, however, is the failure to recognize that many of the parasites of interest may persist across a broad community of hosts, and therefore an important question is whether the parasite could persist even in the absence of dogs (e.g., Prager et al., 2012) because, from the parasite's perspective, a CCS comprised of equally competent taxa allows persistence despite the fact that the CCS is comprised of multiple host taxa.

6.5 Managing interspecies transmission

A variety of management strategies can be applied in an effort to control the diseases that dogs may bring to bear on wildlife. When considering such strategies, it is necessary to address two fundamental questions: (1) are the consequences of transmission of the parasite from dogs to a wildlife population severe enough to consider removing the contribution of that dog population to the maintenance and transmission of the parasite (for instance through vaccination, treatment, population management, or the creation of physical barriers); and (2) if managers are logistically able to do this, what is the likelihood that the parasite will still persist indefinitely in that broader animal community (i.e., what are the roles of other populations in maintaining the parasite)? The latter question in particular requires that an epidemiologic framework is used to address the issue.

Underlying epidemiological approaches for management scenarios can be understood using a conceptual framework to describe the configurations of a host–pathogen community that may lead to disease emergence in a target host. Figure 6.5 provides a schematic diagram for just such an approach. It is a two-host, one-pathogen model in a traditional Susceptible-Infected-Recovered (SIR) framework. Quantitative formulation and discussion of the model in SIR or SI (e.g., for pathogens such as rabies from which recovery does not occur) form can be found in Dobson (2004) and Fenton and Pedersen (2005). The pathogen is endemic within the dog population H_1 such that individual dogs are either susceptible (S_1), infected (I_1), or recovered (R_1). Susceptible and recovered individuals are able to produce new susceptible individuals. Susceptible individual dogs are converted to infected individuals as a function of the per capita rate of contact between the two groups (f_{11}), which can be density dependent.

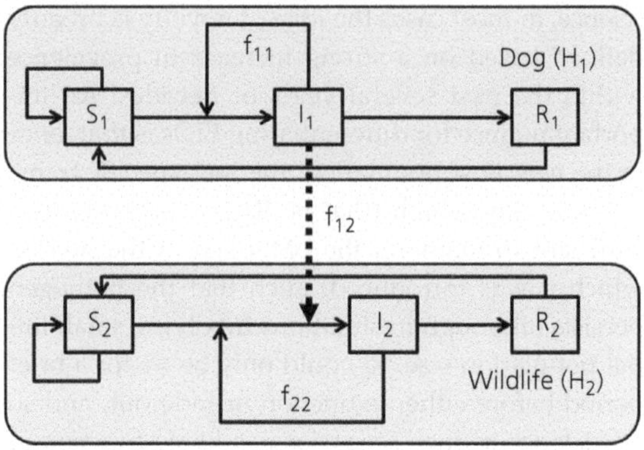

Figure 6.5 Framework for a simplified two host (*H*) species Susceptible–Infected–Recovered (SIR) model. Intraspecific transmission (f_{11}, f_{22}) represents the conversion of susceptible to infected individuals due to conspecific contact. Once infected by a highly pathogenic parasite, infected individuals either mount a sufficient immune response and recover, or die. Interspecific transmission (f_{12}) represents, in this context, contact between an infected dog and a susceptible wildlife individual, with subsequent conversion to an infected stage.

Note that the pathogen may or may not also be endemic in the wildlife population (H_2), whose model mirrors that of the dog population. However given a low density of the wildlife population, f_{22} is likely low. Contact between infected dogs and susceptible wildlife is shown as f_{12}. Because dog populations occur at high density, the density dependent nature of f_{12} implies high rates of transmission and associated conversion of S_2 to I_2. Hypothetically, there might also be transmission of the pathogen from wildlife to dogs (f_{21}), although in most wildlife conservation constructs the importance of this is either ignored or deemed as presumably unimportant in influencing the dynamics of the parasite in dogs compared to f_{11}.

6.5.1 The Laurenson framework

Assuming a target-reservoir system in which a dog population is epidemiologically linked to a target wildlife population, and may therefore pose a potential risk of transmission of pathogens to that target population, Laurenson et al. (2004) propose four general management approaches to reduce the risk of infection and/or spread in the target population: (1) do nothing, (2) target control, (3) blocking control, and (4) reservoir control. The assumptions, advantages, and disadvantages of these approaches, as well as their real-world likelihoods of being enacted and culminating in management success, are discussed at length by Laurenson et al. (2005) and Breed et al. (2009).

It is important to recognize that the first approach—non-intervention—is itself a valid management option (Laurenson et al., 2005). Introduction and spread of a pathogen may not always impact populations. For example, serological evidence of widespread CDV exposure in previously naïve African wild dog packs in the Okavango Delta, Botswana, in 1993–97 was not associated with changes in pup survival or disease-related mortality in adults (Alexander et al., 2010). Even in populations that experience impacts related to episodic invasion of pathogens, conservation efforts might be better directed towards securing large populations or connected metapopulations that are able to absorb or recover from these perturbations (Alexander et al., 2010), rather than attempting to reduce transmission of the pathogen itself. On the other hand, there are circumstances in which the direct threat posed to population viability by a pathogen in a reservoir dog population is sufficient to warrant intervention, which may then take the form of target control, blocking control, or reservoir control.

For target control, knowledge of the reservoir is not required as control efforts are directed within the wildlife target population (for instance, via vaccination or treatment of infected wildlife).[4] For example, following recognition that a subspecies of Arctic foxes was threatened by introduced, dog-associated, *Otodectes* mange, treatment of fox cubs was carried out with anti-parasitic compounds to increase survival (Goltsman et al., 1996). In contrast, blocking control aims to reduce or halt transmission between the source reservoir and target wildlife populations and therefore requires knowledge of source populations within the reservoir, although not necessarily a complete understanding of reservoir infection dynamics. Target control and blocking tactics thus focus on reducing the effects of the pathogen on the wildlife population without necessarily managing the reservoir (dog) population. For infection risks to be fully eliminated, however, control measures must directly address the reservoir host or maintenance populations within the reservoir. Thus, reservoir control focuses on indirectly benefiting the wildlife population by directly managing the reservoir population, through, for example, vaccination or population control, without direct management of the target population. Such approaches attempt to decrease the number of susceptible dogs below the threshold required to sustain the parasite. However, for effective reservoir control an exhaustive identification of all constituent populations of the reservoir is necessary.

The choice of which of the above approaches to apply will depend on the location of the particular host–pathogen system within the community–epidemiology continuum shown in Figure 6.4, as well as the feasibility of particular control options. Assuming that control measures are rarely 100% effective (i.e., that it is not possible to reduce the rate of transmission to zero and maintain it there), the goal of control then becomes to reduce rates of either within- or between-species transmission (or both) so that the host–pathogen system is relocated to the 'spillover' quadrant. In this case, infection in the target wildlife population, possibly followed by short chains of transmission, may still be expected to occur before fade-out of the pathogen, but more extensive population impacts are largely avoided. An exception to this is the case of small populations that may be vulnerable to stochastic extinction following even short chains of pathogen transmission, as discussed above. In these cases, contingency plans should be in place for target intervention in the case of spillover. For example, Haydon et al. (2006) predict that, in the case of rabies spillover from dogs to Ethiopian wolves, 40% of spillover events would fade out with less than ten (and usually less than four) wolves becoming infected. In the remaining 60% of spillover events, the epidemic would almost certainly become large, increasing the probability of catastrophic reduction of the population. Haydon et al. went on to show that strategic reactive vaccination of the wolf population following a spillover event (vaccination of packs in corridors connecting subpopulations, as well as vaccination of even a

[4] Note that Breed et al. (2009) refer to target control as a "targeting the infectious agent." The basic premise is the same: treatment of wildlife prior to exposure or treatment of wildlife to reduce the extent of infection.

small proportion of wolves in a core habitat area) would reduce the probability of catastrophic population reduction to <0.001.

A similar principle would operate in the case of potential EIDs, even in more resilient populations. In these cases, natural impediments occur to the transmission of the pathogen to the target population from the reservoir, for example geographic isolation or restriction in the natural host range of the pathogen. Should these impediments be overcome through either human interference (e.g., translocation of infected individuals) or natural processes (e.g., evolution in pathogen host range), even a single transmission event could have devastating consequences in the target population, because of the inherently high rate of within-species transmission. In this situation, neither blocking control nor reservoir control are likely to be effective, unless the intention is to re-establish the natural impediments to between-species transmission (for example through the extirpation of an introduced, infected, in-contact reservoir population). If these natural impediments cannot be re-established, control in the target population is the only viable option. This could take the form of a prophylactic vaccination campaign of susceptible individuals in the target population before any interspecies transmission, or reactive vaccination after. Treatment of affected individuals in the target population may also be warranted, although this will need to continue until the pathogen fades out naturally or through the effect of other control measures, as treatment alone will not affect the within-species transmission rate markedly (unless implemented very early in the infectious stage, and clearing the animal of the infection).

For an apparent multi-host pathogen, attempting to reduce the rate of within-species transmission in the target population will not have a major impact on the incidence of the disease in this population, as the apparent persistence of the disease in the target is a function of the high between-species transmission rate. Reducing this rate, through either blocking control or reservoir control (both of which have the same intention, namely reducing the force of infection on the target population) will shift the system to the left of the continuum (Figure 6.4), resulting in fewer transmission events of the pathogen into the target population. For example, Rhodes et al. (1998) concluded that rabies in side-striped jackals is not self-sustaining and that the apparent persistence of the disease in this species is due to frequent reintroductions from dog reservoir populations. Following this, it is evident that reducing the force of infection from dogs will reduce the incidence of disease in jackals. There is also strong evidence from the Serengeti ecosystem in Tanzania that rabies in wildlife is not self-sustaining (i.e., is an apparent multi-host pathogen), and that vaccination campaigns in dog populations reduce the incidence of rabies in sympatric wildlife populations (Fitzpatrick et al., 2012; Lembo et al., 2008).

In the case of a pathogen that can independently persist in either a dog or a wildlife population, and where there is substantial between-species transmission (i.e., a true multi-host pathogen), shifting this system to the 'spillover' quadrant of the community–epidemiology continuum would require a reduction in both the within-species transmission in the target population and the between-species transmission rate. If target control is so effective as to eliminate infection in the wildlife population (e.g., through blanket vaccination of the population), or if blocking/reservoir control is so effective as to reduce between-species transmission to zero (e.g., extirpation of a reservoir population), then this becomes a potential re-emerging disease. Such examples can be found in historical and contemporary livestock disease control practices in sub-Saharan Africa (e.g., extirpation of wildlife hosts to eradicate tsetse fly in *Trypanosoma* control campaigns in South Africa, or erection of fences to prevent transmission of foot-and-mouth disease from buffalo reservoir hosts to cattle). As in the case of potential EIDs, contingency plans for target population control still need to be in place in case of any between-species transmission (e.g., vaccination of cattle in the case of a foot-and-mouth disease outbreak).

6.5.2 Reservoir control through vaccination—the Tanzania experience

Vaccination of reservoir host populations may provide a powerful disease management approach for mitigating infectious disease threats to target populations. Well-implemented mass vaccination programs also have the potential benefit of eliminating disease from an area, with implications for wildlife conservation and, in the case of zoonoses, human health. For example, since the second part of the twentieth century, large-scale vaccination of reservoir hosts has been successfully used to eliminate or control rabies in dog and wildlife populations worldwide. The epidemiological theory of infectious disease control based on dog vaccination is described in detail elsewhere (Knobel et al., 2013). In brief, vaccination approaches aim at reaching a critical threshold coverage (P_{crit}) to reduce transmission so that, on average, less than one secondary infection will result from each infected individual.

Given the ethical issues associated with direct vaccination of threatened wildlife (Cleaveland et al., 2006) or the culling of dogs (Box 6.3), the wildlife sector often regards immunization of dog reservoirs as the only acceptable approach to the protection of endangered species. For example, intervention trials involving mass vaccination of dogs against rabies and CDV have been implemented around the Serengeti National Park (north-western Tanzania) since the 1990s (Cleaveland et al., 2007). The initial focus of the campaigns was the higher-density dog population bordering the north-western boundaries of the Serengeti that was considered the main disease threat for wild carnivores within the park (Cleaveland and Dye, 1995; Cleaveland et al., 2000). Vaccination of 60–70% of dogs (the target considered necessary to prevent outbreaks of dog rabies: Hampson et al., 2009) was sufficient to control dog rabies in this area and to reduce bite-injuries from suspected rabid dogs (Cleaveland et al., 2003). However, due to the continuous persistence of rabies in other areas adjacent to the Serengeti, a 'cordon-sanitaire' was subsequently established around the park with mass dog vaccination campaigns still ongoing. Although detailed epidemiological studies demonstrated that, in the Serengeti, dogs are the sole maintenance hosts for rabies, it was also evident that spillover infection from dogs into alternative wildlife hosts occurs, even though these transmission events are not self-sustaining (Lembo et al., 2008). Concerns, however, remained regarding the effectiveness of vaccination targeted at dogs in the presence of intra- and interspecies transmission, especially in areas characterized by greater carnivore species richness. Subsequent modeling approaches demonstrated that vaccination coverage at the level sufficient to control dog rabies is adequate to control rabies also in these multi-host settings (Fitzpatrick et al., 2012). As a result of these vaccination efforts, rabies has now been eliminated in large parts of the Serengeti ecosystem (Lembo et al., 2010), including the Serengeti National Park where no cases of canine rabies have been confirmed for over a decade (Lembo et al., 2008). A substantial conservation outcome of these interventions has been the re-establishment of African wild dog populations in rabies-free areas of the Serengeti (Cleaveland et al., 2012).

With respect to CDV, the effectiveness of dog vaccination as a management approach in the Serengeti remains unclear, partially because the issue of CDV reservoirs in these areas is still unresolved. Epidemiological studies so far have not provided definitive evidence for long-term CDV persistence in either dog or wildlife populations in the Serengeti ecosystem (Lembo, 2007), with the virus continuing to circulate at low levels in both populations, despite the implementation of mass dog vaccination campaigns (Cleaveland et al., 2012). While a clearer understanding of patterns of CDV maintenance in the ecosystem is required, the high costs of CDV vaccines (relative to rabies vaccines) raises questions as to the most cost-effective CDV control approach for the conservation management of wild carnivores in natural ecosystems.

6.5.3 Large-scale oral baiting programs

A significant and repeated hurdle in vaccination efforts is the ability to reach a sufficient proportion of target individuals, whether wildlife or dogs, at a sufficient spatial scale to induce herd immunity (that is, immunity in a sufficiently large proportion of the population that the likelihood of an infected individual contacting a susceptible is low, thus rendering the pathogen unable to remain endemic in the host population by reducing R_0 to <1). In societies where most dogs are owned and receive veterinary care, the ability to gain herd immunity to a pathogen via regular interactions of veterinary workers with dogs is potentially high. However, in most societies a significant portion of dogs do not receive regular veterinary care. Thus large-scale

Box 6.3 Culling dogs to protect wildlife from pathogens?

While culling of dog populations has taken place in an attempt to reduce the risk of pathogen transmission to humans, and dogs are occasionally lethally removed from areas to reduce their direct impacts on wildlife, such events have not to our knowledge occurred over a large spatial scale with a specific goal of reducing the risk of pathogen transmission to wildlife (that is, reducing f_{12}; Figure 6.4). Such approaches are not, however, unfeasible. Culling can be an effective management strategy for particular pathogens of concern (Barlow, 1996). For instance, wildlife is often culled to reduce the risk of pathogen transmission to populations of naïve domestic animals (e.g., Donnelly et al., 2006; White et al., 2011). Recently, the highly endangered Iberian lynx, *Lynx pardinus*, suffered an outbreak of feline leukemia virus, a highly pathogenic retrovirus transmitted mainly through direct contact. The epidemic was linked to contact with infected domestic cats, and management of the outbreak involved trapping and removal of cats in the area of sympatry (López et al., 2009).

However, like all population management options, culling results in altered demographics and population trajectories. While culling, like vaccination, has the potential to reduce interspecific pathogen transmission by reducing the density of both infected and susceptible individuals, such efforts can result in problematic outcomes if the spatial scale of management is insufficient to hinder rapid recolonization, or if density dependence in the reproduction or survival is not fully considered. For example, culling programs might increase birth rates via compensatory recruitment or increase immigration rates, and thereby increase the chance of pathogen persistence (Barlow, 1996; Lloyd-Smith et al., 2005). In addition, the assumption that disease transmission is density-dependent—that disease incidence increases directly with host density, and vice versa—has been shown not to hold true for some important dog diseases, including rabies (Hampson et al., 2009; Morters et al., 2013), rendering culling ineffective as a reservoir control measure for this disease.

The issue is further complicated by the close attachment that people have with dogs, limiting their willingness to allow culls and resulting in the acquisition of new animals should an owned dog be killed (Costa, 2011). For example, Nunes et al. (2008) reported on the outcome of the culling of *Leishmania*-positive dogs as a control mechanism for canine visceral leishmaniasis in a Brazilian endemic area. Within their study area, 61% of the estimated dog population was culled and the mean age of culled animals was 34 months. Owners replaced 39% of dogs, on average within 4 months, and sometimes by ≥2 dogs. Dogs were replaced mostly by puppies (mean age = 7 months) that are more susceptible to canine visceral leishmaniasis, and these recruited dogs rapidly became *Leishmania*-positive. Thus the replacement ratio of culled animals was high, increasing the dog population turnover and leading to a younger population that might be more susceptible to the very pathogen that culling was designed to curtail.

dog vaccination efforts have involved household visitations and vaccination camps, where it is often possible to reach the critical vaccination threshold (e.g., 70% for rabies) that is generally associated with pathogen eradication (Davlin and VonVille, 2012). In some scenarios, however, geographic, cultural, or dog behavior and ownership patterns make reaching critical vaccination thresholds more difficult (e.g., Belsare and Gompper, 2013).

On the other hand, there have been a number of programs in wildlife that have successfully conducted large-scale vaccination programs by combining the technological advances associated with the development of safe (to target and non-target animals), effective virus vaccines embedded within edible baits, which can then be spread across the landscape either by hand or via aerial drops. Such efforts have principally targeted rabies in wildlife through an oral rabies vaccine and have been remarkably successful, resulting in the virtual elimination of fox rabies in western Europe (Cliquet and Aubert, 2004), elimination of several variants of canid rabies in parts of Canada and the United States, and the maintenance of a robust cordon solitaire against raccoon rabies in the eastern United States and Canada (Slate et al., 2009). The use of such large-scale campaigns need not be limited to rabies, however, or indeed solely to vaccination. For instance, oral vaccination campaigns are potentially capable of addressing bovine tuberculosis (*Mycobacterium bovis*) in badgers (*Meles meles*), European wild boars (*Sus scrofa*), and Australian brushtail possums (*Trichosurus vulpecula*) (Garrido et al., 2011; Tompkins et al., 2009; Wilson et al., 2011). Similarly, treatment of wildlife on a large spatial scale using orally-administered compounds as secondary prophylaxes has occurred (Gardner et al., 2013; Jachowski et al., 2011).

6.5 Final thoughts and research needs

Our understanding of the pathogens shared by dogs and wildlife has increased tremendously over the past two decades. This knowledge has facilitated our ability to recognize the potential risks that large unvaccinated populations of dogs may represent to wildlife and, when these concerns materialize into real threats to the persistence of wildlife, to manage the pathogen, dog, or wildlife populations so as to reduce the negative impact on wildlife. Yet, with our increased knowledge of the disease ecology and epidemiology of dogs, wildlife, and their pathogens, has come the concern of overgeneralizing the insights we have gained, failing to recognize where we lack fundamental knowledge, and the hazard of attempting to manage dog–wildlife–pathogen systems without the full knowledge to make completely informed decisions.

Currently, much of our understanding of the community of parasites that infect both dogs and wildlife is based on cross-sectional surveys. Such surveys have the potential to lead to spurious and invalid conclusions about the role of dogs because they are often based on serological surveys of antibodies. Survey results can be notoriously inaccurate if the cross-reactivity and varying virulence of closely related strains is not recognized. Furthermore, such survey results are often indicative of past exposure (and therefore survival) rather than current infection and its associated risk. Perhaps most importantly, there is often a failure to fully consider the epidemiological implications of parasites persisting in a community of multiple hosts. For instance, microparasites may occur at high prevalence in dogs, but knowledge of this in and of itself is insufficient to justify an assumption that dogs are the principal maintenance host of the pathogen of interest, or that elimination of the pathogen from the dog or elimination of the dog from the host community will be sufficient to reduce the prevalence of the pathogen in the target wildlife population. It is likely that a comprehensive systematic review would reveal a lack of evidence for the definitive role of dogs as reservoir hosts for pathogens that represent threats to wildlife in almost all cases. More evidence is needed from well-designed studies that combine a variety of approaches (field ecology, epidemiology, molecular biology, theoretical modeling) if we are to fully test the role of dogs in mediating diseases of concern to wildlife.

It is also worth noting how much attention is dedicated to understanding the ecology and epidemiology of just a handful of parasites. The bulk of attention is focused on just a few taxa (rabies virus, CPV, CDV, *Sarcoptes*), and even for these we often have a poor understanding of basic ecological issues. These focal taxa are undoubtedly important conservation and management concerns. Yet given the diversity of parasitic species that are commonly observed at relatively high prevalences (and in the case of macroparasites, at high intensities) in free-ranging dogs, there is a need to examine other parasitic species that may (or may not) be a threat to wildlife, and examine the possible role of dogs in maintaining these species.

Finally, we have increased our tactical capability to design and undertake management strategies that attempt to address the persistence of pathogens in dog and target wildlife populations. Yet, with such capacity should also come a recognition that pathogens commonly play a direct or indirect role in limiting both dog and wildlife populations. Our increasing ability to manage both parasites and the diseases they cause suggests the need to also develop a more sophisticated understanding of how such management may alter dog and wildlife demographics. If pathogens limit dog populations, then a removal of this limiting factor may result in greater dog survival rates, greater reproductive rates, and increased opportunities for dogs and wildlife to interact in non-disease related ways. By recognizing that the pathogens dogs may transmit are only one of several mechanisms by which dogs may interact with wildlife, we move towards a more holistic understanding of dog–wildlife interactions.

References

Alexander, K. A. and Appel, M.J.G. (1994). African wild dogs (*Lycaon pictus*) endangered by a canine distemper epizootic among domestic dogs near the Masai Mara National Reserve, Kenya. *Journal of Wildlife Diseases*, **30**, 481–85.

Alexander, K.A., Conrad, P.A., Gardner, I.A., Parrish, C., Appel, M., Levy, M.G., Lerche, N., and Kat, P. (1993). Serologic survey of selected microbial pathogens in African wild dogs (*Lycaon pictus*) and sympatric domestic dogs (*Canis familiaris*) in Maasai Mara, Kenya. *Journal of Zoo and Wildlife Medicine*, **24**, 140–44.

Alexander, K.A., Kat, P.W., Wayne, R.K., and Fuller, T.K. (1994). Serologic survey of selected canine pathogens among free-ranging jackals in Kenya. *Journal of Wildlife Diseases*, **30**, 486–91.

Alexander, K.A., McNutt, J.W., Briggs, M.B., Standers, P.E., Funston, P., Hemson, G., Keet, D., and van Vuuren, M. (2010). Multi-host pathogens and carnivore management in southern Africa. *Comparative Immunology, Microbiology and Infectious Diseases*, **33**, 249–65.

Allison, A.B., Harbison, C.E., Pagan, I., Stucker, K.M., Kaelber, J.T., Brown, J.D., Ruder, M.G., Keel, M.K., Dubovi, E.J., Holmes, E.C., and Parrish, C.R. (2012). The role of multiple hosts in the cross-species transmission and emergence of a pandemic parvovirus. *Journal of Virology*, **86**, 865–72.

Arlian, L.G. (1989). Biology, host relations and epidemiology of *Sarcoptes scabiei*. *Annual Review of Entomology*, **34**, 139–61.

Barker, I.K. and Parrish, C.R. (2001). Parvovirus infections. In E.S. Williams and I.K. Barker, Eds., *Infectious Diseases of Wild Mammals*, 3rd edn. Iowa State University Press, Ames, pp. 131–46.

Barlow, N.D. (1996). The ecology of wildlife disease control: simple models revisited. *Journal of Applied Ecology*, **33**, 303–14.

Bartlett, M.S. (1960). The critical community size for measles in the United States. *Journal of the Royal Statistical Society. Series A (General)*, **123**, 37–44.

Belsare, A.V., and Gompper, M.E. (2013). Assessing demographic and epidemiologic parameters of rural dog populations in India during mass vaccination campaigns. *Preventive Veterinary Medicine*, **111**, 139–46.

Benetka, V., Leschnik, M., Affenzeller, N., and Möstl, K. (2011). Phylogenetic analysis of Austrian canine distemper virus strains from clinical samples from dogs and wild carnivores. *Veterinary Record*, **168**, 377.

Bingham, J. (2005). Canine rabies ecology in southern Africa. *Emerging Infectious Diseases*, **11**, 1337–42.

Bingham, J., Foggin, C.M., Wandeler, A.I., and Hill, F.W.G. (1999a). The epidemiology of rabies in Zimbabwe. 1. Rabies in dogs (*Canis familiaris*). *Onderstepoort Journal of Veterinary Research*, **66**, 1–10.

Bingham, J., Foggin, C.M., Wandeler, A.I. Hill, F.W.G. (1999b). The epidemiology of rabies in Zimbabwe. 2. Rabies in jackals (*Canis adustus and Canis mesomelas*). *Onderstepoort Journal of Veterinary Research*, **66**, 11–23.

Blanton J.D., Palmer, D, Christian, K.A., and Rupprecht, C.E. (2008). Rabies surveillance in the United States during 2007. *Journal of the American Veterinary Medical Association*, **233**, 884–97.

Breed, A.C., Plowright, R.K., Hayman, D.T.S., Knobel, D.L., Molenaar, F.M., Gardner-Roberts, D., Cleaveland, S., Haydon, D.T., Kock, R.A., Cunningham, A.A., Sainsbury, A.W., and Delahay, R.J. (2009). Disease management in endangered mammals. In R.J. Delahay, G. Smith and M. Hutchings, Eds., *Management of Disease in Wild Mammals*. Springer Verlag, Berlin, pp. 215–40.

Bryan, H.M., Darimont, C.T., Paquet, P.C, Ellis, J.A., Goji, N., Gouix, M., and Smits, J.E. (2011). Exposure to infectious agents in dogs in remote coastal British Columbia: Possible sentinels of diseases in wildlife and humans. *Canadian Journal of Veterinary Research*, **75**, 11–17.

Butler, J.R.A. (1998). The ecology of domestic dogs (*Canis familiaris*) in the communal lands of Zimbabwe. PhD dissertation, University of Zimbabwe.

Butler, J.R.A., du Toit, J.T., and Bingham, J. (2004). Free-ranging domestic dogs *Canis familiaris* as predators and prey in rural Zimbabwe: threats of competition and disease to large wild carnivores. *Biological Conservation*, **115**, 369–78.

Carpenter, M.A., Appel, M.J.G., Roelke-Parker, M.E., Munson, L., Hofer, H., East, M., and O'Brien, S.J. (1998). Genetic characterization of canine distemper virus in Serengeti carnivores. *Veterinary Immunology and Immunopathology*, **65**, 259–66.

CDC (2007). US declared canine-rabies free. Centers for Disease Control and Prevention. <http://www.cdc.gov/media/pressrel/2007/r070907.htm>.

Chaves, L.F., Hernandez, M-J., Dobson, A.P., and Pascual, M. (2007). Sources and sinks: revisiting the criteria for identifying reservoirs for American cutaneous leishmaniasis. *Trends in Parasitology*, **23**, 311–16.

Cleaveland, S. and Dye, C. (1995). Maintenance of a microparasite infecting several host species: rabies in the Serengeti. *Parasitology*, **111**, S33–47.

Cleaveland, S., Appel, M.G.J., Chalmers, W.S.K., Chillingworth, C., Kaare, M., and Dye, C. (2000). Serological and demographic evidence for domestic dogs as a source of canine distemper virus infection for Serengetti wildlife. *Veterinary Microbiology*, **72**, 217–27.

Cleaveland, S.C., Laurenson, M.K., and Taylor, L.H. (2001). Diseases of humans and their domestic mammals; pathogen characteristics, host range and the risk of emergence. *Philosophical Transactions of the Royal Society of London, B*, **356**, 991–9.

Cleaveland, S., Kaare, M., Tiringa, P., Mlengeya, T., and Barrat, J. (2003). A dog rabies vaccination campaign in rural Africa: impact on the incidence of dog rabies and human dog-bite injuries. *Vaccine*, **21**, 1965–73.

Cleaveland, S., Kaare, M., Knobel, D., and Laurenson, M.K. (2006). Canine vaccination-providing broader benefits for disease control. *Veterinary Microbiology*, **117**, 43–50.

Cleaveland, S., Mlengeya, T., Kaare, M., Haydon, D., Lembo, T., Laurenson, M.K., and Packer, C. (2007). The conservation relevance of epidemiological research into carnivore viral diseases in the Serengeti. *Conservation Biology*, **21**, 612–22.

Cleaveland, S., Keyyu, J., Auty, H., Lembo, T., Hampson, K., Lankester, F.J., and Kazwala, R. (2012). Domestic animal health programmes to protect wildlife. In *Animal Health and Biodiversity: Preparing for the Future*. Compendium of the OIE Global Conference on Wildlife, Paris, France, February 23–25, 2011, pp. 91–100.

Cliquet, F., and Aubert, M. (2004). Elimination of terrestrial rabies in Western European countries. *Developments in Biologicals (Basel)*, **119**, 185–204.

Costa, C.H.N. (2011). How effective is dog culling in controlling zoonotic visceral leishmaniasis? A critical evaluation of the science, politics and ethics behind this public health policy. *Revista da Sociedade Brasileira de Medicina Tropical*, **44**, 232–42.

Courtenay, O., Quinnell, R.J., Garcez, L.M., and Dye, C. (2002). Low infectiousness of a wildlife host of *Leishmania infantum*: the crab-eating fox is not important for transmission. *Parasitology*, **125**, 407–14.

Craft, M.E., Hawthorne, P.L., Packer, C., and Dobson, A.P. (2008). Dynamics of a multihost pathogen in a carnivore community. *Journal of Animal Ecology*, **77**, 1257–64.

Craft, M.E., Volz, E., Packer, C., and Meyers, L.A. (2009). Distinguishing epidemic waves from disease spillover in a wildlife population. *Proceedings of the Royal Society, Series B, Biological Sciences*, **276**, 1777–85.

Crawford, P. C., Dubovi, E. J., Castleman, W. L., Stephenson, I., Gibbs. E. P., Chen. L., Smith, C., Hill, R. C., Ferro, P., Pompey, J., Bright, R. A., Medina, M. J., Johnson, C. M., Olsen, C. W., Cox, N. J., Klimov, A. I., Katz, J. M., and Donis, R. O. (2005). Transmission of equine influenza virus to dogs. *Science*, **310**, 482–5.

Creel, S., Creel, N.M., Munson, L., Sanderlin, D., and Appel, M.J. (1997). Serosurvey for selected viral diseases and demography of African wild dogs in Tanzania. *Journal of Wildlife Diseases*, **33**, 823–32.

Cumming, D.H.M. (1982). A case history of the spread of rabies in an African country. *South African Journal of Science*, **78**, 443–7.

Curi, N.H.D.A., Araújo, A.S., Campos, F.S., Lobato, Z.I.P., Gennari, S.M., Marvulo, M.F.V., Silva, J.C.R., and Talamoni, S.A. (2010). Wild canids, domestic dogs and their pathogens in Southeast Brazil: disease threats for canid conservation. *Biodiversity and Conservation*, **19**, 3513–24.

Daszak, P., Cunningham, A.A., and Hyatt, A.D. (2000). Emerging infectious diseases of wildlife—threats to biodiversity and human health. *Science*, **287**, 443–9.

Davidson, R.K., Romig, T., Jenkins, E., Tryland, M., and Robertson, L.J. (2012). The impact of globalisation on the distribution of *Echinococcus multilocularis*. *Trends in Parasitology*, **28**, 239–47.

Davidson, W.R., Appel, M.J., Doster, G.L., Baker, O.E., and Brown, J.F. (1992). Diseases and parasites of red foxes, gray foxes, and coyotes from commercial sources selling to fox-chasing enclosures. *Journal of Wildlife Diseases*, **28**, 581–9.

Davis, P.L., Rambaut, A., Bourhy, H., and Holmes, E.C. (2007). The evolutionary dynamics of canid and mongoose rabies virus in southern Africa. *Archives of Virology*, **152**, 1251–58.

Davlin, S.L., and Vonville, H.M. (2012). Canine rabies vaccination and domestic dog population characteristics in the developing world: A systematic review. *Vaccine*, **30**, 3492–3502.

Dobson, A. (2004). Population dynamics of pathogens with multiple host species. *American Naturalist*, **164**, S64–78.

Donnelly, C.A., Woodroffe, R., Cox, D.R., Bourne, F.J., Cheeseman, C.L., Wei, G., Gettinby, G., Gilks, P., Jenkins, H., Johnston, W.T., Le Fevre, A.M., McInerney, J.P., and Morrison, W.I. (2006). Positive and negative effects of widespread badger culling on tuberculosis in cattle. *Nature*, **439**, 843–46.

Evermann, J.F., Foreyt, W., Maag-Miller, L., Leathers, C.W., McKiernan, A.J., and LeaMaster, B. (1980). Acute hemorrhagic enteritis associated with canine coronavirus and parvovirus infections in a captive coyote population. *Journal of the American Medical Association*, **177**, 784–6.

Ezenwa, V.O. (2004). Interactions among host diet, nutritional status and gastrointestinal parasite infection in wild bovids. *International Journal for Parasitology*, **34**, 535–42.

Fenton, A. and Pedersen, A.B. (2005). Community epidemiology in theory and practice: A conceptual framework for classifying disease threats in human and wild populations. *Emerging Infectious Diseases*, **11**, 1815–21.

Fiorello, C.V. (2004). Disease ecology of wild and domestic carnivores in Bolivia. PhD dissertation, Columbia University.

Fiorello, C.V., Deem, S.L., Gompper, M.E., and Dubovi, E.J. (2004). Seroprevalence of pathogens in domestic carnivores on the border of Madidi National Park, Bolivia. *Animal Conservation*, **7**, 45–54.

Fitzpatrick, M.C., Hampson, K., Cleaveland, S., Meyers, L.A., Townsend, J.P., and A.P. Galvani. (2012). Potential for rabies control through dog vaccination in wildlife-abundant communities of Tanzania. *PLoS Neglected Tropical Diseases*, **6**(8): e1796.

Foggin, C.M. (1988). Rabies and rabies-related viruses in Zimbabwe: historical, virological and ecological aspects. Ph.D. dissertation, University of Zimbabwe.

Gakuya, F., Rossi, L., Ombui, J., Maingi, N., Muchemi, G., Ogara, W. Soriguer, R.C., and Alasaad, S. (2011). The curse of the prey: *Sarcoptes* mite molecular analysis reveals potential prey-to-predator parasitic infestation in wild animals from Masai Mara, Kenya. *Parasites & Vectors*, **4**, 193.

Garcelon, D.K., Wayne, R.K., and Gonzales, B.J. (1992). A serologic survey of the island fox (*Urocyon littoralis*) on the Channel Islands, California. *Journal of Wildlife Diseases*, **28**, 223–9.

Gardner, C.L., Beckmen, K.B., Pamperin, N.J., and Del Vecchio, P. (2013). Experimental treatment of dog lice infestation in interior Alaska wolf packs. *Journal of Wildlife Management*, **77**, 626–32.

Garrido, J.M., Sevilla, I.A., Beltrán-Beck, B., Minguijón, E., Ballesteros, C., Galindo, R.C., Boadella, M., Lyashchenko, K.P., Romero, B., Geijo, M.V., Ruiz-Fons, F., Aranaz, A., Juste, R.A., Vicente, J., de la Fuente, J., and Gortázar, C. (2011). Protection against tuberculosis in Eurasian wild boar vaccinated with heat-inactivated *Mycobacterium bovis*. *PLoS ONE*, **6**(9): e24905.

Gascoyne, S.C., Laurenson, M.K., Lelo, S., and Borner, M. (1993). Rabies in African wild dogs (*Lycaon pictus*) in the Serengeti region, Tanzania. *Journal of Wildlife Diseases*, **29**, 396–402.

Gese, E.M., Schultz, R.D., Johnson, M.R., Williams, E.S., Crabtree, R.L., and Ruff, R.L. (1997). Serological survey for diseases in free-ranging coyotes (*Canis latrans*) in Yellowstone National Park, Wyoming. *Journal of Wildlife Diseases*, **33**, 47–56.

Goltsman, M., Kruchenkova, E. P., and Macdonald, D. W. (1996). The Mednyi Arctic foxes: treating a population imperiled by disease. *Oryx*, **30**, 151–8.

Gowtage-Sequeira, S., Banyard, A.C., Barrett, T., Buczkowski, H., Funk, S.M., and Cleaveland, S. (2009), Epidemiology, pathology, and genetic analysis of a canine distemper epidemic in Namibia. *Journal of Wildlife Diseases*, **45**, 1008–20.

Haas, L., Hofer, H., East, M., Wohlsein, P., Liess, B., and Barrett, T. (1996). Epizootic of canine distemper virus infection in Serengeti spotted hyenas (*Crocuta crocuta*). *Veterinary Microbiology*, **49**, 147–52.

Hampson, K., Dushoff, J., Cleaveland, S., Haydon, D. T., Kaare, M., Packer, C., and Dobson, A. (2009). Transmission dynamics and prospects for the elimination of canine rabies. *Plos Biology*, **7**, e1000053.

Harder, T.C., Kenter, M., Appel, M.J.G., Roelke-Parker, M.E., Barrett, T., and Osterhaus, A.D.M.E. (1995). Phylogenetic evidence of canine distemper virus in Serengeti's lions. *Vaccine*, **13**, 521–3.

Haydon, D.T., Cleaveland, S., Taylor, L.H., and Laurenson M.K. (2002). Identifying reservoirs of infection: a conceptual and practical challenge. *Emerging Infectious Disease*, **8**, 1468–73.

Haydon, D. T., Randall, D. A., Matthews, L., Knobel, D. L., Tallents, L. A., Gravenor, M. B., Williams, S. D., Pollinger, J. P., Cleaveland, S., Woolhouse, M. E. J., Sillero-Zubiri, C., Marino, J., Macdonald, D. W., and Laurenson, M. K. (2006). Low-coverage vaccination strategies for the conservation of endangered species. *Nature*, **443**, 692–5.

Hoelzer, K. and Parrish, C.R. (2010). The emergence of parvoviruses of carnivores. *Veterinary Research*, **41**, 39.

Hoelzer, K., Shackelton, L.A., Parrish, C.R., and Holmes, E.C. (2008). Phylogenetic analysis reveals the emergence, evolution and dispersal of carnivore parvoviruses. *Journal of General Virology*, **89**, 2280–9.

Hofmeyr, M., Bingham, J., Lane, E. P., Ide, A., and Nel, L. (2000). Rabies in African wild dogs (*Lycaon pictus*) in the Madikwe Game Reserve, South Africa. *Veterinary Record*, **146**, 50–52.

Ikeda, Y., Miyazawa, T., Nakamura, K., Naito, R., Inoshima, Y., Tung, K-C., Lee, W-M., Chen, M-C., Kuo, T-F., Lin, J.A., and Mikami, T. (1999). Serosurvey for selected virus infections of wild carnivores in Taiwan and Vietnam. *Journal of Wildlife Diseases*, **35**, 578–81.

Ikeda, Y., Mochizuki, M., Naito, R., Nakamura, K., Miyazawa, T., Mikami, T., and Takahashi, E. (2000). Predominance of canine parvovirus (CPV) in unvaccinated cat populations and emergence of new antigenic types of CPVs in cats. *Virology*, **278**, 13–19.

Jachowski, D.S., Skipper, S., and M.E. Gompper. (2011). Field evaluation of imidacloprid as a systemic approach to flea control in black-tailed prairie dogs (*Cynomys ludovicianus*). *Journal of Vector Ecology*, **36**, 100–107.

Johnson, N., Mansfield, K.L., Marston, D.A., Wilson, C., Goddard, T., Selden, D., Hemson, G., Edea, L., van Kesteren, F., Shiferaw, F., Stewart, A.E., Sillero-Zubiri, C., and Fooks, A.R. (2010). A new outbreak of rabies in rare Ethiopian wolves (*Canis simensis*). *Archives of Virology*, **155**, 1175–77.

Kameo, Y., Nagao, Y., Nishio, Y., Shimoda, H., Nakano, H., Suzuki, K., Une, Y., Sato, H., Shimojima, M., and Maeda, K. (2012). Epizootic canine distemper virus infection among wild mammals. *Veterinary Microbiology*, **154**, 222–9.

Kapil, S., Rezabek, G., Germany, B., and Johnston, L. (2010). Isolation of a virus related to canine parvovirus type 2 from a raccoon (*Procyon lotor*). *Veterinary Record*, **166**, 24–25.

Kennedy, S., Kuiken, T., Jepson, P.D., Deaville, R., Forsyth, M., Barrett, T., van de Bildt, M.W.G., Osterhaus, A.D.M.E., Eybatov, T., Duck, C., Kydyrmanov, A., Mitrofanov, I., and Wilson, S. (2000). Mass die-off of Caspian seals caused by canine distemper virus. *Emerging Infectious Diseases*, **6**, 637–9.

Knobel, D. L., Fooks, A. R., Brookes, S. M., Randall, D. A., Williams, S. D., Argaw, K., Shiferaw, F., Tallents, L. A., and Laurenson, M. K. (2008). Trapping and vaccination of endangered Ethiopian wolves to control an outbreak of rabies. *Journal of Applied Ecology*, **45**, 109–16.

Knobel, D.L., Lembo, T., Morters, M., Townsend, S.E., Cleaveland, S., and Hampson, K. (2013). Dog rabies and its control. In A.C., Jackson, Ed., *Rabies: Scientific Basis of the Disease and its Management*. 3rd edn. Elsevier/Academic Press, London, pp. 591–615.

Kock, R., Chalmers, W.S.K., Mwanzia, J., Chillingworth, C., Wambura, J., Coleman, P.G., and Baxendale, W. (1998). Canine distemper in lions of the Masai Mara. *Veterinary Record*, **142**, 662–5.

Koski, K.G. and Scott, M.E. (2001). Gastrointestinal nematodes, nutrition and immunity: Breaking the negative spiral. *Annual Review of Nutrition*, **21**, 297–321.

Kuzmin, I.V. and Tordo, N. (2012). Genus *Lyssavirus*. In R.G. Dietzgen and I.V. Kuzmin, Eds., *Rhabdoviruses: Molecular Taxonomy, Evolution, Genomics, Ecology, Host-Vector Interactions, Cytopathology and Control*. Caister Academic Press, UK, pp. 37–58.

Laurenson K., Shiferaw, F., and Sillero-Zubiri, C. (1997). Rabies as a threat to the Ethiopian wolf (*Canis simensis*). In A. King, Ed., *Proceedings of the International Conference on Epidemiology, Control and Prevention of Rabies in Eastern And Southern Africa, Nairobi, Kenya*. Merieux, Lyon.

Laurenson, K., Sillero-Zubiri, C., Thompson, H., Shiferaw, F., Thirgood, S., and Malcolm, J. (1998). Disease as a threat to endangered species: Ethiopian wolves, domestic dogs and canine pathogens. *Animal Conservation*, **1**, 273–80.

Laurenson, M.K., Cleaveland, S., Artois, M., and Woodroffe, R. (2004). Assessing and managing infectious disease threats to canids. In C. Sillero-Zubiri, M. Hoffman and D.W. Macdonald, Eds., *Canids: Foxes, Wolves, Jackals and Dogs. Status Survey and Conservation Action Plan*. IUCN SSC Canid Specialist Group, Cambridge, United Kingdom and Gland, Switzerland, pp. 246–55.

Laurenson, M.K., Mlengeya, T., Shirefaw, F., and Cleaveland, S. (2005). Approaches to disease control in domestic canids for the conservation of endangered wild carnivores. In S. Cleaveland, W.B. Karesh, D. Kock, P.J. Nyhus, L. Starr and A. Yang, Eds., *Conservation and Development Interventions at the Wildlife/Livestock Interface—Implications for Wildlife, Livestock and Human Health*. IUCN Publications Services Unit, Cambridge, United Kingdom and Gland, Switzerland, pp. 141–6.

Lednicky, J.A., Dubach, J., Kinsel, M.J., Meehan, T.P., Bocchetta, M., Hungerford, L.L., Sarich, N.A., Witecki, K.E., Braid, M.D., Pedrak, C., and Houde, C.M. (2004) Genetically distant American canine distemper virus lineages have recently caused epizootics with somewhat different characteristics in raccoons living around a large suburban zoo in the USA. *Virology Journal*, **1**, 2.

Lemberger, K.Y., Gondim, L.F.P., Pessier, A.P., McAllister, M.M., and Kinsel, M.J. (2005). *Neospora caninum* infection in a free-ranging raccoon (*Procyon lotor*) with concurrent canine distemper virus infection. *Journal of Parasitology*, **91**, 960–1.

Lembo, T. (2007). An investigation of disease reservoirs in complex ecosystems: rabies and canine distemper in the Serengeti. PhD dissertation, University of Edinburgh.

Lembo, T., Hampson, K., Haydon, D.T., Craft, M., Dobson, A.P., Dushoff, J., Ernest, E., Hoare, R., Kaare, M., Mlengeya, T., Mentzel, C., and Cleaveland, S. (2008). Exploring reservoir dynamics: a case study of rabies in the Serengeti ecosystem. *Journal of Applied Ecology*, **45**, 1246–57.

Lembo, T., Hampson, K., Kaare, M.T., Ernest, E., Knobel, D., Kazwala, R.R., Haydon, D.T., and Cleaveland, S. (2010). The feasibility of canine rabies elimination in Africa: dispelling doubts with data. *PLoS Neglected Tropical Diseases*, **4**(2):e626.

Lloyd-Smith, J.O., Cross, P.C., Briggs, C.J., Daugherty, M., Getz, W.M., Latto, J., Sanchez, M.S., Smith, A.B., and Swei, A. (2005). Should we expect population thresholds for wildlife disease? *Trends in Ecology and Evolution*, **20**, 511–9.

López, G., López-Parra, M., Fernández, L., Martínez-Granados, C., Martínez, F., Meli, M.L., Gil-Sánchez, J.M., Viqueira, N., Díaz-Portero, M.A., Cadenas, R., Lutz, H., Vargas, A., and Simón, M A. (2009). Management measures to control a feline leukemia virus outbreak in the endangered Iberian lynx. *Animal Conservation*, **12**, 173–82.

Mainka, S.A., Xianmeng, Q., Tingmei, H., and Appel, M.J. (1994). Serologic survey of giant pandas (*Aliuropoda melanoleuca*), and domestic dogs and cats in the Wolong Reserve, China. *Journal of Wildlife Diseases*, **30**, 86–9.

Mamaev, L.V., Denikina, N.N., Belikov, S.I., Volchkov, V.E., Visser, I.K.G., Fleming, M., Kai, C., Harder, T.C., Liess, B., Osterhaus, A.D.M.E., and Barrett, T. (1995). Characterization of morbilliviruses isolated from Lake Baikal seals (*Phoca sibirica*). *Veterinary Microbiology*, **44**, 251–9.

Mansfield, K., McElhinney, L., Hübschle, O., Mettler, F., Sabeta, C., Nel, L.H., and Fooks, A.R. (2006). A molecular epidemiological study of rabies epizootics in kudu (*Tragelaphus strepsiceros*) in Namibia. *BMC Veterinary Research*, **2**, 2.

Martella, V., Cirone, F., Elia, G., Lorusso, E., Decaro, N., Campolo, M., Desario, C., Lucente, M.S., Bellacicco, A.L., Blixenkrone-Møller, M., Carmichael, L.E., and Buonavoglia, C. (2006). Heterogeneity within the hemagglutinin genes of canine distemper virus (CDV) strains detected in Italy. *Veterinary Microbiology*, **122**, 32–42.

McCarthy, A.J., Shaw, M.-A., and Goodman, S.J. (2007). Pathogen evolution and disease emergence in carnivores. *Proceedings of the Royal Society B*, **274**, 3165–74.

McCaw, D.L. and Hoskins, J.D. (2006). Canine viral enteritis. In C.E. Greene, Ed., *Infectious Diseases of the Dog and Cat*, 3rd edn. Elsevier, London, pp. 63–73.

McCue, P.M. and O'Farrell, T.P. (1988). Serological survey for selected diseases in the endangered San Joaquin kit fox (*Vulpes macrotis mutica*). *Journal of Wildlife Diseases*, **24**, 274–81.

Mech, L.D. and Goyal, S. (1993). Canine parvovirus effect on wolf population change and pup survival. *Journal of Wildlife Diseases*, **29**, 330–3.

Mech, L.D. and Goyal, S. (1995). Effects of canine parvovirus on gray wolves in Minnesota. *Journal of Wildlife Diseases*, **59**, 565–70.

Mech, L.D., Kurtz, H.J., and Goyal, S. (1997). Death of a wild wolf from canine parvoviral enteritis. *Journal of Wildlife Diseases*, **33**, 321–2.

Miller, D.S., Covell, D.F., McLean, R.G., Adrian, W.J., Niezgoda, M., Gustafson, J.M., Rongstad, O.J., Schultz, R.D., Kirk, L.J., and Quan, T.J. (2000). Serologic survey for selected infectious disease agents in swift and kit foxes from the western United States. *Journal of Wildlife Diseases*, **36**, 798–805.

Miller, L.L. and Crosbie, P.R. (2011). Canine heartworm (*Dirofilaria immitis*) in Fresno and Madera Counties, California: prevalence differences between foothill and valley habitats. *Veterinary Parasitology*, **175**, 84–91.

Monello, R.J. and Gompper, M.E. (2011). Effects of resource availability and social aggregation on the species richness of raccoon endoparasite infracommunities. *Oikos*, **120**, 1427–33.

Morters, M.K., Restif, O., Hampson, K., Cleaveland, S., Wood, J.L.N., and Conlan, A.J.K. (2013). Evidence-based control of canine rabies: a critical review of population density reduction. *Journal of Animal Ecology*, **82**, 6–14.

Munson, L., Terio, K.A., Kock, R., Mlengeya, T., Roelke, M.E., Dubovi, E., Summers, B., Sinclair, A.R.E. and Packer, C. (2008). Climate extremes promote fatal coinfections during canine distemper epidemics in African lions. *PLoS ONE*, **3**(6): e2545.

Nadin-Davis, S.A. (2007). Molecular epidemiology. In A.C. Jackson and W.H. Wunner, Eds., *Rabies: Scientific Basis of the Disease and its Management*. 2nd edn. Elsevier. London, pp. 69–122.

Nikolin, V.M., Wibbelt, G., Michler, F.-U.F., Wolf, P., and East, M.L. (2012). Susceptibility of carnivore hosts to strains of canine distemper virus from distinct genetic lineages. *Veterinary Microbiology*, **156**, 45–53.

Nunes, C.M., Lima, V.M.F., Paula, H.B., Perri, S.H.V., Andrade, A.M., Dias, F.E.F., and Burattini, M.N. (2008). Dog culling and replacement in an area endemic for visceral leishmaniasis in Brazil. *Veterinary Parasitology*, **153**, 19–23.

Nunn, C.L., and Altizer, S.M. (2005). The global mammal parasite database: An online resource for infectious disease records in wild primates. *Evolutionary Anthropology: Issues, News, and Reviews*, **14**, 1–2.

Oliveira-Sequeira, T.C.G., Amarante, A.F.T., Ferrari, T.B., and Nunes, L.C. (2002). Prevalence of intestinal parasites in dogs from São Paulo State, Brazil. *Veterinary Parasitology*, **103**, 19–27.

Ostfeld, R.S. and Keesing, F. (2000). Biodiversity and disease risk: the case of Lyme disease. *Conservation Biology*, **14**, 722–8.

Papazahariadou, M., Founta, A., Papadopoulos, E., Chliounakis, S., Antoniadou-Sotiriadou, K., and Theodorides, Y. (2007). Gastrointestinal parasites of shepherd and hunting dogs in the Serres Prefecture, Northern Greece. *Veterinary Parasitology*, **148**, 170–3.

Paras, K.L., Little, S.E., Reichard, M.V., and Reiskind, M.H. (2012). Detection of *Dirofilaria immitis* and *Ehrlichia* species in coyotes (*Canis latrans*), from rural Oklahoma and Texas. *Vector-Borne and Zoonotic Diseases*, **12**, 619–21.

Parrish, C.R. and Kawaoka, Y. (2005). The origins of new pandemic viruses: the acquisition of new host ranges by canine parvovirus and influenza A viruses. *Annual Review of Microbiology*, **59**, 553–86.

Parrish C.R., Holmes, E.C., Morens, D.M., Park, E.-C., Burke, D.S., Calisher, C.H., Laughlin, C.A., Saif, L.J., and Daszak, P. (2008). Cross-species virus transmission and the emergence of new epidemic diseases. *Microbiology and Molecular Biology Reviews*, **72**, 457–70.

Pence, D.B., and Ueckermann, E. (2002). Sarcoptic mange in wildlife. *Revue Scientifique et Technique*, **21**, 385–98.

Prager, K.C., Mazet, J.A.K., Munson, L., Cleaveland, S., Donnelly, C. A., Dubovi, E. J., Gunther, M. S., Lines, R., Mills, G., Davies-Mostert, H. T., McNutt, J. W., Rasmussen, G., Terio, K., and Woodroffe, R. (2012). The effect of protected areas on pathogen exposure in endangered African wild dog (*Lycaon pictus*) populations. *Biological Conservation*, **150**, 15–22.

Qin, Q., Loeffler, I.K., Li, M., Tian, K., and Wei, F. (2007). Sequence analysis of a canine parvovirus isolated from a red panda (*Ailurus fulgens*) in China. *Virus Genes*, **34**, 299–302.

Randall, D.A., Williams, S.D., Kuzmin, I.V., Rupprecht, C.E., Tallents, L.A., Tefera, Z., Argaw, K., Shiferaw, F.,

Knobel, D.L., Sillero-Zubiri, C., and Laurenson, M.K. (2004). Rabies in endangered Ethiopian wolves. *Emerging Infectious Diseases*, **10**, 2214–2217.

Randall, D.A., Marino, J., Haydon, D.T., Sillero-Zubiri, C., Knobel, D.L., Tallents, L.A., Macdonald, D.W., and Laurenson, M.K. (2006). An integrated disease management strategy for the control of rabies in Ethiopian wolves. *Biological Conservation*, **131**, 151–62.

Rasero, R., Rossi, L., Maione, S., Sacchi, P., Rambozzi, L., Sartore, S., Soriguer, R., Spalenza, V., and Alasaad, S. (2010). Host taxon-derived *Sarcoptes* mites in European wildlife animals, revealed by microsatellite markers. *Biological Conservation*, **143**, 1269–77.

Reithinger, R., and Davies, C.R. (1999). Is the domestic dog (*Canis familiaris*) a reservoir host of American cutaneous leishmaniasis? A critical review of the current evidence. *American Journal of Tropical Medicine and Hygiene*, **61**, 530–41.

Rhodes, C.J., Atkinson, R.P.D., Anderson, R.M., and Macdonald, D.W. (1998). Rabies in Zimbabwe: reservoir dogs and the implications for disease control. *Philosophical Transactions of the Royal Society B, Biological Science*, **353**, 999–1010.

Roelke-Parker, M.E., Munson, L., Packer, C., Kock, R., Cleaveland, S., Carpenter, M., O'Brien, S.J., Pospischil, A., Hofmann-Lehmann, R., Lutz, H., Mwamengele, G.L., Mgasa, M.N., Machange, G.A., Summers, B.A., and Appel, M.J. (1996). A canine distemper virus epidemic in Serengeti lions (*Panthera leo*). *Nature*, **379**, 441–5.

Salb, A.L, Barkema, H.W., Elkin, B.T., Thompson, R.C., Whiteside, D.P., Black, S.R., Dubey, J.P., and Kutz, S.J. (2008). Dogs as sources and sentinels of parasites in humans and wildlife, northern Canada. *Emerging Infectious Diseases*, **14**, 60–63.

Sillero-Zubiri, C., King, A. A., and Macdonald, D. W. (1996). Rabies and mortality in Ethiopian wolves (*Canis simensis*). *Journal of Wildlife Diseases*, **32**, 80–86.

Slate, D., Algeo, T.P., Nelson, K.M., Chipman, R.B., Donovan, D., Blanton, J.D., Niezgoda, M., and Rupprecht, C.E. (2009). Oral rabies vaccination in North America: opportunities, complexities, and challenges. *PLoS Neglected Tropical Diseases*, **3**(12), e549.

Smith J.S., Orciari, L.A., Yager, P.A., Seidel, H.D., and Warner, C.K. (1992). Epidemiologic and historical relationships among 87 rabies virus isolates as determined by limited sequence analysis. *Journal of Infectious Diseases*, **166**, 296–307.

Soulsbury, C.D., Iossa, G., Baker, P.J., Cole, N.C., Funk, S.M., and Harris, S. (2007). The impact of sarcoptic mange *Sarcoptes scabiei* on the British fox *Vulpes vulpes* population. *Mammal Review*, **37**, 278–96.

Sowemimo, O. A. (2009). The prevalence and intensity of gastrointestinal parasites of dogs in Ile-Ife, Nigeria. *Journal of Helminthology*, **83**, 27–31.

Steinel, A., Munson, L., van Vuuren, M., and Truyen, U. (2000). Genetic characterization of feline parvovirus sequences from various carnivores. *Journal of General Virology*, **81**, 345–50.

Sun, Z., Li, A., Ye, H., Shi, Y.Hu, Z., and Zeng, L. (2010). Natural infection with canine distemper virus in hand-feeding rhesus monkeys in China. *Veterinary Microbiology*, **141**, 374–78.

Tompkins, D.M., Greenman, J.V., and Hudson, P.J. (2001). Differential impact of a shared nematode parasite on two gamebird hosts: implications for apparent competition. *Parasitology*, **122**, 187–93.

Tompkins D. M., Ramsey D. S., Cross M. L., Aldwell F. E., de Lisle G. W., and Buddle B. M. (2009). Oral vaccination reduces the incidence of tuberculosis in free-living brushtail possums. *Proceedings of the Royal Society B*, **276**, 2987–95.

White, P.J., Wallen, R.L., Geremia, C., Treanor, J.J., and Blanton, D.W. (2011). Management of Yellowstone bison and brucellosis transmission risk—expectations and realizations. *Biological Conservation*, **144**, 1322–34.

Williams, E.S., Thorne, E.T., Appel, M.J.G., and Belitsy, D.W. (1988). Canine distemper in black-footed ferrets (*Mustela nigripes*) from Wyoming. *Journal of Wildlife Diseases*, **24**, 385–98.

Williams, E.S., Yuill, T., Artois, M., Fischer, J., and Haigh, S.A. (2002). Emerging infectious diseases in wildlife. *Revue Scientifique et Technique*, **21**, 139–57.

Wilson, G.J., Carter, S.P., and Delahay, R.J. (2011). Advances and prospects for management of TB transmission between badgers and cattle. *Veterinary Microbiology*, **151**, 43–50.

Woldstad, T.M. (2010). *Trichodectes canis*, an invasive ectoparasite of Alaskan wolves: detection methods, current distribution, and ecological correlates of spread. M.S. thesis, University of Alaska.

Woodroffe, R. and Donnelly, C.A. (2011). Risk of contact between endangered African wild dogs *Lycaon pictus* and domestic dogs: opportunities for pathogen transmission. *Journal of Applied Ecology*, **48**, 1345–54.

Woodroffe, R., Prager, K.C., Munson, L., Conrad, P.A., Dubovi, E.J., and Mazet, J.A.K. (2012). Contact with domestic dogs increases pathogen exposure in endangered African wild dogs (*Lycaon pictus*). *PLoS ONE*, **7**(1): e30099.

Zulu, G.C., Sabeta, C.T., and Nel, L.H. (2009). Molecular epidemiology of rabies: focus on domestic dogs (*Canis familiaris*) and black-backed jackals (*Canis mesomelas*) from northern South Africa. *Virus Research*, **140**, 71–78.

CHAPTER 7

Impact of hybridization with domestic dogs on the conservation of wild canids

Jennifer A. Leonard, Jorge Echegaray, Ettore Randi, and Carles Vilà

7.1 Introduction

Dogs (*Canis familiaris*) were domesticated from gray wolves (*C. lupus*) in the late Pleistocene (Vilà et al., 1997; Wayne et al., 2006). Dogs retain the ability to hybridize with gray wolves and produce fertile offspring. However, gray wolves are not the only wild species that can hybridize with dogs. All wolf-like canids (*Canis* spp. and *Lycaon*; Figure 7.1) have the same number of chromosomes (n = 78) and may have the capacity to interbreed and produce viable offspring (MacDonald and Sillero-Zubiri, 2004; Vilà and Wayne, 1999).

Dogs are distributed across the entire range of all wild canids, and are much more abundant than any other canid species. The wild canid with the greatest natural distribution is the gray wolf, which, despite having been extirpated from large parts of its native range, has approximately 115,000 to 139,000 individuals distributed across Europe, Asia, and North America (Mech and Boitani, 2003). On the other end of the scale, only a few small populations remain of the Ethiopian wolf (*C. simensis*), totaling about 500 individuals in the wild (Ethiopian Wildlife Conservation Authority, 2012; MacDonald and Sillero-Zubiri, 2004). For comparison, there are approximately 700 million to 1 billion dogs worldwide (Baker et al., 2010; Hughes and MacDonald, 2013; WHO-WSPA, 1990; Gompper, Chapter 1), and a large fraction of this dog population could be considered free-ranging (reaching up to 75%, according to World Society for the Protection of Animals, 2010 as cited in Massei et al., 2010). Dogs not under control of an owner are considered 'free-ranging,' although some of these animals likely have owners that provide regular care, but allow them to roam for periods of time (Gompper, Chapter 1; Ritchie et al., Chapter 2). The abundance of free-ranging dogs in areas with small and endangered populations of other wolf-like canids could represent a serious conservation threat via risk of hybridization. A special case is that of the dingo, a divergent dog population that hybridizes with dogs of recent arrival to Australia (see Box 7.1).

7.2 Effects of hybridization

Hybridization can affect wildlife in a number of ways, most of them negative but some potentially positive as well (i.e., Grant and Grant, 1992; Larsen et al., 2010; McDonald et al., 2008; Schwenk et al., 2008). The negative impacts are better known and they include loss of reproductive potential, lowered fitness of individuals that hybridize, introduction of maladaptive alleles into wild populations, loss of genetic integrity, potential for disease transfer, and legal consequences that may affect the individual or population's conservation status (i.e., Barilani et al., 2007; Casas et al., 2012; Pierpaoli et al., 2003; Puigcerver et al., 2007; Randi, 2008; Schwartz et al., 2004).

Free-Ranging Dogs and Wildlife Conservation. Edited by Matthew E. Gompper
© Oxford University Press 2014. Published 2014 by Oxford University Press.

Figure 7.1 Some canids that potentially could be affected by hybridization with free-ranging dogs (a) Iberian wolf (*Canis lupus*) in Spain (photo: Jorge Echegaray); (b) coyote (*Canis latrans*) in Montana, USA (photo: Jorge Echegaray); (c) Ethiopian wolf (*Canis simensis*) in the afroalpine highlands of Ethiopia (photo: Martiño Nercellas); (d) African wild dog (*Lycaon pictus*) in South Africa (photo: Martiño Nercellas); (e) golden jackal (*Canis aureus*) in Hungary (photo: Miha Krofel); (f) black-backed jackal (*Canis mesomelas*) in South Africa (photo: Martiño Nercellas).

Box 7.1 Dingoes: mixing of ancient and recent dog populations

Dingoes are free-roaming dogs in Australia that descend from dogs brought to the region by ancient hunter-gatherer societies 3,500–5,000 years BP (Corbett, 1995; Elledge et al., 2006). This population of dogs has been largely isolated from other dogs for thousands of years (Oskarsson et al., 2012) until the arrival of the British with their dogs in 1788. These new arrivals bred with the local dog population, yielding the current mixed population (Corbett, 1995; Dickman et al., 2009). There are still some animals that are considered 'pure' dingo, although, according to skull morphometrics, they have been declining in numbers for the last century (Corbett, 1995). Dingoes have been and are still an important component of native Australian culture.

While the dingo is a protected species in parts of Australia (for example, it is protected in the Northern Territory and is regarded as having important conservation value) in other areas it is considered a pest and there is a legal mandate to decimate or eradicate its populations. Efforts to control the dingo population have grown out of the conflict between

continued

Box 7.1 *Continued*

dingoes and ranchers due to predation on livestock, especially sheep. The best example of this campaign has been the construction of the world's longest fence: the dingo fence, a man-made defense that stretches 5,614 kilometers.

There is a recent and growing body of research that suggests that dingoes may be playing an important role as apex predators (Johnson et al., 2007; Letnic et al., 2009; Wallach et al., 2009; Ritchie et al., Chapter 2). It has been suggested that these free-ranging dogs could be helping to restore the biodiversity in several environments (see e.g., Dickman et al., 2009) by direct predation on introduced competitors of marsupial native communities (i.e., feral goats and pigs), and indirectly benefiting native species through suppression of introduced mesopredators (i.e., fox and cats).

In areas where dingoes are protected, the main threat to their conservation is crossing with dogs of recent arrival to Australia (Elledge et al., 2006). However, quantitatively assessing the extent of this mixture, as well as the extent of genetic transfer between populations, is a difficult task since no diagnostic tools exist that can readily separate the two lineages, and dingoes are heavily admixed over the entire Australia with the possible exception of some remote and protected areas. This implies that it is difficult to find pure dingoes that can be used as reference for the identification of mixes (see Box 7.2). It is not clear how many pure dingoes may exist today and conservation efforts sometimes aim at identifying and removing animals below a specific threshold of dingo ancestry (e.g., quarter or half dingoes, Elledge et al., 2006).

In small populations, hybridization may result in a loss of intraspecific mating opportunities and a subsequent reduction in the effective number of breeders. This implies a loss in the population's reproductive potential and may reduce the population growth rate to a level that is below what is needed for its long-term survival. In this way, hybridization may present a demographic threat to a species or population even if hybrids die and do not lead to genetic admixture through introgression (introduction of genes of one species into the gene pool of another, see Box 7.2).

Introgression following hybridization can introduce maladapted genes into wild populations. The gene pool of each population and species is the result of thousands to millions of years of selection and adaptation to specific environmental conditions. A rapid influx of genes from another species or population, that evolved under different conditions and with a different set of co-evolved genes, usually precipitates a decline in fitness. Despite strong negative selection, these genes have a real chance of achieving high frequencies or even becoming fixed in small populations through random genetic drift. These genes could reduce the population's chances of long-term survival. In cases of high frequency of hybridization and introgression, a major conservation threat is the potential for the loss of the genetic integrity of a species.

Several of these negative effects of hybridization in the wild are observed in the Ethiopian wolf. The distribution of Ethiopian wolves has been reduced and fragmented to a handful of very small populations, some of which are in populated areas with high numbers of free-ranging dogs. In at least one of these populations, Web Valley, these endangered canids hybridized with dogs, and the hybrids have integrated into the population and backcrossed (Gottelli et al., 1994). The level of hybridization and backcrossing is high enough to threaten the integrity of this population, and dog alleles were also identified in another population. Currently, small size, strong population structuring with low gene flow, demographic stochasticity, and hybridization with dogs are considered the main threats for the survival of this species (Gottelli et al., 2013). The contact during successful or attempted intercrossing with dogs has also further threatened Ethiopian wolves by spreading rabies (Laurenson et al., 1998).

Hybridization with dogs can threaten wild populations of canids even when the hybrids do not backcross into the population, by changing people's perception of the population and, in some cases, changing their legal conservation status. Popular perception of wildlife, expectations of wildlife management, and feelings about hybridization involving wildlife are drastically different across the range of all species in the genus *Canis*, even between North America and Europe. For example, a female gray wolf from the highly endangered Scandinavian population hybridized with a male dog in Norway (Vilà et al., 2003). Concern about the presence of hybrids led the authorities to allow killing any wolf-like canid in the area by government officials and hunters despite it being outside the hunting season. In other countries such as Finland, some hybridization events were reported in areas geographically distant from the known established wolf packs. The suspected wolf-dog individuals inspired fear in local people with their apparent lack of shyness, aggressive behavior, and some livestock damage possibly caused by them (Ministry of Agriculture and Forestry, 2005). In Spain, the existence of some suspected hybrids has recently fanned persecution of both wolves and dogs in areas where hybridization may have taken place. Wolf conservation and management programs across Europe emphasize the eradication of hybrids as a priority (for example, in Estonia, Lohmus, 2001; Italy, Genovesi, 2002; Finland, Ministry of Agriculture and Forestry, 2005), and this is also emphasized in the *Action plan for the conservation of wolves* (Canis lupus) *in Europe* (Boitani, 2001). Although this priority has been widespread, the scientific basis is not clear.

Hybridization and introgression are generally assumed to be negative. However, they can act as a source of genetic variation that could be beneficial in some cases, promoting rapid diversification or even speciation, and introduction of alleles with adaptive value (i.e., Cadieu et al., 2009; Grant and Grant, 2008; Larsen et al., 2010; Pardo-Díaz et al., 2012). When populations are large, hybridization and subsequent introgression could provide additional genetic material for selection to act upon. An example of this seems to have happened when a gene coding for melanistic coloration was transferred to wild populations of both gray wolves and coyotes (*C. latrans*) in North America from dogs through ancient hybridization (Anderson et al., 2009). Analysis of the gene and surrounding sequence (haplotype) in wolves, coyotes, and dogs showed that the causative mutation arose in dogs and then entered the wolf and coyote gene pools. The mutation likely arose many thousands of years ago in dogs and was more recently transferred, likely in pre-Columbian times, to gray wolves from Native American dogs. The very low diversity in the wolf haplotypes and the high frequency of this allele in certain populations suggests that this locus had been positively selected for in some populations of wild wolves. In this case, the populations of wolves and coyotes were large enough that the positive selection for the introgressed gene was able to separate it from other linked genes, and it increased in frequency in some populations. This would have been less likely if the populations were small.

Box 7.2 Identifying hybrids and assessing introgression

The terms 'hybridization' and 'introgression' are related, but not the same. Hybridization, interbreeding between two different species, even when it results in viable, fertile offspring does not always lead to introgression. Introgression is the transfer of genetic material from one species to another. Introgression starts with hybridization, but then those hybrid offspring must backcross with the parent species for the DNA from one species to be incorporated into the gene pool of the second species (Allendorf and Luikart, 2007).

Although morphological irregularities can indicate a hybrid origin of an individual (Ciucci et al., 2003), there are no unambiguous traits that definitively distinguish dog–wild canid hybrids. The difficulty in identifying hybrids is especially great in this case because of the huge phenotypic diversity present in the domestic species (Boyko and Boyko, Chapter 8). For this reason, the unambiguous identification of hybrids and of individuals with signs of introgression demands the application of genetic tests.

There are thousands of genetic markers that are useful for the study of gray wolves and dogs. The markers inform about variants (alleles) in very small portions of the genome of these species. When investigating hybridization, researchers look for genetic markers (such as short repetitive sequences called microsatellites, or single nucleotides that differ from one sequence to another, called SNPs) that are present in the

continued

Box 7.2 *Continued*

two species with different frequencies. The genetic identification of hybrids involves typing a panel of genetic markers (often between 10 and 20 microsatellite loci) in the suspect individuals and comparing the alleles observed with the frequencies previously found in dogs and in the wild canid. Since the two species are relatively well separated, the allele frequencies tend to be different. Genetic markers showing fixed or almost fixed differences between the two species are the most useful ones because they are expected to generate heterozygote genotypes in the first hybrid generation, following Mendel's laws. Alleles that are present with high frequency in dogs, but are rare in the wild canid, can thus be considered 'dog alleles.' First generation (F1) hybrids will tend to have a 'dog allele' and a 'wild canid allele' for each one of the genetic markers studied. Even if species-specific alleles cannot be identified, the genotype found for a hybrid should have a very low probability of being found in each one of the parental species and should fit with the expectation of an admixed genotype.

Even though the genetic identification of hybrids is relatively easy (except for dingoes; see Box 7.1), the identification of cases of introgression is much more difficult. If a hybrid mates with the wild parental species, only half of the genetic markers would be expected to show a mix of dog and wild canid alleles. This proportion will continue to decrease in subsequent generations. This implies that the robust identification of individuals with admixed ancestry requires a much larger number of genetic markers in order to increase the chances of unambiguously finding regions of the genome originating in the other species. The number of markers required will exponentially increase as one tries to identify older introgressions. This is further complicated by the difficulty of characterizing true parental populations and because hybridization may be episodic or fluctuating, particularly in anthropogenic cases of hybridization. Only more sophisticated genomic approaches, currently being developed, promise to provide this robust identification of old introgression events (Anderson et al., 2009; vonHoldt et al., 2011).

7.3 Hybridization is often directional

Hybridization between wild canids and dogs typically seems to be directional. Vilà and Wayne (1999) surveyed all of the maternally inherited mitochondrial DNA data available for gray wolves at that time and found no instance of introgressed dog mitochondrial DNA in any gray wolf population. Since then, thousands more mitochondrial DNA sequences have been published (i.e. Koblmüller et al., 2009, 2012; Leonard et al., 2005; Muñoz-Fuentes et al., 2009; Musiani et al., 2007; Pilot et al., 2010; Randi et al., 2000; Vilà et al., 1999) and very few instances of introgression of a dog mitochondrial haplotype into a wild canid population have been reported, each of which is most likely explained by a single hybridization event (Adams et al., 2003; Muñoz-Fuentes et al., 2010). This implies that hybrid matings involving a female dog and a male wolf are extremely rare, or that those hybrids fail to integrate themselves in the wolf populations.

On the other hand, almost all confirmed hybridization events between these two species involved the mating of female wolves and male dogs. Thus the hybrids carry wolf mitochondrial DNA (Andersone et al., 2002; Godinho et al., 2011; Khosravi et al., 2013; Randi, 2011; Randi and Lucchini, 2002; Verardi et al., 2006; Vilà et al., 2003; but see Hindrikson et al., 2012). These cases could not be identified through mitochondrial DNA analyses because this marker is maternally inherited, so paternally or bi-parentally inherited nuclear markers must be employed (Iacolina et al., 2010; Vilà et al., 2003). A similar pattern was also observed in Ethiopian wolves (Gottelli et al., 1994).

In the case of gray wolves, the asymmetric pattern of hybridization has been explained by the lack of synchrony in breeding time with that of dogs (Vilà and Wayne, 1999). In most wild canids, females have a single estrus per year, and males are only sexually active at that time of year. This differs fundamentally from dogs, where females have two non-synchronized estrus cycles per year and males are ready to breed throughout the year. This means that whenever a female, wild or domestic, is ready to breed, male dogs are also ready. However, many female dogs will not be in estrus during the short window that wild male canids are able to breed (Vilà and Wayne, 1999).

In some of the few cases where it has been documented that a male wild canid hybridized with a female dog, those males were the first or among the first individuals of their species to (re)colonize an area, and there were either few, or possibly no, conspecific females in the area (Adams et al., 2003; Muñoz-Fuentes et al., 2010). This specific social/ecological condition is an extreme example of an Allee effect (Allee, 1931) where it is not only difficult, but potentially impossible, for an individual to find an appropriate mate. These conditions are probably rare in less disturbed systems. However, in both of these cases this condition was anthropogenically facilitated. In one case young coyote males were trapped, transported outside the range of the species, and released, so there was no possibility of them finding female coyotes with which to mate (Adams et al., 2003). In the second case the local wolf population in Vancouver Island was eradicated, and the first natural recolonizers to reach it found a large population of dogs, but no conspecific potential mates (Muñoz-Fuentes et al., 2010).

7.4 Hybridization, but limited introgression

Theoretically, the arrival of just a few effective migrants—individuals contributing to the reproduction and to the population's gene pool—per generation is enough to erase the differentiation between two populations (Fisher, 1941; Mills and Allendorf, 1996; Wright, 1978). The frequency of hybridization between dogs and wolves in small, fragmented European populations where hybrids have been reported (Figure 7.2) appears to be close to 3–5% (Andersone et al., 2002; Godinho et al., 2011; Randi, 2011; Randi and Lucchini, 2002; Verardi et al., 2006; Vilà et al., 2003). Gottelli et al. (1994) reported that 8–17% of the Bale Mountains Ethiopian wolf population had hybrid ancestry. Therefore, reported levels of hybridization should quickly lead to the complete admixture and loss of differentiation of the two species if the hybrids are incorporated into the wild canid populations and do not have a lower fitness. These cannot be taken as species-wide estimates of hybridization rates. Genetic studies of hybridization are costly and are not likely to be undertaken unless there is reasonable chance (lots of dogs and few wild canids) or indirect evidence of hybridization. However, these rate estimates show that hybridization can be locally frequent.

Nevertheless, all these studies show that the hybridizing species remain genetically distinct, suggesting that the genetic admixture may be lower than that estimated by the estimates rates of hybridization. In part this may be due to the fundamentally different behavior of dogs as compared to wild canids (Miklósi, 2007). Many of these behaviors and behavioral complexes have been shown to be genetically inherited, and not a product of the environment. For this reason it is expected that hybrid offspring will share some of their behaviors with each parental species (Box 7.3). Some dog behaviors, such as a tendency to promiscuous matings without forming stable pairs and a low level of contribution by males to raising pups, may seriously impede or

Figure 7.2 Gray wolf–domestic dog hybrids (a) born and living in the wild in Asturias, Spain (photo: Xurde Gayol); (b) born and living in the wild in Italy (photo: Lorenzo Rigacci).

Box 7.3 Behavior of hybrids

Many behavioral characteristics are genetically determined, and have been differentially selected for in wild and domestic canids. Because hybrids are a combination of two genetic lineages, their behavioral characteristics are likely to be a mix of both. Nevertheless, our knowledge about inheritance of behavior is limited.

The modern advances in molecular genetics have found links between genes, brain function, and a wide range of social behaviors on diverse organisms (Moon-Fanelli, 2011; Robinson et al., 2008). For example, in silver foxes, a naturally occurring melanistic variant of the red fox (*Vulpes vulpes*), an extensive study including genetics, pedigree, biochemical and morphological data, found that some behavioral phenotypes have a genetic basis (i.e., Kukekova et al., 2012). In this sense, each hybrid could show very different personality traits, depending on the degree of admixture and also on chance inheritance of specific alleles. In the course of domestication, dogs have been selected for traits that enable them to understand some human signals and to produce a wider range of communication signals (Hare et al., 2002; Miklósi, 2007). The development of these skills likely has a genetic basis (i.e., Kukekova et al., 2012), as it has also been shown through training experiments with socialized wolves and dogs (Miklósi, 2007; Miklósi et al., 2004). This ability does not seem shared with either wolves or wolf–dog hybrids.

In general, wolf–dog hybrids are less predictable and manageable than dogs. They generally exhibit unpredictable behavior, often including absence of tameness, lack of fear, anti-social behavior, and aggressiveness towards both animals and humans, even if they have been raised with them. Wildlife and livestock may be much more vulnerable to predation by hybrids than dogs because of their higher prey drive instinct (Fritts et al., 2003). There are even reports of fatal attacks by wolf–dog hybrids on unattended children (Hope, 1994).

Dogs are the result of thousands of generations of selection for tame animals. Hybridization represents the disruption of this co-selected set of genes, altering the behavioral repertoire in an unpredictable way. This makes hybrids poor pets in almost all circumstances.

even prevent hybrids from integrating into the wild canid society and achieving breeder status (Vilà and Wayne, 1999). This factor may be especially important in the more social canids, such as the gray wolf and the Ethiopian wolf.

In some cases, physiological differences between the species can contribute to reduce the fitness of the hybrids and their chances of integrating into the wild populations. The timing of estrus in females is genetically based. In dog X coyote crosses, female F1 and F2 hybrids have an estrus cycle that is shifted by two months, from February to December (Mengel, 1971; Silver and Silver, 1969). If these females are able to find mates and breed, their offspring will be born in winter instead of spring, which is likely to increase mortality in colder portions of their range (Adams et al., 2003). A similar effect on the physiology of hybrids with other wild species is probable.

7.5 Cases of introgression

Despite the lower fitness of hybrids, there are a few clear cases of introgression of dog alleles into wild canid populations, including both the Vancouver Island gray wolf population and the south-east United States population of coyotes examples mentioned above (Adams et al., 2003; Muñoz-Fuentes et al., 2010). These examples involve the rare introgression of mitochondrial DNA, often considered to be selectively neutral. Selectively neutral autosomal microsatellite alleles have also been reported to have introgressed from domestic dogs to wild canids, for example in the case of the Ethiopian wolf (Gottelli et al., 1994). Neutral loci should not have any major impact on the phenotype or fitness of the carriers of these introgressed alleles.

Alleles at functional genes associated with coat color appear also to have introgressed from domestic dogs to wild canids. The best-documented case is that of the introgression of a *k* locus allele, which causes a melanistic coat color in dogs, into wild populations of both wolves (Figure 7.3) and coyotes in North America (Anderson et al., 2009), as described above (Section 7.2). Once this mutation was in the wolf population, it provided some selective advantage and spread, although the specific advantage incurred is not known. The black color

Figure 7.3 Pack of wolves comprising gray and black individuals in Yellowstone National Park, USA (photo: Dan Stahler/NPS photo).

in North American wolves has been correlated with ecological factors, and its high frequency in some areas is suggested to be an adaption to specific ecological conditions (Coulson et al., 2011; Musiani et al., 2007). Thus, modern black wolves carry in their DNA evidence of an ancient introgression of dog genes, but this does not imply that they are hybrids or that they have a recent hybrid ancestor.

Similarly, results from Italian wolves suggest that introgression of portions of the dog genome in the gene pool of wild wolves can be selected for, affecting only some parts of the genome, while other genomic regions remain virtually untouched (Randi, 2011). In both of these cases the functional or putatively functional gene or region seems to be very discrete and not to have 'carried along' much else. This suggests strong selection against most dog DNA in wild canids.

Due to the strong effect of selection in wolves, the rate of gene introgression cannot be directly related to the frequency of hybridization. If hybrids fail to reproduce or to integrate themselves into the wild canid population, these crosses do not result in the transfer of dog genes, reducing the threat posed by hybridization. As mentioned above, the observation that the species remain differentiated indicates that hybrids often do not transfer their genome into the wild populations. This, together with the observation that most of the genome of the wild canids remains virtually untouched (see vonHoldt et al., 2011 for genome-wide analyses involving multiple populations of dogs and wolves), suggest that introgression in nature might be strongly counteracted by selection (Randi and Lucchini, 2002; Vilà and Wayne, 1999). In very small populations, however, random genetic drift can overwhelm selection and even deleterious alleles can rise in frequency or even become fixed. For example, recessive deleterious alleles likely cause the observed inbreeding depression in the Scandinavian wolf population (Liberg et al., 2005), which was highly inbred because of its extremely small founder population of just three individuals (Vilà et al., 2003), and removing those maladaptive alleles from the gene pool has proven difficult because the effects of selection are likely offset by genetic drift (Hagenblad et al., 2009). This may have impacted the population recovery.

7.6 When does hybridization occur?

The factors driving hybridization between wild and domestic canids are multiple and likely complex (Randi, 2008; Vilà and Wayne, 1999). Hybridization has been primarily documented in cases where the wild species exists as a small population, on the edge of its distribution, or when dispersing individuals fail to finds suitable mates. Allowing wild canids to exist in populations that are large enough to be demographically and genetically self-sustaining would greatly alleviate this threat (i.e., Wayne and Brown, 2001; Gottelli et al., 2013).

Additionally, hybridization between wild canid species has been shown to increase under conditions of anthropogenic disturbance, such as changes in land-use and high human-caused mortality, likely due to social structure disruption (Hailer and Leonard, 2008; Koblmüller et al., 2009; Rutledge et al., 2010; Stronen et al., 2012; vonHoldt et al., 2011). The same conditions that facilitate hybridization between wild canids may also facilitate hybridization between wild and domestic canids. In this case, the hunting or removal for management of wolves or other wild canids, especially in small, isolated, or 'edge' populations, may be an important factor resulting in hybridization (Andersone et al., 2002).

7.7 Conservation implications

Despite the huge capacity of large canids, such as gray wolves, Ethiopian wolves, and coyotes, to disperse, these species have been shown to have a strong population structure likely associated with local adaptation (Carmichael et al., 2001; Geffen et al., 2004; Muñoz-Fuentes et al., 2009; Musiani et al., 2007; Pilot et al., 2006; Sacks et al., 2004, 2005; Gottelli et al., 2013). This pattern is probably common to all wild canids distributed across multiple habitats. Hybridization has the potential to break down both species-level and population-level adaptations.

Most of the documented hybridization between wild and domestic canids has been between gray wolves and dogs (i.e., Randi, 2008). This is primarily a reflection of the amount of effort that has gone into studying the population genetics of gray wolves versus other wild canids. However, hybridization has also been documented between dogs and coyotes (Freeman and Shaw, 1979; Gipson et al., 1974) and Ethiopian wolves (Gottelli et al., 1994). It seems likely that, as more studies are undertaken with other large canids, more examples of hybridization will be identified. The golden jackal (*C. aureus*) is a widespread species that inhabits many human-dominated areas, and has expanded its distribution in some European countries (Arnold et al., 2012). Recently, two jackals with anomalous phenotypes were collected in Dalmatia (Croatia). Preliminary genetic analyses suggested that they are probably first generation hybrids with dogs (E. Randi, pers. obs.). Hybridization between dogs and common, widespread species such as coyotes and gray wolves (and probably golden jackals) does not usually threaten the wild species due to loss of breeding opportunities. However, hybridization between dogs and rarer wild species, such as the Ethiopian wolf, can reduce its effective number of breeders (MacDonald and Sillero-Zubiri, 2004) and so contribute to the erosion of its genetic diversity (Randall et al., 2010).

At the same time, dog-transmitted diseases have caused dramatic die-offs and local extinctions (Knobel et al., Chapter 6), such as rabies in Ethiopian wolves (Laurenson et al., 1998; MacDonald and Sillero-Zubiri, 2004) and canine parvovirus in gray wolves in Isle Royale (Peterson, 2007). This may constitute an anthropogenic 'edge effect' originating from the large and uncontrolled presence of dogs in many environments (Woodroffe and Ginsberg, 1998), which could also favor hybridization between dogs and some wild canids (e.g., Vilà and Wayne, 1999).

A less biological, but still important, threat to wild canids resulting from hybridization with dogs can be a change in social or legal status. Many populations of canids are protected by regional, national, and/or international laws and treaties. A few individuals hybridizing in these populations may remove the legal protection from either those individuals, their social group, or the whole population. Ironically, the populations under the highest conservation threat, and thus most in need of legal protection, are also those most likely to be involved in hybridization, as discussed above. Perhaps the observation of hybridization in wild populations of canids should be an indicator of poor population health, and a reason for more, not less, protection.

A common response to the presence of wolf–dog hybrids in an area is trying to remove the individuals to minimize the impact of those hybrids on the native population. However, most studies suggest that hybrids often fail to introduce themselves in the wild population and reproduce (see Section 7.4). Consequently, those management measures may not be necessary. Increased hunting pressure and disturbance of social groups created during the hybrid removal efforts could translate into increased chances of hybridization, and thus amplify rather than contain the problem. Allowing populations to grow and to reach stability could be a more efficient (and cheaper) management measure to reduce the impact of hybrids.

The risk of hybridization between wild and domestic canids could be further diminished by reducing contact between them (Randi, 2008). This would require the control of free-ranging dog populations and, most importantly, allowing an increase in the numbers of wild canids where they exist as small populations so they can form functional populations. This is simple in principle, but complex in practice (Randi, 2011; Young et al., 2011). Fertility control through immunocontraception and surgical sterilization has been shown as an efficient alternative for the long-term reduction of dog numbers (Woodroffe et al., 2004; Young et al., 2011). Direct lethal measures focused on dogs and presumed hybrids are very costly, complex, difficult in practice, and ineffective to implement in the long term. A more effective way to control dog populations would be intense public education and reduced access to anthropogenic food resources such as carrion and garbage dumps (Butler et al., 2004; Woodroffe and Donnelly, 2011; Woodroffe et al., 2004).

7.8 Conclusion

Introgression of dog genes into wild canid populations does not seem to be a large problem for widespread wild species, as selection against the hybrids seems generally to be strong enough to remove them from wild populations. The ability of selection to favor adaptation and remove unfit alleles is dependent on large population sizes. In cases where populations are small, selection will not be able to counteract drift and deleterious genes may increase in frequency or even become fixed. For this reason, it is important to allow small, isolated populations of canids to increase in numbers. This includes both rare species, such as the Ethiopian wolf, and isolated populations of more common species, such as the Mexican wolf and the Scandinavian wolf populations (Hagenblad et al., 2009; Hailer and Leonard, 2008).

7.9 Future research

The frequency, distribution, and impact of hybridization between domestic and wild species are just now being realized, largely with the use of increasingly sensitive genetic and genomic tools (Randi, 2008; vonHoldt et al., 2010, 2011). Many of these same patterns of hybridization are also being identified in other taxonomic groups that include both wild and domestic species (Alemayehu et al., 2011; Allendorf et al., 2001; Halbert et al., 2005; Scandura et al., 2011; Silbermayr et al., 2010).

In addition to documenting and characterizing hybridization, we need to better understand the conditions that promote it. This will require an integration of field ecology and genetic approaches. Europe is a good place to study these interactions because of high human and dog density, overlapping with fragmented and relatively small wolf populations. In many regions, conservation guidelines and laws aim at establishing a functional system of connected landscapes and wilderness areas to allow natural dispersal of wildlife. This could support the conservation of marginal or small wild canid populations (Boitani and Ciucci, 2009; MacDonald and Sillero-Zubiri, 2004; Gottelli et al., 2013).

Most of the documented hybridization and introgression has been between dogs and their wild ancestor, the gray wolf. However, this may primarily be a reflection of effort. Very little work has been done on interactions between other *Canis* species and dogs. In particular, golden jackals could be prone to hybridization. Golden jackals may live in humanized areas, are opportunistic, and are likely increasing their distribution in Europe (Arnold et al., 2012). Other species of jackals and wild dogs also co-occur with free-ranging dogs and could potentially hybridize.

Hybridization in *Canis* species is likely not a novel phenomenon. Understanding the history of hybridization over the course of the dog domestication process and over the course of the past several thousand years of change in human societies would be informative. Such information (which could be achieved through the ancient DNA analysis of Pleistocene and Holocene populations of co-distributed dogs and wild canids) would provide insight into the conditions under which hybridization is most common, and the longer-term effect(s) of that hybridization on natural populations.

Acknowledgments

The Programa de Captación del Conocimiento para Andalucía (Spain) supported CVA. The Spanish National Research Council (CSIC) JAE Predoctoral fellowship supported JE. We want to thank Xurde Gayol, César Alonso Guzmán, Martiño Nercellas, Angel Nuño, Lorenzo Rigacci, and Dan Stahler for sharing their pictures and field experience with wolves and hybrids in the wild.

References

Adams, J.R., Leonard, J.A., and Waits, L.P. (2003). Widespread occurrence of a domestic dog mitochondrial DNA haplotypes in southeastern US coyotes. *Molecular Ecology*, **12**, 541–46.

Alemayehu, K., Gizaw, S., Mekasha, Y., Haile, A., and Dessie, T. (2011). The probable genetic introgression between walia ibex (*Capra walie*) and domestic goats (*Capra hircus*) at Simien Mountains National Ppark (SMNP) in Ethiopia. *African Journal of Agricultural Research*, **6**, 856–65.

Allee, W.C. (1931). *Animal Aggregations*. University of Chicago Press, Chicago.

Allendorf, F.W., Leary R.F., Spruell, P., and Wenburg, J.K. (2001). The problems with hybrids: setting conservation guidelines. *Trends in Ecology and Evolution*, **16**, 613–22.

Allendorf, F.W. and Luikart, G. (2007). *Conservation and the Genetics of Populations*. Blackwell Publishing Ltd, Oxford.

Anderson, T.M., vonHoldt, B.M., Candille, S.I., Musiani, M., Greco, C., Stahler, D.R., Smith, D.W., Padhukasahasram, B., Randi, E., Leonard, J.A., Bustamante, C.D., Ostrander, E.A., Tang, H., Wayne, R.K., and Barsh, G.S. (2009). Molecular and evolutionary history of melanism in North American gray wolves. *Science*, **323**, 1339–43.

Andersone, Z., Lucchini, V., Randi, E., and Ozolins, J. (2002). Hybridisation between wolves and dogs in Latvia as documented using mitochondrial and microsatellite DNA markers. *Mammalian Biology*, **67**, 79–90.

Arnold, J., Humer, A., Heltai, M., Murariu, D., Spassov, N., and Hackländer, K. (2012). Current status and distribution of golden jackals *Canis aureus* in Europe. *Mammalian Review*, **42**, 1–11.

Baker, P.J., Soulsbury, C.D., Iossa, G., and Harris, S. (2010). Domestic cat (*Felis catus*) and domestic dog (*Canis familiaris*). In S.D. Gehrt, S.P.D. Ridely and B.L. Cypher, Eds., *Urban Carnivores. Ecology, Conflict, and Conservation*. The Johns Hopkins University Press, Baltimore, pp. 156–71.

Barilani, M., Bernard-Laurent, A., Mucci, N., Tabarronic, C., Kark, S., Garridoe, J.A.P., and Randi, E. (2007). Hybridization with introduced chukars (*Alectoris chukar*) threatens the gene pool integrity of native rock (*A. graeca*) and red-legged (*A. rufa*) partridge populations. *Biological Conservation*, **137**, 57–69.

Boitani, L. (2001) *Action plan for the conservation of wolves (Canis lupus) in Europe*. Council of Europe. Strasburg, France.

Boitani, L. and Ciucci, P. (2009). Wolf Management across Europe: Species conservation without boundaries. In M. Musiani, P.C. Paquet and L. Boitani, Eds., *A New Era for Wolves and People. Wolf Recovery, Human Attitudes, and Policy*. University of Calgary Press. Energy, Ecology and Environment Series, Calgary, pp. 15–39.

Butler, J.R.A., Du Toit, J.T., and Bingham, J. (2004). Free-ranging domestic dogs (*Canis familiaris*) as predators and prey in rural Zimbabwe: threats of competition and disease to large wild carnivores. *Biological Conservation*, **115**, 369–78.

Cadieu, E., Neff, M.W., Quignon, P., Walsh, K., Chase, K., Parker, H.G., vonHoldt, B.M., Rhue, A., Boyko, A., Byers, A., Wong, A., Mosher, D.S., Elkahloun, A.G., Spady, T.C., André, C., Lark, K.G., Cargill, M., Bustamante, C.D., Wayne, R.K., and Ostrander, E.A. (2009). Coat variation in domestic dog is governed by variants in three genes. *Science*, **326**, 150–3.

Carmichael, L.E., Nagy, J.A., Larter, N.C., and Strobeck, C. (2001). Prey specialization may influence patterns of gene flow in wolves of the Canadian Northwest. *Molecular Ecology*, **10**, 2787–98.

Casas, F., Mougeot, F., Sánchez-Barbudo, I., Dávila, J.A., and Viñuela, J. (2012). Fitness consequences of anthropogenic hybridization in wild red-legged partridge (*Alectoris rufa*, Phasianidae) populations. *Biological Invasions*, **14**, 295–305.

Ciucci, P., Lucchini, V., Boitani, L., and Randi, E. (2003). Dewclaws in wolves as evidence of admixed ancestry with dogs. *Canadian Journal of Zoology*, **81**, 2077–81.

Corbett, L. (1995). *The Dingo in Australia and Asia*. University of New South Wales Press, Sydney.

Coulson, T., MacNulty, D.R., Stahler, D.R., vonHoldt, B., Wayne, R.K., and Smith, D.W. (2011). Modeling effects of environmental change on wolf population dynamics, trait evolution, and life history. *Science*, **334**, 1275–8.

Dickman, C.R., Glen, A.S., and Letnic, M. (2009). Reintroducing the dingo: Can Australia's conservation wastelands be restored? In M.W. Hayward and M.J. Somers,

Eds., *Reintroduction of Top-Order Predators*. Blackwell Publishing, Oxford, pp. 238–69.

Elledge, A.E., Leung, L.K.-P., Allen, L.R., Firestone, K., and Wilton, A.N. (2006). Assessing the taxonomic status of dingoes *Canis familiaris dingo* for conservation. *Mammal Review*, **36**, 142–56.

Ethiopian Wildlife Conservation Authority (2012). <http://www.ethiopianwolf.org/>.

Fisher, R.A. (1941). Average excess and average effect of a gene substitution. *Annals of Human Genetics*, **11**, 53–63

Freeman, C.R. and Shaw, J.H. (1979). Hybridization in *Canis* (Canidae) in Oklahoma. *Southwestern Naturalist*, **24**, 485–99.

Fritts, S.H., Stephenson, R. O., Hayes, R., and Boitani, L. (2003). Wolves and humans. In L.D. Mech and L. Boitani, Eds., *Wolves. Behavior, Ecology and Conservation*. The University of Chicago Press, Chicago, pp. 289–316.

Geffen, E., Anderson, M.J., and Wayne, R.K. (2004). Climate and habitat barriers to dispersal in the highly mobile grey wolf. *Molecular Ecology*, **13**, 2481–90.

Genovesi, P. (ed.) (2002). *National action plan for wolf (Canis lupus) conservation in Italy*. Nature Conservation Report 13, Ministry of Environment-ISPRA, Bologna.

Gipson, P.S., Sealander, J.A., and Dunn, J.E. (1974). The taxonomic status of wild *Canis* in Arkansas. *Systematic Zoology*, **23**, 1–11.

Godinho, R., Rodríguez-Llaneza, L., Blanco, J.C., Lopes, S., Álvares, F., García, E.J., Palacios, V., Cortés, Y., Talegón, J., and Ferrand, N. (2011). Genetic evidence for multiple events of hybridization between wolves and domestic dogs in the Iberian Peninsula. *Molecular Ecology*, **20**, 5154–66.

Gottelli, D., Sillero-Zubiri, C., Applebaum, G.D., Roy, M.S., Girman, D.J., Garcia-Moreno, J., Ostrander, E.A., and Wayne, R.K. (1994). Molecular genetics of the most endangered canid: the Ethiopian wolf *Canis simensis*. *Molecular Ecology*, **3**, 301–12.

Gottelli, D., Sillero-Zubiri, C., Marino, J., Funk, S.M., and Wang, J. (2013). Genetic structure and patterns of gene flow among populations of the endangered Ethiopian wolf. *Animal Conservation* **16**, 234–47.

Grant, P.R. and Grant, B.R. (1992). Hybridization of bird species. *Science*, **256**, 193–7.

Grant, B.R. and Grant, P.R. (2008). Fission and fusion of Darwin's finches populations. *Philosophical Transactions of the Royal Society B, Biological Sciences*, **363**, 2821–29.

Hagenblad, J., Olsson, M., Parker, H.G., Ostrander, E.A., and Ellegreen, H. (2009). Population genomics of the inbred Scandinavian wolf. *Molecular Ecology*, **18**, 1341–51.

Hailer, F. and Leonard. J.A. (2008). Hybridization among three native North American *Canis* species in a region of natural sympatry. *PLoS ONE*, **3**, e3333.

Halbert, N.D., Ward, T.J., Schnabel, R.D., Taylor, J.F., and Derr, J.N. (2005). Conservation genomics: disequilibrium mapping of domestic cattle chromosomal segments in North American bison populations. *Molecular Ecology*, **14**, 2343–62.

Hare, B., Brown, M., Wiliamson, C., and Tomasello, M. (2002). The domestication of social cognition in dogs. *Science*, **298**, 1634–36.

Hindrikson, M., Männil, P., Ozolins, J., Krzywinski, A., and Saarma, U. (2012). Bucking the trend in wolf-dog hybridization: first evidence from Europe of hybridization between female dogs and male wolves. *PLoS ONE*, **7**, e46465.

Hope, J. (1994). Wolves and wolf hybrids as pets are big business, but a bad idea. *Smithsonian*, **25**, 34–44.

Hughes, J. and MacDonald, D.W. (2013). A review of the interactions between free-roaming domestic dogs and wildlife. *Biological Conservation*, **157**, 341–51.

Iacolina, L., Scandura, M., Gazzola, A., Cappai, N., Capitani, C., Mattioli, L., Vercillo, F., and Apollonio, M. (2010). Y-chromosome microsatellite variation in Italian wolves: A contribution to the study of wolf-dog hybridization patterns. *Mammalian Biology*, **75**, 341–7.

Johnson, C.N., Isaac, J.L., and Fisher, D.O. (2007). Rarity of a top predator triggers continent-wide collapse of mammal prey: Dingoes and marsupials in Australia. *Proceedings of the Royal Society of London, B*, 274: 341–46.

Khosravi, R., Rezaei, H.R., and Kaboli, M. (2013). Detecting hybridization between Iranian wild wolf (*Canis lupus pallipes*) and free-ranging domestic dog (*Canis familiaris*) by analysis of microsatellite markers. *Zoological Science*, **30**, 27–34.

Koblmüller, S., Nord, M.,Wayne, R.K., and Leonard, J.A. (2009). Origin and status of the Great Lakes wolf. *Molecular Ecology*, **18**, 2313–16.

Koblmüller, S., Wayne, R.K., and Leonard, J.A. (2012). Impact of Quaternary climatic changes and interspecific competition on the demographic history of a highly mobile generalist carnivore, the coyote. *Biology Letters*, **8**, 644–7.

Kukekova, A.V., Temnykh, S.V., Johnson, J.L., Trut, L.U., and Acland, G.M. (2012). Genetics of behavior in the silver fox. *Mammalian Genome*, **23**, 164–77.

Larsen, P.A., Marchan-Rivadeneira, M.R., and Baker, R.J. (2010). Natural hybridization generates mammalian lineage with species characteristics. *Proceedings of the National Academy of Sciences*, **107**, 11447–52.

Laurenson, M.K., Sillero-Zubiri, C., Thompson, H., Shiferwa, F., Thirgood, T., and Malcolm, J.R. (1998). Disease threats to endangered species: Ethiopian wolves, domestic dogs, and canine pathogens. *Animal Conservation*, **1**, 273–80.

Leonard, J.A, Vilà, C., and Wayne, R.K. (2005). Legacy lost: Genetic variability and population size of extirpated US gray wolves (*Canis lupus*). *Molecular Ecology*, **14**, 9–17.

Letnic, M., Koch, F., Gordon, C., Crowther, M.S., and Dickman, C.R. (2009). Keystone effects of an alien top-predator stem extinctions of native mammals. *Proceedings of the Royal Society of London B*, **276**, 3249–56.

Liberg, O., Andrén, H., Pedersen, H.-C., Sand, H., Sejberg, D., Wabakken, P., Åkesson, M., and Bensen, S. (2005). Severe inbreeding depression in a wild wolf *Canis lupus* population. *Biological Letters*, **1**, 17–20.

Lohmus, A. (2001). *Large Carnivore Control and Management Plan for Estonia (2002–2011)*. Convention of the Conservation of European Wildlife and Natural Habitats, Strasbourg.

Macdonald, D.W. and Sillero-Zubiri, C. (2004). *Biology and Conservation of Wild Canids*. Oxford University Press, Oxford.

Massei, G., Miller, L.A., and Killian, G.J. (2010). Immunocontraception to control rabies in dog populations. *Human–Wildlife Interactions*, **4**, 155–7.

McDonald, D.B., Parchman, T.L., Bower, M.R., Hubert, W.A., and Rahel, F.J. (2008). An introduced and a native vertebrate hybridize to form a genetic bridge to a second native species. *Proceedings of the National Academy of Sciences*, **105**, 10837–42.

Mech, L.D. and Boitani, L. (Eds.) (2003). *Wolves: Behavior, Ecology and Conservation*. University of Chicago Press, Chicago.

Mengel, R.M. (1971). A study of dog-coyote hybrids and implications concerning hybridization in *Canis*. *Journal of Mammalogy*, **52**, 316–36.

Miklósi, Á., Topál, J., and Csányi, V. (2004). Comparative social cognition: what can dogs can teach us?. *Animal Behaviour*, **67**, 995–1004.

Miklósi, A. (2007). *Dog Behaviour, Evolution, and Cognition*. Oxford University Press. Oxford.

Mills, L.S. and Allendorf, F.W. (1996). The one-migrant-per-generation rule in conservation and management. *Conservation Biology*, **6**, 1509–18.

Ministry of Agriculture and Forestry (2005). *Management Plan for the Wolf Population in Finland*. Ministry of Agriculture and Forestry 11b/2005, Helsinki, Finland.

Moon-Fanelli, A. (2011). The ontogeny of expression of communicative genes in coyote-beagle hybrids. *Behavior Genetics*, **41**, 858–75.

Muñoz-Fuentes, V., Darimont, C.T., Wayne, R.K., Paquet, P.C., and Leonard, J.A. (2009). Ecological factors drive differentiation in wolves from British Columbia. *Journal of Biogeography*, **36**, 1516–31.

Muñoz-Fuentes, V., Darimont, C.T., Paquet, P.C., and Leonard, J.A. (2010). The genetic legacy of extirpation and re-colonization in Vancouver Island wolves. *Conservation Genetics*, **11**, 547–56.

Musiani, M., Leonard, J.A., Cluff, H.D, Gates, C.C., Mariani, S., Paquet, P.C., Vilà, C., and Wayne, R.K. (2007). Differentiation of tundra/taiga boreal coniferous forest wolves: genetics, coat color and association with migratory caribou. *Molecular Ecology*, **16**, 4149–70.

Oskarsson, M.C.R., Klütsch, C.F.C., Boonyaprakob, U., Wilton, A., Tanabe, Y., and Savolainen, P. (2012). Mitochondrial DNA data indicate an introduction through mainland southeast Asia for Australian dingoes and Polynesian domestic dogs. *Proceedings of the Royal Society of London, B*, **279**, 967–74.

Pardo-Diaz, C., Salazar, C., Baxter, S.W., Merot, C., Figueiredo-Ready, W., Joron, M., McMillan, W.O., and Jiggins, C.D. (2012). Adaptive Introgression across species boundaries in *Heliconius* butterflies. *PLoS Genetics*, **8**, e1002752.

Peterson, R.O. (2007). *The Wolves of Isle Royale: A Broken Balance*. University of Michigan Press, Ann Arbor.

Pierpaoli, M., Biró, Z.S., Herrmann, M., Hupe, K., Fernandes, M., Ragni, B., Szemethy, L., and Randi, E. (2003). Genetic distinction of wildcat (*Felis silvestris*) populations in Europe, and hybridization with domestic cats in Hungary. *Molecular Ecology*, **12**, 2585–98.

Pilot, M., Jedrzejewski, O., Branicki, W., Sidorovich, V.E., Jedrzejewska, B., Stachura, K., and Funk, S.M. (2006). Ecological factors influence population genetic structure of European grey wolves. *Molecular Ecology*, **15**, 4533–53.

Pilot, M., Branicki, W., Jędrzejewski, W., Goszczyński, J., Jedrzejewska, B., Dykyy, I., Shkvyrya, M., and Tsingarska, E. (2010). Phylogeographic history of grey wolves in Europe. *BMC Evolutionary Biology*, **10**, 104.

Puigcerver, M., Vinyoles, D., and Rodríguez-Teijeiro, J.D. (2007). Does restocking with Japanese quail or hybrids affect native populations of common quail *Coturnix coturnix*?. *Biological Conservation*, **136**, 628–35.

Randall, D.A., Pollinger, J.P., Argaw, K., and Wayne, R.K. (2010). Fine-scale genetic structure in Ethiopian wolves imposed by sociality, migration, and population bottlenecks. *Conservation Genetics*, **11**, 89–101.

Randi, E. and Lucchini, V. (2002). Detecting rare introgression of domestic dog genes into wild wolf (*Canis lupus*) populations by Bayesian admixture analyses of microsatellite variation. *Conservation Genetics*, **3**, 31–45.

Randi, E., Lucchini, V., Christensen, M.F., Funk, S.M., Dolf, G., and Loeschke, V. (2000). Mitochondrial DNA variability in Italian and east European wolves: detecting the consequences of small population size and hybridization. *Conservation Biology*, **14**, 464–73.

Randi, E. (2008). Detecting hybridization between wild species and their domesticated relatives. *Molecular Ecology*, **17**, 285–93.

Randi, E. (2011). Genetics and conservation of wolves *Canis lupus* in Europe. *Mammal Review*, **41**, 99–111.

Robinson, G.E., Fernald, R.D., and Clayton, D.F. (2008). Genes and social behavior. *Science*, **322**, 896–900.

Rutledge, L.Y., Garroway, C.J., Loveless, K.M., and Patterson, B.R. (2010). Genetic differentiation of eastern wolves in Algonquin Park despite bridging gene flow between coyotes and grey wolves. *Heredity*, **105**, 520–31.

Sacks, B.N., Brown, S.K., and Ernest, H.B. (2004). Population structure of California coyotes corresponds to habitat-specific breaks and illuminates species history. *Molecular Ecology*, **13**, 1265–75.

Sacks, B.N., Mitchell, B.R., Williams, C.L., and Ernest, H.B. (2005). Coyote movements and social structure along a cryptic population genetic subdivision. *Molecular Ecology*, **14**, 1241–9.

Scandura, M., Iacolina, L., and Apollonio, M. (2011). Genetic diversity in the European wild boar *Sus scrofa*: Phylogeography, population structure and wild x domestic hybridization. *Mammal Review*, **41**, 125–37.

Schwartz, M.K. Pilgrim, K.L., McKelvey. K.S., Lindquist, E.L., Claar, J.J., Loch, S., and Ruggiero, L.F. (2004). Hybridization between Canada lynx and bobcats: genetic results and management implications. *Conservation Genetics*, **5**, 349–55.

Schwenk, K., Brede, N., and Streit, B. (2008). Introduction. Extent, processes and evolutionary impact of interspecific hybridization in animals. *Philosophical Transactions of the Royal Society, Biological Sciences*, **363**, 2805–11.

Silbermayr, K., Orozco-terWengrl, P., Charruau, P., Enkhbileg, D., Walzer, C., Vogl, C., Schwarzenberger, F., Kaczensky, P., and Burger, P.A. (2010). High maternal differentiation levels between wild and domestic Bactrian camels: a basis for rapid detection of maternal hybridization. *Animal Genetics*, **41**, 315–18.

Silver. H., and Silver, W.T. (1969). Growth and behavior of the coyote-like canids of northern New England with observations on canid hybrids. *Wildlife Monographs*, **17**, 1–41.

Stronen, A.V., Tessier, N., Jolicoeur, H., Paquet, P.C., Hénault, M., Villemure, M., Patterson, B.R., Sallows, T., Goulet, G., and Lapointe, F.J. (2012). Canid hybridization: contemporary evolution in human-modified landscapes. *Ecology and Evolution*, 2, 2128–40.

Verardi, A., Lucchini, V., and Randi, E. (2006). Detecting introgressive hybridisation between free-ranging domestic dogs and wild wolves (*Canis lupus*) by admixture linkage disequilibrium analysis. *Molecular Ecology*, **15**, 2845–55.

Vilà, C. and Wayne. R.K. (1999). Hybridization between wolves and dogs. *Conservation Biology*, **13**, 195–8.

Vilà, C., Savolainen, P., Maldonado, J.E., Amorim, I.R., Rice, J.E., Honeycutt, R.L., Crandall, K.A., Lundeberg, J., and Wayne, R.K. (1997). Multiple and ancient origins of the domestic dog. *Science*, **276**, 1687–9.

Vilà, C, Amorim, I.R., Leonard, J.A., Posada, D., Castroviejo, J., Petrucci-Fonseca, F., Crandall, K.A., Ellegren, H., and Wayne, R.K. (1999). Mitochondrial DNA phylogeography and population history of the grey wolf *Canis lupus*. *Molecular Ecology*, **8**, 2089–2103.

Vilà, C., Walker, C., Sundqvist, A.-K., Flagstad, Ø., Andersone, Z., Casulli, A., Kojola, I., Valdmann, H., Halverson, J., and Ellegren, H. (2003). Combined use of maternal, paternal and bi-parental genetic markers for the identification of wolf-dog hybrids. *Heredity*, **90**, 17–24.

vonHoldt, B.M., Pollinger, J.P., Lohmueller, K.E., Han, E., Parker, H.G., Quignon, P., Degenhardt, J.D., Boyko, A.R., Earl, D.A., Auton, A., Reynolds, A., Bryc, K., Brisbin, A., Knowles, J.C., Mosher, D.S., Spady, T.C., Elkahloun, A., Geffen, E., Pilot, M., Jedrzejewski, W., Greco, C., Randi, E., Bannasch, D., Wilton, A., Shearman, J., Musiani, M., Cargill, M., Jones, P.G., Qian, Z., Huang, W., Ding, Z., Zhang, Y., Bustamante, C.D., Ostrander, E.A., Novembre, J., and Wayne, R.K. (2010). Genome-wide SNP and haplotype analyses reveal a rich history underlying dog domestication. *Nature*, **464**, 898–902.

vonHoldt, B.M., Pollinger, J.P., Earl, D.A., Knowles, J.C., Boyko, A.R., Parker, H., Geffen, E., Pilot, M., Jedrzejewski, W., Jedrzejewski, B., Sidorovich, V., Greco, C., Randi, E., Musiani, M, Kays, R., Bustamante, C.D., Ostrander, E.A., Novembre, J., and Wayne, R.K. (2011). A genome-wide perspective on the evolutionary history of enigmatic wolf-like canids. *Genome Research*, **21**, 1294–1305.

Wayne, R.K. and Brown, D.M. (2001). Hybridization and conservation of carnivores. In J.L. Gittleman, S.M. Funk, D.W. Macdonald and R.K. Wayne, Eds., *Carnivore Conservation*. Cambridge University Press, Cambridge, pp. 145–62.

Wayne, R.K., Leonard, J.A., and Vilà, C. (2006). Genetic analysis of dog domestication. In M.A. Zeder, D.G. Bradley, E. Emshwiller and B.D. Smith, Eds., *Documenting Domestication: New Genetic and Archaeological Paradigms*. University of California Press, Los Angeles, pp. 279–93.

Wallach, A.D., Ritchie, E.G., Read, J., and O'Neill, A.J. (2009). More than mere numbers: The impact of lethal control on the social stability of a top-order predator. *PLoS ONE*, **4**, e6861.

WHO-WSPA. (1990). *Guidelines for Dog Population Management*. World Health Organization and World Society for the Protection of Animals, Geneva.

Woodroffe, R., Cleaveland, S, Courtneay, O., Laurenson, M.K., and Artois, M. (2004). Infectious disease. In D.W. Macdonald and C. Sillero-Zubiri, Eds., *Biology and Conservation of Wild Canids*. Oxford University Press, Oxford, pp. 123–42.

Woodroffe, R. and Donnelly, C.A. (2011). Risk of contact between endangered African wild dogs *Lycaon pictus* and domestic dogs: Opportunities for pathogen transmission. *Journal of Applied Ecology*, **48**, 1345–54.

Woodroffe, R. and Ginsberg, J.R. (1998). Edge effects and the extinction of populations inside protected areas. Science, **280**, 2126–8.

Wright, S. (1978). *Evolution and the Genetics of Populations. Volume 4. Variability Within and Among Natural Populations*. University of Chicago Press, Chicago.

Young, J.K., Olson, K.A., Reading, R.P., Amgalanbaatar, S., and Berger, J. (2011). Is wildlife going to the dogs? Impacts of feral and free-roaming dogs on wildlife populations. *BioScience*, **61**, 125–32.

CHAPTER 8

Dog conservation and the population genetic structure of dogs

Ryan H. Boyko and Adam R. Boyko

8.1 Introduction

The domestication of dogs likely began 12,500–30,000 years ago, giving dogs more time to evolve and diversify than any other domesticated species (Clutton-Brock, 2012). Over the course of just 5,000–10,000 generations, dogs adapted to a variety of environments and niches, a process accelerated in many populations by artificial selection. The wide assortment of shapes, sizes, temperaments, and behaviors in modern dogs testifies to the power with which human-directed selection can transform the dog genome to produce novel and desirable phenotypes suited to diverse tasks and predilections. The ubiquitous distribution of dogs across the globe testifies to the dogs' own ability to adapt to a wide array of anthropogenic niches.

In this chapter, we summarize what is known about the genetic and phenotypic distinctiveness of modern breed dogs and free-breeding dog populations, both truly feral populations (like dingoes) and the more common 'village' dog populations that are found throughout much of the world (see Box 8.1 for an explanation of terms). Because of the relatively recent origin (in evolutionary time-scales) of the dog, no dog population can fairly be described as a separate biological species. In fact, dogs can freely interbreed with wolves (*Canis lupus*) and coyotes (*C. latrans*) (Leonard et al., Chapter 7), and hybridization and introgression within the genus can make it difficult to neatly apply traditional species concepts (vonHoldt et al., 2011). Nevertheless, isolation and local adaptation created genetically distinct village dog populations, some of which are now threatened by the encroachment of non-indigenous dogs. On top of this, even genetically similar modern breed dogs demonstrate substantial phenotypic diversity that could interest conservation biologists. In this chapter, we begin by addressing the questions of what one might want to conserve and why. We then proceed to summarize the current state of dog diversity. Finally, we suggest ways to determine which populations should be conserved and present ideas on how to conserve them.

8.1.1 What are we conserving?

As wolves transformed into dogs, they arguably became integrated into our lives in a deeper and more complex manner than any other animal. We maintain working relationships of all sorts with dogs, using them to help us hunt, herd, guard, carry burdens, clear landmines, find missing persons, assist disabled individuals, find illicit substances, and detect cancers (e.g., VerCauteren et al., Chapter 9; Woollett et al., Chapter 10; Koster and Noss, Chapter 11). Depending on the culture of an area, dogs are also used as food and companions. Given this diversity of uses, it is unsurprising that specific kinds of dog are bred to have phenotypic and, perhaps, genetic advantages in performing one or another of these functions. For example, Poodles seem to contain more transcribed olfactory genes than Boxers, probably due to stronger selection on Poodles' ability to smell game and truffles (Tacher et al., 2005). In these cases, conserving dogs with unique abilities will conserve the genetics underpinning them and allow for their continued use and study.

Free-Ranging Dogs and Wildlife Conservation. Edited by Matthew E. Gompper
© Oxford University Press 2014. Published 2014 by Oxford University Press.

Box 8.1 Terminology

To clarify our use of terms and to distinguish our use of these terms from other, sometimes conflicting, uses of the same terms, we provide the following guide to terminology used throughout this chapter.

Dogs: Canis familiaris including modern breed dogs, village dogs, New Guinea singing dogs, and dingoes, but not including wolves or coyotes despite their ability to occasionally, albeit rarely, interbreed with dogs.

Breed/Purebred dogs: Dogs that have restrictive breed books and are generally recognized by kennel clubs (groups of dog owners that collectively focus on the breeding, maintenance, and promotion of particular breeds of dogs). Most dog breeds underwent a bottleneck during breed formation with some breeds encountering subsequent bottlenecks and/or inbreeding. *Modern dog breeds* come predominantly from Europe (see Figure 8.2) and developed closed breeding populations sometime during or after the Victorian era of the mid–late 1800s. Boxers and Poodles are examples.

Ancient breed dogs: In ancient times, some dogs were deliberately bred for certain characteristics, although not necessarily with the rigorously maintained pedigree records of modern purebreds. Ancient breed dogs today are purebred dogs with genetic signatures inherited from those dogs, signatures that are identifiably separate from the modern European breeds. Basenjis and Salukis are examples.

Land races: Dogs that exhibit physical traits and behavioral tendencies characteristic of dogs originating in a particular place. These characters have developed over hundreds or thousands of years though adaptation to the local environment, possibly with breeding interference by humans (artificial selection), but without official studbooks (and thus despite interbreeding with sympatric or parapatric dog populations). In many ways they are similar to ancient breed dogs but their breeding is less closely controlled and in most cases (e.g., the Africanis) it seems like the original land races were mostly or completely genetically swamped by modern breed dogs brought to these areas. In other cases these land races may just be local village dogs that happen to comport to a certain physical appearance (e.g., the Indog).

Village dogs: Dogs that live relatively free-breeding and oftentimes partially free-ranging existences as human commensals or mutualists in many places around the world. These dogs are not usually undergoing strong programs of human-directed breeding, but people may preferentially feed, shelter, or cull certain individuals. These dogs' relationship with the local humans and other animals varies greatly depending on cultural and ecological context. They tend to show a genetic signature of their place of origin and tend not to be closely related to major European dog breeds, although in some places (e.g., Central Namibia and much of the Western Hemisphere) they show significant admixture with European-derived dogs (Boyko et al., 2009; Castroviejo-Fisher et al., 2011). We use the term *indigenous village dog* to refer to a village dog that has little admixture with non-native dog breeds and *admixed village dog* to refer to a village dog that has significant admixture with non-native (usually European) dog breeds. Compared to land races, village dogs have a much wider variety of physical appearances within a location. Free-breeding city-dwelling dogs in Russia and India fit this definition, as well as dogs living at the margins of Egyptian society or living in rural villages in Uganda and elsewhere. Populations of admixed village dogs may be consistently replenished by new stray dogs while indigenous village dog populations are usually self-perpetuating, not requiring newly released dogs to maintain their populations.

Feral dogs: Dogs living completely or nearly completely free from human-derived resources (such as trash), for example dingoes. For our purposes of identifying conservation targets based primarily on genetics, we differentiate populations of feral dogs from village dogs based on the interactions most individuals have with people.

Free-breeding dogs: Dog populations with a substantial proportion of dogs that often choose mating partners for themselves, including village dogs and feral dogs. While we acknowledge that there is a range of dog breeding and husbandry practices across the globe, in general village dog breeding involves more sexual/natural selection and less artificial selection than modern breed dog breeding practices. This difference has important implications for the level of genetic and phenotypic diversity found in these populations, and for the diversity found between different populations and breeds. We prefer this term to *semi-feral dogs* because it encapsulates the most important difference between village dogs and breed dogs from a conservation standpoint, which is their mating system and its effects on genetic diversity and adaptation. It is also a more accurate term, as some village dog populations contain individuals that have nearly no interaction with people (truly semi-feral) while others contain mostly individuals that interact extensively with a human owner, but in general most bitches in these populations are either allowed to breed freely with other local dogs or are bred with locally available sires in such a way

continued

Box 8.1 *Continued*

as to not overly skew the variance in reproductive success between males and females or quickly diminish the population's genetic variation.

Introgression: The incorporation of portions of the genome from individuals of one species/population to another through admixture or hybridization and back-crossing.

Species: For sexually reproducing organisms, the biological unit consisting of similar individuals capable of interbreeding and reproductively isolated from other such groups. In dogs, some extreme breeds (Chihuahua and Great Dane) may be physically incapable of interbreeding, but are still genetically compatible and therefore considered the same species. Conversely, although wolves, coyotes, and dogs are all capable of interbreeding and producing fertile hybrids, they are often considered separate species on the basis that hybridization under natural conditions is rare. Where these species are sympatric, they remain genetically distinct even though some hybridization may occur (Leonard et al., Chapter 7).

Artificial selection: Human-controlled selective breeding of individuals for particular traits. In this chapter, we generally use the term to refer to directed breeding of particular individuals or the intentional killing or spaying/neutering of certain individuals, as opposed to the more subtle selection that occurs by favoring some individuals with higher quality resource provisioning.

Nevertheless, for the most part the phenotypic diversity of modern dog breeds is decoupled from the diversity of roles dogs can fulfill. Some phenotypes, like skin wrinkling in Shar Peis or brachycelphaly in Bulldogs, became more extreme during the last century as fewer dogs fulfilled working roles and breeding was driven more by aesthetics. For many of these visible morphological traits, artificial selection for novelty itself, accelerated by careful breeding with managed populations, has generated spectacular phenotypic diversity through the selection and fixation of a small number of genetic variants with major phenotypic effects (Figure 8.1; Boyko et al., 2010). In these cases, conserving these dogs would conserve the unique products of extreme artificial selection, which could help elucidate biological pathways and evolutionary processes.

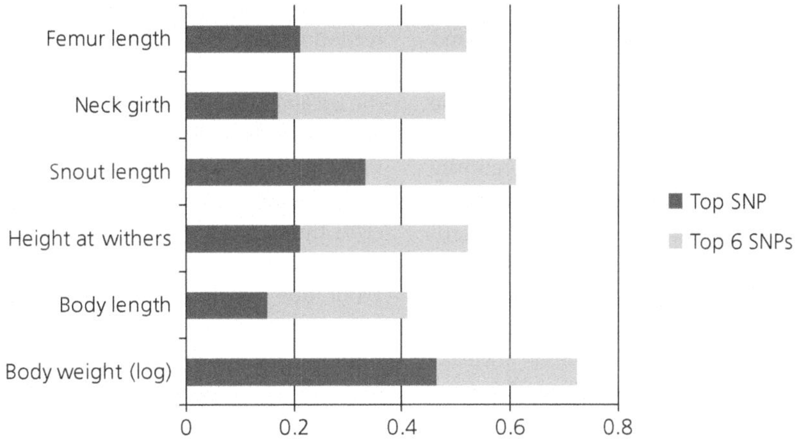

Figure 8.1 Mean proportion of between-breed phenotypic variance in various traits explained by the single nucleotide polymorphism (SNP) with greatest effect and the top six SNPs by effect size. Breed dog phenotypic traits are largely determined by a few SNPs of great effect. This figure shows the proportion of phenotypic variance between 80 breeds of dog (breed average phenotypic values derived from 890 dogs) explained by the SNP with the highest explanatory power and the top six SNPs in terms of explanatory power. For most traits, the top SNP explains about 20% of the variance and the top six SNPs explain more than 40% of the variance for all traits. Except for body size, all of these traits were allometrically scaled against ln (body size). Data are from Boyko et al. (2010).

Beyond morphological differences, dogs vary phenotypically in other ways, most notably in behavior. Surely genetics plays a large role in the distinct aptitudes of herders, pointers, and retrievers, but the genes underlying these traits have not yet been discovered. Still, in many cases various breeds of dogs can perform functions equally well (e.g., markedly reducing depression and negative health outcomes through companionship with nursing home residents, acquired immunodeficiency syndrome patients, and other groups; Nimer and Lundahl, 2007; Perelle and Granville, 1993; Siegel et al., 1999). Conserving a variety of dogs with different abilities and temperaments will give science time to better understand the genetic underpinnings of mental processes and behavioral traits before that remarkably diverse study system is lost forever.

In many regions, village dogs perform jobs such as guarding crops and livestock. For example, the presence of village dogs has been shown to reduce attacks on livestock grazing in northern Kenya by 63% (Treves and Karanth, 2003; Woodroffe et al., 2007). In at least two societies in Ethiopia, 'nurse dogs' help raise babies and small children, cleaning the children and providing warmth and companionship (Fuller and Fuller, 1981). Even free-ranging dogs scavenging human-derived foods might perform valuable roles for human communities. Evidence from India suggests that village dogs consume most of the available human-derived foods in and around agricultural areas, excluding native foxes from the agricultural areas and thus, perhaps, mitigating the potential conflict between foxes and farmers (Vanak and Gompper, 2009). Village dogs could theoretically reduce populations of pest species such as rodents. However, leftover dog meals could also attract rodents (Masi et al., 2010) and dogs themselves carry or transmit some human parasites (Macpherson, 2005). The degree to which genetics has adapted village dogs to perform their various duties is unclear.

At the very least, many of these village dog populations contain genetic adaptations for survival in their local environment. Desert dogs are almost universally lanky, presumably facilitating heat dissipation. Other populations likely contain unique genetic variants to help them survive harsh winters, food shortages, high altitudes, unique diets, parasitic infections, and other biotic and abiotic stresses. Only by careful study of indigenous populations of village dogs fulfilling their natural roles in intact human communities will we be able to discover the genetic basis of their adaptation to these various niches over thousands of years. Village dogs may also perform important sociocultural functions in many societies, and may contain important genetic and behavioral clues for improving our understanding of the evolutionary history of dogs and the process of domestication. Conserving these village dogs will conserve any local adaptations and preserve the selective and demographic history written into their genes.

8.2 An overview of dog diversity

Dogs have diversified in size, shape, and behavior perhaps more than any other mammal (Figure 8.2). This diversification recently accelerated as dog breeders established closed populations for various breeds and deliberately selected some lines for novel or exaggerated phenotypes according to the distinctive standards of each breed. Depending on one's viewpoint, the 400 or so modern breeds of dog persisting today represent either the perfection or the perversion of the canine form, drastically expanding the range of phenotypic diversity present in the dog's wild progenitor, the gray wolf.

Genetic analysis of purebred dogs and wild canids shows that most breeds trace back relatively recently with only a few breeds—the Basenji and a smattering of Asian, Middle Eastern, and Nordic breeds—showing more ancient roots or unique signatures of wolf admixture (Larson et al., 2012; Parker et al., 2004; vonHoldt et al., 2010). Certainly, distinct 'kinds' of dogs were present in ancient times, but most of these either died out (e.g., the English Turnspit dog, Morris, 2002; the Salish Wool dog of the Pacific Northwest, Crockford, 1997) or admixed with other dogs sufficiently to destroy much of their ancient or localized heritage (e.g., Rhodesian Ridgebacks and Pharaoh Hounds; Boyko et al., 2009; Parker et al., 2004, 2007).

Neolithic dogs likely had similar relationships to the humans that lived with them as present-day village dogs do, having fulfilled varied roles in the human communities they associated with. It seems unlikely that they were bred in the same manner as current breed dogs, with closed breed books or

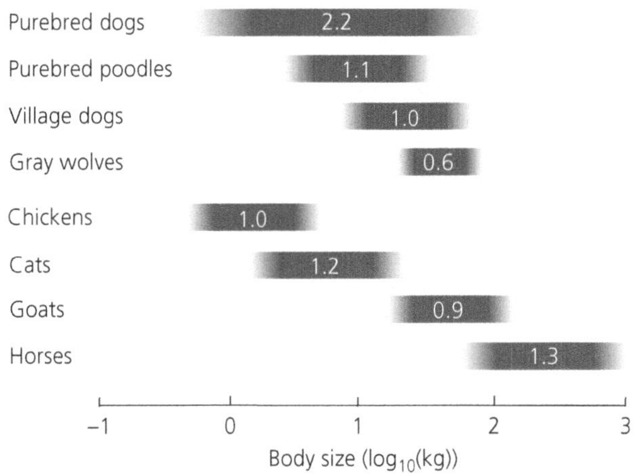

Figure 8.2 Size variation within various groups of dogs, gray wolves, and several domesticated species. Size variation within purebred dogs varies over 2 orders of magnitude, from Chihuahuas weighing less than 1 kg to Great Danes weighing 80 kg. Even within single breeds formed within the last two centuries, size variation can be extreme and similar to the variation found in other domesticated animals (e.g., Poodles vary across 1.1 orders of magnitude). In contrast to breed dogs, the order of magnitude variation in size in free-ranging village dogs is similar to that observed in other domesticated animals. This variation still exceeds that observed across all extant gray wolf subspecies. Data taken from Carroll and Huntingtom (1988), Wayne and Ostrander (1999), Galal (2005), Brooks et al. (2010), Hunter (2011), Henderson (2012), and Boyko et al. (unpublished data).

similar strict protocols guarding the line's purity. Ancient dog populations or breeds that could not be kept isolated from the emerging 'modern' European breeds lost their genetic distinctiveness, a process accelerated in populations with close proximity to populations of modern breeds or with attributes such as small body size that made them easy to transport (Larson et al., 2012; Pires et al., 2009). Deliberate interbreeding of ancient breeds with modern stock also occurred in some lineages, particularly those with breed-defining dominant mutations like the Rhodesian Ridgeback or the Mexican Xoloitzcuintli (Fox, 2003), or those facing dwindling numbers as their utility waned (e.g., Irish Wolfhounds and Finnish Spitzes).

Yet, most dogs throughout history and even today are not breed dogs in any sense, but are free-breeding human commensals (Coppinger and Coppinger, 2001). The population history of these village dogs is potentially much richer than that of modern breeds, which largely reflect genetic variation present in a few dogs in Europe several centuries ago. Village dogs have a nearly global distribution, with most continental populations first established millennia ago. Notably, these village dogs reflect the ancestral stock for all dog breeds, and may represent an important genetic resource for reinvigorating some purebred lineages using outbred individuals related to the breed founders.

Like many modern breeds, some populations of village dogs are also genetic mixtures of several modern European breed dogs that were relatively recently imported to those areas (e.g., Puerto Rican and central Namibian village dogs; Boyko et al., 2009). These dogs resumed a scavenging, free-breeding existence (they are 'secondarily free-breeding'), but they retain little or no localizable genetic signature and do not contain unique genes resulting from local adaptation over millennia. We refer to these dogs as admixed village dogs. Other village dog populations, however, have much more ancient roots and are likely to be very informative for deciphering the origin of dogs and the movement of early dog populations across the globe (e.g., Ugandan village dogs; Boyko et al., 2009). These indigenous village dogs also represent unique genetic resources for understanding local adaptation and may provide unique services to the humans that live with them.

In many ways, indigenous village dogs are intermediate between purebred dogs and wolves. Village dogs, living off human scraps, are mostly freed from the demands of needing to hunt prey and thus have reduced selective pressure on many functional traits. However, without strict breeding controlled by humans, they still must compete for mating opportunities. Even in cases where humans control breeding for some village dogs, sympatric scavenging dogs that are not under human control also contribute to the dog population. Further, these dogs are generally selected for functional traits like greater hunting aptitude, which tends to decrease genetic diversity less than breeding for conformation (Pedersen et al., 2013). Given this, village dogs exhibit more diversity in their behavior and morphology than do wolves, but nothing like what could be seen in an afternoon at the Westminster Kennel Club Dog Show (but see de Caprona and Savolainen, 2013, who argue that a high level of phenotypic diversity co-occurs with a high level of genetic diversity in southern Chinese village dogs). Likewise, even though all dogs (village dogs and purebred dogs) descend from the same ancestral stock, the lack of strong artificial selection in most village dog populations means they have more genetic variants and genome characteristics (e.g., a high level of heterozygosity) in common with the first domestic dogs (and also modern wolves) than purebred dogs, which rapidly lost their genetic diversity in the last few decades or centuries (Calboli et al., 2008). Finally, whereas wolves are a keystone species and clearly an important conservation target from an ecological perspective (Fortin et al., 2005) and purebred dogs are not generally ecologically important (e.g., a keystone species), free-breeding dogs, because they interact with both humans and the natural environment, present an interesting intermediate case. They can potentially mediate the interactions between humans, other domestic animals, and wildlife (Woodroffe et al., 2007; Ritchie et al., Chapter 2; Vanak et al., Chapter 3, Butler et al., Chapter 5) and, at least in some animal communities, act as an important predator species (e.g., dingoes, Johnson et al., 2007; Zimbabwean village dogs, Butler et al., 2004).

Dogs are the only domesticated species that pre-dates the origin of agriculture, and rural free-breeding dog populations likely live a similar lifestyle to that of the very first dogs, mostly choosing their own mating partners while relying on scavenging food from humans (Coppinger and Coppinger, 2001). Whether dogs 'pre-adapted' humans for the Neolithic revolution or not, the fact remains that village dogs have filled an important niche (guard/companion/scavenger) ever since farming communities first existed. As human populations expanded and diversified, so did dog populations, with dogs serving as hunters, sentries, shepherds, warriors, and food animals. Thus, genetic analysis of village dog populations could shed light on theories of dog origins and also yield unique anthropological insights and improve our understanding of the genetic basis of natural and artificial selection.

As dogs spread across the globe, they encountered different geographical features, ecological contexts, and historical events. These led to different selection regimes and demographic histories of the dog populations in different areas. Due to this, the dogs on each continent are not equally useful for preserving the genetic diversity of dogs as a whole. In the following sections we will examine extant dog genetic diversity on each continent, which will inform the discussion of dog conservation that follows.

8.3 Africa

The prototypical image of the proud, independent Basenji of Central Africa evokes a sense of rugged independence and hunting prowess maintained since ancient times. For many Africans, however, a more typical image would be dogs foraging on trash, waste, and animal carcasses on the periphery of human settlements. African dogs have a complex relationship with the humans and wildlife with which they share the continent and an equally complex genetic background. Because of this, there is no simple answer to the question of which African dog populations are especially worthy of conserving.

8.3.1 History of dogs in Africa

Mummified dogs have been found in Egyptian tombs, sometimes sleeping curled at their master's feet, dating from around 4,500 years ago (Ikram, 2005). With deserts, dense forests, and tsetse fly infested savanna to cross, it took about 3,000 more years for dogs to make their way to South Africa (Larson et al., 2012). Thus, no southern African dog has a truly ancient distinctive genetic makeup in the context of the 15,000-plus year history of the dog. However, the diseases and terrain that slowed dogs' initial advance across the African continent also served as a buffer against the subsequent intermixing with European dogs that overwhelmed the local diversity in many places across the globe (Diamond, 1997). This allowed some African dogs to maintain relatively distinctive genetic lineages that provide a glimpse of some of the dog genetic diversity that existed prior to the formation of European breed clubs that instituted closed breed books and ultimately sharply reduced the genetic diversity of European dog populations (Calboli et al., 2008; Larson et al., 2012).

By the time Europeans first visited the Cape of Good Hope in 1652, indigenous people were using dogs to assist in hunting, guarding, and herding throughout the continent (Gallant, 2002). Ridged dogs were present in southern Africa as well as Basenjis north of them in the Congo basin (Gallant, 2002). Both Basenji fanciers and southern African breed (Rhodesian Ridgeback and Africanis) enthusiasts today claim ancient breed status, but recent genetic studies only back-up the claim for Basenjis (Bannasch et al., 2005; Boyko et al., 2009; Larson et al., 2012).

8.3.2 A case study from Namibia

To understand why some ancient dog populations maintained their distinctive genetic signatures while others now appear genetically identical to modern European breed dogs, we consider the distribution of dogs in Namibia. Namibia provides a particularly instructive example in how climate and geography interact with chance historical events to influence dog population histories.

In the late nineteenth century, European immigrants displaced the native peoples and established ranches in the most productive and easily exploited land in German South-west Africa (present-day Namibia). European settlement covered much of the southern 80% or so of the country while the northern area of the country experienced colonial administration without large immigrant-owned ranches (Meischer, 2012). European and South African authorities limited the movement of farm animals from the North to the South of Namibia to prevent livestock disease from spreading to European-owned ranches, eventually building a physical chain-link fence across the country after the Second World War (Meischer, 2012).

The Namibian fence (also called the Red Line) did nothing to prevent dogs from moving freely about the country prior to its physical substantiation and did not actually prohibit their crossing after its construction. However, the fence created a sharp delineation between tropical Africa, with its agriculturally poorer soils and high tropical disease burdens that ethnic Europeans (and their dogs) were not accustomed to, and the more temperate southern lands that were suitable for ranching and harbored fewer tropical ailments.

Although there are now a few modern European breed dogs south of the Red Line, most dogs today on both sides of the fence appear to be 'typical' village dogs, similar to those found throughout much of rural Africa (Figure 8.3): tan, prick ears, short hair, and about 15 kg (Boyko et al., 2009). Since canids naturally have large home ranges, high gene flow, and low genetic differentiation among populations, one would expect Namibian dogs, which are not prevented from crossing the Red Line, to show low genetic differentiation between populations north and south of the Red Line (Wayne et al., 1992). This is especially true given the phenotypic similarity and small geographic distance between dogs on either side of the fence. However, dogs north of the fence averaged 87% indigenous African dog ancestry while those south of the fence had

Figure 8.3 A young bitch (about 1 year old) in Boende, Democratic Republic of the Congo, July 2012. This dog has a standard village dog appearance. Photo credit: Julia A. Randall.

only 9% indigenous African ancestry on average, the rest coming from recent imports of European dogs (Figure 8.4; Boyko et al., 2009). This result is confirmed by other studies that have found southern African dog breeds (e.g., Rhodesian Ridgeback and Africanis) to have significant recent European ancestry and low genetic diversity (Bannasch et al., 2005; Larson et al., 2012). Clearly if one cares about preserving indigenous genetic lines, African dogs north of the Red Line represent a good conservation target. However, within that area, less is known about which populations are genetically distinct; we explore that question below.

8.3.3 Current status of dog diversity in Africa

Outside of southern Africa, relatively little data exist to determine which village dog populations have primarily indigenous ancestry. East Africa had fairly intensive European settlement in some areas and the dogs there may have significant European ancestry, though this is not addressed by any studies to date. Dogs in Giza, Egypt have some European ancestry, though not nearly as much as dogs from southern Namibia (Boyko et al., 2009). A Y chromosome study is consistent with indigenous ancestry for Basenjis as well as some Middle Eastern dogs, supporting the notion that North African village dogs may be primarily indigenous (Bannasch et al., 2005). Mitochondrial DNA evidence shows high levels of diversity in Moroccan dogs as well, although there is likely some European admixture with these dogs given their proximity to Iberia (Pires et al., 2006). Thus, these dogs may be diverse, but that diversity is likely partially due to having a mix of mitochondrial DNA from the African village dog line and from modern European breeds. Given the abundance of nearby modern European breed dogs, these populations are unlikely, at first glimpse, to be the most useful conservation targets for preserving African village dog lineages. Mitochondrial studies have also confirmed that Malagasy village dogs are closely related to indigenous African village dogs and show a higher genetic diversity than other island populations that have been sampled (Oskarsson, 2012). Dogs on the island of Madagascar thus represent another viable African indigenous village dog population.

Outside of the periphery of Africa (southern Africa, Madagascar and other offshore islands, the Mediterranean Coast, and part of East Africa that had colonization featuring European emigrant owned ranches), African dogs may be a generally panmictic population with some fairly small

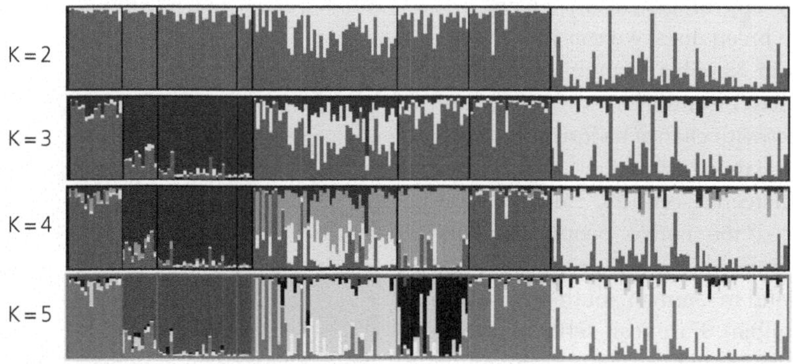

Figure 8.4 STRUCTURE analysis across 389 SNP and microsatellite loci in African village and American mixed breed dogs. Each column represents an individual dog, with dogs grouped by population. Each color represents one of *k* populations, and individuals are colored according the proportion of their genome assigned to each population by the program. Despite being separated by only a few kilometers, central Namibian dogs do not cluster genetically with northern Namibian dogs but rather with European breed-admixed street dogs from Puerto Rico and elsewhere. This figure is based on a STRUCTURE analysis across 389 SNP and microsatellite loci in 223 unrelated African village dogs and 17 American mixed breed dogs (from Boyko et al., 2009).

variations due to natural dispersal barriers such as lakes and deserts. Supporting this idea, dogs from northern Namibia and Uganda, 2,900 km apart, varied little (Fst = 0.025, Boyko et al., 2009). However, dogs from islands in Lake Victoria did vary some from the Ugandan mainland dogs 10–20 km away (Fst = 0.038). Similarly, dogs from the Kharga Oasis in Egypt showed some differentiation from the dogs 230 km away in Luxor (Fst = 0.09). Still, it seems that this variation is most likely due to founder effects and genetic drift and does not represent any lineages distinct from the ones inhabiting most of sub-Saharan Africa above the Red Line. Given the low coverage of genetic studies on African dogs to date, isolated populations representing unique lineages may still be found in remote regions there.

Of course, genetic heritage is only one factor to consider when determining populations to target for conservation. Ridged dogs in southern and western Africa have distinctive appearances and many dedicated enthusiasts and Basenji lovers hold special esteem for rural Congolese dog populations. In southern Africa, just as in more northern sub-Saharan Africa, dogs are used for hunting and may be locally adapted. Dogs in urban environments are often larger and have different temperaments than dogs in rural environments, which enable them to physically compete against other dogs and animals while remaining fearful of, and keeping their distance from, people (R. Boyko and A. Boyko, pers. obs.). These dogs may benefit local people by reducing the number of trash-eating and disease-carrying small animals living in the cities and villages, although little research has been done on the overall effect of dog populations on disease aside from rabies. Some research has shown that having guard dogs may mitigate human–wildlife conflict near the borders of national parks (Saj et al., 2001). Conversely, dogs in some areas can kill wildlife and spread disease to wildlife and people (Ritchie et al., Chapter 2; Knobel et al., Chapter 6).

8.4 Oceania and Island South-east Asia

8.4.1 Dingoes and New Guinea Singing Dogs

Dingoes and New Guinea Singing Dogs (NGSDs) are well-known examples of truly feral dog populations, and are already given some status as conservation targets (Koler-Matznick et al., 2007; Letnic et al., 2012). Genetically, these groups are sister taxa, clearly descended from domestic dogs, but separated from other dog populations for over 4,000 years (Ardalan et al., 2012; Fillios et al., 2012; Oskarsson et al., 2012; Savolainen et al., 2004). No archeological evidence for these dogs exists before this time, so it is likely they were introduced sometime after Australia and New Guinea were separated by rising sea levels approximately 8,000 years ago.

These feral dogs share many 'primitive' characteristics, including annual estrus and a lack of barking, suggesting they retain (or, less likely, have regained) characters found in pre-Neolithic and early Neolithic dogs that have been subsequently lost in modern mainland populations. Both dingoes and NGSDs show relatively low levels of genetic diversity, likely due to strong founder effects or low population sizes, and they are at extreme risk of genetic contamination from interbreeding with modern dogs (Corbett, 1995). A recent study found that only 12.5% of the 24 sampled dingoes in south-east Australia had <25% modern European breed dog ancestry (Claridge et al., 2009), though earlier studies using morphological instead of genetic measures suggest that dingoes may be less mixed with modern breed dogs elsewhere on the Australian continent (Stephens, 2011). Indeed, a microsatellite study involving nearly 4,000 dingoes across Australia revealed that a majority of dingoes in central and western Australia, including 87% of dingoes in the Northern Territory, were pure dingo and not hybrid (Stephens, 2011). NGSDs are extremely rare in the wild, limited to elevations above 4,000 m, and captive populations are small and at high risk for inbreeding (Koler-Matznick et al., 2007). Genome-wide analysis of 48,000 single nucleotide polymorphism (SNP) markers showed that dingoes and NGSDs are highly diverged from other dogs (vonHoldt et al., 2010) despite some admixture from European-derived dogs, at least in dingoes. Although distinguishing dingoes/NGSDs from other dogs based on genetic markers is relatively simple due to their strong divergence, so far no studies have identified genetic differences underlying unique dingo and NGSDs traits.

8.4.2 Other dogs in Oceania and Island South-east Asia

In contrast to the truly feral and highly diverged dingo and NGSD, the village dogs found throughout Oceania are behaviorally and genetically much closer to other dog lineages (Irion et al., 2005; Runstadler et al., 2006). Even village dogs in the highlands of Papua New Guinea share more genetic affinity with mainland village dogs than they do with NGSDs (Boyko et al., unpublished data), suggesting perhaps multiple waves of dog migration through Oceania, with the isolation of NGSDs and dingoes prior to the most recent migrations.

The urban street dogs on the island of Bali were one of the first village dog populations to be analyzed genetically, and were found to be intermediate between mainland Asian dogs and dingoes based on microsatellite data (Irion et al., 2005). Despite living on an island of approximately 5,600 km^2 containing fewer than 1 million dogs, Bali street dogs had much more mitochondrial, Y chromosome, microsatellite, and dog leukoctye antigen (DLA; a series of genes involved in dogs' immune function) diversity than the dingoes of 7.6×10^6 km^2 Australia, and harbored several unique haplotypes not found in modern dog breeds (Brown et al., 2011; Irion et al., 2005; Runstadler et al., 2006). These data show that dogs were introduced to Bali over 3,000 years ago and have subsequently been isolated from other dog populations (Brown et al., 2011).

Because of their isolation, indigenous island dogs are potentially highly informative for ancestral dog diversity and also human migration patterns and trade routes. Recent analysis of mtDNA shows that modern-day Polynesian street dogs are most closely related to Indonesian and Melanesian dogs, and not to dogs from Taiwan or the Philippines (Oskarsson et al., 2012). However, reaching definitive conclusions about the spread of early dogs in the region based on this relatedness is complicated since some of these mtDNA haplotypes were likely introduced in modern times. Island dog ancestry has implications for understanding the spread and trade networks of Polynesians, although studies using genomic markers will be required to determine whether contamination with modern breeds needs to be taken into account when estimating colonization history. Beyond ancestry analysis, genome-wide datasets from indigenous island dog populations will be particularly useful for detecting signatures of selection that may underlie genetic adaptations to local conditions. Thus far, few island dogs have been analyzed to this resolution, and many island dog populations are still completely uncharacterized.

8.5 Mainland Eurasia

Dogs evolved from Eurasian gray wolves (Vilà et al., 1997; Wayne, 1993). This continent is clearly the cradle of dog origins, and likely contains the oldest free-breeding dog populations. These dogs may carry important clues regarding the evolutionary process and population history of the dog. Mitochondrial and chromosome Y haplotypes in East Asian village dogs, particularly those in southern China, are especially diverse, making this region a diversity hotspot and perhaps the center of origin for the species (Ding et al., 2011; Pang et al., 2009; Savolainen et al., 2002). Southern Chinese village dogs may also exhibit high phenotypic diversity for village dogs (de Caprona and Savolainen, 2013), but systematic, quantitative comparisons with other village dog populations to demonstrate this have not been attempted thus far. Because village dogs are found throughout South-east Asia but Asian dog breeds disproportionately hail from China and Japan (and some of these, such as the Chinese Crested and Pekinese, have mixed Asian–European ancestry; Larson et al., 2012), genetically analyzing village dogs will be particularly valuable for providing a finer-scale geographic pattern to this East Asian center of diversity. Indeed, Brown et al. (2011) recently found mtDNA and Y chromosome diversity as high in village dog populations in far South-east Asia as in southern China, extending the geographic area of known high diversity in Asian dogs. Many potentially important areas (e.g., Myanmar and Bangladesh) have not yet been studied and most other populations have only been studied with uniparentally inherited markers (chromosome Y and the mitochondrion), so there is still much to learn about them.

8.5.1 Asian dogs

Genetic clustering of Asian village dog populations reveals two major groupings: South-east Asian dogs and Middle Eastern dogs (Brown et al., 2011; Ding et al., 2011). The diverse South-east Asian dogs show some affinity with the dogs of Oceania (including dingoes and NGSDs) whereas the Middle Eastern populations, home of the oldest archeological evidence for dogs, share some affinity with European and African dogs (Larson et al., 2012) (although Y chromosome evidence supports a closer relationship between Asian and European dogs than between Middle Eastern and European dogs; Brown et al., 2011). Between these clusters, India has large populations of village dogs (sometimes referred to as 'pariah dogs') that have been studied in terms of anatomy and behavior, as well as a diverse assortment of indigenous breeds. These Indian dog populations have yet to be well characterized genetically. Genetic analysis of Middle Eastern dogs revealed lower overall levels of diversity than in East Asia, but also evidence of localized mtDNA haplotypes (Ardalan et al., 2011; Brown et al., 2011; Pang et al., 2009; but see vonHoldt et al., 2010 who found similar levels of nuclear DNA variation between Middle Eastern and East Asian dogs).

Genome-wide analysis of Middle Eastern dogs and wolves shows that they clearly interbred in the past, and that genes from these wolves may have been critical for the evolution of some dog traits like small body size or limb dwarfism (Gray et al., 2010; Parker et al., 2009; vonHoldt et al., 2010). Thus, Middle Eastern dog populations represent an important genetic resource for understanding dog evolution. The Canaan dog, a land race from the eastern Mediterranean area around Israel and Lebanon, clusters genetically with Middle Eastern purebreds (Afghan Hounds and Salukis) but with lower genetic diversity in the imported stock, suggesting that genetic analysis on dogs in the Middle East will be highly informative (Shiboleth, 2004; vonHoldt et al., 2010). Y chromosome studies also show Canaan Dogs have relatively high haplotype diversity and Canaan Dogs and Salukis have deeply rooted Y chromosome haplotypes, supporting a lengthy evolutionary history with significant population size for these dogs (Bannasch et al., 2005).

In addition to Asian village dogs, Asian Spitz-type dogs, such as the Akita and Chow Chow, also contain some haplotypes not seen in most modern breeds (Larson et al., 2012; Parker et al., 2004). These 'ancient' breeds tend to be strongly diverged from other breeds, which could be a consequence of long-maintained genetic separation from other dogs or simply a product of strong inbreeding (Parker et al., 2004). Dogs appeared in the fossil record over 12,000 years ago in northern China and the Russian Far East, so current Asian Spitz-type dogs and other northern Asian dogs may have a lengthy history apart from other dogs (Cui and Zhou, 2008; Dikov, 1996; Jing 2010a, b). Scientists have not yet tested whether or not any free-breeding populations of northern Asian dogs retain genetic signatures of local, ancient heritage, though this seems likely given the area's social and geographic separation from Europe. Likewise, dogs living in relatively inaccessible places like the high altitudes of Tibet have admixed little with modern breed dogs and exhibit high genetic diversity (Li and Zhang, 2012).

8.5.2 European dogs

Village dogs also occur in many European countries, presenting a possible conservation problem by interbreeding with endangered gray wolf populations (Verardi et al., 2006; Vilà et al., 2003; Leonard et al., Chapter 7). Early European village dog populations were likely some of the founder stock for many of our modern dog breeds. But as the popularity of purebred dogs grew, homogenization of these village dog populations through interbreeding with purebred dogs likely greatly reduced European village dog populations' genetic diversity and distinctiveness, especially in urban areas. Nevertheless, unstudied pockets of ancestral genetic diversity may exist, with isolated free-breeding dog populations and indigenous working dog breeds the most likely candidates to harbor that diversity. Although modern breed dogs with European ancestry continue some ancient European dog genetic lineages, some regions of the continent have few if any representatives in modern kennel clubs. For these regions, studying intact village dog populations or ancient DNA samples are the only methods available to assess their early dogs' genetic history.

The Arctic region of Europe was also important in creating some modern dog lineages. Spitz-type dogs were likely developed here thousands of years ago, in part through accidental or deliberate interbreeding with local wolves (Klütsch et al., 2011; Parker, 2012; Parker et al., 2004; Savolainen, 2006; vonHoldt et al., 2010). In fact, the modern breed descendents of these dogs carry clear mtDNA signatures of this interbreeding with local wolves, having a private haplogroup found almost exclusively in Spitzes (Klütsch et al., 2011; Savolainen, 2006). These village dog populations essentially disappeared as tribal cultures were replaced with modern societies in this region, but through the extraordinary efforts of some individuals, some of their genetic legacy lives on in breeds such as the Finnish Spitz (Morris, 2002).

8.6 The Americas

Before its discovery by Europeans, the American continents teemed with village dogs, including some land races with distinctive phenotypes, such as the hairless Xoloitzcuintli (Morey, 2010; Schwartz, 1998). These dogs were not independently domesticated from North American gray wolves, but were instead brought from Asia by early Americans (Leonard et al., 2002). European colonization not only destroyed great American tribes and empires, but also led to the extinction of nearly every single Native American dog breed, including extremely unique breeds like the Salish Wool Dog of British Colombia (Crockford, 1997). Many dogs likely disappeared as their niches at the feet and trash heaps of Native American peoples collapsed. In other cases, the local dogs may have bred with European-derived dog stock to the point where the pre-Colombian American dog genetic signature was completely lost. The Mexican hairless (Xoloitzcuintli) and its hairless Peruvian counterpart live on, but since hairlessness is a dominant mutation, it is likely that hairlessness survived by introgression of hairless dogs with European stock, leaving the modern American hairless breeds' genomes primarily derived from European breed dogs (Vilà et al., 1999). While one study found no evidence of pre-Colombian American dog mtDNA in 19 Xoloitzcuintli (Leonard et al., 2002), another found some evidence for pre-Colombian American dog mtDNA in the Xoloitzchintli, Chihuahua, and Peruvian hairless (Oskarsson, 2012). Further research will be required to quantify the amount of ancient American dog heritage in these breeds.

8.6.1 Current state of dog diversity in the Americas

The human population of Central and South America today has approximately 10–50% Native American ancestry depending on the country analyzed, with the rest of its genetic heritage coming from European or African ancestors. Thus, in a sense, even tribes that have been lost since European contact, like the Taíno of Puerto Rico, live on in the genomes of the modern population (Bryc et al., 2010; Young, 2011). The Native American dogs, however, did not fare so well. Diagnostic mitochondrial haplotypes discovered through ancient DNA analysis of Native American dog burials are almost completely absent from modern populations, with perhaps the exception of some Arctic sled dogs and possibly a few dogs around the Yucatán Peninsula (Brown et al., 2013; Castroviejo-Fisher et al., 2011; Leonard et al., 2002). At most, 5% of the surveyed dogs descend from ancient American dogs, and the true number is likely much lower (possibly zero). While the 2011 study of Castroviego-Fisher et al. analyzed 400 modern dogs from several isolated areas across the Americas, the sampling emphasized some geographic areas over others and the study only included 13 ancient Latin American samples. It is still possible that certain areas with relatively few modern dogs sampled (such as the south-eastern USA, where the 'American Basenji,' or Carolina Dog, is found) may yield greater levels of Native American dog DNA, or that increased sampling of ancient American dogs will lead to reinterpretation of the study's results. Indeed, a recent study found that all tested Carolina Dog mtDNA haplotypes belonged to East Asian or universal clades, including 37% private haplotypes not found in any other dogs (Oskarsson, 2012). This study lends credence to the hypothesis that feral, free-breeding Carolina Dogs are remnant populations of pre-Colombian American dogs (Brisbin and Risch, 1997). Once again, thorough analysis of isolated dog populations, including the use of genome-wide DNA markers to detect admixture proportions, will be key for determining how much, if any, genetic legacy from this profoundly interesting evolutionary branch still survives.

The Native American Arctic dogs may have fared better than their more southern counterparts. The Alaskan Malamute shows evidence of ancient heritage, forming, along with the Siberian Husky, a distinct clade separate from modern European breeds (Parker et al., 2004; vonHoldt et al., 2010). Modern Malamutes contain significant admixture with modern breeds, but likely retain some ancestry from early American dog lineages such as the Pre-Columbian working dogs of the native Iñupiat people in Alaska's far north-west (Brown et al., 2013; Cummins, 2002). Competitive sled dogs derive over 50% of their DNA from this Malamute–Husky lineage that likely includes some ancient Inuit sled dog ancestry (Huson et al., 2010, 2012). A new study of modern and ancient mtDNA in Arctic dogs shows that modern Eastern Arctic Inuit sled dogs, to a much greater extent than even Malamutes, retain the mtDNA haplotypes found in the pre-Columbian dogs of region (Brown et al., 2013). Most likely, they will serve as a useful population for understanding the history of the dog.

8.7 Conserving dog diversity

As noted in the above sections, the patterns of genotypic and phenotypic diversity in dogs around the world do not match. Phenotypic diversity—at least for the most obvious traits like size, shape, and specialized behavior—is concentrated in breed dogs generally held in developed countries, whereas genotypic diversity is concentrated in indigenous village dogs living mostly in developing countries. This pattern is not unusual for domestic animals. For example, goats are far more numerous in the Asia-Pacific and African regions than in Europe (75% of the goats live in the Asia-Pacific and African regions versus 4% which live in Europe), but Europe has the largest number of described breeds of any region (33%) (Galal, 2005). It does appear, however, that the disconnect between population size (and likely high genetic diversity) and number of breeds (and likely high phenotypic diversity) is particularly large in the case of the dog, as Europe boasts over 80% of the Fédération Cynologique Internationale-recognized breeds while probably containing less than 10% of all individuals (Figure 8.5).

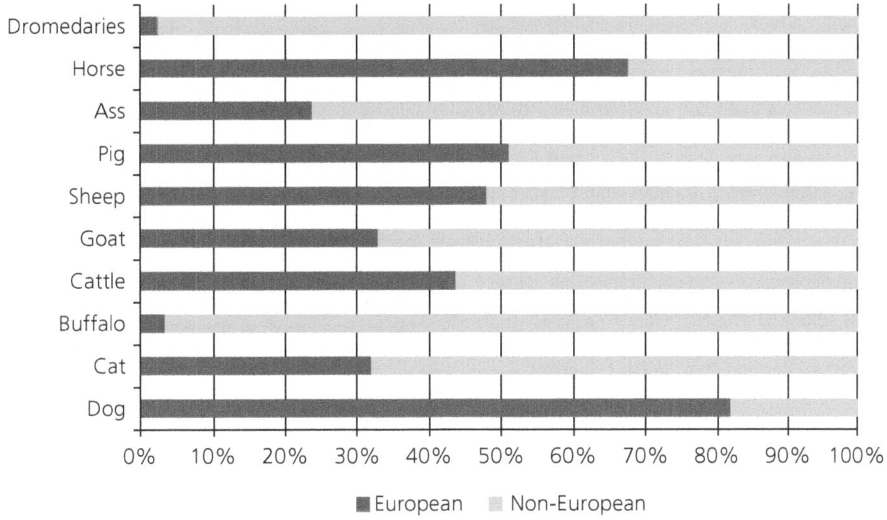

Figure 8.5 Percentage of breeds in each species or species group with European origin. Despite the fact that Europe has <25% of the individual animals of each of these species (Coppinger and Coppinger, 2001; Galal, 2005), it is often the birthplace of half or nearly half of the breeds. In fact, in every one of these domesticated species, Europe has more breeds for every million animals than any other continent. However, that rate is particularly high for dogs, where Europe is home to 82% of the dog breeds recognized by the Fédération Cynologique Internationale. Sources: Galal (2005), The International Cat Association, Fédération Cynologique Internationale.

8.7.1 Conserving village dog diversity

Indigenous village dogs have locally distinct genotypes that are the result of successive founder events, genetic drift, and probably local adaptation, although little progress has been made in identifying signatures of genetic adaptation in these populations. To the extent that these dogs adapted to their different environments, preserving their genetic diversity is important for maintaining healthy populations of well-adapted village dogs across their current geographic range. In fact, undiscovered genetic variants in village dogs could be used to breed disease resistance and other traits into established dog breeds in response to new ecological, economic, or sociocultural factors, as has been considered for other plants and animals of agricultural importance (Oldenbroek, 1999). Separately, there would be some benefit to preserving many modern breeds of dog as well, for the genetic structure of breeds makes it easier to identify the genetics underlying various traits and diseases, including some genetic diseases affecting humans (see Box 8.2).

Box 8.2 The value of dogs as genetic resources

In addition to preserving dogs for intrinsic reasons or for their ecological, economic, and sociocultural value, dogs provide useful genetic resources for studying people and evolutionary processes. Some examples of recent work using dogs to study genetic processes in people and other organisms include:

- *Mapping canine disorders*: Because modern dog breeds are highly inbred with successful males siring dozens of litters, their genomes have long runs of linkage disequilibrium and low heterozygosity (Lindblad-Toh et al., 2005). Taking advantage of this genomic architecture, scientists have mapped the genetic underpinnings of a number of canine disorders much more simply than they could map similar human diseases. In a number of cases, such as narcolepsy and epilepsy, similar human diseases have proven to be caused by similar mutations in loci that correspond to the canine disease variants (Lin et al., 1999; Lohi et al., 2005; Seppälä et al., 2011).
- *Acting as models for gene therapy*: For example, dogs with one form of progressive retinal atrophy that causes blindness were successfully treated using a recombinant virus that then proved successful in treating the human form of the disease (Acland et al., 2001; Ostrander, 2012).
- *Understanding complex inheritance patterns*: Given the genomic resources available for studying dogs, including published genomes, microsatellites and SNP genotyping arrays, as well as the breed structure and extreme phenotypic diversity of dogs, dogs are a model system for understanding the genetic architecture of traits with complex inheritance (Boyko et al., 2010; Wayne and Ostrander, 2007). Multigenic traits in particular are much easier to map in dogs than many other animals (Ostrander, 2012).
- *Understanding artificial selection*: Village dogs are a group of geographically widespread domesticated animals that do not undergo strong artificial selection. Studying their genetics could elucidate the differences between natural and artificial selection, as has been done for selection on size. Between dog breeds, one locus, IGF1, explains 50% of the variation in size (Boyko et al., 2010). The situation is similar in other domestic animals (e.g., four loci explain 83% of the size variation in horses; Makvandi-Nejad et al., 2012). In village dogs, however, the top three SNPs explain only 38% of the variation in body mass (Boyko et al., 2010). While this does not approach the complexity of size determination in humans (697 genes are estimated to explain 15.7% of the variation in human height; Allen et al., 2010), it does suggest that village dogs may be something of an 'in between' in terms of artificially and naturally selected animals. After undergoing an initial domestication event bottleneck they have since been impacted by both human and natural selective pressures. Thus village dogs could elucidate the genomic changes associated with domestication bottlenecks versus those that are the result of continuing, strong artificial selection.
- *Understanding local adaption processes*: Dogs underwent dietary and lifestyle changes during and after domestication. They inhabit nearly every place humans do and share similar food to humans. Genetic analysis of diverse dog populations might help researchers find local adaptations to diet and physical conditions, such as altitude. It may also help us understand complex disease

continued

Box 8.2 *Continued*

processes. For example, dogs are the only animal besides humans to regularly suffer prostate cancer as they age. This likely has to do with their dietary overlap of red meat and fats as well as phytoestrogens in Western food and environmental factors (Coffey, 2001). High fat diets can also cause pancreatitis that can lead to diabetes in dogs (Rand et al., 2004). Studying obesity and metabolic syndrome in dogs helped point out directions for human research into diabetes and related conditions and could do so for other conditions (Kaiyala et al., 2000; Kim et al., 2003).

- *Understanding genomic integrity*: Canids are the only mammals that do not have functional PRDM9 genes which, for all other mammals including humans, localize recombination hotspots in their genomes (Oliver et al., 2009). Dog genomes also contain a highly active, canine-specific SINE_Cf transposable element. Disruptive SINE insertions underlie many important dog phenotypes and possibly contribute to structural instability in some genomic regions (Kirkness et al., 2003). Studying dog genomes will further our understanding of evolutionary processes involved in genomic integrity.

Conserving village dog populations represents a unique challenge. Unlike other canids, village dogs are often found at much higher densities and thrive in urban environments. Traditional canid conservation strategies tend to focus more on keeping population numbers healthy than on preventing admixture between native free-breeding animals and nonlocal purebred or admixed animals (e.g., Ginsberg and Macdonald, 1990). Additionally, attempts at preserving breeds of other domestic animals take advantage of their artificially created population structure and controlled breeding, meaning village dogs require different methods to conserve them. For example, studies on protecting cattle show that native cattle breeds have far smaller population ranges than village dogs, enabling conservation efforts over smaller geographic areas using different tools than required for village dog conservation (e.g., 49 African cattle breeds, Reist-Marti et al., 2003; 20 North European cattle breeds, Kantanen et al., 2000).

Given the inadequate models of conservation available, how should we approach village dog conservation? Because dogs vary so much phenotypically and maintain the ability to interbreed with each other and even with some other canids, recent literature has widely varied in their classification schemes (e.g., in 2012 alone, dingoes have been called *Canis lupus dingo* (Ardalan et al., 2012), *Canis dingo* (Smith et al., 2012), and *Canis familiaris dingo* (Kutt, 2012)). Given the relatively short evolutionary time-scale of the dog and the disconnection between genetic and phenotypic diversity in dogs, we cannot rely on traditional taxonomic units to target which populations to conserve. Systems incorporating ecological differentiation (e.g. Crandall et al., 2000), extinction risk to domestic breeds (e.g., Reist-Marti et al., 2003; Simianer et al., 2003), or evolutionarily significant units (Moritz, 1994) are potentially more viable. However, the amount of ecological differentiation between village dog populations and the factors that increase extinction risk (other than widespread contact with Western breed dogs) are poorly understood. Which ecological interactions need to be considered when conserving dogs is also unknown. In some places dogs may play important roles in their ecosystems. For example, 6% of the dogs in one area in Zimbabwe fell prey to leopards over the course of a year, possibly increasing leopard density and affecting the density of traditional leopard prey (Butler et al., 2004). However, when the alternatives are indigenous village dogs or admixed village dogs living in an area, it is not yet established that a change in the dogs' genetic profiles would have an important effect on the ecosystem.

At this time, we do not have enough genetic, ecological, or sociocultural data to know the geographic extent of various indigenous village dog populations, the number of individuals in the population, the ecological and phenotypic distinctiveness of each population, or the external extinction risk factors for the population. We also do not have enough experience conserving village dog populations to know how expensive this would be, even if we could decide what constituted a single population and which populations we wanted to conserve. Simple answers, such as, 'conserve the ancestral population' do not work, since 'ancient' dog breeds and village dog populations generally derive from isolated areas, not areas with ancient (i.e., archeological) dogs or areas where the dogs' progenitor, the gray wolf, lives (Larson et al., 2012).

With the above caveats in mind, we will try to make some suggestions regarding populations that deserve consideration from conservationists. It seems clear that in the postcolonial era, countries geographically closest to developed countries are the most at risk for losing their indigenous village dog populations to swamping gene flow (e.g., northern Egyptian dogs had relatively high levels of admixture with European dogs; Boyko et al., 2009). Urban dog populations found in large cities that maintain global trade networks are also likely highly admixed. In addition to having a high volume of global trade, urban areas often have spay-and-neuter programs aimed at preventing dog overpopulation and the accompanying public health and safety problems. These programs likely hasten the replacement of indigenous dogs with admixed ones as they prevent many indigenous street dogs from reproducing while allowing owned dogs with nonlocal ancestry to contribute disproportionately to the next generation of street dogs. Isolated or peripheral populations such as those on islands in the Indonesian archipelago (Irion et al., 2005), those on the Tibetan plateau (Larson et al., 2012), those in and around Iran (Brown et al., 2011), those in eastern Arctic North America (Brown et al., 2013; Darwent 2013), and those in Central Africa (Boyko et al., 2009) are most likely to be genetically distinct from modern European breed dogs. However, many village dog populations, like gray wolf populations, are interconnected over large ranges (e.g., African village dogs in Uganda and northern Namibia, separated by >2,000 km, are very similar genetically; Boyko et al., 2009). More work on dog abundance, phenotypic distinctiveness, and genetic relatedness in free-breeding populations is needed to determine where conservation considerations are warranted. Clearly, though, there are some possible target populations on most, if not all, continents.

Deciding which particular populations in each geographic area to conserve and how to do so depends on the goals and resources available. Targeting the most vulnerable populations could have the biggest impact in terms of maximizing the number of distinct dog populations remaining on Earth, but would require an immediate investment of significant resources to fight the forces currently threatening those populations. Conserving currently unthreatened populations would presumably be cheaper and require fewer resources, at least for the moment. Determining which populations are most threatened requires population abundance estimates for dog populations around the world, which currently do not exist, but could be accomplished with reasonable effort. Regardless of which populations are targeted, conserving them could focus on conserving their genetic lineage, conserving most of their extant genetic diversity in a viable population, or conserving them across much of their current range.

The cheapest and easiest solution for 'saving' a population is to choose several non-admixed individuals in the population to form a new dog breed, and then begin to breed them in the traditional manner. To slow the loss of genetic diversity, one could allow some outbred crosses with other non-admixed individuals from the population for as long as the free-breeding population survives. This would preserve some of the unique local genetic information and adaptations, but could result in a highly inbred population and would certainly cause the loss of much of the current population's genetic variation. It would also prevent that lineage from continuing to evolve with changing ecological conditions as dogs have done quite successfully for millennia.

Preserving a viable free-breeding population somewhere within its current range is somewhat more complex, but would enable much of the population's genetic variability to be maintained and allow the population to continue to evolve in response to ecological changes. Such efforts would require the buy-in of the local human population and active efforts to prevent non-local dogs from entering the local breeding population. To achieve that would require public education on the value of local dogs over imported ones, which might also improve the welfare of local animals. Depending on the location, it might involve significant trade-offs regarding spay-and-neuter dog control programs and other conservation and public health and safety projects.

Saving a dog population across most or all of its current range seems prohibitively difficult and expensive given current interest in dog conservation. It would require extensive genetic surveillance and programs tailored to the local social and ecological context across wide geographic areas. In many places, local efforts are already underway to try to conserve indigenous village dog populations (which are often referred to as land races), but current efforts are primarily focused on preserving just a sliver of the present genetic diversity through local breed formation. However, conservation requires effort and money, and these are potential quick and easy wins for conservation if saving particular genetic lineages is an important goal. For example, the Indian Indog, the South African Africanis, and the Indonesian Kintamani dog (Puja et al., 2005) make attractive conservation targets because local people are willing to put the time and effort into conserving them although, in the case of the Africanis, finding enough non-admixed individuals to form a viable breed may be challenging. Geneticists and conservationists could work with these groups to identify dogs that have a high proportion of indigenous DNA and, to the extent possible, represent the range of genetic and phenotypic diversity of local dogs. These dogs could then form the basis of a closed or partially closed breeding group. However, the history of numerous breeds demonstrates that conserving a particular phenotype in a breeding line is much simpler and cheaper than conserving high genetic diversity. It is not clear that dog fanciers would be willing to pay the higher costs and go through the additional effort to conserve a large enough breeding population to encompass most of an indigenous breed's genetic diversity. If long-term conservation of dog populations is important, efforts to save viable populations of free-breeding dogs will need to be undertaken and will most likely need to be led by conservationists coordinating with local dog enthusiasts (Box 8.3).

8.7.2 Conserving breed dog diversity

While conserving purebred lineages is beyond the scope of this chapter, we note that some such lineages (particularly those of ancient or indigenous dog breeds) may be the only genetic descendants left for some endangered or extinct free-ranging or free-breeding dog populations. For example, in the absence of Latin American dog populations with significant Native American dog ancestry, preserving Xoloitzcuintlis or Chihuahuas might be the only way to perpetuate at least some of the early American dog gene pool. Even this would likely preserve only a small portion of the pre-Colombian American dog genetic diversity, mostly that centered on the genes responsible for the phenotypic traits that distinguish these dogs (Leonard et al., 2002; but see Oskarsson, 2012 who argues that there may be more ancient American dog ancestry in today's free-breeding populations than previously thought). In general, genetic distinctiveness is a good measure for identifying breeds that are potential targets for conservation, though extinction risk, phenotypic distinctiveness, and other criteria could also be used as supplemental criteria in identifying breeds for conservation efforts.

Fortunately, there are many individual dog fanciers and breed organizations that can afford to undertake the great efforts necessary to conserve dog breeds. For example, one man, Hugo Roos, single-handedly rescued the Finnish Spitz from extinction in the late nineteenth century, carefully breeding some of the last remaining native Spitz-type individuals (Morris, 2002). The Portuguese Water Dog was also saved largely through the work of one man, Vasco Bensaude, in the 1930s (Braund, 1997). More recently, the Basenji Club of America has opened its studbook twice (in 1990 and 2009–13) to newly imported dogs from the Congo to preserve the breed in the face of genetic disorders very common in American Basenjis due to founder effects (Bell, 2007). These efforts have included difficult and expensive trips to the Democratic Republic of the Congo to bring back native Basenjis. Clearly, there is money available to conserve at least some breeds, though care must be taken not to conserve merely one particular phenotype at the expense of genetic diversity, or worse, attempt to reconstruct a phenotype by the selective breeding of unrelated stock.

8.7.3 Conserving dingoes and New Guinea Singing Dogs

Dingoes serve an important role as top-predator across much of Australia (Johnson et al., 2007; Letnic et al., 2009, 2012; Ritchie et al., Chapter 2). They are genetically distinct from all other dog lineages except NGSDs (Savolainen et al., 2004; vonHoldt et al., 2010). They are also culturally important to Australian aboriginal populations (Meggitt, 1965). In parts of Australia, they are at risk from pest control measures such as baiting and hunting, but in many areas they are most at risk from admixture with local, modern breed dogs (Elledge et al., 2006). Current conservation efforts focus on identifying and removing dingo–dog hybrids as well as mitigating the impacts of human and livestock encroachment on dingo lands (Claridge and Hunt, 2008; Elledge et al., 2006). Understanding dingo behavior and ecology better will also improve dingo conservation. For example, their large home ranges suggest that larger dingo conservation areas are warranted (Claridge et al., 2009).

Although NGSDs are genetically distinct from other non-dingo dog lineages, less work has been completed on NGSD conservation (Koler-Matznick et al., 2007; vonHoldt et al., 2010). These dogs have unique behavioral, ecological, and cultural significance, and are threatened by small population size and hybridization with dogs (Koler-Matznick et al., 2003, 2007). At this point little is known about their population size, except that it is extremely small, probably shrinking, and limited to high altitudes (Koler-Matznick et al., 2007). Indeed, the last confirmed sighting of an NGSD by a scientist occurred in 1989, though there have been a number of reports of sightings by Papuans in remote mountain areas since then (Koler-Matznick et al., 2007). More research is clearly needed.

Box 8.3 Future directions for free-breeding dog conservation

Dogs are a well-studied model species for a number of genetic conditions, but relatively little is known about many of the most threatened dog populations, hampering efforts to conserve non-breed dog populations. The following are some areas in which more study and action could prove particularly useful for conserving dogs.

- More studies are needed characterizing genetic diversity across diverse populations of free-breeding and indigenous breed dogs, particularly for the majority of the genome that is biparentally inherited (unlike the mitochondrion and chromosome Y). Whereas mitochondrial and Y chromosome studies have contributed greatly to our understanding of dog population history, genome-wide studies can detail the role of selection in shaping genetic diversity and provide finer resolution for parameterizing models of demographic history and identifying distinct free-breeding dog populations versus populations that are amalgamations of imported, non-native dog lineages.
- More archeological DNA studies of ancient dogs, particularly in the Americas, are required to better understand what diversity existed and to perhaps start to understand how displacement with modern breed dogs occurred and how that can be avoided in the future.
- Studies of gene expression in various village dog populations could uncover specific phenotypic adaptations these dogs harbor that might make certain populations higher priority conservation targets (e.g., adaptations to specialized diet, extreme altitudes, etc.). These studies may also prove useful for understanding human and other animal adaptations to similar environmental stressors.
- Breeding programs that increase the genetic diversity in certain breeds would better protect the health of individual animals and enable the long-term survival of currently highly inbred breeds.
- Studies are needed to identify breeds facing the threat of extinction and determine which are worthy targets of conservation based on genetic and phenotypic distinctiveness, particularly in breeds that were founded from free-ranging dog populations that have subsequently gone extinct (e.g., Finnish Spitz and Xoloitzcuintli).
- Continued efforts should be made to improve dingo and New Guinea Singing Dog (NGSD) conservation. For dingoes, this involves improving the ability to identify and remove hybrids as well as addressing human encroachment. For NGSDs this involves more basic research in identifying their distribution and numbers and identifying

continued

Box 8.3 *Continued*

threats to the population's survival in any pockets of territory in which it still persists. If no viable populations are found in the wild, zoo breeding programs and possible reintroduction to the wild could be considered.

- Studies demonstrating what benefits local people obtain from keeping indigenous village dogs could be used to convince local people to prefer those dogs over imported ones and could also help galvanize outside individuals and groups to contribute to indigenous village dog conservation.
- More research is required to determine the positive and negative effects that village dogs and feral dogs have on other species' populations (e.g., by fulfilling a top-predator niche or by competing with endangered or threatened carnivore species). Without this, conservation efforts could cause unintended consequences and ultimately fail if, for example, an increase in village dogs in an area reduces populations of the endangered African wild dog, *Lycaon pictus*.
- Investments in vaccination programs in cities and research into better vaccine delivery systems would reduce the negative impact of village dogs, which is a prerequisite for any large-scale attempt to conserve them. Designing and implementing spay/neuter programs that do not result in decreased indigenous dog representation in the next generation is also important to mitigate the conflict between public health and safety and indigenous village dog conservation.

8.8 Conclusions

Over the past several millennia, village dogs spread across the globe and diversified as genetic drift and selection acted upon isolated populations. Particularly in areas with relatively large populations that avoided breeding with imported dogs in recent times, village dog populations maintained a genetic diversity that was lost in modern breeds. These populations bear unique genetic signatures that arose in geographic areas not represented by modern breeds. Unfortunately, many important village dog populations have already been lost, including nearly all of the dog populations that lived in the New World and Polynesia prior to European contact and colonization.

Currently, village dogs still inhabit much of the globe. Some, such as the street dogs in many large cities, are undoubtedly mixes of various indigenous and imported dogs, but many others still occupy traditional niches in the community and retain localized genetic signatures and physical features. The patterning of genetic diversity in these 'indigenous' populations is largely based on geographic separation over centuries or millennia. Particularly in isolated and peripheral dog populations, these dogs likely harbor important undiscovered genetic variants contributing to local adaptation. These adaptations that are not present in modern breeds warrant further study before these populations are lost in our increasingly urban and interconnected world.

Breed dogs also deserve conservation consideration. While most modern breeds are closely related to each other, many harbor unique genetic variants underlying extreme phenotypic differences. Additionally, several ancient breeds from peripheral areas of the dog's range harbor distinct genes derived from older, isolated dog populations or local wolf populations. Some breeds are also particularly useful for biomedical research on the genetic components of disease processes.

Dingoes and NGSDs are free-ranging dogs that represent distinct genetic lineages from all other modern dogs. These dogs require different kinds of conservation efforts than other dogs, as they have been completely free-ranging for millennia. Dingoes, and possibly NGSDs, also play important ecological roles in their environments. Given this, conservation of these dogs is important for maintaining healthy ecosystems and can be accomplished using more traditional conservation methods than will be required for the conservation of other dog populations.

Effective dog conservation requires better definition of the goals of such efforts. Merely maintaining the range of genetic lineages found in modern dogs requires identifying distinct village dog populations and genetically distinct breeds and then conserving a large enough representative sample of these populations. This could potentially be done through the formation of 'new' internationally recognized breeds of dogs from indigenous village dog populations, as is being done in India with the Indog.

Preserving viable free-breeding populations that would conserve genetic diversity, evolutionary potential, and the ecological and sociocultural roles of indigenous village dogs where they are found now would require more intensive conservation efforts aimed at keeping modern breed dogs out of areas with indigenous village dogs. It would also require working with local human populations to balance dog conservation with the conservation of nearby wildlife and with public health and safety. This would be most helpful in areas still practicing traditional hunting, farming, or ranching, where indigenous village dogs directly assist people and are not at such high population sizes that they constitute a significant risk to public safety.

However, in some areas individuals may prefer to breed their dogs with modern breed dogs that are often larger or have other desirable traits. In these areas, local support for conservation of indigenous village dogs may be harder to achieve and might crucially depend on trying to change the hearts and minds of individual dog owners to believe that owning local dogs is preferable to owning imported ones. If this cannot be achieved, we note that rapid evolution and successive waves of colonization have been the norm for dogs as long as they have existed. The continuation of free-breeding populations in areas where they currently exist, even if those populations are already admixed or become admixed in the future, would allow dogs to continue to adapt to these local conditions. Given the demonstrated ability of dogs to adapt to a wide variety of human-associated niches, the conservation of indigenous village dogs' ecological and sociocultural roles may not necessarily require the conservation of indigenous village dogs themselves. However, indigenous village dogs adapted to local conditions over many centuries or millennia, and their ancestral populations presumably started with more genetic diversity than modern breeds have. Therefore, conserving indigenous village dogs does seem preferable wherever possible.

Ultimately, conservationists must decide if saving those indigenous populations is feasible, or if they should instead focus on conserving dogs' roles in a local community. If conservationists do not protect indigenous free-breeding populations, some benefit might come from creating a new breed by saving a few representative members of an indigenous village dog population before it is swamped by modern breed dogs.

Alternatively, a focus on preserving phenotypic variation in dogs would require a greater focus on conserving modern breed dogs as well as efforts to find genes responsible for local adaptation and other distinctive traits in indigenous village dog populations. Such a focus would preserve specific alleles of large effect and allow dogs to continue to serve as a model system for understanding artificial selection and for discovering the genes underlying many interesting traits. It would, however, reduce dog genetic diversity and likely limit their potential to evolve new adaptations and traits of interest in the future. It would also make it nearly impossible to use genetics to detail the history of the dog. Further, many of the breeds with the most distinctive traits are not currently in danger of extinction and conservation efforts for most breeds are probably unnecessary.

As a broader point, biologists should consider dogs' remarkable adaptability when planning conservation efforts. Dogs followed humans across the globe and thrive in a remarkable number of niches in every environment humans live in. They exhibit enormous phenotypic variation. Conservation efforts aimed at merely maintaining the status quo, such as by forming local breeds with closed breed books and strict phenotypic conformation standards, short circuit dogs' hallmark trait of adaptability. The extent to which conservationists should prevent continuing genetic change, or even prevent the spread of non-native dog genetic material in indigenous populations, is an open question without a clear objective answer. Interventions aimed at mitigating human impacts, such as reducing the number of modern breed dogs breeding with indigenous village dogs, are probably more defensible than those aimed at reducing the effects of more localized or 'natural' population processes, but even these need to consider the wishes of the local human population and the fact that dogs have been colonizing and recolonizing territories for many millennia to produce the present-day distribution of dogs (e.g., Li and Zhang, 2012).

Regardless of what people choose to conserve, more research is needed to fully catalogue the genetic diversity found in modern dogs and to understand the range of uses people in different areas of the world have for their dogs. Once that is completed, it will be easier to identify distinct populations and choose appropriate conservation goals. A closer working relationship between dog enthusiasts, conservationists, and scientists would make it easier to coordinate in making these decisions and allocate funds to the most pressing conservation issues.

Dogs will undoubtedly live on in the world for centuries to come. However, most of the genetic diversity found in dogs will be lost if the diverse indigenous village dog populations alive today vanish or are replaced with non-native imported modern breed dogs. Without a concerted effort, some populations will certainly be lost as globalization brings more competition from non-native imported dogs and changes the structure of the human communities to which native dogs have adapted. In many cases the imported dogs and their purebred or admixed offspring may benefit from human breeding and dog control practices, such as spay/neuter programs. Given the sentimental, economic, and sociocultural value of dogs to many people, and their value as a model organism for studying the genetic underpinnings of biological processes, it is clear that humanity would lose much if we did not save at least some of these unique populations.

Acknowledgments

We wish to thank all the scientific collaborators, assistants, financial backers, and dog owners that have helped us sample and study village dog populations from across the world. We also thank all the dog breeders, dog enthusiasts, scientists, and laypeople who have questioned our tentative hypotheses and helped push our thinking about dogs in all directions over the past few years. We would particularly like to thank the insightful comments of Jan Koler-Matznick, Ben Sacks, Julia Randall, and one anonymous reviewer.

References

Acland, G.M., Aguirre, G.D., Ray, J., Zhang, Q., Aleman, T.S., Cideciyan, A.V., Pearce-Kelling, S.E., Anand, V., Zeng, Y., Maguire, A.M., Jacobson, S.G., Hauswirth, W.W., and Bennett, J. (2001). Gene therapy restores vision in a canine model of childhood blindness. *Nature Genetics*, **28**, 92–5.

Allen, H.L., Estrada, K., Lettre, G., Berndt, S.I., Weedon, M.N., Rivandeneira, F., Willer, C.J., Jackson, A.U., Vedantam, S., Raychaudhuri, S., et al. (2010). Hundreds of variants clustered in genomic loci and biological pathways affect human height. *Nature*, **467**, 832–8.

Ardalan, A., Kluetsch, C.F.C., Zhang, A.-b, Erdogan, M., Uhlén, M., Houshmand, M., Tepeli, C., Reza, S., Ashtiani, M., and Savolainen, P. (2011). Comprehensive study of mtDNA among Southwest Asian dogs contradicts independent domestication of wolf, but implies dog-wolf hybridization. *Ecology and Evolution*, **1**, 373–85.

Ardalan, A., Oskarsson, M., Natanaelsson, C., Wilton, A.N., Ahmadian, A., and Savolainen, P. (2012). Narrow genetic basis for the Australian dingo confirmed through analysis of paternal ancestry. *Genetica*, **140**, 65–73.

Bannasch, D.L., Bannasch, M.J., Ryun, J.R., Famula, T.R., and Pedersen, N.C. (2005). Y chromosome haplotype analysis in purebred dogs. *Mammalian Genome*, **16**, 273–80.

Bell, J.S. (2007). American breed clubs and health initiatives. In E.A. Ostrander, U. Giger and K. Lindblad-Toh, Eds., *The Dog and its Genome*. Cold Spring Harbor Laboratory Press, New York, pp. 31–45.

Boyko, A.R., Boyko, R.H., Boyko, C.M., Parker, H.G., Castelhano, M., Corey, L., Degenhardt, J.D., Auton, A., Hedimbi, M., Kityo, R., Ostrander, E.A., Schoenebeck, J., Todhunter, R.J., Jones, P., and Bustamante, C.D. (2009). Complex population structure in African village dogs and its implications for inferring dog domestication history. *Proceedings of the National Academy of Sciences*, **106**, 13903–8.

Boyko, A.R., Quignon, P., Li, L., Schoenebeck, J.J., Degenhardt, J.D., Lohmueller, K.E., Zhao, K., Brisbin, A., Parker, H.G., vonHoldt, B.M., Cargill, M., Auton, A., Reynolds, A., Elkahloun, A.G., Castelhano, M., Mosher, D.S., Sutter, N.B., Johnson, G.S., Novembre, J., Hubisz, M.J., Siepel, A., Wayne, R.K., Bustamante, C.D., and Ostrander, E.A. (2010). A simple genetic architecture underlies morphological variation in dogs. *PLoS Biology*, **8**, e1000451.

Braund, K. (1997). *The New Complete Portuguese Water Dog*. 2nd edn. Howell Book House, New York.

Brisbin, Jr., I. L. and Risch, T. S. (1997). Primitive dogs, their ecology and behavior: unique opportunities to study the early development of the human-canine bond. *Journal of the American Veterinary Medical Association*, **210**, 1122–6.

Brooks, S.A., Makvandi-Nejad, S., Chu, E., Allen, J.J., Streeter, C., Gu, E., McCleery, B., Murphy, B.A., Bellone, R., and Sutter, N.B. (2010). Morphological variation in the horse: defining complex traits of body size and shape. *Animal Genetics*, **41**, 159–65.

Brown, S.K., Pedersen, N.C., Jafarishorijeh, S., Bannasch, D.L., Ahrens, K.D., Wu, J-T, Okon, M., and Sacks, B.N. (2011). Phylogenetic distinctiveness of Middle Eastern and Southeast Asian village dog Y chromosomes illuminates dog origins. *PLoS ONE*, **6**, e28496.

Brown, S.K., Darwent, C.M., and Sacks, B.N. (2013). Ancient DNA evidence for genetic continuity in arctic dogs. *Journal of Archaeological Science*, **40**, 1279–88.

Bryc, K., Velez, C., Karafet, T., Moreno-Estrada, A., Reynolds, A., Auton, A., Hammer, M., Bustamante, C.D., and Ostrer, H. (2010). Colloquium paper: Genome-wide patterns of population structure and admixture among Hispanic/Latino populations. *Proceedings of the National Academy of Sciences*, **107 Suppl 2**, 8954–61.

Butler, J.R.A., du Toit, J.T., and Bingham, J. (2004). Free-ranging domestic dogs (*Canis familiaris*) as predators and prey in rural Zimbabwe: Threats of competition and disease to large wild carnivores. *Biological Conservation*, **115**, 369–78.

Calboli, F.C.F., Sampson, J., Fretwell, N., and Balding, D.J. (2008). Population structure and inbreeding from pedigree analysis of purebred dogs. *Genetics*, **179**, 593–601.

Carroll, C.L. and Huntington, P.J. (1988). Body condition scoring and weight estimation of horses. *Equine Veterinary Journal*, **20**, 41–5.

Castroviejo-Fisher, S., Skoglund, P., Valadez, R., Vilà, C., and Leonard, J. (2011). Vanishing native American dog lineages. *BMC Evolutionary Biology*, **11**, 73.

Claridge, A.W. and Hunt, R. (2008). Evaluating the role of the dingo as a trophic regulator: Additional practical suggestions. *Ecological Management and Restoration*, **9**, 116–9.

Claridge, A.W., Mills, D.J., Hunt, R., Jenkins, D.J., and Bean, J. (2009). Satellite tracking of wild dogs in south-eastern mainland Australian forests: implications for management of a problematic top-order carnivore. *Forest Ecology and Management*, **258**, 814–22.

Clutton-Brock, J. (2012). *Animals as Domesticates*. Michigan State University Press, Ann Arbor.

Coffey, D.S. (2001). Similarities of prostate and breast cancer: Evolution, diet, and estrogens. *Urology*, **57**, 31–38.

Coppinger, R. and Coppinger, L. (2001). *Dogs: A Startling New Understanding Of Canine Origin, Behavior and Evolution*. Scribner, New York.

Corbett, L.K. (1995). *The Dingo in Australia and Asia*. University of New South Wales Press, Sydney.

Crandall, K.A., Bininda-Emonds, O.R.P., Mace, G.M., and Wayne, R.K. (2000). Considering evolutionary processes in conservation biology. *Trends in Ecology and Evolution*, **15**, 290–5.

Crockford, S.J. (1997). *Osteometry of Makah and Coast Salish Dogs*. Publication No. 22, Archaeology Press, Simon Fraser University, Burnaby, Canada.

Cui, J. and Zhou, S. (2008) Zhongguo xin shi qi shi dai wen hua yan bian dong li fen xi (An analysis of the evolutionary forces of Neolithic period cultures in China). *Journal of Chinese Historical Geography*, **23**, 33–45.

Cummins, B.D. (2002). *First Nations, First Dogs: Canadian Aboriginal Ethnocynology*. Detselig, Calgary.

de Caprona, M.D.C. and Savolainen, P. (2013). Extensive phenotypic diversity among South Chinese dogs. *ISRN Evolutionary Biology*, **2013**, 621836.

Diamond, J. (1997). *Guns, Germs and Steel*. W.W. Norton and Company, New York.

Dikov, N.N. (1996). The Ushki sites, Kamchatka Peninsula. In F.H. West, Ed., *American Beginnings: The Prehistory and Palaeoecology of Beringia*. University of Chicago Press, Chicago, pp. 244–50.

Ding, Z.L., Oskarsson, M., Ardalan, A., Angleby, H., Dahlgren, L-G, Tepeli, C., Kirkness, E., Savolainen, P., and Zhang, Y.-P. (2011). Origins of domestic dog in southern East Asia is supported by analysis of Y-chromosome DNA. *Heredity*, **108**, 507–14.

Elledge, A.E., Leung, L.K-P, Allen, L.R., Firestone, K., and Wilton, A.N. (2006). Assessing the taxonomic status of dingoes *Canis familiaris dingo* for conservation. *Mammal Reviews*, **36**, 142–56.

Fillios, M., Crowther, M.S., and Letnic, M. (2012). The impact of the dingo on the thylacine in Holocene Australia. *World Archaeology*, **44**, 118–34.

Fortin, D., Beyer, H.L., Boyce, M.S., Smith, D.W., Duchesne, T., and Mao, J.S. (2005). Wolves influence elk movements: Behavior shapes a trophic cascade in Yellowstone National Park. *Ecology*, **86**, 1320–30.

Fox, S. (2003). *Pet Manual: Rhodesian Ridgebacks (A Complete Pet Owner's Manual)*. Barron's Educational Series Inc., New York.

Fuller, G.K. and Fuller, D.C. (1981). Part II: Hydatid disease in Ethiopia: epidemiological findings and ethnographic observations of disease transmission in southwestern Ethiopia. *Medical Anthropology: Cross-Cultural Studies in Health and Illness*, **5**, 293–312.

Galal, S. (2005). Biodiversity in goats. *Small Ruminant Research*, **60**, 75–81.

Gallant, J. (2002). *The Story of the African Dog*. University of KwaZulu-Natal Press, Durban.

Ginsberg, J.R. and Macdonald, D.W. (1990). *Foxes, Wolves, Jackals and Dogs: An Action Plan for the Conservation of Canids*. International Union for the Conservation of Nature, Gland, Switzerland.

Gray, M.M., Sutter, N.B., Ostrander, E.A., and Wayne, R.K. (2010). The IGF1 small dog haplotype is derived from Middle Eastern grey wolves. *BMC Biology*, **8**, 16.

Henderson, J. (2012). Henderson's Chicken Breed Chart. Ithaca College. <http://www.ithaca.edu/staff/jhenderson/chooks/chooks.html>.

Hunter, L. (2011). *Carnivores of the World*. Princeton University Press, New Jersey.

Huson, H.J., Parker, H.G., Rundstadler, J., and Ostrander, E.A. (2010). A genetic dissection of breed composition and performance enhancement in the Alaskan sled dog. *BMC Genetics*, **11**, 71.

Huson, H.J., vonHoldt, B.M., Rimbault, M., Byers, A.M., Runstadler, J.A., Parker, H.G., and Ostrander, E.A. (2012). Breed-specific ancestry studies and genome-wide association analysis highlight an association between the MYH9 gene and heat tolerance in Alaskan sprint racing sled dogs. *Mammalian Genome*, **23**, 178–94.

Ikram, S. (2005). *Divine Creatures: Animal Nummies in Ancient Egypt*. The American University in Cairo Press, Cairo.

Irion, D.N., Schaffer, A.L., Grant, S., Wilton, A.N., and Pedersen, N.C. (2005). Genetic variation analysis of the Bali street dog using microsatellites. *BMC Genetics*, **6**, 6.

Jing, Y. (2010). Zhongguo gu dai jia yang dong wu de dong wu kao gu xue yan jiu (The zooarchaeology of domesticated animals in ancient China). *Quaternary Sciences*, **30**, 298–306.

Jing, Y. (2010). Research report of the faunal remains from Nanzhuangtou site, XuShi of Hebei Province. (Translated from Chinese). *Acta Archaeology Sinica*, **2010**, 385–91.

Johnson, C.N., Isaac, J.L., and Fisher, D.O. (2007). Rarity of a top predator triggers continent-wide collapse of mammal prey: dingoes and marsupials in Australia. *Proceedings of the Royal Society B*, **274**, 341–46.

Kaiyala, K.J., Prigeon, R.L., Kahn, S.E., Woods, S.C., and Schwartz, M.W. (2000). Obesity induced by a high-fat diet is associated with reduced brain insulin transport in dogs. *Diabetes*, **49**, 1525–33.

Kantanen, J., Olsaker, I., Holm, L-E, Lien, S., Vikki, J., Brusgaard, K., Eythorsdottir, E., Danell, B., and Adalsteinsson, S. (2000). Genetic diversity and population structure of 20 north European cattle breeds. *Journal of Heredity*, **91**, 446–57.

Kim, S.P., Ellmerer, M., van Citters, G.W., and Bergman, R.N. (2003). Primacy of hepatic insulin resistance in the development of the metabolic syndrome induced by an isocaloric moderate-fat diet in the dog. *Diabetes*, **52**, 2453–60.

Kirkness, E.F., Bafna, V., Halpern, A.L., Levy, S., Remington, K., Rusch, D.B., Delcher, A.L., Pop, M., Wang, W., Fraser, C.M., and Venter, J.C. (2003). The dog genome: survey sequencing and comparative analysis. *Science*, **301**, 1898–1903.

Klütsch, C. F. C., Seppälä, E. H., Fall, T., Uhlén, M., Hedhammer, Å., Lohi, H., and Savolainen, P. (2011). Regional occurrence, high frequency but low diversity of mitochondrial DNA haplogroup d1 suggests a recent dog-wolf hybridization in Scandinavia. *Animal Genetics*, **42**, 100–3.

Koler-Matznick, J., Brisbin Jr., I.L., Feinstein, M., and Bulmer, S. (2003). An updated description of the New Guinea singing dog (*Canis hallstromi*, Troughton 1957). *Journal of Zoology*, **261**, 109–18.

Koler-Matznick, J., Yates, B.C., Bulmer, S., and Brisbin, I.L.J. (2007). The New Guinea singing dog: Its status and scientific importance. *Australian Mammalogy*, **29**, 47–56.

Kutt, A.S. (2012). Feral cat (*Felis catus*) prey size and selectivity in north-eastern Australia: implications for mammal conservation. *Journal of Zoology*, **287**, 292–300.

Larson, G., Karlsson, E.K., Perri, A., Webster, M.T., Ho, S.Y.W., Peters, J., Stahl, P.W., Piper, P.J., Lingaas, F., Fredholm, M., Comstock, K.E., Modiano, J.F., Schelling, C., Agoulnik, A.I., Leegwater, P.A., Dobney, K., Vigne, J-D, Vilà, C., Andersson, L., and Lindblad-Toh, K. (2012). Rethinking dog domestication by integrating genetics, archeology, and biogeography. *Proceedings of the National Academy of Sciences*, **109**, 8878–83.

Leonard, J.A., Wayne, R.K., Wheeler, J., Valadez, R., Guillén, S., and Vilà, C. (2002). Ancient DNA evidence for Old World origin of New World dogs. *Science*, **298**, 1613–16.

Letnic, M., Koch, F., Gordon, C., Crowther, M.S., and Dickman, C.R. (2009). Keystone effects of an alien top-predator stem extinctions of native mammals. *Proceedings of the Royal Society B*, **276**, 3249–56.

Letnic, M., Ritchie, E.G., and Dickman, C.R. (2012). Top predators as biodiversity regulators: the dingo *Canis lupus dingo* as a case study. *Biological Reviews of the Cambridge Philosophical Society*, **87**, 390–413.

Lin, L., Faraco, J., Li, R., Kadotani, H., Rogers, W., Lin, X., Qiu, X., de Jong, P.J., Nishino, S., and Mignot, E. (1999). The sleep disorder canine narcolepsy is caused by a mutation in the hypocretin (orexin) receptor 2 gene. *Cell*, **98**, 365–76.

Lindblad-Toh, K., Wade, C.M., Mikkelsen, T.S., Karlsson, E.K., Jaffe, D.B., Kamal, M., Clamp, M., Chang, J.L., Kulbokas, E.J. III, Zody, M.C., et al. (2005). Genome sequence, comparative analysis and haplotype structure of the domestic dog. *Nature*, **438**, 803–19.

Li, Y., and Zhang, Y. (2012). High genetic diversity of Tibetan Mastiffs revealed by mtDNA sequences. *Chinese Science Bulletin*, **57**, 1483–7.

Lohi, H., Young, E.J., Fitzmaurice, S.N., Rusbridge, C., Chan, E.M., Vervoort, M., Turnbull, J., Zhao, X-C, Ianzano, L., Paterson, A.D., Sutter, N.B., Ostrander, E.A., André, C., Shelton, G.D., Ackerley, C.A., Scherer, S.W., and Minassian, B.A. (2005). Expanded repeat in canine epilepsy. *Science*, **307**, 81.

Macpherson, C.N.L. (2005). Human behaviour and the epidemiology of parasitic zoonoses. *International Journal for Parasitology*, **35**, 1319–31.

Makvandi-Nejad, S., Hoffman, G.E., Allen, J.J., Chu, E., Gu, E., Chandler, A.M., Loredo, A.I., Bellone, R.R., Mezey, J.G., Brooks, S.A., and Sutter, N.B. (2012) Four loci explain 83% of size variation in the horse. *PLoS ONE* 7(7): e39929.

Masi, E., Pino, F.A., Santos, M.G.S., Genehr, L., Albuquerque, J.O.M., Bancher, A.M., and Alves, J.C.M. (2010). Socioeconomic and environmental risk factors for urban rodent infestation in Sao Paulo, Brazil. *Journal of Pest Science*, **83**, 231–41.

Meischer, G. (2012). *Namibia's Red Line: The History of a Veterinary and Settlement Border*. Palgrave Macmillan, New York.

Meggitt, M. (1965). The association between Australian Aborigines and dingoes. In A. Leeds and A. Vayda, Eds., *Man, Culture, And Animals: The Role of Animals in Human Ecological Adjustments*. American Association for the Advancement of Science, New York, pp. 7–26.

Morey, D.F. (2010). *Dogs: Domestication and the Development of a Social Bond*. Cambridge University Press, Cambridge.

Morris, D. (2002). *Dogs: The Ultimate Dictionary of Over 1000 Dog Breeds*. Trafalgar Square Publishing, North Pomfret, Vermont, USA.

Moritz, C. (1994). Defining "evolutionary significant units" for conservation. *Trends in Ecology and Evolution*, **9**, 373–5.

Nimer, J. and Lundahl, B. (2007). Animal-Assisted Therapy: A Meta Analysis. *Anthrozoos*, **20**, 225–38.

Oldenbroek, J.K. (1999). Introduction. In J.K. Oldenbroek, Ed., *Genebanks and the Conservation of Farm Animal Genetic Resources*. Institute for Animal Science and Health, Lelystad, Netherlands, pp. 1–9.

Oliver, P.L., Goodstadt, L., Bayes, J.J., Birtle, Z., Roach, K.C., Phadnis, N., Beatson, S.A., Lunter, G., Malik, H.S., and Ponting, C.P. (2009). Accelerated evolution of the Prdm9 speciation gene across diverse metazoan taxa. *PLoS Genetics*, **5**, e1000753.

Oskarsson, M. (2012). Analysis of the origin and spread of the domestic dog using Y-chromosome, DNA, and mtDNA sequence data. PhD dissertation, KTH Royal Institute of Technology.

Oskarsson, M.C.R., Klütsch, C.F.C., Boonyaprakob, U., Wilton, A., Tanabe, Y., and Savolainen, P. (2012). Mitochondrial DNA data indicate an introduction through Mainland Southeast Asia for Australian dingoes and Polynesian domestic dogs. *Proceedings of the Royal Society B*, **279**, 967–74.

Ostrander, E.A. (2012). Both ends of the leash: the human links to good dogs with bad genes. *The New England Journal of Medicine*, **367**, 636–46.

Pang, J-F, Kluetsch, C., Zou, X-J, Zhang, A., Luo, L-Y, Angleby, H., Ardalan, A., Ekström, C., Sköllermo, A., Lundeberg, J., Matsumura, S., Leitner, T., Zhang, Y-P and Savolainen, P. (2009). mtDNA data indicate a single origin for dogs south of Yangtze River, less than 16300 years ago, from numerous wolves. *Molecular Biology and Evolution*, **26**, 2849–64.

Parker, H.G., Kim, L.V., Sutter, N.B., Carlson, S., Lorentzen, T.D., Malek, T.B., Johnson, G.S., DeFrance, H.B., Ostrander, E.A., and Kruglyak, L. (2004). Genetic structure of the purebred domestic dog. *Science*, **304**, 1160–4.

Parker, H.G., Kukekova, A.V., Akey, D.T., Goldstein, O., Kirkness, E.F., Baysac, K.C., Mosher, D.S., Aguirre, G.D., Acland, G.M., and Ostrander, E.A. (2007). Breed relationships facilitate fine-mapping studies: A 7.8-kb deletion cosegregates with Collie eye anomaly across multiple dog breeds. *Genome Research*, **17**, 1562–71.

Parker, H.G., vonHoldt, B.M., Quignon, P., Marguiles, E.H., Shao, S., Mosher, D.S., Spady, T.C., Elkahloun, A., Cargill, M., Jones, P.G., Maslen, C.L., Acland, G.M., Sutter, N.B., Kuroki, K., Bustamante, C.D., Wayne, R.K., and Ostrander, E.A. (2009). An expressed *Fgf4* retrogene is associated with breed-defining chondrodysplasia in domestic dogs. *Science*, **325**, 995–8.

Parker, H.G. (2012). The history and relationships of dog breeds. In E.A. Ostrander and A. Ruvinsky, Eds., *Genetics of the Dog*. 2nd edn. CABI, Wallingford, UK. pp. 38–56.

Pedersen, N., Lui, H., Theilen, G., and Sacks, B. (2013). The effects of dog breed development on genetic diversity and the relative influences of performance and conformation breeding. *Journal of Animal Breeding and Genetics* **130**, 236–248.

Perelle, I.B. and Granville, D.A. (1993). Assessment of the effectiveness of a pet facilitated therapy program in a nursing home setting. *Society and Animals*, **1**, 91–100.

Pires, A.E., Ouragh, L., Kalboussi, M., Matos, J., Petrucci-Fonseca, F., and Bruford, M.W. (2006). Mitochondrial DNA sequence variation in Portuguese native dog breeds: Diversity and phylogenetic affinities. *The Journal of Heredity*, **97**, 318–30.

Pires, A.E., Amorim, I.R., Ginja, C., Gomes, M., Godinho, I., Simões, F., Oom, M., Petrucci-Fonseca, F., Matos, J., and Bruford, M.W. (2009). Molecular structure in peripheral dog breeds: Portuguese native breeds as a case study. *Animal Genetics*, **40**, 383–92.

Puja, I.K., Irion, D.N., Schaffer, A.L., and Pedersen, N.C. (2005). The Kintamani dog: genetic profile of an emerging breed from Bali, Indonesia. *Journal of Heredity*, **96**, 854–9.

Rand, J.S., Fleeman, L.M., Farrow, H.A., Appleton, D.J., and Lederer, R. (2004). Canine and feline diabetes mellitus: nature or nurture? *Journal of Nutrition*, **134**, 2072S–80S.

Reist-Marti, S.B., Simianer, H., Gibson, J., Hanotte, O., and Rege, J.E.O. (2003). Weitzman's approach and conservation of breed diversity: an application to African cattle breeds. *Conservation Biology*, **17**, 1299–1311.

Runstadler, J.A., Angles, J.M., and Pedersen, N.C. (2006). Dog leucocyte antigen class II diversity and relationships among indigenous dogs of the island nations of Indonesia (Bali), Australia and New Guinea. *Tissue Antigens*, **68**, 418–26.

Saj, T.L., Sicotte, P., and Paterson, J.D. (2001). The conflict between vervet monkeys and farmers at the forest edge in Entebbe, Uganda. *African Journal of Ecology*, **39**, 195–9.

Savolainen, P. (2006). MtDNA studies of the origin of dogs. In E.A. Ostrander, U. Giger and K. Lindblah-Toh, Eds., *The Dog and its Genome*. Cold Spring Harbor Laboratory Press, New York, pp. 119–40.

Savolainen, P., Zhang, Y-P, Luo, J., Lundeberg, J., and Leitner, T. (2002). Genetic evidence for an East Asian origin of domestic dogs. *Science*, **298**, 1610–3.

Savolainen, P., Leitner, T., Wilton, A.N., Matisoo-Smith, E., and Lundeberg, J.A. (2004). A detailed picture of the origin of the Australian dingo, obtained from the study of mitochondrial DNA. *Proceedings of the National Academy of Sciences*, **101**, 12387–90.

Seppälä, E.H., Jokinen, T.S., Fukata, M., Fukata, Y., Webster, M.T., Karlsson, E.K., Kilpinen, S.K., Steffen, F., Dietschi, E., Leeb, T., Eklund, R., Zhao, X., Rilstone, J.J., Lindblad-Toh, K., Minassian, B.A., and Lohi, H. (2011). LGI2 truncation causes a remitting focal epilepsy in dogs. *PLoS Genetics*, **7**, e1002194.

Schwartz, M. M. (1998). *A History of Dogs in the Early Americas*. Yale University Press, New Haven.

Shiboleth, M. (2004). The Canaan dog: expanding the gene pool from the desert. *Israel Journal of Veterinary Medicine*, **59**, 1–3.

Siegel, J.M., Angulo, F.J., Detels, R., Wesch, J., and Mullen, A. (1999). AIDS diagnosis and depression in the multicenter AIDS cohort study: the ameliorating impact of pet ownership. *AIDS Care: Psychological and Socio-medical Aspects of AIDS/HIV*, **11**, 157–70.

Simianer, H., Marti, S.B., Gibson, J., Hanotte, O., and Rege, J.E.O. (2003). An approach to the optimal allocation of conservation funds to minimize loss of genetic diversity between livestock breeds. *Ecological Economics*, **45**, 377–92.

Smith, B.P., Appleby, R.G., and Litchfield, C.A. (2012). Spontaneous tool-use: An observation of a dingo (*Canis dingo*) using a table to access an out-of-reach food reward. *Behavioural Processes*, **89**, 219–24.

Stephens, D. (2011). The molecular ecology of Australian wild dogs: hybridization, gene flow and genetic structure at multiple geographic scales. PhD dissertation, University of Western Australia.

Tacher, S., Quignon, P., Rimbault, M., Dreano, S., Andre, C., and Galibert, F. (2005). Olfactory receptor sequence polymorphism within and between breeds of dogs. *Journal of Heredity*, **96**, 812–6.

Treves, A. and Karanth, K.U. (2003). Human-carnivore conflict and perspectives on carnivore management worldwide. *Conservation Biology*, **17**, 1491–99.

Vanak, A.T. and Gompper, M.E. (2009). Dietery niche separation between sympatric free-ranging domestic dogs and Indian foxes in Central India. *Journal of Mammology*, **90**, 1058–65.

Verardi, A., Lucchini, V., and Randi, E. (2006). Detecting introgressive hybridization between free-ranging domestic dogs and wild wolves (*Canis lupus*) by admixture linkage disequilibrium analysis. *Molecular Ecology*, **15**, 2845–55.

Vilà, C., Savolainen, P., Maldonado, J.E., Amorim, I.R., Rice, J.E., Honeycutt, R.L., Crandall, K.A., Lundeberg, J., and Wayne, R.K. (1997). Multiple and ancient origins of the domestic dog. *Science*, **276**, 1687–9.

Vilà, C., Maldonado, J.E., and Wayne, R.K. (1999). Phylogenetic relationships, evolution, and genetic diversity of the domestic dog. *Journal of Heredity*, **90**, 71–7.

Vilà, C., Walker, C., Sundqvist, A-K, Flagstad, Ø., Andersone, Z., Casulli, A., Kojola, I., Valdmann, H., Halverson, J., and Ellegren, H. (2003). Combined use of maternal, paternal and bi-parental genetic markers for the identification of wolf-dog hybrids. *Heredity*, **90**, 17–24.

vonHoldt, B.M., Pollinger, J.P., Lohmueller, K.E., Han, E., Parker, H.G., Quignon, P., Degenhardt, J.D., Boyko, A.R., Earl, D.A., Auton, A., Reynolds, A., Bryc, K., Brisbin, A., Knowles, J.C., Mosher, D.S., Spady, T.C., Elkahloun, A., Geffen, E., Pilot, M., Jedrzejewski, W., Greco, C., Randi, E., Bannasch, D., Wilton, A., Shearman, J., Musiani, M., Cargill, M., Jones, P.G., Qian, Z., Huang, W., Ding, Z., Zhang, Y., Bustamante, C.D., Ostrander, E.A., Novembre, J., and Wayne, R.K. (2010). Genome-wide SNP and haplotype analyses reveal a rich history underlying dog domestication. *Nature*, **464**, 898–902.

vonHoldt, B.M., Pollinger, J.P., Earl, D.A., Knowles, J.C., Boyko, A.R., Parker, H., Geffen, E., Pilot, M., Jedrzejewski, W., Jedrzejewski, B., Sidorovich, V., Greco, C., Randi, E., Musiani, M., Kays, R., Bustamante, C.D., Ostrander, E.A., Novembre, J., and Wayne, R.K. (2011). A genome-wide perspective on the evolutionary history of enigmatic wolf-like canids. *Genome Research*, **21**, 1294–1305.

Wayne, R.K. (1993). Molecular evolution of the dog family. *Trends in Genetics*, **9**, 218–24.

Wayne, R.K. and Ostrander, E.A. (1999). Origin, genetic diversity, and genome structure of the domestic dog. *BioEssays*, **21**, 247–57.

Wayne, R.K. and Ostrander, E.A. (2007). Lessons learned from the dog genome. *Trends in Genetics*, **23**, 557–67.

Wayne, R.K., Lehman, N., Allard, M.W., and Honeycutt, R.L. (1992). Mitochondrial DNA variability of the gray wolf: Genetic consequences of population decline and habitat fragmentation. *Conservation Biology*, **6**, 559–69.

Woodroffe, R., Frank, L.G., Lindsey, P.A., ole Ranah, S.M.K., and Romañach, S. (2007). Livestock husbandry as a tool for carnivore conservation in Africa's community rangelands: a case-control study. *Vertebrate Conservation and Biodiversity*, **16**, 1245–60.

Young, S. (2011). Rebuilding the genome of a hidden ethnicity. *Nature*, doi:10.1038/news.2011.592.

CHAPTER 9

Dogs as mediators of conservation conflicts

Kurt VerCauteren, Michael Lavelle, Thomas M. Gehring, Jean-Marc Landry, and Laurie Marker

9.1 Introduction

There are many positive aspects to human–wildlife interactions, such as wildlife viewing and hunting. However, human safety and economic well-being can be adversely impacted by wildlife, for example, by aircraft and vehicle collisions with wildlife, depredation of livestock by predators, and wildlife-borne pathogens that can infect humans and livestock. These and other conflicts are of global importance and are increasing in magnitude. When such conflicts occur, wildlife, humans, and other resources—primarily livestock—can suffer. Historically, humans have been quick to resolve conflicts with wildlife, especially wild carnivores, using lethal means. Advancing technologies associated with firearms and poisons, coupled with establishment of bounties, often resulted in highly efficient carnivore removal efforts. Thus, in many developed regions of the world human populations increased while populations of conflict-associated species, such as carnivores, decreased, often to the point of extirpation. Increasing human populations also put severe pressure on populations of other wildlife species—directly via harvest for food and other resources, and indirectly via competition from land development and introduced domestic livestock on habitat formerly used by wildlife alone. Today, however, many societies have come to value wildlife more highly, necessitating development of management strategies that preserve human health and allow human commercial interests to succeed in the presence of broad and thriving communities of wildlife species.

As a result of recently increased public interest in wildlife, and large carnivores in particular, dwindling predator populations have gained protection and extirpated populations have been re-established. For example, gray wolf (*Canis lupus*) populations in North America and Europe have successfully recovered following generations of persecution by humans and this has resulted in an increase in livestock depredations (Mech et al., 1995). Conflicts revolving around shared pathogens of wildlife and livestock are also increasing with particular concern over wildlife diseases, such as bovine tuberculosis and brucellosis that have moved from livestock into wild ungulates, which may in turn serve as reservoirs of diseases and continually transmit infection back to livestock (Frölich et al., 2006; Schmitt et al., 2006; Walter et al., 2012). These challenges as well as others have the potential to be addressed through the strategic development and deployment of livestock protection dogs (LPDs). Worldwide, the recent recovery of large carnivore populations and their recolonization of human-dominated landscapes has prompted resurgence in the use of LPDs. This resurgence began in North America and followed in particular regions of Europe where large carnivores had been extirpated. During the 1970s, LPD use in North America was prompted at least partly by the banning of predator toxicants (Linhart et al., 1979). Use of dogs has

Free-Ranging Dogs and Wildlife Conservation. Edited by Matthew E. Gompper
© Oxford University Press 2014. Published 2014 by Oxford University Press.

also expanded for a variety of other conservation-specific practices to address increasing wildlife–human conflict challenges.

This chapter is a review of past and present use of dogs for mediating wildlife–human conflict. An underlying assumption herein is that current and future use of dogs in wildlife management benefits wildlife conservation if social pressure to implement lethal control of wildlife is reduced or eliminated. Historically, such pressure most frequently resulted from depredation to crops and livestock. This chapter also highlights other past and present uses of dogs and future areas of research that are needed to more effectively and extensively use dogs to address conservation conflicts. There are significant unexplored avenues that deserve attention in terms of the use of dogs for resolving conservation conflicts. Humans are ingenious and dogs are malleable; as such, we are still determining how broadly and specifically dogs can be employed to play major roles in resolving conflicts between humans and wildlife.

9.2 History of use of dogs in conservation and management

Livestock protection dogs, as pastoral protectors of livestock, have long been used in the context of wildlife conservation and management. Protecting livestock is one of the oldest anthropogenic functions of dogs (Coppinger and Coppinger, 2001; Landry and Valensi, 2011). The use of dogs in agriculture appears to have originated concurrently with the domestication of sheep and goats in western Asia as early as 9,000 to 10,000 ybp (Gehring et al., 2010a; Landry, 1999). Archeological sites dating to 5,585 ybp provide physical evidence of dogs and sheep together (Olsen, 1985). Livestock protection dogs were historically developed to protect small stock from predators and their use was common around the world (Coppinger and Coppinger, 2001; de la Cruz, 1995) though their current use worldwide is less common (Landry, 2010).

The use of LPDs has continued uninterrupted for hundreds of years in areas where predators have persisted over millennia, such as the Iberian Peninsula, Italy, the Balkans, the Carpathian Mountains, Turkey, the Caucasian Mountains, Russia, Central Asia, the Himalaya Mountains, and the Atlas Mountains of North Africa (Landry, 1999; Rigg, 2001). Conversely, regional predator extirpations resulted in many societies ceasing to employ LPDs as they were no longer essential. As a result, many local LPDs were maintained through the creation of breeds (e.g., Great Pyrenees) and persist only as pets. Thus the knowledge of how to raise and train them to protect livestock has not been passed on to the current generation of livestock producers in many regions of the world.

The versatility and adaptability of LPDs has contributed to the resurgence in their use where wolves are recovering (e.g., the Alps, the Jura Mountains, the Iberian Peninsula, eastern Germany, Finland) or have been reintroduced (Yellowstone) and where bear (*Ursus* spp.) populations have re-established (e.g., the Pyrenean Mountains). The use of LPDs in North America began more recently as there were no livestock in the New World until European introductions. Among the first employments in North America was the involvement of mixed-breed dogs that were raised and kept with sheep by Navajo people who learned and adopted the processes from Spanish missionaries (Black and Green, 1985; Coppinger and Coppinger, 2001).

Protecting livestock through the implementation of LPDs is also relatively new in Nordic countries, and their recent use and results are being documented across Finland (Otstavel et al., 2009), Norway (Hansen and Smith, 1999; Hansen, 2005), and Sweden (Levin, 2005). Similar exploration into the use of LPDs is occurring in response to growing predator populations and increases in livestock predation, combined with the demand for non-lethal solutions in Australia (van Bommel and Johnson, 2012), Poland (Nowak and Myslajek, 2004; Śmietana, 2005), Slovakia (Rigg, 2005), Switzerland (Landry et al., 2005), and Portugal (Ribeiro and Petrucci-Fonseca, 2005). Predator movements into new areas of Spain, such as Avila, are being monitored by researchers and they are also working closely with livestock producers using LPDs to evaluate outcomes.

The resurgence of LPD use has been facilitated by the creation of governmental and private organizations (Table 9.1). These organizations, as well as

Table 9.1 Organizations from around the world that are promoting and evaluating the use of livestock protection dogs to alleviate damage to resources by wildlife.

Organization	Country	Website
Cheetah Conservation Fund	Namibia	<http://www.cheetah.org/>
Invasive Animals Cooperative Research Centre	Australia	<http://www.invasiveanimals.com>
United States Department of Agriculture, Animal and Plant Health Inspection Service, Wildlife Services	USA	<http://www.aphis.usda.gov/wildlife_damage/nwrc/>
Livestock Guarding Dog Project at Hampshire College	USA	
Wind River Bear Institute	USA	<http://www.beardogs.org/programs/wrkbd.html>
Association for Nature WOLF	Poland	<http://www.polishwolf.org.pl>
Association Chiens de protection des troupeaux Suisse	Switzerland	<http://www.cpt-ch.ch/fr/association-cpt-ch/>
Pôle Grands Prédateurs (Jura Mountains)	France	<http://www.polegrandspredateurs.org>
La Pastorale Pyrénéenne (Pyrenees)	France	<http://www.pastoralepyreneenne.fr>
Grupo Lobo	Portugal	<http://lobo.fc.ul.pt/>
The Slovak Wildlife Society	Slovakia	<http://www.slovakwildlife.org>
BBPS Semperviva	Bulgaria	<http://www.save-foundation.net/semperviva/dog.htm>
Arcturos	Greece	<http://www.arcturos.gr>

others from around the world, are conducting research and promoting the use of LPDs to protect livestock and ensure sustainable agricultural practices continue while also easing conflict associated with predators. For example, multiple programs in the USA have conducted research into the use of LPDs and have facilitated deployment of LPDs to alleviate predation issues (Coppinger and Coppinger, 1978; Green and Woodruff, 1999; Lorenz, 1985; Sims and Dawydiak, 1990). Further, novel uses for LPDs are being explored and evaluated worldwide and are discussed in this chapter.

9.3 Reducing predation in agricultural systems

Efficacious tools that agricultural producers can adapt into their normal husbandry practices are needed to reduce economic losses associated with damage due to wildlife. Lethal control, as a management tool, can be effective (Conover, 2002). However, livestock depredations commonly recur annually after individual predators are removed lethally following depredation (Fritts et al., 1992; Gehring and Potter, 2005) and lethal control does not appear to reduce depredations on a regional scale (Musiani et al., 2005). Alternatively, non-lethal management tools are regarded by society as more humane than lethal means and deserve evaluation to determine efficacy (Reiter et al., 1999; Reynolds and Tapper, 1996). Numerous non-lethal management options exist; however, few provide reliable or long-term protection (Shivik, 2004). Livestock protection dogs may be the best preventive method for addressing predation on livestock because they act as a disruptive-stimulus tool (Gehring et al., 2010a) that remains with the mobile flock, as opposed to other stationary tools (e.g., audio and visual repellents). The use of LPDs is, however, context dependent. For instance, LPDs are generally regarded as effective in reducing livestock depredation caused by coyotes (*C. latrans*; Andelt and Hopper, 2000; Andelt, 1992; Green et al., 1984; Smith et al., 2000), but their effectiveness against wolves is sometimes more tenuous (Gehring et al., 2010a).

Overall, effectiveness of LPDs against predators has been the research objective of a few studies, and more recently their efficacy for repelling wild ungulates has begun to be assessed. Among predator-focused studies, most have relied on producer-based reporting and surveys rather than field experimentation (Gehring et al., 2010a). A study evaluating LPD efficacy with wolves suggested that LPDs displayed protective behavior against free-ranging wolves and defended experimenter-created bait stations (Coppinger et al., 1987, 1988). Linhart et al. (1979) demonstrated that LPDs reduced sheep depredation by coyotes on three ranches over a 20-day period, and coyotes appeared to be displaced from ranches for an additional 20 days after LPDs were removed. Gehring et al. (2010b) documented almost no use of LPD-guarded livestock pastures by wolves and coyotes, with visitation indices declining to zero. This suggested that LPDs can be effective for reducing the risk of livestock depredations by these predators on pastures associated with small- and medium-sized cattle farms.

A significant reduction in predation often results following the introduction of LPDs into a flock or herd. For example, reductions in wolf-caused mortality of up to 75% have been documented in protected sheep herds in Portugal (Ribeiro and Petrucci-Fonseca, 2005). Based on a recent survey of livestock producers that employ LPDs, 90% of survey respondents reported reductions in predation with an average decrease of 64% in predation rates observed associated with the use of LPDs in western USA (M. Marlow, USDA APHIS WS, pers. comm.). The presence of LPDs with a flock does not always prevent wolves from attacking, but can reduce the number of livestock killed per attack and can inhibit surplus killing behavior (Rigg, 2005), especially when several LPDs are present within the flock and working together to protect it. When using two or more LPDs within a flock, dogs appear more confident and efficient than a solo dog in protecting and confronting a threat. The use of LPDs has also proven effective against attacks by other predators including lynx (*Lynx lynx*), wolverines (*Gulo gulo*), bears, black-backed jackals (*C. mesomelas*), golden jackals (*C. aureus*), leopards (*Panthera pardus*), cheetahs (*Acinonyx jubatus*), lions (*Panthera leo*), Chacma baboons (*Papio ursinus*), and even free-ranging dogs (Hansen, 2005; Landry and Raydelet, 2010; Marker et al., 2005; Rigg, 2005).

In general, LPD pups should be raised with the species of livestock they are to protect. Though not ideal, it is possible to raise pups with sheep and then to introduce them into a flock of goats or cattle, even as adults (Landry, 2011; Marker et al., 2005). Oftentimes, integrating a pup into a herd can be facilitated by introducing it into a herd that already contains at least one established adult LPD. The number of LPDs to employ should be based on likely adversaries and characteristics of the surrounding environment (Landry and Raydelet, 2010). Individual dogs may demonstrate particular behaviors, thus selecting individuals that complement one another based on their strengths can be advantageous. For specific information on selecting, raising, and implementing LPDs for protecting livestock see Lorenz and Coppinger (1986); Sims and Dawydiak (1990), and Dohner (2007).

9.4 Livestock protection dog breed selection

Dogs used for protecting resources fall into distinctive groups that are typically identified by organizations such as the American Kennel Club and the Fédération Cynologique Internationale; oftentimes including working dogs and herding dogs. Working dogs include, but are not limited to, LPD breeds as well as breeds thought of as 'sled dogs.' There are over 40 breeds of LPDs throughout the world today, the most common being Akbash, Anatolian Shepherd, Great Pyrenees, Komondor, Maremma, and Kangal. Beyond purebred breeds, there are also unlimited and unrecognized mixed-breeds and mongrels, sometimes called 'landrace' dogs, which are tasked with protecting livestock, often with positive results. Other groupings that may include breeds mentioned herein include utility or pastoral groups. Commonalities of these groups are the inclusion of large, strong, and intelligent dogs that are engaged in physically active work. Working dog breeds are typically motivated to respond to stimuli (e.g., approaching wolves) by protective instincts. Although LPDs are more adept at protecting a flock, they demonstrate proficiency in protecting property as well, though to a lesser degree. Conversely, herding-dog breeds are instinctually motivated to move (i.e., herd) livestock by approaching and pushing them in a variety of styles.

Breeds commonly used within a region are frequently those that were developed there. For example, in western Turkey, LPDs commonly used are of the Akbash breed while in eastern Turkey one finds more Kangals. Anatolian Shepherds are also from Turkey and may be indistinguishable from Kangals, though they are recognized as a separate breed. Both have been exported across the globe and are gaining popularity in places such as Africa and the USA. Anatolian Shepherds and Kangals were selected for use in Africa, as they are independent in their thinking, can move long distances with their herds each day, and can be left unattended.

There are apparent physical differences between breeds. However, differences among individuals within a breed are often more notable than those between breeds. Rearing and bonding processes used with pups apparently have a greater effect on the development of an effective dog than the breed itself. These differences emphasize the importance of selecting individuals from proven breeders. General differences among breeds described by varying combinations of attributes, including attentiveness, trustworthiness, and aggressiveness, further facilitate breed selection based upon needs in a particular situation. Additionally, breed-specific physical attributes, such as dominance display, agility, and strength are important considerations as well. For example, Karakachan dogs of Bulgaria are known for their aggression and determination in pursuing predators, which may be necessary in situations where the possibility of predation by large and persistent predators is high.

Herding breeds such as Border Collies, Australian Shepherds, and heelers were evaluated for excluding ungulates such as deer from crops and did not perform well (VerCauteren et al., 2005). Characterization of herding breed styles is frequently attempted but is discouraged due to the high level of variation between individuals within breeds. Characteristics such as strong-eyed dogs versus loose-eyed dogs oftentimes can be the result of upbringing or even the basics of a particular task at hand. Border Collies have been a popular breed for wildlife-hazing jobs and were used successfully to haze elk (*Cervus canadensis*) in Canada and white-tailed deer (*Odocoileus virginianus*) in Missouri, USA, (Beringer et al., 1994; Kloppers et al., 2005). Additionally, Border Collies diminished hazards through hazing birds from airports and communities (Ball, 2000; Holevinski et al., 2007, respectively).

Mixed or crossbreeds have also been utilized and have excelled in protecting stationary resources such as orchards and organic farms in the USA, suggesting that choosing a dog with the ability to withstand weather extremes is possibly more important than selecting a specific breed for some jobs (Curtis and Rieckenberg, 2005; VerCauteren et al., 2005). The use of mixed breeds in livestock protection is relatively common, sometimes with the intention of benefiting from desired characteristics possessed by contributing breeds. However, it has been claimed that crossing breeds can actually disrupt these desired heritable traits, potentially creating unintended results (Dohner, 2007). The dogs used as LPDs to protect sheep and goats within the Navajo Reservation in the USA, and in Turkey and Uruguay, may be of no single identifiable breed but the result of crossing several breeds (Black and Green, 1985).

Less common breeds, such as the Karelian Bear Dog breed that was developed for hunting aggressive game, have become valuable in specialized roles. Karelian Bear Dogs have been used and proven effective in hazing bears, thus mitigating bear–human conflict in several locales in North America (see Section 9.5.3). This is one particular role in which very specific breed characteristics are important and selection of the 'wrong' breed may have dire consequences. Breed selection based on traits commonly exhibited by that breed is a good starting point in the selection process. Refinement of a particular bloodline by breeders and trainers plays an equally important role in development that must also be taken into consideration. Beyond these considerations in selecting a dog for a particular purpose, and possibly the most important factor, is the training provided to the individual dog from an early age which will essentially determine the potential for success of that individual in its desired role.

9.5 Non-traditional uses in other conservation conflicts

The demonstrated adaptability of dogs has prompted the use of various breeds of dogs in research and management contexts for reducing a variety of conservation conflicts. Researchers have demonstrated the abilities of dogs to protect various resources of value to humans from wildlife species other than predators, primarily white-tailed deer and Canada geese (*Branta canadensis*). Dogs have been used against deer to protect cattle (from the threat of bovine tuberculosis; VerCauteren et al., 2008; Box 9.1), forest plantations (Beringer et al., 1994), orchards (Curtis and Rieckenberg, 2005), and vegetable farms (VerCauteren et al., 2005). They have been shown to be effective in deterring geese from golf courses (Woodruff and Green, 1995) and landscapes around office complexes (Castelli and Sleggs, 2000). Further, use of LPDs in protecting non-traditional animals such as penguins in Australia is being explored (Lustig, 2011). Accounts of chickens, geese, and even pigs benefiting from farmyard LPDs are commonly heard.

9.5.1 LPDs for reducing transmission of wildlife-borne diseases

Along with endemic infectious disease-causing pathogens, a growing number of emerging diseases continue to manifest and become established in wildlife. Wildlife serves as hosts, reservoirs, and vectors, transmiting pathogens to livestock and humans. The disease threats to livestock posed by wildlife are analogous to predation and in many cases infectious agents have larger and further reaching implications. Wolves, for example, may greatly impact the profit of local livestock producers (Gehring et al., 2006), while bovine tuberculosis (TB) established in free-ranging deer that repeatedly transmit it to cattle could lead to regional, state, national, and international movement and trade restrictions that impact the economic viability of not only individual producers but entire industries. Brucellosis, foot and mouth disease, *Escherichia coli*, and keratoconjunctivitis are other wildlife-mediated parasites and diseases of livestock that LPDs could help curtail. Pathogens, of course, may also be transmited from livestock to wildlife

Box 9.1 TB dogs: novel disease management strategy in Michigan, USA

Disease transmission between wildlife and livestock is a worldwide issue. Culling potential reservoirs is a common strategy for mitigating potential spread. However, there is a need for additional tools to address disease issues. VerCauteren et al. (2008) theorized that LPDs raised and bonded with cattle could be employed to reduce the risk of bovine tuberculosis (*Myobacterium bovis*; TB) transmission between white-tailed deer and cattle by minimizing contact between the two species as well by reducing the use of cattle feed by deer (Figure 9.1). Due to the numerous modes of

Figure 9.1 In Michigan, USA, research is evaluating the use of livestock protection dogs to minimize potential for transmission of bovine tuberculosis from white-tailed deer (*Odocoileus virginianus*) to cattle. Photo credit: Kurt VerCauteren.

continued

Box 9.1 *Continued*

transmiting causative agents of disease (i.e., direct via contact with infectious hosts, indirect via contaminated fomites, etc.), mitigation tools must be adaptable and versatile. In this particular situation, there was both risk via potential physical contact and close proximity, as well as a more likely potential for transmission through contaminated feed in concentrated form (i.e., hay bales) and dispersed feed (i.e., available forage in pastures).

Researchers evaluated four Great Pyrenees LPDs and found that they were highly effective in preventing deer from using cattle feed (likely the greatest risk factor of TB transmission on farms). Dogs also prevented deer from approaching cattle in core areas of pastures and were very effective throughout pastures. Direct observations documented 79 events in which deer approached to within 5 m of cattle in pastures not protected by LPDs compared to only 3 events in LPD-protected pastures. Further, researchers observed 113 events during which deer consumed concentrated feed in unprotected pastures and no events in LPD-protected pastures.

In this situation, LPD pups were bonded with calves beginning at 8 weeks of age and demonstrated their versatility and potential in the novel role of protecting cattle from non-traditional threats. Researchers concluded that LPDs may be a practical tool to minimize potential for livestock to contract TB from infected deer in small-scale cattle operations. Operationally, producers that rotate pastures frequently have developed strategies to make it easy to provide resources for LPDs that are inaccessible to livestock (Figure 9.2).

Figure 9.2 Mobile livestock protection dog station used to minimize time away from livestock by providing a cattle-free loafing and feeding area for a working dog. Photo credit: National Wildlife Research Center.

and thus threaten conservation efforts (Ward et al., 2009). Die-offs of wild bighorn sheep (*Ovis canadensis*), for example, have been caused by pneumonia that they may have contracted through contact with domestic sheep (e.g., Clifford et al., 2009; Schommer and Woolever, 2008). The presence of LPDs, though, may serve to deter wildlife from coming in contact with livestock and thus reduce the risk of disease transmission to wildlife.

Wildlife carrying diseases that threaten livestock are often not species we traditionally consider employing LPDs to manage. Furthermore the motivation of these species is not uniform. Wild ungulates can be attracted to areas where livestock are present due to the availability of food resources, which can be natural vegetation, standing agricultural forage, stored agricultural feedstuffs, and mineral supplements such as salt. Wild ungulates may also be attracted to livestock with the intent of procreation. Bighorn sheeprams, for example, may be attracted to domestic ewes in heat (Singer et al., 2001) as feral boar swine (*Sus scrofa*) are to domestic sows (Wyckoff et al., 2012).

Though regulated recreational hunting is the primary means of managing populations of wild ungulates, non-lethal means to aid in reducing disease threats from wildlife are needed and an integrated approach that employs a combination of multiple strategies will enhance the likelihood of success. Traditional non-lethal strategies for deterring wild ungulates from livestock and crops include various applications of fencing and frightening devices (see VerCauteren et al., 2006 and Gilsdorf et al., 2002 for reviews). Most fencing strategies could be considered physical deterrents while most frightening devices function as psychological deterrents (VerCauteren et al., 2006). Dogs have the potential to function first as psychological deterrents that can turn into physical deterrents when called for by especially persistent individuals or groups of wildlife. Dogs belong in a category of deterrent we define as 'biological control.' Non-lethal biological control strategies for protecting resources from wildlife are generally considered more 'green' and acceptable to various publics than alternative methods.

Livestock protection dogs offer many advantages over traditional physical and psychological strategies. Unlike fences or frightening devices, LPDs are mobile, moving with the herds or flocks they are protecting on open range or within fenced pastures. They are dynamic, adjusting real-time to varying settings, situations, and threats. Livestock protection dogs also provide 24-hour protection 7 days per week, as they are with livestock and vigilant at all times, poised and ready to position themselves to repel intruding wildlife. The presence of LPDs alone is often psychological deterrent enough to prevent wild ungulates from approaching areas, and thus livestock, where LPDs are present, since they represent and purvey aggression and predation risk. We have observed, and have numerous anecdotes from livestock producers who employ LPDs, that wild ungulates quickly learn to completely avoid areas inhabited by LPDs (e.g., Gehring et al., 2010b; VerCauteren et al., 2008). Once the routine of local wildlife using an area is broken, they and subsequent generations are less apt to attempt to enter the area, thus making the job of the LPD easier.

Although LPD breeds have been developed and employed by humans for centuries, their use to deter wild ungulates is in its infancy. We predict that their potential efficacy and ability to deter wildlife without requiring lethal actions will lead to their widespread adaptation and employment in a variety of situations. Livestock protection dogs may either be motivated to pursue wildlife in a predatory manner or chase them off as a protective response. Regardless, the result of excluding the animal being targeted is still the same. Hansen and Smith (1999) reported that LPDs chased and repelled moose (*Alces alces*) and roe deer (*Capreolus capreolus*) in Norway. Similarly, LPDs repeatedly proved effective in deterring white-tailed deer (Beringer et al., 1994; Coppinger et al., 1988; Curtis and Rickenburg, 2005). VerCauteren et al., (2008) demonstrated that LPDs reduced transmission potential of bovine tuberculosis from free-ranging deer to cattle in a controlled experimental setting (see Box 9.1). Livestock protection dogs prevented deer from coming in direct contact with cattle, from coming in contact with and potentially contaminating hay and grain meant for cattle, and greatly reduced their use of pastures in general. Lending credence to the concept, Gingold et al. (2009) reported that mountain gazelles (*Gazella gazelle*) in large enclosures that contained cattle herds with LPDs avoided cattle and were more vigilant and active than gazelles in enclosures that contained cattle herds without LPDs. In the Alps, LPDs are also sometimes used to prevent red deer (*Cervus elaphus*) from grazing new spring grass in pastures or destroying sheep enclosures.

As LPDs are used more extensively for the purpose of reducing likelihood of disease transmission between wildlife and livestock, potential exists for development of new lines or 'breeds' of LPDs. Dog breeders and members of livestock industries employing dogs for keeping wild ungulates from contacting livestock or feed meant for livestock will undoubtedly pair their best performing dogs, over time developing genetic lines excelling at and specifically suited to this purpose. We have already observed this occurring in the Great Lakes region of the USA, where Great Pyrenees have been employed for about 10 years to reduce contact between deer potentially infected with TB and cattle. The LPDs in this area are being called 'TB Dogs.'

Other potential applications include employing LPDs in north-western USA, where *Brucella* is established in free-ranging elk and bison (*Bison bison*) and these species have the potential to transmit it to cattle. In this setting, cattle are often on open range or in very large pastures. The presence of LPDs with cattle has the potential to keep elk and bison from using areas occupied by cattle, thus reducing the risk of disease transmission while allowing them access to graze in other areas. Similarly, in the Alps, LPDs may also be used to prevent contact between domestic sheep and chamois (*Rupicapra rupicapra*), ibex (*Capra ibex*), or red deer, thus reducing potential for transmission of keratoconjunctivitis.

Additionally, LPDs could benefit conservation by decreasing disease transmission risk in the other direction, from livestock to wildlife. Domestic sheep, as mentioned above, may pose health risks to wild bighorn sheep in western USA or chamois and ibex in the Alps, through transmission of viral and bacterial diseases. Anecdotal evidence suggests LPDs can function to deter bighorn sheep from approaching and coming in contact with domestic sheep (C. Urbigkit, pers. comm.) and researchers are initiating efforts to rigorously evaluate their efficacy.

Although disease transmission by wild ungulates has been the focus to this point; dogs could also serve to reduce disease transmission potential mediated by other species of wildlife. Raccoons (*Procyon lotor*), Virginia opossums (*Didelphis virginiana*), and Eurasian badgers (*Meles meles*), for example, may play a role in TB persistence (Atwood et al., 2009; Böhm et al., 2009; Gallagher and Clifton-Hadley, 2000) and dogs may serve both as non-lethal deterrents and lethal measures to keep them from coming in direct contact with livestock or contaminating feed destined for consumption by livestock.

9.5.2 Border Collies for reducing avian conflicts

In some cases, conflict between humans and large bird species is a major wildlife management concern. The presence of large populations of gull species near airports could lead to an increase in aircraft–bird strikes. Additionally, there are aesthetic and human health concerns with the congregation of large populations of bird species at localized sites. For example, the ring-billed gull (*Larus delawarensis*) population in the Great Lakes region of the USA has shown a dramatic, exponential increase in numbers, estimated at 10% per year since the early 1970s (Solman, 1994). After being nearly extirpated by the early 1900s, this population rebounded to 27,000 nesting pairs in Lake Michigan and Lake Huron by 1960 and by the mid 1980s had reached over 700,000 nesting pairs (Greenlaw and Sheehan, 2003). Conflict arises with ring-billed gulls at public beaches due to complaints that large numbers of gulls are not aesthetically pleasing. Further, gull feces at beaches may be an important source of *E. coli* and other pathogens that could jeopardize human health and safety (Converse et al., 2012).

Since 2008, Border Collies have been used for displacing gulls in some areas near the Great Lakes of the USA and Canada (Hartman et al. 2009; Hiller, 2009; Toronto Beaches Plan, 2009). Though not yet optimized or rigorously evaluated, the concept shows promise for lessening conflict associated with beach recreation and public safety concerns. Koski and Kinzelman (2010) indicated that Border Collies appeared to be effective at displacing gulls from beaches but the use of dogs was costly and required trained dog handlers to constantly supervise dogs. Converse et al. (2012) found that average daily gull counts were reduced from 665 to 17 gulls when Border Collies were deployed on one study beach. They also found that *Enterococcus* spp. and *E. coli* densities were reduced when gulls were excluded by dogs. Although the Converse et al. (2012) study included only 1 beach site where dogs were used for only 16 days to exclude gulls, it does highlight dogs as a possibly important management tool to reduce human–gull conflict at public beaches. E. Alm and T. Gehring (unpublished data) currently are conducting more rigorous evaluations with experimental design evaluating the efficacy of Border Collies on public beaches to reduce gull use and microbial contamination and have thus far demonstrated reduced gull use and lower *E. coli* densities at Border Collie-protected beaches.

Woodruff and Green (1995) reported anecdotal accounts of the successful use of Border Collies for dispersing Canada geese from golf courses in New York, Oregon, and Idaho, USA, and from alfalfa fields in Oregon, USA, when alternative methods were ineffective or impractical. Border Collies, rather than traditional LPDs, are often used for dispersing bird species due to their innate pursuit behavior of groups of targeted animals. Castelli and Sleggs (2000) conducted an *a posteriori* examination of the efficacy of Border Collies for reducing goose conflicts on a corporate office complex in New Jersey, USA, that included manicured lawns and a pond. Interviews with property managers indicated that goose feces, general annoyance, and helicopter safety were the primary conflicts with geese. Dogs were contained by an invisible electric fencing system and were allowed to chase geese 24 hrs per day every day, with minor exceptions for special events. The original two Border Collies purchased were not from working stock, exhibited little herding instinct, and were not effective at hazing geese. Subsequently, another pair of Border Collies (from working stock) replaced the original pair. Thereafter, property managers observed an abrupt decline in on-site goose numbers and within three years geese were rarely observed on the property. Property owners believed this success justified the costs of the program. A caveat is that data from aerial and ground counts appear somewhat contradictory to owners' observations and authors' results and discussion. Aerial counts were in sharp decline for several years before dogs were introduced and ground counts the year prior to dog introduction had already dropped to an extremely low level, additionally both aerial and ground counts remained near zero after dogs were implemented. Although strong evidence of efficacy in the use of Border Collies for reducing conflict with geese was lacking in the Castelli and Sleggs (2000) review, it did highlight the possible future application of dogs for addressing this problem and the need for more rigorous study.

Holevinski et al. (2007) found that Border Collies deployed by handlers during daylight hours were successful at displacing geese (i.e., >90% of geese present displaced) during 94% of hazing events (n = 113) at three locations (hazing sites) in New York, USA. In this setting the geese had access to aquatic habitats that dogs couldn't efficiently access. Border Collies were present on individual hazing sites only a small proportion of the time, however, and geese readily returned to areas when dogs were absent. The use of Border Collies coincident with small, remote-controlled boats increased the success rate of displacing geese during 97% of hazing events (n = 37). Swift (2000) reported >67% reduction in numbers of geese encountered after four weeks of patrolling with Border Collies several times per day at each of two locations in another region of New York. Researchers recorded eventual reductions of 80–100% at established molting and feeding areas as long as regular patrols were maintained. Preusser et al. (2008) evaluated Border Collies, remote-controlled boats, and Border Collies in conjunction with remote-controlled boats; they found the dog–boat combination to be most effective.

Results of these studies suggest that frequent and persistent hazing by Border Collies can displace geese from particular locations. Hazing alone will not reduce goose populations, so problems across larger landscapes may not be solved as birds may simply shift their use to adjacent areas. Additional management actions such as egg addling and goose roundup and removals may be required to reduce area-wide goose numbers (Preusser et al., 2008; Swift, 2000). Preusser et al. (2008) suggested that such integrated approaches could reduce human–goose conflicts across large landscapes but extensive coordination of local projects, inclusion of public involvement processes, and intensive long-term commitment of resources would be required.

Airports are a particular area of emphasis for use of Border Collies for hazing geese and other birds that may pose risk of bird–aircraft collision. Civil and military airports have implemented handler–dog teams to augment existing bird-strike management operations. Although reports of initial results at some airports suggest promise for reducing avian abundance and numbers of bird–aircraft collisions (e.g., Carter, 2000; O'Rick, 2000; Patterson, 2000; Froneman and van Rooyen, 2003), to date there have been no rigorous, long-term evaluations of the efficacy of using dogs as bird deterrents at airports.

9.5.3 Karelian Bear Dogs for reducing conflicts with bears

The Karelian Bear Dog breed was developed from hunting stock existing in the Karelian region of north-eastern Europe (astride the current Finnish–Russian border) during the 1930s and 1940s. These were versatile dogs used for hunting predators, cervids, and smaller mammalian and avian species. In North American wildlife conservation applications, Karelian Bear Dogs are primarily used for managing (and protecting people from) their namesake species, although they have been used for other applications where a well-trained versatile hunting dog would be beneficial. Although not typically aggressive toward humans, they were bred for strong hunting and fighting instincts and a very confident and independent temperament. Thus, extensive and highly skilled training is required for wildlife management and conservation applications.

In Canada, the Alberta Ministry of Environment and Sustainable Resource Development (ASRD) initiated the Karelian Bear Dog Program in 2001 to protect people and property while maintaining viable black bear (*U. americanus*) populations in the province (ASRD, 2009). Four Karelian Bear Dogs were deployed with ASRD personnel with a goal of interacting with bears such that bears learn to recognize and avoid humans and human-occupied areas through aversive conditioning. They are also used to track bears and mountain lions (*Felis concolor*) in residential areas, and to locate tranquilized bears. In addition to bear management, they are used to assist with public education, detect wildlife carcasses or parts, improve public and officer safety at response sites, and respond to conflicts involving other wildlife including moose, bighorn sheep, elk, wolf, and mountain lion. As of March 2009, Karelian Bear Dogs had been employed by ASRD for 1,643 bear-related conflicts, 350 ungulate-related conflicts, 190 mountain lion-related conflicts, and 13 wolf-related conflicts in Alberta.

In the USA, Washington Department of Fish and Wildlife (WDFW) first used a Karelian Bear Dog in 2003 and as of 2012 had four experienced adults in service and two pups in training, with each Karelian Bear Dog assigned to a specific wildlife officer/handler (WDFW, nd a; Grimley, 2012). Through the use of Karelian Bear Dogs for addressing bear–human conflicts, the WDFW has reduced their need for lethally removing black bears because of their success in training bears to avoid humans. Their original Karelian Bear Dog was involved in more than 50 bear-related captures and releases and another has helped capture over 50 mountain lions and 100 bears (WDFW, nd b). The dogs are also used to track animals, find carcasses, and will potentially be used in search and rescue.

Karelian Bear Dog programs in both Alberta and Washington were developed in partnership with Wind River Bear Institute (Table 9.1; WDFG, nd a; ASRD, 2009). The WRBI developed an approach for teaching problem bears to avoid areas used by humans instead of destroying the bears or relocating them outside their home ranges (returns are common). Specially selected and trained Karelian Bear Dogs are integral to the process. The WRBI breeds Karelian Bear Dogs and matches individual dogs to prospective owners (including agency personnel and private citizens), trains dogs and owners in bear education and deterrence, and provides community education to help people avoid creating conditions leading to human–bear conflicts.

The efficacy of Karelian Bear Dogs in reducing human–wildlife conflict has not been rigorously evaluated and reported in the scientific literature, although both ASRD and WDFW appear convinced that they provide real value for managing problem wildlife. On-line sources stress the importance of identifying and selecting for those dogs with the inherent tendency to stand up to large mammals such as bears and mountain lions, as not all Karelian Bear Dogs have that trait. Equally important, is matching individual people with a particular Karelian Bear Dog—and not all people have the proper temperament to train and handle them. As with nearly all types of deterrence approaches, Karelian

Bear Dogs are likely not ideal for all applications and their optimal use requires careful planning, training, and prudent deployment in management scenarios.

9.5.4 Dogs to reduce deer and other wildlife damage to crops

Coppinger et al. (1988) demonstrated the potential of a dog that naturally chased deer, and whose movements were spatially restricted by invisible electronic containment fencing, for protecting a heavily damaged apple orchard in Missouri, USA. Beringer et al. (1994) further explored the potential of dogs for protecting crops from deer browsing in a multi-year study examining the protection of white pine (*Pinus strobes*) seedlings. In the first year of study, plots were randomly assigned one of three treatments: electronically contained dogs, a chemical deer deterrent, or a control with no form of deer deterrence. Treatments were rotated among plots each year so that each plot received each treatment type. Over three years, the mean percentage of total seedlings with evidence of browsing was 13, 37, and 56% for dog, chemical, and no protection treatments, respectively. Dog-protected plots retained higher seedling biomass in the first two years of study and for the three-year mean, but these treatments had equivalent seedling biomass in the last year. An economic evaluation (see Box 9.2) of net present value of crop over a typical eight-year rotation suggested a clear superiority of dog protection (high returns) compared to chemical protection (losses or small returns) or no protection (large losses).

VerCauteren et al. (2005) compared crop protection dogs, contained with shock-collar based electric dog-containment systems, to double-strand electric polytape fence for protecting organic crops from deer browsing during three growing seasons in Wisconsin, USA (Figure 9.3). Dogs were randomly applied to one of three fields (1.4 ha) and polytape fencing was applied to the remaining fields (1.2 and 3.7 ha, respectively). Early on, five dogs were rejected as behaviorally unsuitable, before a Siberian Husky and a Siberian Husky–Malamute mix were identified as having suitable temperament. After these two dogs were placed in service, no crop damage was observed in the dog-protected field.

9.5.5 Dogs, mesopredators, and grassland bird conservation

Hansen and Smith (1999) noted that LPDs excluded and/or killed mesopredators in protected pastures

Figure 9.3 Organic crop-protection dog on duty in Wisconsin, USA. Such dogs were evaluated and proven effective in their ability to reduce wildlife damage to crops. Photo credit: National Wildlife Research Center.

Box 9.2 Economics of using livestock protection dogs

In comparison to the costs of other techniques for addressing predation issues with livestock, costs associated with LPDs are relatively low. Results from a 2009 survey in western USA indicated that average investment for non-lethal techniques such as shed lambing (US$18,000) and fencing (US$8,000) were four to nine times greater than using LPDs (<US$2,000; M. Marlow, pers. comm.). Further, the effectiveness of LPDs in mitigating depredation (mean decrease of 64%) was deemed nearly as great as that of shed lambing and greater than fencing. Survey respondents also provided information pertaining to start-up costs associated with incorporating LPDs into their operations. Purchase prices averaged US$413 per puppy with an additional average of $618 spent on that puppy during the first year of use. As adults, LPDs reportedly cost an average of US$115/month. Researchers have estimated annual costs associated with LPDs at US$937, $850, and $1,040/year (Landry et al., 2005; VerCauteren et al., 2008; and Gehring et al., 2010b, respectively).

Little effort has been put into evaluating savings realized by employing LPDs. A study in South Africa showed that of 70 LPDs placed in South Africa between May 2005 and July 2011, producers saved an average of US$3,189 ± $302 per farm annually due to a reduction in depredation for all livestock species (Rust et al., in press). In VerCauteren et al.'s (2005) comparison of crop protection dogs (Figure 9.3) to fencing for protecting organic crops from deer browsing in Wisconsin, USA, prior to introducing dogs the study field sustained US$3,762–5,200/year of damage. Fields protected by polytape fencing experienced browsing losses estimated at US$638–3,797 during the study. Protection by dogs cost US$3,575 for the first year, including fence installation and materials, dogs and related supplies, with annual maintenance of US$650 thereafter. The estimated annual average cost of dogs over a 25-year period was US$767 including initial costs. The annual cost for a 2.4m-tall woven-wire fence would have been approximately US$650 and would likely have provided a similar level of protection (assuming proper installation, maintenance, and consistent gate closure). Thus dogs may be a preferred alternative for those who prefer not to fence fields due to aesthetic or other practical reasons.

and VerCauteren (unpublished data) documented fewer small mammals in pastures protected by LPDs, and observed LPDs to occasionally capture and consume them. We also documented the killing of ≥10 Virginia opossums per year on one farm, although population abundance and the number of unharmed opossums was not measured. The presence of free-ranging dogs interfered with Indian fox (*Vulpes bengalensis*) use of areas, causing foxes to shift their use of resources (Vanak and Gompper, 2010). Vanak et al. (2009) also noted that Indian foxes modified their foraging behavior by being more vigilant and consuming less food in the presence of dogs. Similarly, Gehring et al. (2010b) noted a slight decrease in mesopredator visitation to livestock pastures following deployment of LPDs. Ground-nesting bird nests were also more abundant in the presence of LPDs, possibly because of greater rates of nest predation from mesopredators in non-LPD-protected pastures (Gehring et al., 2010b). Similarly, in western USA LPDs are employed in areas inhabited by sage grouse (*Centrocercus urophasianus*), a species of conservation concern, and it is thought the presence of LPDs reduces impacts of predators on the grouse (C. Urbigkit, pers. comm.). Individual LPDs, though, could depredate nests themselves or adversely influence wildlife indirectly (Weston and Stankowich, Chapter 4).

A parallel phenomenon is observed with wild apex predators; wolves, for example, may limit the presence and thus impact of mesopredators on small prey (Ritchie and Johnson, 2009). Thus, apex predators and LPDs play similar roles in mediating mesopredators (Vanak et al., Chapter 3), and the latter might also serve as a more general tool for wildlife conservation objectives, such as reducing mortality of ground-nesting birds and nest predation as stated above.

9.5.6 Dogs and conservation of declining wildlife species

The image of traditional uses for LPDs surrounded by numerous sheep in a high European mountain meadow may still be a reality. However, societal demands for alternative means for protecting resources without sacrificing the existence of wildlife species is expanding that image to diverse landscapes accompanied by various species. For example, Maremma LPDs originating from Italy are being used on the beaches of Australia to protect little penguins (*Eudyptula minor*) from predation by red foxes (*V. vulpes*) (Lustig, 2011). Additionally, sage grouse in the western USA appear to be inadvertent benefactors of LPDs used to protect sheep from predators (Urbigkit, pers. comm.). The potential for the use of dogs in various roles for protecting resources is virtually unlimited, especially when a well thought out strategy is implemented.

Even large predatory species can be indirectly afforded protection, through the implementation of LPDs to minimize conflict over predation. Protecting livestock in some areas like the Trans-Himalayan region is a key step toward predator conservation of species such as the Himalayan wolf (*C. l. himalyensis/chanco*) and the snow leopard (*Uncia uncia*; Namgail et al., 2007). Tigers (*Panthera tigris*) in Bangladesh are experiencing reduced levels of persecution by humans partly due to the implementation of dogs to alert citizens (i.e., potential prey) to the presence of tigers, thus reducing the potential for attack on humans (Kerley, 2010; Khan, 2009). Dogs from local pastoral communities are also functioning similarly to traditional LPDs in the southern part of Africa (South Africa, Lesotho, and Botswana). For example, efforts of the Cheetah Conservation Fund (CCF) have afforded protection to cheetahs by using LPDs as a tool enabling livestock owners to reduce both perceived threats and real predation, thus minimizing retributive killing of cheetahs and other potential predators (Marker et al., 2005). Once widespread across Africa, Asia, and the Middle East, the cheetah has undergone a serious decline over the past century with population estimates falling from around 100,000 animals in 1900 to less than 10,000 by 2007 (Bartels et al., 2001). One of the few remaining strongholds for cheetahs is in Namibia. Namibia contains the largest remaining population of free-ranging cheetahs in the world, estimated at 3,000 adult animals, of which 95% occur on private rangeland (Marker, 2002). As such, a high degree of conflict exists with producers who perceive cheetahs as posing a threat to their livestock and farmed game (Marker-Kraus et al., 1996). Although there is minimal evidence to support this perception (Marker et al., 2003), there has been widespread killing and capture of cheetahs on rangelands. Almost 7,000 cheetahs were reportedly removed from Namibian rangelands during the 1980s (CITES, 1992), halving Namibia's cheetah population between 1975 and 1987 (Morsbach, 1987), and the conflict continues (Marker et al., 2007). In an effort to understand and resolve this conflict that imperils the cheetahs' Namibian stronghold, the CCF explored diverse techniques to lessen actual or perceived depredation on livestock (Marker et al., 2003). The strategy that seemed likely to have most relevance to the Namibian situation was the use of specialized LPDs.

In 1994 the CCF began their LPD program, exploring the use of LPDs in an African system where livestock (cattle, goats, and sheep) mostly range untended over vast areas amongst a multitude of predators, including cheetahs, leopards, caracals (*Felis caracal*), and black-backed jackals. Research and experience has led CCF to employ primarily Anatolian Shepards and Kangals. These breeds were chosen in preference to other available LPD breeds as they are short-coated, well adapted to working in a hot, arid climate, and are heavy, imposing dogs that outweigh the majority of Namibian farmland predators (Richardson, 1994). The dogs are placed primarily with small stock like goats and sheep, which typically roam over large areas in the day (Figure 9.4), are sometimes accompanied by a herder, and are usually corralled at night (Marker-Kraus et al., 1996). Producers in the area have used a variety of techniques aimed at reducing livestock depredation, including employing human herders, donkeys, and even baboons (*Papio ursinus*, Marker-Kraus et al., 1996). Local dogs, called 'pavement specials' were often kept with herds. However, the majority of these dogs were small to medium sized and showed herding behaviors using the eye–stalk behavior to move livestock. It is believed that when a predator approaches the herd, the dog instinctively begins to herd the livestock. This stimulates the predatory motor pattern of the predator (eye–stalk–chase–trip–bite–consume), stimulating it to chase and kill

Figure 9.4 Anatolian Shepherd and flock of goats in Namibia. Photo credit: Laurie Marker.

livestock. Herds that had dogs with these behaviors actually had higher losses than those with no dog (Marker-Kraus et al., 1996).

The performance of LPDs placed with 117 producers in Namibia between 1994 and 2002 were evaluated through surveys. Nearly 75% of producer responses indicated a large decline in the numbers of livestock lost since using a LPD (Marker et al., 2005). A follow up survey in 2009 of 164 producers showed that >90% of the dogs reduced or eliminated livestock losses (Potgieter, 2011; Potgieter et al., in press). The majority of producers felt they had benefited economically from employing LPDs (Potgieter et al., in press).

Overall, this long-term case study has shown that the relatively simple strategy of placing LPDs with Namibian livestock can be an effective tool for local producers. The dogs reduced livestock losses and 89% of the farmers surveyed in 2009 perceived their LPDs as economically beneficial, thus resulting in reduced conflict with predators. Importantly, illegal cheetah removals and lethal predator control efforts have dropped (Marker et al., 2003; Potgieter, 2011), though it is hard to assess the extent to which these changes were due to LPDs versus other factors, such as education and changes in cheetah population size. The success of CCF's work in Namibia has encouraged the use of LPDs in other African countries, including South Africa (Rust et al., in press), Botswana, Kenya, and Tanzania (Stannard, 2006; L. Marker, unpublished data).

9.6 Potential limitations, conflicts, and problems

As with any wildlife damage management strategy, an integrated approach utilizing multiple techniques improves the overall potential for success. For example, the effectiveness of LPDs can be maximized when used in conjunction with night penning and with the presence of a human herder (Espuno et al., 2004). Use of LPDs is not without its problems or limitations. Each LPD is an individual, and one may perform excellently in the same situation where another fails. Every situation that an LPD is put into is unique as well, and it is the responsibility of LPD owners to deploy and prepare each LPD in a way that maximizes the potential for success. For example, LPDs protecting livestock on the open range are exposed to greater risk of predation and frequently more LPDs are necessary than with livestock in fenced pastures (Andelt and Hopper, 2000). Each dog is also an investment in time and money for the producer, they mature at about 2 years of age and there is no guarantee that given individuals will perform to expectations and remain healthy. Most have a working lifespan of about 7 years. Some individuals tend to roam, leaving their livestock unprotected. Others can show aggressive behaviors toward livestock, pets, and humans they are not familiar with, which can be especially problematic on open range and public lands. Husbandry practices also impact the application of LPDs. For example, in some regions such as Jura and Vosges of France, flocks of sheep are dispersed in small groups (from 2 to 15) making it logistically impractical to deploy a LPD with each group (Landry and Raydelet, 2010). In open mountain rangelands in Norway, systematic patrolling of small dispersed flocks of sheep by herders with LPDs has been evaluated and found to be only moderately effective (Hansen et al., 2002).

9.6.1 Behavioral problems of LPDs impacting livestock and wildlife

A variety of behavioral problems may arise between LPDs and livestock or LPDs and non-target wildlife. These problems need to be detected, addressed, and eliminated early and rapidly to ensure the development and maintenance of an effective LPD. Occasionally, when introduced into a flock with young animals (especially lambs), LPDs might demonstrate unacceptable behaviors while being playful, such as chasing and biting or pulling at wool, ears, and tails (Landry et al., 2005). Juvenile livestock may not be able to defend themselves, and may become frightened or injured. Therefore, dog owners need to intervene promptly to correct dogs and eliminate the development of inappropriate habits or aggression.

The CCF surveyed Namibian producers and monitored behaviors and successes of dogs placed through their program, evaluating the benefits and drawbacks. Three primary LPD behaviors first identified by Coppinger and Coppinger (1980) were examined. These behaviors included: (1) attentiveness—the tendency of the dog to stay with the flock; (2) trustworthiness—the lack of predatory or other inappropriate behaviors towards the flock; and (3) protectiveness—the tendency of the dog to display protective behaviors. Almost all the evaluated dogs demonstrated inappropriate behaviors at some stage of their development, most often when they were young and formative (Marker et al., 2005). The three most common problems were: (1) chasing non-target wildlife (see Box 9.3), which sometimes resulted in the dogs killing and even feeding on wildlife such as kudu (*Tragelaphus strepsiceros*) or chasing warthogs (*Phacochoerus africanus*) which could gore an LPD; (2) staying at home instead of going out with livestock; and (3) harassing or even killing livestock. In Namibia, the majority of problems were corrected through training under direction of the CCF, with dogs becoming attentive, trustworthy, and protective. Seldom was it necessary to resort to transferring a dog into a pet situation or culling it. To prevent such behavioral problems, it is recommended that young, unproven dogs be monitored by a herder who has the skills to correct the dog's behaviors (Marker et al., 2005; Schumann, 2003). In addition, the CCF has developed a farmer training course and dog training guide that presents to farmers predicted behaviors and ages at which to closely monitor the dogs during growth and development (Schumann, 2003). Dogs and farmers are monitored regularly by CCF during these key times (Marker et al., 2005; Potgieter et al., in press). The satisfaction of producers with their LPDs was dependent primarily on how attentive the dogs were to their flocks, followed by trustworthiness and protectiveness.

Organized LPD breeding and development programs provide consistency and expertise that maximizes the potential for success in deploying effective dogs. For example, in Switzerland and Turkey the breeding of LPDs is regulated by breeding centers, which should guarantee the quality of the LPDs they produce. In Namibia, the CCF serves as a breeding center, as farmers do not want to be burdened with the downtime of females with pups. Puppies are born at CCF's model farm and raised with the flock until they are placed with herds. All puppies are neutered prior to placement and farmers attend a mandatory training day prior to placement, which provides farmers with guidelines for management and training. Training and management strategies to overcome traditional challenges and broaden the application of LPDs are being examined (e.g., VerCauteren et al., 2012). Such strategies include raising pups in a training setting until at least 12 months of age before introducing them into a new flock. It is theorized that the dog will have passed through its problem stages by this age and will thus integrate and be accepted into its new setting more seamlessly.

Box 9.3 Impacts of dogs on wildlife

Although the goal of placing trained dogs on the landscape is directed at protecting resources such as livestock, crops, or even wildlife species, unintentional and detrimental outcomes toward local wildlife species also may occur. The nature of the job of LPDs leaves them in a role with great independence; producers rely on them to take appropriate action when non-target species of wildlife approach the livestock. An approaching animal may be perceived as a threat, and thus confronted with potential for over-reactive aggression by the LPD. Particular breed characteristics, as well as training and experience, all contribute to how a dog will respond in such a situation and thus the outcome.

Several studies have examined the effects of feral and free-ranging dogs and quantified results (e.g., Young et al., 2011; Ritchie et al., Chapter 2). Free-ranging LPDs with livestock have been shown to chase, kill, and even eat local wildlife such as mountain gazelle (*Gazella gazella*; Gingold et al., 2009). One study even suggested that free-roaming dogs consumed more livestock than wolves (Echegaray and Vilà, 2010). Occasionally, feral and free-ranging dog categorizations are combined, thus LPDs are implicated with the impacts of truly feral dogs. Livestock protection dogs used in Namibia were reportedly observed chasing wildlife by 19% of survey respondents (Potgieter, 2011) and this misbehavior was easily corrected (L. Marker, unpublished data). Furthermore, aside from the potential negative direct impacts of dogs on wildlife, there is the potential for the spread of canine pathogens such as canine distemper virus, adenovirus, and parvovirus (e.g., Laurenson et al., 1998). Thus all LPDs should have all appropriate vaccinations.

Examples of LPDs demonstrating pursuit behavior toward non-predator species initially motivated researchers to develop and evaluate the use of LPDs for the purpose of excluding wildlife from livestock-related resources in attempts to alleviate transmission of disease (Gehring et al., 2010b; VerCauteren et al., 2008). Results showed that LPDs effectively reduced the potential for disease transmission by excluding deer; however, the presence of other wildlife, such as mesopredators and rodents, was also reduced concurrently. The goal of using LPDs is to allow them to work independently with livestock with no or little supervision from a herder or handler. In attempts to minimize the negative impacts of LPDs on non-target wildlife species, emphasis has been placed on targeted training to deter predators, the importance of containing LPDs with protected livestock, and reprimanding offending dogs when negative behaviors are exhibited. The CCF's research has shown that corrective training early will correct most behaviors (Marker et al., 2005, Potgieter et al., submitted). In some instances, Namibian farmers have witnessed their LPDs fighting with predators, and the dogs have been documented killing black-backed jackal, leopards, and Chacma baboons that were threatening livestock (Marker et al., 2005). Eurasian badgers, red foxes (*V. vulpes*), marmots (*Marmota marmota*), young wild boars, and wolves have reportedly been killed by LPDs (J.-M. Landry, unpublished data). In the USA, LPDs have been known to kill coyotes and mesopredators (Gehring et al., 2010b; C. Urbigkit, pers. comm.). The impact of LPDs on large prey is unsubstantiated and documented cases involve only a few particular dogs (Lapeyronie and Moret, 2003; Potgieter, 2011). The CCF recommends corrective training immediately and has successfully used a dangle stick, if a herder is available to monitor, and has had success in stopping dogs from chasing wildlife (Schumann, 2003).

9.6.2 Conflicts with various public interest groups

Current societal changes toward recreational land uses create new challenges for agricultural uses, including the use of LPDs in proximity to recreational lands. For the most part, this is the first time in history that producers employing LPDs have had to worry about conflicts with recreationists such as backcountry hikers and mountain bikers. Remote alpine areas have always been destinations for recreational purposes, but today many recreationists are not familiar with agriculture and agricultural practices. Thus, conflicts can occur when recreationists, unwittingly or otherwise, disrespect the work of producers by disturbing livestock (e.g., passing through idle flocks). Situations like this can be challenging for LPDs because of their alarming behaviors (e.g., rapid approach, barking) and because they are often unsupervised and roaming freely with the herd. Unfortunately, some LPDs have bitten people and injured or killed companion dogs in proximity to their flocks. Conflicts occasionally arise when LPDs are deployed near the residences of humans, usually due to barking or dissuasive behavior that may be frightening. To avoid conflicts, some producers forgo deploying LPDs in proximity to human dwellings or communities. Evaluations of relative levels of aggression of different LPD breeds have been conducted to determine particular breeds that could potentially be more or less dangerous towards people passing by a herd (Durand and LePape, 1998; Green and Woodruff, 1988; Hansen and Bakken, 1999; Landry, 2004; Landry and Raydelet, 2010). Of the breeds evaluated, Great Pyrenees have proven to be the least aggressive toward humans. The presence of a companion dog with recreationists increases the probability of approach by LPDs (Landry, 2004), and thus increases the probability of conflict. Additionally, hunters may complain that LPDs disturb wildlife and even render hunting more difficult. Further, some may also be concerned that LPDs may attack and kill their hunting dogs during hunting activities. In the French Jura Mountains, 30% of interviewed LPD owners commented on conflicts with hunters (Landry and Raydelet, 2010). Such conflicts could lead to the death of LPDs (shot or poisoned), but fortunately these events seem uncommon. As part of their seasonal husbandry, French producers often remove their livestock and LPDs from rangelands by the time hunting seasons begin, which serves to lessen the potential for conflicts. To help address these issues, entities in several countries (e.g., Namibia, France, Switzerland, USA) have established educational campaigns, enlisting various media, to inform the public about how to behave when meeting LPDs and to educate them about pastoral agriculture and the role of LPDs in protecting livestock from predators in a non-lethal manner (Table 9.1; Figure 9.5).

Another issue that can create conflict with local humans involves LPDs that leave their livestock and begin to roam. One or a variety of containment strategies can be implemented to encourage LPDs to remain with livestock and minimize potential roaming. To determine the best and most cost-effective option for containing LPDs and livestock in pastures, producers need to consider their existing infrastructure and management practices. Existing livestock fences provide a visible boundary that may facilitate training LPDs to remain within the perimeter. When existing fences prove insufficient for containment, adding strands of electrified wire can serve to help contain LPDs. Supplemental training may be required to ensure individual LPDs maintain respect for electric fences and to deter escape behavior, by setting up scenarios where LPDs come in contact with electric fences and thus learn not to test them. While training LPDs for electric fences, producers must ensure that when dogs receive negative stimuli (electrical shock) that they associate it with the fence and not the individual doing the training. Shock-collar based electric dog-containment systems facilitate the establishment and maintenance of a LPD's respect for a perimeter. They also minimize the potential for negative association with handlers, since the handler need not be near the dog or even present (Schilder and van der Borg, 2004). Successful containment not only reduces the potential for conflict but also ensures safety of LPDs. For more detailed information on fencing options see VerCauteren et al. (2008) and Gehring et al. (2010c).

As a result of the types of conflicts described above, local authorities could restrict or even ban the use of LPDs in particular areas (Landry et al., 2005). For example, since 2004 a division of Switzerland has maintained a list of 'dangerous' breeds that includes the Spanish Mastiff, a commonly used LPD in Spain. Such breeds must always be muzzled and leashed, thus they are not allowed to function as guardians of livestock. In some regions, associations of producers who employ LPDs have formed to serve as references for local and federal authorities, to help educate the public, and to aid in mitigating conflicts by overseeing LPD use.

In some areas where wolves have re-established or been re-introduced, LPDs have become entangled in political controversy. Although LPDs should serve as a tool to help allow livestock and wild carnivores to inhabit common areas, some argue that LPDs are a danger to people and pets, and thus recreational activities, and that livestock should not be protected by this means; as such, wolves should be controlled with lethal methods. Others pressure producers who employ LPDs, claiming that "working with LPDs means accepting the wolf" (Landry et al., 2005). It is essential that conflicts such as these be identified and understood because there is concern they could lead

Figure 9.5 Sign used in the USA to educate land users about the role of livestock protection dogs and how to respond to their presence to avoid conflict. Courtesy of the USDA APHIS WS.

to the malicious deaths of LPDs or that LPDs could be banned from protecting livestock in some areas. Because of this, research into human dimensions associated with the employment of LPDs in various settings for various purposes is needed, as are educational efforts so that various segments of the public can understand and appreciate LPDs and the job they do.

9.6.3 Mortality of dogs

Although free-ranging dogs can adversely affect wildlife, they can also serve as prey for other wild carnivores (Vanak and Gompper, 2009; Athreya, 2006; Butler et al., Chapter 5). Predation of livestock, and now LPDs, in western USA has become more commonplace over the last 15 years due to the reintroduction of the gray wolf (see Box 9.4). Between 1987 and 2005, 18 LPDs were reportedly killed by wolves in western USA. In contrast, from 2005 to 2010, 28 LPDs were reported to have been killed by wolves and another 30 injured (M. Marlow, pers. comm.). In one area of Romania, 157 adult LPDs were killed by wolves from January 2001 to October 2002, of which 77% were killed near the flock, which was left unhurt. Nearly all LPDs were consumed (Mertens and Schneider, 2005). One hypothesized cause for this situation was a lack of training and uncontrolled breeding among LPDs and stray dogs, a lack of selection, and resultant smaller and less-effective offspring. Additionally, the use of too few dogs per flock may be leaving LPDs susceptible to attack by packs of wolves.

Occasionally, wolves and LPDs have been observed in proximity apparently tolerating one another, even following depredation events. In one particular situation, following the removal of a depredating wolf that had apparently befriended an LPD, the LPD also preyed on sheep and had to be removed as well (Bangs et al., 2005). Other predators, including bear, coyote, and mountain lion have occasionally been reported to kill LPDs. Livestock protection dogs have been killed by wolves in Portugal and France, but compared to other causes of death these events are rare (Riberio and Petrucci-Fonseca, 2005; J.-M. Landry, unpublished data).

Organized LPD breeding and deployment programs have enabled researchers to follow and evaluate the success and other aspects of individual dogs. Causes of mortality in LPDs are quite diverse and often not due to old age. For example, in the CCF program in Namibia, only dogs that died as pets (18%) and those that worked on commercial farms (6%) were reported to die of old age (Figure 9.7). On average, dogs in the CCF program had a working lifespan of 4.3 years, similar to that reported in the USA where fewer than 50% of working LPDs lived that long and 36% lived to 6 years of age (Green et al., 1994; Lorenz et al., 1986). The leading cause of death in Namibia was field accidents (36%, 77 dogs) including dogs that were killed by snakes, baboons, other predators, lost in the veld, and other accidents (e.g., drowned in a river, kicked by a horse, killed in a dog fight); 41 dogs (19%) died of unknown causes and 39 (18%) died of medical or health related issues (Figure 9.7). Culling by owners also accounted for a substantial proportion of working dog deaths, particularly on commercial farms, usually as a result of the dog chasing or harassing stock. In Portugal, 97 LPDs were deployed and 75% were still alive after 7.5 years of life, the main causes of mortality were disease (e.g., leishmaniasis, leptospirosis, hip dysplasia) and accidents (Ribeiro and Petrucci-Fonseca, 2005).

Mortality from field accidents is to be expected under the dangerous circumstances that working dogs are exposed to, especially for young dogs, which are likely to be relatively inattentive (Lorenz and Coppinger, 1986; Lorenz et al., 1986). Inattentive dogs were found to be more likely to be lost or killed in the USA (Lorenz et al., 1986) and in Namibia, with dogs that were ultimately removed, by either death or transfer, being significantly less attentive than other dogs (Marker, 2002). As CCF's program has progressed, with lessons learned, the lifespan of the working dogs has been increasing. Also, contrary to situations related to wolves in North America and Europe, CCF reported no LPDs to have been killed by the primary species they were protecting livestock from (i.e., cheetahs or leopards). This may be attributed to differences between canid and felid predators, for example wolves may view LPDs as competitors.

Box 9.4 Providing protection to the protector

The use of spiked collars (Figure 9.6) to protect LPDs in the event of attack by predators is relatively common and considered a necessity in areas across Europe populated with wolves. Spiked collars are just beginning to be employed in western USA (C. Urbigkit, pers. comm.). These collars not only provide protection against wolves (which target the neck region when attacking) but may also become a weapon for experienced LPDs. There have been observations of wolves being wounded by spiked collars being worn by LPDs, which gave the LPD the advantage in the conflict (J.-M. Laundry, unpublished data). It has also been observed that LPDs wearing spiked collars appear more self-confident when interacting with wolves.

Figure 9.6 Spiked collars used on dogs such as this Anatolian Shepherd provide both protection and an offensive tool against aggressive predators. Photo credit: Cat Urbigkit.

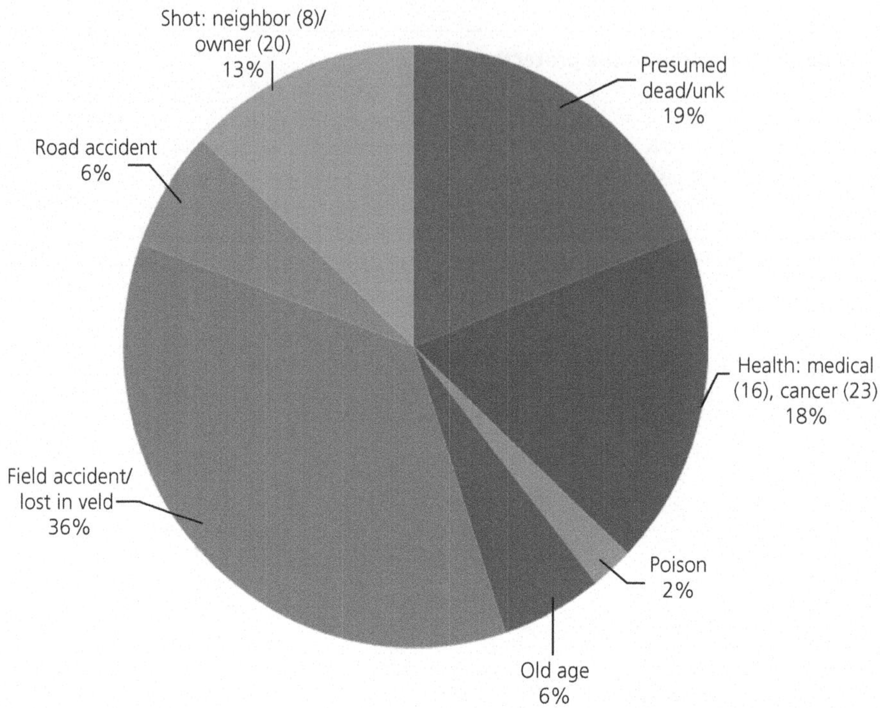

Figure 9.7 Cause of death of Cheetah Conservation Fund's livestock guarding dogs (including adults and immature animals) in Namibia ($n = 216$; 1994–2012).

9.7 Conclusions and future directions

Many challenges undoubtedly lie ahead at the interface between wildlife and human interests, especially as societies continue to become more urbanized and disconnected from natural systems. Past experience has shown that dogs can be effective intermediaries for helping people, livestock, and wildlife coexist. Dogs can be quite flexible and versatile and can be applied to a wide variety of conservation conflicts. Dogs alone, though, may not be able to permanently alleviate damages. Thus, integrated strategies that employ a variety of non-lethal and lethal management tools must be considered (Gehring et al., 2010a).

To date, studies on LPDs were mainly from North America, Europe, and Africa. Yet LPDs originated from Central Asia, where they have been used for centuries without interruption, and thus there are likely lessons and strategies to be learned by studying dogs and their people in this region. Use of LPDs there seems to differ from the occidental way, rather than instilling a strong bond with the animal to be protected LPDs are often chained (Subba, 2012), lack training, and even become feral (Namgail et al., 2007). Better understanding of how well these strategies work with LPDs could lead to new ways of applying them in other areas. Establishing a mechanism for international information exchange, like the former newsletter Carnivore Damage Prevention News (www.lcie.org/res_damage.htm), could facilitate improvements in training and using dogs and accelerate their use for mitigating conservation conflicts.

Many research questions need to be addressed relative to the use of dogs to resolve conservation conflicts, and well-designed experimental studies with large sample sizes of dogs are required. Basic questions related to prescribing the appropriate dog-based solution to the challenge at hand include:

- How many LPDs are needed to protect herds of various species and sizes from various predators in various settings?
- How do dynamics of sex, age, relationship, experience, and so on impact LPDs that are working together?
- Are social interactions between LPDs and flock members enough to protect a flock?
- What is the role of aggression and other behaviors in communicating with wildlife and resulting in protection?
- How do groups of LPDs cooperate, are there synergies and conflicts, and how do they influence the protective ability of the group?
- What are the optimal ways to raise and train dogs for specific uses?
- What new, innovative uses of dogs can be explored?

Extirpation of large carnivores in many areas of Europe likely had detrimental effects on some breeds as fewer dogs were needed to protect flocks, potentially leading to genetic bottlenecks (e.g., Great Pyrenees population in France). Today's gene pools may therefore be impoverished and produce a high proportion of dogs unsuited to livestock protection or conservation applications. Therefore, it would be valuable to study breed-specific questions relevant to contemporary issues, such as:

- Which breeds are most applicable for which use?
- Which breeds have the potential to be further developed and suited for novel purposes?
- Do we need more aggressive LPD breeds to be effective against large predators, such as wolves?
- Do LPDs that are aggressive towards predators also show aggression towards humans?
- As LPDs from Asia are selected from a large gene pool, how do behaviors (e.g., protective behavior) of these dogs differ within breeds, how are they selected, and how effective are they in comparison to LPDs elsewhere in the world?

Ultimately, there will be few simple or universal answers to these questions. Though LPDs have been bred for thousands of years to fulfill the niche of protecting livestock, they are all individuals, varied and versatile, and the humans who train and employ them must also be versatile as well as innovative. Understanding the behavior of dogs as well as that of the wildlife species they are working against will require continued research—and achieving more widespread and successful use of dogs for mediating conservation conflicts will occur as researchers, managers, agricultural producers, and the public at large begin to further unleash the potential of dogs for these important purposes.

References

Andelt, W.F. (1992). Effectiveness of livestock guarding dogs for reducing predation on domestic sheep. *Wildlife Society Bulletin*, **20**, 55–62.

Andelt, W.F. and Hopper, S.N. (2000). Livestock guard dogs reduce predation on domestic sheep in Colorado. *Journal of Range Management*, **53**, 259–67.

ASRD. (2009). *Karelian BearDog program*. Alberta Sustainable Resource Development <http://www.srd.alberta.ca/FishWildlife/WildlifeManagement/BearManagement/KarelianBearDogProgram.aspx>.

Athreya, V. (2006). Is relocation a viable management option for unwanted animals? The case of the leopard in India. *Conservation and Society*, **4**, 419–23.

Atwood, T.C., Deliberto, T.J., Smith, H.J., Stevenson, J.S., and VerCauteren, K.C. (2009). Spatial ecology of raccoons related to cattle and bovine tuberculosis in northeastern Michigan. *Journal of Wildlife Management*, **73**, 647–54.

Ball, D. (2000). Wildlife control at Vancouver International Airport: Introducing Border Collies. *Bird Strike Committee-USA/Canada Meeting, Minneapolis, Minnesota*, **2**, 8–10.

Bangs, E.E., Fontaine, J.A., M.D. Jimenez, M.D, Meier, T.J., Bradley, E.H., Niemeyer, C.C., Smith, D.W., Mack, C.M., Asher, V., and Oakleaf, J.K. (2005). Managing wolf/human conflict in the northwestern United States. In R. Woodroffe, S. Thirgood and A. Rabinowitz. Eds., *People and Wildlife: Coexistence or Conflict?* Cambridge University Press, Cambridge, pp. 340–56.

Bartels, P., Bouwer, V., Crosier, A., Cilliers, D., Durant, S., Grisham, J., Marker, L., Mulama, M., Venter, L., Wildt, D., and Freidmann, Y. (2001). *Global Cheetah Action Plan Final Workshop Report*. Conservation Breeding Specialist Group SSC/IUCN, Pretoria.

Beringer, J., Hansen, L.P., Heinen, R.A., and Giessman, N.F. (1994). Use of dogs to reduce damage by deer to a white pine plantation. *Wildlife Society Bulletin*, **22**, 627–32.

Black, H.L., and Green, J.S. (1985). Navajo use of mixed-breed dogs for management of predators. *Journal of Range Management*, **38**, 11–15.

Böhm, M., Hutchings, M.R., and White, P.C.L. (2009). Contact networks in a wildlife-livestock host community: identifying high-risk individuals in the transmission of bovine TB among badgers and cattle. *PLoS ONE*, 4:e5016.

Carter, N.B. (2000). The use of Border Collies in avian and wildlife control programs. In M.C. Brittingham, J. Kays, and R. McPeake, Eds. *The Ninth Wildlife Damage Management Conference Proceedings*. State College, Pennsylvania, pp. 70–6.

Castelli, P.M. and Sleggs, S.E. (2000). Efficacy of Border Collies to control nuisance Canada geese. *Wildlife Society Bulletin*, **28**, 385–92.

CITES. (1992). Quotas for trade in specimens of cheetah. *Convention of International Trade in Endangered Species of Wild Fauna and Flora*, **8**, 1–5.

Clifford, D.L., Schumaker, B.A., Stephenson, T.R., Bleich, V.C., Cahn, M.L., Gonzales, B.J., Boyce, W.M., and Mazet, J.A.K. (2009). Assessing disease risk at the wildlife-livestock interface: a study of Sierra Nevada bighorn sheep. *Biological Conservation*, **142**, 2559–68.

Conover, M. (2002). *Resolving Human-Wildlife Conflicts: The Science of Wildlife Damage Management*. CRC Press, Boca Raton, Florida.

Converse, R.R., Kinzelman, J., Sams, E.A, Hudgens, E., Dufour, A.P., Ryu, H., Santo- Domingo, J.W., Kelty, C.A., Shanks, O.C., Siefring, S.D., Haugland, R.A., and Wade, T.J. (2012). Dramatic improvements in beach water quality following gull removal. *Environmental Science & Technology*, **46**, 10206–13.

Coppinger, R. and Coppinger, L. (1978). *Livestock Guarding Dogs for U. S. Agriculture*. Livestock Dog Project, Hampshire College, Massachusetts.

Coppinger, R. and Coppinger, L. (1980). Livestock-guarding dogs: an Old World solution to an age-old problem. *Country Journal*, **7**, 68–77.

Coppinger, R. and Coppinger, L. (2001). *Dogs: A startling New Understanding of Canine Origin, Behavior, and Evolution*. Scribner, New York.

Coppinger, R., Lorenz, J., and Coppinger, L. (1987). New uses of livestock guarding dogs to reduce agriculture/wildlife conflicts. *Proceedings of the Wildlife Damage Control Conference*, **3**, 253–59.

Coppinger, R., Coppinger, L., Langeloh, G., Gettler, L., and Lorenz, J. (1988). A decade of use of livestock guarding dogs. *Proceedings of the Vertebrate Pest Conference*, **13**, 209–14.

Curtis, P. and Rieckenberg, R. (2005). Use of dogs for reducing deer damage in apple orchards. *Proceedings of the Wildlife Damage Management Conference*, **11**, 149–58.

de la Cruz, C. (1995) Another View of Livestock Guarding Dog History. *AKC Gazette* <http://www.lgd.org/library/origins.htm>.

Dohner, J.V. (2007). *Livestock Guardians: Using Dogs, Donkeys, and Llamas to Protect Your Herd*. Storey Publishing, Massachusetts.

Durand, C. and Le Pape, G. (1998). Interactions entre promeneurs et les chiens de protection de troupeaux dans le massif du Mercantour. *Report of Life—loup Program* <http://www.loup.developpement-durable.gouv.fr>.

Echegaray, J. and Vilà, C. (2010). Noninvasive monitoring of wolves at the edge of their distribution and the cost of their conservation. *Animal Conservation*, **13**, 157–61.

Espuno, N., Lequette, B., Poulle, M., Migot, P., and Lebreton, J. (2004). Heterogeneous response to preventive sheep husbandry during wolf recolonization of the French Alps. *Wildlife Society Bulletin*, **32**, 1195–1208.

Fritts, S.H., Mech, L.D., Paul, W.J., and Scott, D.P. (1992). Trends and management of wolf-livestock conflicts in Minnesota. *U.S. Department of Interior Fish Wildlife Services Resource Publication*, **181**, 27.

Frölich, K., Thiede, S., Kozikowski, T., and Jakob, W. (2006). A review of mutual transmission of important infectious diseases between livestock and wildlife in Europe. *Annals of the New York Academy of Sciences*, **969**, 4–13.

Froneman, A. and Van Royen, M. (2003). The successful implementation of a Border Collie bird scaring program at Durban International Airport, South Africa. *Twenty-sixth International Bird Strike Committee Conference* <http://www.int-birdstrike.org/Warsaw_Papers/IBSC26%20WPAE1.pdf>.

Gallagher, J. and Clifton-Hadley, R.S. (2000). Tuberculosis in badgers; a review of the disease and its significance for other animals. *Research in Veterinary Science*, **69**, 203–17.

Gehring, T.M., and Potter, B.A. (2005). Wolf habitat analysis in Michigan: an example of the need for proactive land management for carnivore species. *Wildlife Society Bulletin*, **33**, 1237–44.

Gehring, T.M., Hawley, J.E., Davidson, S.J., Rossler, S.T., Cellar, A.C., Schultz, R.N., Wydeven, A.P., and VerCauteren, K.C. (2006). Are viable non-lethal management tools available for reducing wolf-human conflict? Preliminary results from field experiments. *Proceedings of the Vertebrate Pest Conference*, **22**, 2–6.

Gehring, T.M., VerCauteren, K.C., and Landry, J.M. (2010a). Livestock protection dogs in the 21st century: is an ancient tool relevant to modern conservation challenges? *BioScience*, **60**, 299–308.

Gehring, T.M., VerCauteren, K.C., Provost, M.L., and Cellar, A.C. (2010b). Utility of livestock protection dogs for deterring wildlife from cattle farms. *Wildlife Research*, **37**, 715–21.

Gehring, T.M., Cellar, A.C., and VerCauteren, K.C. (2010c). Good fences make good neighbors: implementation of electric fencing for establishing effective livestock protection dogs. *Human-Wildlife Interactions*, **5**, 106–11.

Gilsdorf, J.M., Hygnstrom, S.E., and VerCauteren, K.C. (2002). Use of frightening devices wildlife damage management. *Integrated Pest Management Reviews*, **7**, 29–45.

Gilsdorf, J.M., Hygnstrom, S.E., VerCauteren, K.C., Clements, G.M., Blankenship, E.E., and Engeman, R.M. (2004). Evaluation of a deer-activated bio-acoustic frightening device for reducing deer damage in cornfields. *Wildlife Society Bulletin*, **32**, 515–23.

Gingold, G., Yom-Tov, Y., Kronfeld-Schor, N. and Geffen, E. (2009). Effect of guard dogs on the behavior and reproduction of gazelles in cattle enclosures on the Golan Heights. *Animal Conservation*, **12**, 155–62.

Green, J.S. and Woodruff, R.A. (1988). Breed comparisons and characteristics of use of livestock guarding dogs. *Journal of Range Management*, **41**, 249–51.

Green, J.S. and Woodruff, R.A. (1999). Livestock guarding dogs: protecting sheep from predators. *Agricultural Information Bulletin 588*. U. S. Department of Agriculture.

Green, J.S., Woodruff, R.A., and Tueller, T.T. (1984). Livestock-guarding dogs for predator control: costs, benefits, and practicality. *Wildlife Society Bulletin*, **12**, 44–50.

Green, J.S., Woodruff, R.A., and Andelt, W.F. (1994). Do livestock guarding dogs lose their effectiveness over time? *Vertebrate Pest Conference*, **16**, 41–4.

Greenlaw, J.S. and Sheehan, W.J. (2003). The current status and history of ring-billed gulls (*Larus delawarensis*) in Maine. *Northeastern Naturalist*, **10**, 55–66.

Grimley, B. (2012). *Kitsap's newestwildlife officer: Karelian bear dog Spencer* <http://www.kitsapsun.com/news/2012/mar/25/kitsaps-newest-wildlife-officer-karelian-bear/>.

Hansen, I. (2005). Use of livestock guarding dogs in Norway: a review of the effectiveness of different methods. *Carnivore Damage Prevention News*, **8**, 2–8.

Hansen, I. and Bakken, M. (1999). Livestock-guarding dogs in Norway: Part I. Interactions. *Journal of Range Management*, **52**, 2–6.

Hansen, I. and Smith, M.E. (1999). Livestock-guarding dogs in Norway: Part II. Different working regimes. *Journal of Range Management*, **52**, 312–6.

Hansen, I., Staaland, T., and Rings, A. (2002). Patrolling with livestock guarding dogs: a potential method to reduce predation on sheep? *Acta Agriculture Scandinavia, Section A, Animal Science*, **52**, 38–43.

Hartmann, J., Beckerman, S., Seamans, T., Engeman, R., and Abu-Absi, S. (2009). *Report to the City of Chicago on Conflicts With Ring-Billed Gulls and the 2008 Integrated Ring-Billed Gull Damage Management Project*. Prepared for Chicago Department of Environment.

Hiller, J. (2009). Treatment of Stormwater Runoff at Public Bathing Beaches. *2009 Joint Conference, Lake Michigan: State of the Lake and Great Lakes Beach Association*. <http://aqua.wisc.edu/SOLM/LinkClick.aspx?fileticket=ycDBc7nSb_Y%3d&tabid=74>.

Holevinski, R.A., Curtis, P.D., and Malecki, R.A. (2007). Hazing of Canada geese is unlikely to reduce nuisance populations in urban and suburban communities. *Human-Wildlife Conflicts*, **1**, 257–64.

Kerley, L.L. (2010). Using dogs for tiger conservation and research. *Integrative Zoology*, **5**, 390–95.

Khan, M.M.H. (2009). Can domestic dogs save humans from tigers *Panthera tigris*? *Oryx*, **43**, 44–47.

Kloppers, E.L., St. Clair, C.C., and Hurd, T.E. (2005). Predator-resembling aversive conditioning for managing habituated wildlife. *Ecology and Society*, **10(1)**, 31.

Koski, A. and Kinzelman, J. (2010). *A review of best management practices benefiting Great Lakes recreational waters: current success stories and future innovations*. A review paper published on behalf of the International Joint Commission <http://meeting.ijc.org/sites/default/files/workgroups/Kinzelman_2.pdf>.

Landry, J.M. (1999). The use of guard dogs in the Swiss Alps: A first analysis. *KORA Report no. 2*. <http://www.kora.ch/pdf/reports/rep2_e.pdf>.

Landry, J.M. (2004). Livestock guarding dogs and hikers in Bas-Valais (Switzerland). *Report for the Swiss Agency for the Environment, Forest and Landscape. IPRA sàrl, Chemin, Switzerland [in French]*. <http://www.kora.ch/en/proj/cdpnews/publications/Landry2004Chientourisme.pdf>.

Landry, J.M. (2010). Essais de l'utilisation du chien de protection et d'autres outils de prévention pour protéger un groupe de veaux contre la prédation de loups sur l'estive de Champlaitier (74). *Report for French Agriculture and Fish Ministry*. Institute for the Promotion and Research on Guarding Animals, Martigny, Switzerland.

Landry, J.M. and Raydelet, P. (2010). Efficacité des chiens de protection contre la prédation du lynx dans le Massif jurassien. Report for the French Ministry of Environment. *Pôle Grand Prédateur*. (http://www.polegrandspredateurs.org).

Landry, J.M. and Valensi, P. (2011). Un regard croisé entre l'éthologie, la biologie moléculaire et l'archéozoologie pour expliquer les conditions nécessaires à la domestication du loup en chien. In *Prédateurs dans tous leurs états. Évolution, biodiversité, interactions, mythes, symboles. XXXIe rencontres internationales d'archéologie et d'histoire d'Antibes*. pp. 113–125.

Landry J.M., Burri, A., Torriani, D., and Angst, C. (2005). Livestock guarding dogs: a new experience for Switzerland. *Carnivore Damage Prevention News*, **8**, 40–8.

Lapeyronie, P. and Moret, A. (2003). Chiens de troupeaux dans le Parc national du Mercantour (chiens de travail et chiens de protection): Étude comportementale, impact sur la faune sauvage des estives. *Programme Life: Le Retour du Loup dans les Alpes Francaises*, France.

Laurenson, K., Sillero-Zubiri, C., Thompson, H., Shiferaw, F., Thirgood, S., and Malcom, J. (1998). Disease as a threat to endangered species: Ethiopian wolves, domestic dogs and canine pathogens. *Animal Conservation*, **1**, 273–80.

Levin, M. (2005). Livestock guarding dogs in Sweden: a preliminary report. *Carnivore Damage Prevention News*, **8**, 8–9.

Linhart, S.B., Sterner, R.T., Carrigan, T.C., and Henne, D.R. (1979). Komondor guard dogs reduce sheep losses to coyotes: a preliminary evaluation. *Journal of Range Management*, **32**, 238–41.

Lorenz, J.R. (1985). Introducing livestock-guarding dogs. *Oregon State University Extension Service, Extension Circular 1224.*

Lorenz, J.R., and Coppinger, L. (1986). Raising and training a livestock-guarding dog. *Oregon State University Extension Service, Extension Circular 1238.*

Lorenz, J.R., Coppinger, R.P., and Sutherland, M.R. (1986). Causes and economic effects of mortality in livestock guarding dogs. *Journal of Range Management*, **39**, 293–95.

Lustig, D. (2011). Maremma sheepdogs keep watch over Little Penguins. *The Bark*, **65**, 1–4.

Marker, L.L. (2002). Aspects of cheetah (*Acinonyx jubatus*) biology, ecology and conservation strategies on Namibian farmlands. PhD dissertation, University of Oxford.

Marker, L.L., Mills, M.G.L., and Macdonald, D.W. (2003). Factors influencing perceptions of conflict and tolerance toward cheetahs on Namibian farmlands. *Conservation Biology*, **17**, 1290–98.

Marker, L.L., Dickman, A.J., and Macdonald, D.W. (2005). Perceived effectiveness of livestock guarding dogs placed on Namibian farms. *Rangeland Ecology and Management*, **58**, 329–36.

Marker, L.L., Dickman, A.J., Wilkinson, C., Schumann, B., and Fabiano, E. (2007). The Namibia cheetah: status report. *Cat News Special Issue*, **3**, 5–13.

Marker-Kraus, L.L., Kraus, D., Barnett, D., and Hurlburt, S. (1996). *Cheetah survival on Namibian farmlands.* Cheetah Conservation Fund, Windhoek, Namibia.

Mech, L.D., Fritts, S.H., and Wagner, D. (1995). Minnesota wolf dispersal to Wisconsin and Michigan. *American Midland Naturalist*, **133**, 368–70.

Mertens, A. and Schneider, H. (2005). What is wrong with Romanian livestock guarding dogs? A discussion. *Carnivore Damage Prevention News*, **9**, 9–14.

Morsbach, D. (1987). Cheetahs in Namibia. *Cat News*, **6**, 25–6.

Musiani, M., Muhly, T., Gates, C.C., Callaghan, C., Smith, M.E., and Tosoni, E. (2005). Seasonality and reoccurrence of depredation and wolf control in western North America. *Wildlife Society Bulletin*, **33**, 876–87.

Namgail, T.J., Fox, L., and Bhatnagar, Y.V. (2007). Carnivore-caused livestock mortality in Trans-Himalaya. *Environmental Management*, **39**, 490–96.

Nowak, S. and Mysłajek, R.W. (2004). Livestock guarding dogs in the western part of the Polish Carpathians. *Carnivore Damage Prevention News*, **January**, 13–17.

Olsen, J.W. (1985). Prehistoric dogs in mainland East Asia. In Olsen, S. J., Ed., *Origins of the Domestic Dog: The Fossil Record*. University of Arizona Press. Tucson, pp. 47–70.

O'rick, B.D. (2000). The use of Border Collies to disperse birds at Southwest Florida International Airport. In *Proceedings of the Second Bird Strike Committee-USA/Canada Meeting*. Minneapolis, Minnesota.

Otstavel, T., Vuori, K., Sims, D.E., Valros, A., Vainio, O., and Saloniemi, H. (2009). The first experience of livestock guarding dogs preventing large carnivore damages in Finland. *Estonian Journal of Ecology*, **58**, 216.

Patterson, B. (2000). Wildlife control at Vancouver International Airport: introducing Border Collies. *Twenty-fifth International Bird Strike Committee Conference* <http://www.int-birdstrike.org/Amsterdam_Papers/IBSC25%20WPA6.pdf>.

Preusser, S.E., Seamans, T.W., Gosser, A.L., and Chipman, R.B. (2008). Evaluation of an integrated non-lethal Canada goose management program in New York (2004–2006). In R.M. Timm and M.B. Madon, Eds., *Proceedings of the Twenty-Third Vertebrate Pest Conference*. University of California, Davis, pp. 66–73.

Potgieter, G.C. (2011). The effectiveness of livestock guarding dogs for livestock production and conservation in Namibia. MS thesis, Nelson Mandela Metropolitan University, South Africa.

Potgieter, G.C., Marker, L. L. Avenant, N.L., and Kerley, G. I.H. (in press). Why Namibian farmers are satisfied with the performance of their livestock guarding dogs. *Human Dimensions of Wildlife*.

Reiter, D.K., Brunson, M.W., and Schmidt, R.H. (1999). Public attitudes toward wildlife damage management and policy. *Wildlife Society Bulletin*, **27**, 746–58.

Reynolds, J.C. and Tapper, S.C. (1996). Control of mammalian predators in game management and conservation. *Mammalian Review*, **26**, 127–56.

Ribeiro S. and Petrucci-Fonseca, F. (2005). The use of livestock guarding dogs in Portugal. *Carnivore Damage Prevention News*, **9**, 27–33.

Richardson, W. E. (1994). Introduction of guarding dogs to Namibia. *DogLog*, **3**, 2–5.

Ritchie, E.G. and Johnson, C.N. (2009). Predator interactions, mesopredator release and biodiversity conservation. *Ecology Letters*, **12**, 982–8.

Rigg, R. (2001). *Livestock guarding dogs: their current use worldwide*. IUCN/SSC Canid Specialist Group Occasional Paper no 1 <http://www.canids.org/occasionalpapers/livestockguardingdog.pdf>.

Rigg, R. (2005). Livestock depredation and livestock guarding dogs in Slovakia. *Carnivore Damage Prevention News*, **January**, 17–27.

Rust, N. A., Whitehouse-Tedd, K. M., and MacMillan, C. (in press). Perceived efficacy of livestock guarding dogs in South Africa. *Wildlife Society Bulletin*.

Schilder, M.B.H. and van der Borg, J.A.M. (2004). Training dogs with help of the shock collar: short and long term behavioural effects. *Applied Animal Behaviour Science*, **85**, 319–34.

Schmitt, S.M., O'Brien, D.J., Bruning-Fann, C.S., and Fitzgerald, S.D. (2006). Bovine tuberculosis in Michigan wildlife and livestock. *Annals of the New York Academy of Sciences*, **969**, 262–8.

Schommer, T. and Woolever, M. (2008). A review of disease related conflicts between domestic sheep and goats and bighorn sheep. *General Technical Report RMRS-GTR-209*, U. S. Department of Agriculture, Forest Service, Rocky Mountain Research Station, Fort Collins, Colorado.

Schumann, M. (2003). *Guide to Integrated Livestock and Predator Management*. Cheetah Conservation Fund, Windhoek, Namibia.

Shivik, J.A. (2004). Non-lethal alternatives for predation management. *Sheep and Goat Research Journal*, **19**, 64–71.

Sims, D.E. and Dawydiak, O. (1990). *Livestock Protection Dogs: Selection, Care and Training*. OTR Publications. Centreville.

Singer, F.J., Zeigenfuss, L.C., and Spicer, L. (2001). Role of patch size, disease, and movement in rapid extinction of bighorn sheep. *Conservation Biology*, **15**, 1347–54.

Śmietana, W. (2005). Use of Tatra Mountains shepherd dog in the Bieszczady Mountains and the Bieszczady Foothills, Poland. *Carnivore Damage Prevention News*, **8**, 10–12.

Smith, M.E., Linnell, J.D.C., Odden, J., and Swenson, J.E. (2000). Review of methods to reduce livestock depredation: I. guardian animals. *Acta Agriculture Scandinavia, Section A, Animal Science*, **50**, 279–90.

Solman, V.E.F. (1994). Gulls. In S.E. Hygnstrom, Timm, R.M., and G.E. Larson, Eds., *Prevention and Control of Wildlife Damage*. U.S. Department of Agriculture, APHIS-WS. pp. E49–E52.

Stannard, C. (2006). Livestock protection dogs. In B. Daly, H. Davies-Mostert, W. Davies-Mostert, S. Evans, Y. Friedmann, N. King, N., T. Snow, and H. Stadler, Eds., *Prevention is the Cure. Proceedings of a Workshop on Holistic Management of Human-Wildlife Conflict in the Agricultural Sector of South Africa*. Endangered Wildlife Trust, Johannesburg, pp. 43–45.

Subba, S. A. (2012). Assessing the genetic status, distribution, prey selection and conservation issues of Himalayan wolf (*Canis himalayensis*) in Trans-Himalayan Dolpa, Nepal. MS thesis, Lund University.

Swift, B. L. (2000). Suburban goose management: insights from New York state. In M.C. Brittingham, J. Kays and R. McPeake, Eds., *The Ninth Wildlife Damage Management Conference Proceedings*. State College, Pennsylvania, pp. 307–21.

Toronto Beaches Plan. (2009). <http://www.toronto.ca/legdocs/mmis/2009/ex/bgrd/backgroundfile-18573.pdf>.

Vanak, A.T. and Gompper, M.E. (2009). Dogs *Canis familiaris* as carnivores: their role and function in intraguild competition. *Mammal Review*, **39**, 265–83.

Vanak, A.T. and Gompper, M.E. (2010). Interference competition at the landscape level: the effect of free-ranging dogs on a native mesocarnivore. *Journal of Applied Ecology*, **47**, 1225–32.

Vanak, A.T., Thaker, M., and Gompper, M.E. (2009). Experimental examination of behavioural interactions between free-ranging wild and domestic canids. *Behavioral Ecology and Sociobiology*, **64**, 279–87.

van Bommel, L. and Johnson, C.N. (2012). Good dog! Using livestock guardian dogs to protect livestock from predators in Australia's extensive grazing systems. *Wildlife Research*, **39**, 220–29.

VerCauteren, K., Seward, N., Hirchert, D., Jones, M., and Beckerman, S. (2005). Dogs for reducing wildlife damage to organic crops: a case study. *Proceedings of the Wildlife Damage Management Conference*, **11**, 286–93.

VerCauteren, K.C., Lavelle, M.J., and Hygnstrom, S.E. (2006). Fences and deer-damage management: a review of designs and efficacy. *Journal of Wildlife Management*, **34**, 191–200.

VerCauteren, K.C., Lavelle, M.J., and Phillips, G.E. (2008). Livestock protection dogs for deterring deer from cattle and feed. *Journal of Wildlife Management*, **72**, 1443–48.

VerCauteren, K.C., Lavelle, M.J., Gehring, T.M., and Landry, J. (2012). Cow dogs: use of livestock protection dogs for reducing predation and transmission of pathogens from wildlife to cattle. *Applied Animal Behaviour Science*, **140**, 128–36.

Walter, W.D., Anderson, C.W., Smith, R., Vanderklok, M., Averill, J.J., and VerCauteren, K.C. (2012). On-farm mitigation of transmission of tuberculosis from white-tailed deer to cattle: literature review and recommendations. *Veterinary Medicine International*, **2012**, 616318.

Ward, A.I., VerCauteren, K.C., Walter, W.D., Gilot-Fromont, E., Rossi, S., Edwards-Jones, G., Lambert, M.,

Hutchings, M.R., and Delahay, R.J. (2009). Options for the control of disease 3: targeting the environment. In R.J. Delahay, G.C. Smith and M.R. Hutchings, Eds., *Management of Disease in Wild Mammals*. Springer, Tokyo, pp. 147–68.

WDFW. (no date *a*). *Karelian Bear Dog program*. Washington Department of Fish and Wildlife <http://www.wdfw.wa.gov/enforcement/kbd/index.html>.

WDFW. (no date *b*). *Karelian Bear Dog program: Karelian Bear Dogs at work*. Washington Department of Fish and Wildlife <http://www.wdfw.wa.gov/enforcement/kbd/kbd_dogs.html>.

Woodruff, R.A. and Green, J.S. (1995). Livestock herding dogs: A unique application for wildlife damage management. *Proceedings of the Great Plains Wildlife Damage Control Workshop*, **12**, 43–45.

Wyckoff, C., Henke, S., Campbell, T., Hewitt, D., and VerCauteren, K.C. (2012). Movement and habitat use of feral swine near domestic swine facilities. *Wildlife Society Bulletin*, **36**, 130–8.

Young, J.K., Olson, K.A., Reading, R.P., Amgalanbaatar, S., and Berger, J. (2011). Is wildlife going to the dogs? Impacts of feral and free-roaming dogs on wildlife populations. *BioScience*, **61**, 125–32.

CHAPTER 10

The current and future roles of free-ranging detection dogs in conservation efforts

Deborah A. (Smith) Woollett, Aimee Hurt, and Ngaio L. Richards

10.1 A brief history of detection dogs

In a working relationship, the acute olfactory abilities of dogs and their unique capacity to engage with humans have been harnessed to accomplish myriad tasks. Whether as partners and recognized assets in hunting, law-enforcement, and search and rescue missions, or more recently in the detection of chemicals, animal and plant pests, and diseases, dogs have been trained to perceive and distinguish a most impressive array of odors. Missing living and deceased persons (Button, 1990), accelerants (Katz and Midkiff, 1998), explosives (Gazit and Terkel, 2003), cancer cells (McCulloch et al., 2006; Willis et al., 2004), blood sugar levels (Chen et al., 2000), termites (Brooks et al., 2003), bed bugs (Pfiester et al., 2008), and contraband agricultural products (Cross, 2006) comprise only a fraction of the target scents that dogs have successfully learned to detect for our benefit.

Though typically less well known than traditional scent dog applications or professions, the use of dogs in wildlife and conservation efforts nonetheless boasts an extensive history. Over the past 100 years, applications of dogs in wildlife research and conservation studies have expanded substantially, and their involvement has become increasingly common worldwide (see Figure 10.1 and Table 10.1). The earliest conservation canines were hunting dogs bred to point to and/or retrieve birds, in this case enlisted to help recover field data for wildlife and conservation studies. As early as the 1890s, these dogs were selected to perform scent work in New Zealand, where they were used to locate the flightless and hence highly vulnerable kiwi (*Apteryx spp.*) and kakapo (*Strigops habroptilus*) as part of efforts to relocate them to a predator-free island (Hill and Hill, 1987).

The wealth of conservation benefits that have been reaped over time using dogs to search for instinctual targets (i.e. birds) for conservation purposes, is discussed in Dahlgren et al. (2012). When statistical and laboratory techniques were fine tuned and the amount of information that could be obtained from scat rose sharply, so too did its value to conservation studies. Although dogs were already being fielded to seek non-instinctual targets, this development in techniques provided the impetus to further expand on their use to seek the scat of a specific species of interest (discussed in Section 10.2.1), among other targets. In this expansion, trainers began to focus solely on selecting dogs with a keen interest in receiving a reward, and that could readily be taught to detect a target that is inherently meaningless to them in order to elicit the desired reward outcome (Hurt and Smith, 2009). This particular training approach has further broadened the use of dogs to facilitate the collection of data on imperiled species and to assist in the location and eradication of invasive species. Drawing on the training principles and practices from narcotics, cadaver, and search-and-rescue work (extensively detailed in Bulanda, 1994; Button, 1990; Pearsall and Verbruggen, 1982; Rebmann et al., 2000; Robicheaux and Jons, 1996) helped to further refine the search for non-instinctual targets.

Free-Ranging Dogs and Wildlife Conservation. Edited by Matthew E. Gompper
© Oxford University Press 2014. Published 2014 by Oxford University Press.

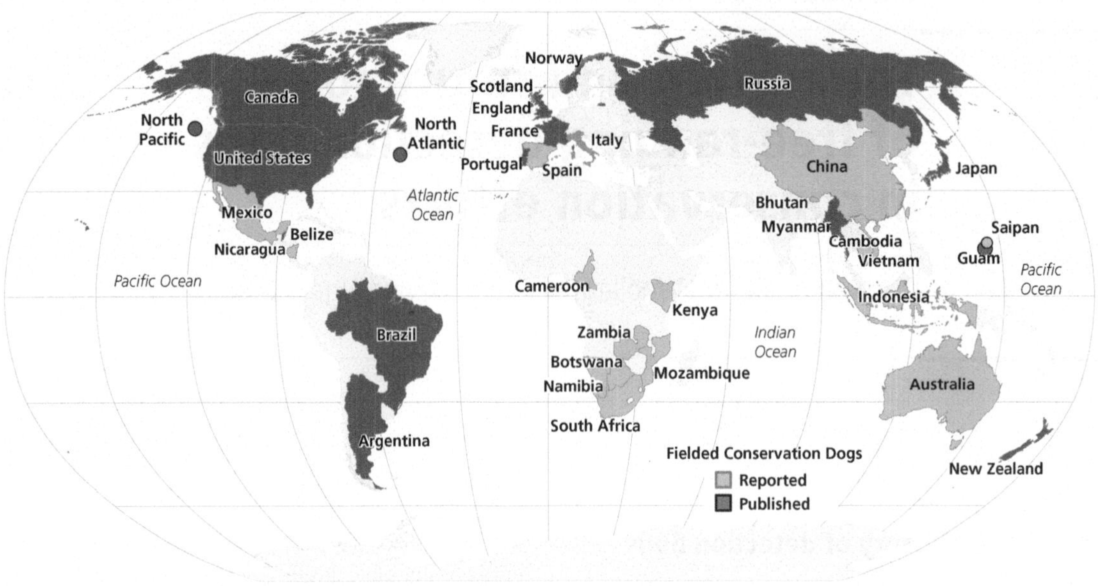

Figure 10.1 There is an ever-increasing use of conservation dogs worldwide. At the time of writing, published accounts in peer-reviewed journals detail dogs being fielded in 14 countries and on 2 oceans. Dogs have reportedly been fielded in at least 18 other countries (based on news reports and personal communications). Figure courtesy W. E. Thogmartin, United States Geological Survey.

Table 10.1 Species detected with or assisted by conservation dogs. This table cites field-based studies in which dogs have been used to recover data about specified taxa. Many additional or novel applications involving laboratory-based dogs, dogs in training and awaiting field deployment, unpublished endeavors not yet appearing in peer-reviewed publications or otherwise unavailable to the authors at the time of writing, may not appear in this table.

Class, Order	Common Name	Taxon	Target sought	Location	Citation
Aves, Galliformes, Chardriiformes, Anseriformes	Grouse, ptarmigan, quail, plover, ducks, geese	Phasianidae, Odontophoridae, Charadriidae, Anatidae	Carcass/Live animal	USA, UK, Canada, Europe	extensive use, see Dahlgren et al. (2012)
Aves, Struthioniformes	Kiwi	*Apteryx sp.*	Live animal	New Zealand	Hill and Hill (1987); Robertson and Fraser (2009)
Mammalia, Artiodactyla	Moose	*Alces alces*	Scat	Canada	Wasser et al. (2011); Kretser and Glennon (2011)
	Woodland caribou	*Rangifer tarandus caribou*	Scat	Canada	Wasser et al. (2011)
Mammalia, Carnivora	Coyote	*Canis latrans*	Scat	USA	Tom and Cox (in preparation)
	Gray wolf	*Canis lupus*	Scat	USA, Canada	Sturdivan (1993); Beckmann (2006); Reed et al. (2011); Wasser et al. (2011); Beckmann et al. (in preparation)
	Maned wolf	*Chrysocyon brachyurus*	Scat	Brazil	Vynne et al. (2011)
	Small Indian mongoose	*Herpestes javanicus*	Live animal	Japan	Fukuhara et al. (2010)
	River otter	*Lontra canadensis*	Scat	USA	Alexander (2005)

continued

Table 10.1 *Continued*

Class, Order	Common Name	Taxon	Target sought	Location	Citation
	Bobcat	*Lynx rufus*	Scat	USA	Harrison (2006); Long et al. (2007a, 2007b, 2011); Reed et al. (2011); Tom and Cox (in preparation)
	Fisher	*Martes pennanti*	Scat	USA	Long et al. (2007a, 2007b, 2011); Thompson et al. (2010, 2012)
	Ermine	*Mustela erminea*	Den	New Zealand	Theobald and Coad (2002)
	Long-tailed weasel	*Mustela frenata*	Scat	USA	Steury (2011)
	Black-footed ferret	*Mustela nigripes*	Scat/Live animal	USA	Dean (1979); Winter (1981); Reindl-Thompson et al. (2006)
	Jaguar	*Panthera onca*	Scat	Brazil, Belize	Wultsch (2009); Vynne et al. (2011)
	Amur tiger	*Panthera tigris altaica*	Scat	Russia	Kerley (2010)
	Ringed seal	*Phoca hispida*	Lair	Canada, Norway	Lydersen and Gjertz (1986)
	Cougar	*Puma concolor*	Scat	USA, Brazil	Beckmann (2006); Reed et al. (2011); Vynne et al. (2011); Beckmann et al. (in preparation)
	Bush dog	*Speothos venaticus*	Sign	Argentina	DeMatteo et al. (2009)
	Gray fox	*Urocyon cinereoargenteus*	Scat	USA	Tom and Cox (in preparation)
	Black bear	*Ursus americanus*	Scat	USA, Canada	Wasser et al. (2004); Long et al. (2007a, b, 2011); Beckmann et al. (in preparation); Steury (in preparation)
	Grizzly bear	*Ursus arctos*	Scat	USA, Canada	Wasser et al. (2004); Beckmann (2006); Beckmann et al. (in preparation)
	San Joaquin kit fox	*Vulpes macrotis mutica*	Scat	USA	Smith et al. (2001, 2006); Ralls and Smith (2004); Ralls et al. (2010); Reed et al. (2011)
	Red fox	*Vulpes vulpes*	Scat	USA	Boydston (2005); Smith et al. (2006); Reed et al. (2011)
Mammalia, Cetacea	North Atlantic right whale	*Eubalaena glacialis*	Scat	North Atlantic Ocean	Rolland et al. (2006)
	Killer whale	*Orcinus orca*	Scat	Pacific Ocean	Ayres et al. (2012)
Mammalia, Chiroptera	Bats	Multiple sp.	Carcass	USA, Portugal	Arnett (2006); Weller (2008); Paula et al. (2011)
Mammalia, Cingulata	Giant armadillo	*Priodontes maximus*	Scat	Brazil	Vynne et al. (2011)
Mammalia, Diprotodontia	Brushtail possum	*Trichosurus vulpecula*	Live animal	New Zealand	Cowan (1992)
Mammalia, Pilosa	Giant anteater	*Myrmecophaga tridactyla*	Scat	Brazil	Vynne et al. (2011)

continued

Table 10.1 *Continued*

Class, Order	Common Name	Taxon	Target sought	Location	Citation
Mammalia, Rodentia	House mouse	*Mus musculus*	Live animal	New Zealand	Gsell et al. (2010)
	Franklin's ground squirrel	*Poliocitellus franklinii*	Live animal/Scat	USA	Duggan et al. (2011)
	Norway rats	*Rattus rattus*	Live animal	New Zealand	Gsell et al. (2010)
Reptilia, Squamata	Brown tree snake	*Boiga irregularis*	Live animal	USA/Guam	Engeman et al. (1998a, b, 2002); Vice and Engeman (2000); Savidge et al. (2010)
	Eastern indigo snake	*Drymarchon couperi*	Live animal/Shed skin	USA	Stevenson et al. (2010)
	Burmese python	*Python molurus bivittatus*	Live animal	USA	Romagosa et al. (in preparation)
	North African python	*Python sebae*	Live animal	USA	Romagosa et al. (in preparation)
Reptilia, Testudines	Burmese star tortoise	*Geochelone platynota*	Live animal	Myanmar	Platt et al. (2003)
	Desert tortoise	*Gopherus agassizii*	Live animal	USA	Cablk and Heaton (2006); Cablk et al. (2008); Nussear et al. (2008)
	Florida box turtle	*Terrapene carolina bauri*	Live animal	USA	Liu et al. (2004)
	Eastern box turtle	*Terrapene carolina carolina*	Live animal	USA	Kapfer et al. (2012)
	Three toed box turtle	*Terrapene carolina triunguis*	Live animal	USA	Schwartz and Schwartz (1974)
Reptilia, Thynchocephalia	Tuatara	*Sphenodon spp.*	Live animal	New Zealand	Browne (2005)
Insecta, Hymenoptera	Bumblebee	*Bombus sp.*	Nest	UK	Waters et al. (2010); O'Connor et al. (2012)
	Red imported fire ant	*Solenopsis invicta*	Nest/Live animal	Taiwan	Lin et al. (2011)
Plantae, Asterales	Spotted knapweed	*Centaurea stoebe*	Plant	USA	Goodwin et al. (2010)
Plantae, Poales	Salt marsh grass	*Spartina patens*	Plant	USA	Milne (2007)
Fungi, Ophiostomatales, Russulales	Invasive tree root fungi (pine decline, little leaf disease, Annosum root rot)	*Leptographium sp., Grosmannia sp., Heterobasidion irregulare sp.*	Fungus	USA	Eckhardt and Steury (2011)

This chapter focuses on the use of free-ranging detection dogs in conservation and management efforts as an overview of the benefits (and drawbacks) of this unique survey approach. Throughout the chapter, they are referred to as conservation dogs, conservation detection dogs and scat detection dogs. In Section 10.2, selected case studies highlight the versatility of free-ranging detection dogs and some of their most notable contributions to date. Discussions that encompass both free- *and* non-free-ranging dogs in wildlife research and conservation are beyond the scope of this chapter, but can be found in MacKay et al. (2008); Hurt and Smith (2009), and Dahlgren et al. (2012), and are briefly considered in

Box 10.1 Non-free-ranging conservation dogs

Non-free-ranging dogs, for example on-leash and/or in a building, assist federal and state or provincial agencies during inspections to detect concealed endangered animals, living or dead, and their harvested parts (Parry-Jones, 2009) as well as stowaway non-native species, such as brown treesnakes (*Boiga irregularis*) (Engeman et al., 1998a, b) and dreissenid (e.g., zebra) mussels (California Department of Fish and Game, unpublished data). The scent discrimination abilities of dogs have been harnessed to differentiate the scat of species or individuals in a controlled setting (Harrison, 2006; Kerley and Salkina, 2007; Wasser et al., 2009). Such an effort can greatly minimize erroneous and expensive genetic analysis of non-target scat samples (see also Dalhgren et al., 2012). One of the most interesting non-free-ranging dog applications has been in the realm of cetacean studies. From the bow of a boat, dogs have been able to orient the search for, and detect, fecal samples from North Atlantic right whales (*Eubalaena glaciali*) and killer whales (*Orcinus orca*) for endocrine, disease, genetic, nutritional, and/or biotoxin studies (Ayres et al., 2012; Rolland et al., 2006).

Box 10.1. Section 10.3 addresses limitations inherent in the use of free-ranging detection dogs in the field and provides recommendations for working dogs with, optimally, little or no impact. Lastly, Section 10.4 summarizes possible avenues for the involvement and integration of free-ranging detection dogs in future research and conservation efforts and outlines areas where safety and performance can be enhanced.

10.2 Use of free-ranging detection dogs in conservation: selected case studies

Identifying conservation priorities and planning appropriate management and monitoring strategies requires high quality data. In a conservation context, detection dogs are above all a unique and efficient data collection tool that researchers and managers have at their disposal. Therefore, the primary goal of incorporating trained dogs in conservation studies is to increase the size and quality of datasets so as to better understand factors such as population size and demographics, presence or absence in an area, habitat selection, disease and parasite levels, and diet. Free-ranging dogs have most often been deployed in natural environments by wildlife researchers and managers to seek the sign (e.g., scat, hair, dens) of low density and/or secretive animal species, study multiple species simultaneously, and increase sample sizes. In some instances, the key information necessary to make conservation decisions can only, or best, be obtained using detection dogs due to their proficiency in detecting samples (see Box 10.2). Given modern developments discussed in this section, such key information can now be obtained from sign rather than the animal itself, making dogs one of the least invasive survey tools available. Here, we highlight the array of targets sought by free-ranging conservation dogs, from scat to reclusive or invasive animals, insects, carcasses, and plants. The case studies provided reflect the fact that the analysis of dog-collected samples augmented the quality and breadth of the information available, which either led to key recommendations or to tangible conservation impacts.

10.2.1 Detection and recovery of scat

Scat has long been recognized as a rich source of information about wild populations (Kohn and Wayne, 1997; Putnam, 1984). With this in mind, dogs have been trained to identify and locate the scat of species of interest for wildlife studies since the 1970s (see Dean, 1979; MacKay et al., 2008; Winter, 1981; Paquet, unpublished data; Breitenmoser and Breitenmoser-Wursten, unpublished data; Table 10.1). Once it became possible to extract DNA from scat in the 1990s, wildlife researchers began to formally integrate dog-handler teams in a more systematic search for scat samples (Wasser, Parker and Davenport, unpublished data and see MacKay et al., 2008; Smith et al., 2003; Wasser et al., 2004; Wasser, Parker, and Davenport, unpublished data). The possibility of gaining additional and better information from scat and increasing sample size by using trained dogs has made it a compelling focal sample in conservation studies and monitoring efforts. The advancement in both the training and survey application also served as a significant stepping stone for the role of detection dogs in conservation work to be recognized and further amplified.

Box 10.2 Effectiveness and cost: dogs versus other survey methods

Detection dogs can be used as a stand-alone data collection tool or in conjunction with other survey methods (e.g., live trapping, radio tracking, remote cameras, hair collection—i.e., hair snares). Several studies have focused on comparing these techniques (Duggan et al., 2011; Harrison, 2006; Long et al., 2007b), and numerous other studies have contrasted the performance of detection dogs for particular uses with an alternative survey method (Table 10.2).

Capable of searching large areas in a variety of habitats and detecting single or multiple species simultaneously, detection dog-handler teams generally find more samples than people searching alone. Harrison (2006) reported that a dog trained to find bobcat (*Lynx rufus*) scats produced nearly ten times the number of detections as automatic cameras, hair snares, and scent stations combined. He noted that although it was most expensive (and required more field time) to deploy the dog in the field, it was also most efficient, and only one visit to each survey site was required. Smith et al. (2001) demonstrated that a trained dog found up to four times as many kit fox (*Vulpes macrotis*) scats along transects as an experienced surveyor visually searching in the same area. Likewise, when searching for avian carcasses in dense vegetation, dogs showed more than twice the search efficiency of human searchers, providing compelling evidence of their ability to improve field-based assessments of avian mortality rates (Homan et al., 2001).

Table 10.2 Selected comparisons between the performance of conservation dogs and other survey methods.

Animal or plant sought	Factor compared	Dog search, type	Other search type		Citation
		Seeking scat	Camera	Hair snare	
Fisher presence	Probability of detection, if present	84%	28%	0%	Long et al. (2007b)
	Percent of time this method produced the only detection of the species	75%	2%	0%	
Black bear presence	Probability of detection, if present	87%	33%	8%	Long et al. (2007b)
	Percent of time this method produced the only detection of the species	65%	2%	4%	
Bobcat presence			Automated cameras, scent stations, hair snares		
	Number of detections	Dog produced 10 times more detections than other 3 methods combined			Harrison (2006)
	Days required to complete search across study area	5 to 10	3 to 5		
			Camera	Hair snare	
Bobcat presence	Probability of detection, if present	27%	13%	0%	Long et al. (2007b)
	Percent of time this method produced the only detection of the species	79%	14%	0%	

continued

Box 10.2 *Continued*

Table 10.2 *Continued*

Animal or plant sought	Factor compared	Dog search, type	Other search type	Citation
North Atlantic right whale scat	Scats/hr		Opportunistic recovery	Rolland et al. (2006)
		1.10	0.25	
		Seeking animal or burrow	Spotlighting	
Black-footed ferret, presence	Find rate	79%/86%	100%[a]	Reindl-Thompson et al. (2006)
	Time to detection (mins)	21	208	
	Search speed (ha/hr)	26	1.60	
			Live trapping	
Franklin's ground squirrel, presence	Find rate: 1 dog team/1 day trapping	59%	61%	Duggan et al. (2011)
	Find rate: 2 dog teams/2 days trapping	83%	84%	
	Days required to complete search across study area	9	18	
			Visual surveys	
Eastern box turtles	Number of turtles located	25	22	Kapfer et al. (2012)
	Time searched (hrs)	9	316.50	
Desert tortoises	Probability of detecting tortoises on surface, in:	70%	70%	Nussear et al. (2008)
	-burrows	Performed same		
	-tortoises in shrubs	Dogs found more		
	-males v. female tortoises	Performed same		
	-time searched (hrs)	5.92	8.52	
Desert tortoises	Number of tortoises located	50	40	Cablk and Heaton (2006)
	Size of tortoises located	30–280 mm	110–280 mm	
Brown treesnakes	Snakes located	35%*	7%#	*Savidge et al. (2010); #Christy et al. (2010)
Bumble bee nests		Seeking nests		O'Connor et al. (2012)
	Nests per ha^{-1}	1.41	1.44	
Bat carcasses		Seeking carcasses		Arnett (2006)
	Percent carcasses located	71%/81%	42%/14%	

continued

Box 10.2 *Continued*

Table 10.2 *Continued*

Animal or plant sought	Factor compared	Dog search, type	Other search type	Citation
		Seeking plants		
Spotted knapweed	Plants located—all sizes combined	81%	59%	Goodwin et al. (2010)
	Small plants	67%	35%	
	Medium plants	94%	78%	
	Large plants	100%	94%	

[a]Spotlighting was used to establish ferret presence, against which dog performance was compared. It is, however, possible that some ferrets went undetected by spotlighting.

In a Vermont, USA, study, scat detection dogs yielded the highest probability of detection for each of three target species (black bears *Ursus americanus*, fishers *Martes pennanti*, and bobcats) over remote cameras and hair snares, and the greatest number of unique detections or occasions when only one method detected the target species (Long et al., 2007b). Overall, the efficiency of scat detection dogs made them the most cost-effective survey method assessed in this study (Long et al., 2007b). Detection/non-detection data amassed from this study provided valuable insights into habitat use at both the regional and landscape scale (Long et al., 2011). Here, the use of detection dogs in conjunction with the other non-invasive survey methods facilitated the gathering of information and prediction of occupancy, based on land type (e.g., forested, human developed).

Comparisons of the effectiveness and cost of distribution surveys using live trapping versus detection dog-handler teams were made for a cryptic rodent, Franklin's ground squirrel (*Poliocitellus franklinii*), a species that occurs at low densities in the Midwestern United States and, depending on its location, is listed as threatened, endangered, or a species of concern (Duggan et al., 2011). Using conservation dogs provided a rapid means of determining the distribution of ground squirrels. Twice as many sites could be surveyed per season using dogs than live trapping, without excessively increasing the cost. However, as a management tool for larger scale monitoring, and to ensure that determination of population status be as accurate as possible, Duggan et al. (2011) suggested that detection dogs be applied *in conjunction* with live trapping, during periods when dogs might be likely to alert—not incorrectly per se—to residual scent even after an individual animal may have vacated its burrow.

There remains some debate as to the efficacy and cost-effectiveness of using dogs. For example, O'Connor et al. (2012) concluded that it was not cost-effective to use detection dogs to seek out bee (*Bombus* spp.) nests. The value placed on a dog's work or the dog-directed search and the quality and quantity of samples may vary substantially according to the conservation goal, the target, the rarity or elusiveness of the focal species, the novelty of the work underway, and the complexity of the survey habitat, among others. If other minimally-invasive techniques are available for collecting data on these species (e.g., cameras or track plates with baits or attractants that can be used for extended periods of time), it may be worth pursuing those alternatives instead of, or in addition to, deploying conservation dogs. However, in many cases, just a few samples from rare and elusive animals may be deemed of inestimable value, and conservation dogs may still offer the best (or even only) option for their recovery. DeMatteo et al. (2009) maintained that rare bush dog (*Speothos venaticus*) locations could not have been detected by humans alone without the use of a dog. Similarly, Browne and Stafford (2003) reported that it would have been difficult and impractical to implement many of the protected species management programs now underway in New Zealand without using dogs.

In this section we outline some of the ways in which scat detection dogs have played an integral role in monitoring and conservation efforts, primarily by facilitating the recovery of samples that would otherwise be hard to come by. The first such example concerns a 4-year study conducted in the Centennial Mountains along the Idaho and Montana border in the Greater Yellowstone

Ecosystem, USA. With an overall aim of examining resource selection within this critical linkage zone, which is positioned between vast protected areas, the study investigated the role that habitat parameters, public land management, and temporal changes in land use might have on landscape usage by four large carnivore species (Beckmann, 2006; Beckmann et al., In preparation). Four dogs were trained to detect the scat of black bears, grizzly bears (*U. arctos*), wolves (*Canis lupus*), and puma (*Puma concolor*). Subsequent DNA analysis and sample location data were then used to develop a resource selection function model to understand the way this carnivore suite was using the Centennial range and both man-made and natural features relevant to that use. This improved understanding helped guide the closure of more than 40% of roads in the western half of the mountain range and played a role in suspending development of a 1,200-home housing development and 18-hole golf course within this critical area (Beckmann et al., In preparation).

Similarly, a multi-year study initiated in 2009 using conservation dogs to find evidence of the endangered San Joaquin kit fox (*V. m. mutica*) via detection of their scats ultimately cemented the purchase of new protected area land parcels in the Ciervo-Panoche Natural Area in the San Joaquin Valley, California, USA (Westphal et al., unpublished data). In a separate study, also conducted in the San Joaquin Valley, analysis of kit fox scat recovered using detection dogs revealed that robust populations occur in only a few locations within this region (Smith et al., 2006). The existence of such discrete core populations led to the recommendation that conservation efforts must be focused on the areas where these populations occur and, to the extent possible, the lands that connect them should be protected (Smith et al., 2006).

Dogs have also proven to be a useful tool to locate alternative, and supplemental, samples and obtain new data on moose (*Alces alces*) to help steer future management efforts. In New York State, USA, 141 Adirondack moose scats recovered by dog-handler teams (considerably more samples than were expected in one survey season) complemented 293 moose tissue and hair samples also gathered from surrounding areas, increasing the overall sample size for DNA analysis and facilitating an assessment of regional subpopulations (Kretser and Glennon, 2011). These data suggested the possibility of a meta-population and the need to pinpoint potential barriers to moose movements. They also provided a concrete basis upon which researchers prioritized maintaining linkages between populations. Kretser and Glennon (2011) highlighted that improperly placed or ill-conceived development could affect the permeability of a landscape to wildlife such as moose, and may restrict or challenge their movement—especially if temperatures at their southern range were to become inhospitable as a result of climate change.

During a multi-year study in and around Emas National Park in the Brazilian cerrado, detection dogs were used to locate scat samples to evaluate the presence and resource selection of five large mammals threatened with extinction by habitat loss: maned wolf (*Chrysocyon brachyurus*), jaguar (*Panthera onca*), puma, giant anteater (*Myrmecophaga tridactyla*), and giant armadillo (*Priodontes maximus*) (Vynne et al., 2010, 2011). Analyses of the locations where scat samples were recovered and their implications about land use revealed a strong presence of all species both within the park and also occupying areas outside its boundaries. These findings were consistent with prior research by Woodroffe and Ginsberg (1998), which had repeatedly shown that protected areas are often too small for species to thrive within their confines. Based on these results, Vynne et al. (2011) noted that supplementary conservation efforts (e.g., maintaining federally-mandated habitat remnant set-asides as corridors and buffer zones) on private lands adjacent or near protected areas were as important as managing the primary protected area itself. They also recommended establishing a more formal system through which to offer special conservation incentives to private land owners (Vynne et al., 2010).

Analysis of scat sometimes yields unexpected results (Box 10.3). In the Alberta oil sands region of western Canada, detection dogs were used over the course of four years by Wasser et al. (2011) to recover scat from caribou (*Rangifer tarandus*), moose, and wolves. Results obtained from analysis of scat were then used to estimate resource selection, measure physiological stress, and provide individual genetic identification for precise mark–recapture abundance estimates (Wasser et al., 2011). Instead of the projected declines in caribou abundance, the data obtained revealed higher individual caribou numbers than predicted, and no significant changes in the abundance of any of the three target species. Interestingly, and contrary to expectation, wolves in this region were found to primarily prey upon deer rather than caribou. Of the possible contributory factors investigated, the degree of human activity on the landscape was determined to have a far greater potential to negatively affect the habitat and health of caribou (Wasser et al., 2011). Detection dogs considerably facilitated the recovery of samples, and the results of their analyses enabled the authors of the study to make detailed and alternative recommendations focusing on addressing human uses and impacts on the landscape prior to initiating and/or focusing on predator removal, and to suggest a new long-term caribou monitoring approach.

Box 10.3 What information can be gained from scat?

- Habitat usage: The GPS coordinates of each scat sample detected by a dog are recorded and mapped to provide a sense of movement and occupancy within a habitat.
- Diet: Analysis of key nutrients can provide insight into resource use, forage quality or preference, and habitat quality.
- DNA: Analyzed to identify the species and the sex and unique genetic profile of individuals, in conjunction with mapped scat points, provides confirmation of a species' presence, individual habitat usage, group dynamics, and population densities.
- Physiological stress level: For example, corticosteroids can be derived from fresh scat.
- Diseases, viral and parasite loads: To assess ecosystem and population health as well as intraspecies transmission, which can provide an important indicator of interaction between humans and non-wildlife species.

In some cases, detection dogs offer the best and perhaps only way to survey a given rare species. In the Upper Paraná Atlantic Forest of Misiones, Argentina, a trained dog found 11 new and previously unrecorded locations of bush dogs, showing evidence of their presence and habitat use (DeMatteo et al., 2009). The authors emphasized the complexity of the environment and maintained that these locations simply could not have been detected by humans alone. They also stated that further use of detection dogs would greatly facilitate the recovery of new presence and habitat information critical to the development of conservation strategies, the design of wildlife corridors and biological crossings, and to examining overall species' distributions.

In addition to working in difficult terrain, dogs have also been of inestimable value in locating the scats of elusive species. In the Sierra National Forest of California, USA, scat detection dogs were deployed on surveys of the Pacific fisher to fulfill an immediate need to identify and fill in gaps in current understanding of fisher ecology and the habitat requirements of this little known population (Thompson et al., 2010). By using dog surveys in combination with live trapping and telemetry, researchers were able to provide more robust estimates of population density, and retrieve information on habitat use, diet, and disease (Thompson et al., 2010; Thompson et al., 2012). Upon this project's completion, researchers and managers, armed with more precise estimates of fisher abundance and other important population data, will be better able to address the uncertainty surrounding the effects of timber harvest and fuel treatments on fishers and their habitat, and ultimately better able to inform future management actions (Thompson et al., 2010).

Scat detection dogs have also been incorporated into efforts to eradicate invasive species. For example, Fukuhara et al. (2010) investigated the ability of dog teams to survey for scats of the small Indian mongoose (*Herpestes javanicus*) in the Yambaru forest region of Okinawa Island, Japan (Fukuhara et al., 2010). Dogs were highly successful in locating these scats, thereby confirming mongoose presence and elucidating their habitat use. Consequently, the authors of this study concluded that detection dogs showed strong merit as an additional eradication tool and could potentially assist in future efforts by helping to detect nesting places, obtain new habitat use information, and even determine the best subsequent trap positions.

10.2.2 Detection of native and invasive live animals

Dogs have also been successfully used to locate living animals, thereby assuming another important role in conservation, monitoring, and eradication efforts. An extensively researched and tested example of detection dogs locating live animals for conservation purposes is that of the endangered Mojave desert tortoise (*Gopherus agassizii*) in Nevada and California, USA (Cablk and Harmon, 2011; Cablk and Heaton, 2006; Nussear et al., 2008). This application of dogs represents a significantly improved survey method for locating smaller age and size classes of tortoises, notable because hatchling individuals can be just two centimeters long (Cablk and Harmon, 2011). In essence, the refined method facilitates the collection of new demographic information and considerably enhances the detection of population trends. These developments can be drawn from to directly support environmental regulatory compliance and indirectly support delisting the Mojave desert tortoise due to more accurate population demographic assessments (Cablk and Harmon, 2011).

The dog-directed search has also been evaluated as a method to monitor populations of endangered black-footed ferrets (*Mustela nigripes*) in the black-tailed prairie dog (*Cynomys ludovicianus*) colonies they inhabit (Reindl-Thompson et al., 2006). After two dogs were assessed in their ability to indicate the presence of live ferrets, and with neither dog falsely indicating presence when ferrets were absent, the authors concluded that well-trained detection dogs show promise as an additional tool for monitoring reintroduction sites, especially those that are remote and difficult to access (Reindl-Thompson et al., 2006).

Dogs have also shown great potential in locating invasive species, thereby lending further support to control efforts. Trained dogs have been used in conjunction with other methods (e.g., trapping, aerial poisoning) to assist in the eradication of introduced Australian brushtail possums (*Trichosurus vulpecula*) from Kapiti Island, a New Zealand nature reserve (Cowan, 1992). A separate study conducted by Gsell et al. (2010) experimentally tested the ability of trained dogs to find rodents or their scent trails at known, very low population densities, indicative of incursions (i.e., rodent invaders) or survivors of late-stage eradications in conservation sanctuaries. In this study, dogs were highly successful in locating (deliberately released, radio-tagged) mice and rats in a pest-free forested area, further validating their adeptness at detecting rodent survivors and invaders (Gsell et al., 2010). Given such results, it may even be preferable to use dogs as eradication agents over more conventional means of pest control, such as rodenticides. Alternately, dogs may be used to pinpoint 'hotspots' that will then be selected for pesticide applications, thus containing the application area and decreasing the likelihood of direct or secondary exposure to non-target wildlife.

In another noteworthy example, the viability of detection dogs was explored as a potential tool for finding wild invasive brown treesnakes (*Boiga irregularis*) on Guam, where these snakes have decimated the avifauna, eliminated several species of lizards, and severely impacted Mariana fruit bats (*Pteropus mariannus*) (Savidge et al., 2010). Although the jungle environment was dense and complex (Figure 10.2), the dog-led teams were still able to find five times the number of snakes previously reported for visual surveys conducted by humans alone. In light of these results, conservation dogs were considered to show promise in locating treesnakes in a spatially complicated forest environment, and to be particularly useful in supplementing human search efforts during the day, when they are otherwise largely undetectable by humans (Savidge et al., 2010). Similarly, dogs are currently being assessed as a potential tool for locating introduced Burmese and North African python (*Python* spp.) in and around Everglades National Park in Florida, USA (Romagosa et al., In preparation), where these non-native species could severely undermine native wildlife populations. Dogs have also been evaluated as a means to locate native, threatened eastern indigo snakes (*Drymarchon couperi*), which are difficult to locate in their natural habitat because of the significant proportion of time they spend concealed in tortoise burrows and other subterranean

Figure 10.2 Field work in densely vegetated areas, such as the jungles of Guam, would not be typically possible working a dog on-leash. Here, conservation dog Sam is shown searching for arboreal brown treesnakes. In this study, dog-led teams exhibited five times the search success reported for human surveys. Photo courtesy Working Dogs for Conservation.

refugia (Stevenson et al., 2010). Results from this pilot effort suggest that dogs have value as a field survey method for live snakes or their shed skins and that the effectiveness of canine surveys may be enhanced if conducted in concert with other techniques (e.g., visual encounter surveys of gopher tortoise burrows), enabling researchers to better determine distribution, and ultimately more effectively conserve this species (Stevenson et al., 2010).

10.2.3 Detection of carcasses

Carcasses retrieved from the field provide essential data, and often irrefutable evidence, that can be used to assess risks and estimate mortality within a given habitat or landscape. The use of dogs has been investigated as an alternative tool to gather data to more accurately quantify bat fatalities at wind turbines (Arnett, 2006). Dog and handler teams significantly outperformed human surveyors in finding bat carcasses (see Table 10.2), even when the height and density of vegetation increased. The author further noted that wind energy programs are increasing and the results of this assessment demonstrated that dogs are a promising tool for gauging potential impacts of wind farms to wildlife. Dog-collected samples may also be analyzed as part of efforts to develop long-term mortality mitigation strategies and solutions (Arnett, 2006). Another study demonstrated the usefulness of dogs in field surveys to improve bird-strike mortality estimates at wind farms and other anthropogenic structures implicated in bird fatalities worldwide (Paula et al., 2011). Here again, the authors reported that a dog–handler team was found to be more accurate than human searchers, independently of vegetation density, supporting their application in field surveys for bird carcass searches.

As part of anti-poaching initiatives, dogs have also been specially trained to recover the carcasses of poisoned wildlife (and bait matter) during investigations into alleged intentional wildlife poisoning incidents in southern Spain (Fajardo et al., 2011). The evidence recovered from these carcasses, collected during routine patrols in targeted problem areas, has been used to secure fines and penalties. As such, this use of dogs has also had a preventive effect, reducing the number of wildlife poisonings within the region by discouraging prospective poisoners.

10.2.4 Detection of nests and dens

Current or recently used refuges and breeding sites may hold information about population size, habitat selection, predation, and reproductive success. One example of harnessing these sites is the extensive use of dogs to find the nests of Galliformes (see Dahlgren et al., 2012). Dogs have also been called upon to seek nests and dens of a different sort. On the Hebridean island of Tiree (northern Scotland, UK), researchers demonstrated that detection dogs could play a key role in future studies examining nesting habitat and nest survival of rare bumblebees (Waters et al., 2010; but see O'Connor et al., 2012 for a contrasting view with focus on the cost-effectiveness of this method). Bee nests tend to occur underground and as such are not readily discernible by people. In this study, a trained dog proved 100% accurate in locating (intentionally) buried nests during trials. The authors reported the same dog was then used in actual field searches, locating 33 wild nests from 4 species including 2 (*Bombus distinguendus* and *B. muscorum*) that are rare and declining across much of Europe.

In a study to detect red imported fire ants (*Solenopsis invicta*), trained dogs were highly competent in pinpointing emerging and smaller nests in ant-infested areas (Lin et al., 2011). The authors recommended detection dogs to supplement other typically used inspection methods, noting their potential to significantly increase the efficacy of red imported fire ant control by providing more accurate identifications of hidden ant nests, especially in low density areas. Finally, since 2011, a dog trained in nest detection has assisted in obtaining data on the last remaining little penguin (*Eudyptula minor*) colony on mainland New South Wales, Australia. As new nests are located, they can be monitored, and mapped in relation to protected areas. This same dog has also been trained to locate red fox (*V. vulpes*) dens, scat, and urine to enable managers to control this primary predator of the little penguin (New South Wales Government, unpublished data).

10.2.5 Detection of native and invasive plants

Using the same training methods and principles as employed for other immobile conservation targets, detection dogs can be trained to seek out rare native or invasive plant species. Browne and Stafford (2003) report that using a dog to find specimens of the scarce wood rose (*Dactylanthus taylorii*) in New Zealand reduced the search time six fold. Search dogs were able to find early- to mid-growth stages of invasive spotted knapweed (*Centaurea maculosa*) at almost twice the rate of human surveyors (Goodwin et al., 2010; see Table 10.2). Similarly, the results following control efforts of the introduced saltmeadow cordgrass (*Spartina patens*) in Washington State, USA, suggest that searches by trained dogs may greatly improve plant discovery rates (Milne, 2007).

Sometimes plant surveys can also reveal information about habitat quality and resource availability which in turn may be relevant to the monitoring or conservation of species of interest. Field experiments in the Willamette Valley of Oregon, USA, demonstrated that conservation dogs located the rare, native Kincaid's lupine (*Lupinus sulphureus kincaidii*) with a high degree of accuracy (Vesely et al., unpublished data). Interestingly, Kincaid's lupine is the primary host plant to the endangered Fender's blue butterfly (*Icaricia icarioides fenderi*). Thus, by successfully locating plants upon which this butterfly relies, dogs may indirectly provide indicators of its presence and distribution within a given habitat.

By contrast, invasive weeds, which can rapidly spread and out-compete native plant communities and permanently change ecosystems, represent one of the most serious problems that conservation managers face today. In response to the severity of this threat, researchers in Montana, USA, are using dog-based surveys to eradicate invasive weeds. In conjunction with human surveyors, trained weed detection dogs are being used to locate every remaining individual plant of Dyer's woad (*Isatis tinctoria*) over 200 acres in a prominently used open space, so that even the plants that are the hardest to visually detect can be found and hand pulled before there is any chance of them flowering or going to seed (Goodwin et al., unpublished data). In the first season they were deployed, the dogs' ability to find root remnants, and plants that had sprouted from remnant roots, led to the understanding that more extensive (i.e., complete) root removal and targeted herbicide application is necessary to exterminate the weed.

Finally, we note promising research that has successfully tested dogs in their ability to locate pathogenic fungus involved in 'pine decline' disease, demonstrating their viability in assisting land managers in formulating management strategies, and lessening effects on endangered species habitat in the future (Eckhardt and Steury, 2011). As dogs proved highly accurate in positively identifying the presence of fungi growing in pine tree roots, the authors stated that detection dogs could serve as an early detection system, allowing the identification of infected stands before symptoms are evident above ground.

10.3 Maximizing safety and success when using free-ranging conservation detection dogs

The previous sections have elaborated on the many ways that conservation detection dogs have meaningfully contributed within the framework of species and wild land protection and management. Throughout Section 10.2 it was also noted that, under the right circumstances, detection dogs are a non- or minimally invasive monitoring tool, often the least invasive of those available. Nonetheless, by their very nature, free-ranging dogs have the potential to act as agents of disturbance (Weston and Stankowich, Chapter 4 and see Sections 10.3.1 to 10.3.3). As such, the use of detection dogs is advised only when the benefits clearly outweigh any potential undesirable outcomes. Appropriately using and drawing the most benefit from conservation dogs requires a thorough understanding of both their behavior and limitations. It also necessitates identifying and proactively addressing the potential risks that may arise when intentionally bringing dogs into proximity with wildlife species, sensitive or otherwise. This section identifies key traits required of conservation dogs and their handlers, factors for determining appropriate search targets and uses of dogs, and considerations for study design and dog deployment. Though the discussion focuses on free-ranging dog applications, many of the principles considered also apply to non-free ranging dogs.

10.3.1 Fundamental attributes of a conservation detection dog

To be selected for conservation detection work, a dog must exhibit the strong inherent attributes that are needed to determinedly seek a target odor. When dogs are to seek instinctual targets by applying traditional hunting skills such as pointing or flushing, as is typical for locating Galliformes and Anseriformes, it behooves a practitioner to use a breed reared for that purpose (although Dahlgren et al. (2012) point out that individual attributes—not just breed generalities—are still paramount). However, when seeking non-instinctual targets, possession of a requisite suite of traits is favored over breed. These traits include an unceasing compulsion to search and a fervent and inexhaustible interest in receiving a reward, either a toy or food. From the onset, the dog is taught to associate the search target with the reward by immediately receiving it when that target is found, i.e. smelled. Thus, the dog learns to persistently seek the target, which has no inherent value, in order to repeatedly earn this reward. To be deemed 'field-ready,' a dog must demonstrate the ability to remain focused during hours of search time even when there are few targets to locate, and hence less opportunities to be rewarded. The dog must also be willing and able to communicate the location of the target to the handler by performing a trained behavior that does not involve interacting with or compromising it in any way. Referred to as a 'passive alert,' this trained behavior typically takes the form of sitting or lying down near the target (Figure 10.3).

The process of training dogs to detect non-instinctual live animal targets (see also Section 10.2.2) does not differ materially from that for inanimate targets, but extra precautionary measures are necessary. For example, the dog must be taught to shift their visual focus from the target to the handler when they passively alert—from a greater distance than for an inanimate target. Barking and/or performing a 'refind' alert, where the dog repeatedly moves between target and handler until the latter arrives at the target, are both strongly discouraged when seeking live animals. In some instances additional measures will be taken, such as having dogs work on leash and wear muzzles (Cablk and

Figure 10.3 Conservation dogs communicate the location of the target to their handler passively, that is without interfering or compromising it, by sitting or lying down near it. (Left) Conservation dog Lily waits for her reward after finding invasive Chinese bush clover (*Lespedeza cuneata*). (Right) Conservation dog Rio alerts to kit fox scat within a canid latrine. Both photos courtesy Working Dogs for Conservation.

Box 10.4 Deciding when to use (and not use) a leash

Working dogs while leashed can add a measure of safety and will sometimes be a necessity. However, leashing a dog does not mean they are completely safe, either to or from wildlife. At best, a leash should be seen as an extra precaution for a dog that is already well-suited to the task and target(s) rather than as a tool that enables the use of an inappropriate dog. Indeed, a dog working on-leash may give the appearance of being a 'restrained menace' but should not be regarded as such; most of the time on-leash work is actually about optimizing search coverage for smaller targets (e.g., emerging weed rosettes) or simply fulfilling stipulations for permission to work in areas that would ordinarily be off-limits to dogs.

A great majority of the time, however, working on-leash will not be an option. Exceedingly dense vegetation (such as shown in Figure 10.2) tangles a leash, and steep terrain makes it unsafe for the dog and handler to pull on one another while traversing a slope. Working a dog off-leash is advantageous in that it *may* enable greater coverage of an area and frees the handler for added agility and trekking safety. Although the majority of off-leash searches occur with the dog working within sight of the handler, dense vegetation can frequently impede visibility, and hence visual contact between the team. Also, a dog might move out of sight when it catches scent and follows it to its source faster than the handler can keep up. Reed et al. (2011) suggested that free-ranging dogs have unfettered mobility, which may allow them to compensate for wind conditions that could otherwise impact their rate of target detection (Box 10.2). Beckmann (2006) reported that while humans walked 365 km in a mountainous environment, the accompanying off-leash conservation dogs covered 767 km, more than doubling the linear distance covered by the handlers and expanding both the search area and the number of scats they had access to detect within it.

Harmon, 2011; see Box 10.4). Such provisions are further discussed in Section 10.3.2.

All field-deployed dogs must consistently demonstrate impeccable functional obedience and be responsive when the handler verbally issues a command (e.g., to stop a pursuit). Fitting a dog with an electronic collar, which can be engaged when necessary (e.g., a verbal command is not heard or heeded), secures an additional level of protection for both the dog and wildlife. However, even a well-selected and trained dog may not be suited to every target of interest, particularly living animals. Some dogs naturally have a heightened interest in the pursuit and killing of quarry. When this is the case neither training nor implementation of additional safety provisions or equipment will be sufficient to deter such dogs from pursuing wildlife when the opportunity presents itself. Even when the target itself is inanimate, wild (and sometimes domestic) animals will be present within the search environment. For all these reasons, a professional trainer must assess a dog's level of interest in pursuing prey, and how readily it can be curbed in a controlled setting, before advanced conservation training is undertaken (see Section 10.4.2).

10.3.2 Fundamental attributes of a conservation detection dog handler

A well-trained conservation dog can only flourish if paired with a correspondingly well-trained handler, and if the two work cohesively as a team. Otherwise, the dog's detection performance may be hampered or adversely impacted (see Box 10.5). A conservation dog handler must have the ability to interpret canine behavior, i.e. exhibited behaviors that indicate the dog smells the target. The handler is responsible for maintaining the dog's fidelity to the target by making split-second decisions and responding appropriately when the dog alerts to a putative target. At the same time, he or she must have the capacity to assess how environmental conditions (see Box 10.5) might be affecting the dispersal of the scent and assist the dog in troubleshooting to the source accordingly. Acting as an anchor point, the handler creates the line off of which the dog searches. When vegetation density permits, the dog works in front of the handler and 'quarters'—continually sniffing while moving to the left and the right of the handler and the invisible line on which the handler progresses (Figure 10.4). To support this search behavior, the handler must effectively offer verbal praise and corrections, and issue obedience and directional commands. An in-depth discussion of handler attributes required for similar detection work can be found in Matre (2003).

A handler is responsible for more than facilitating the performance of their dog. While the search team often comprises additional members who navigate, assist in sample collection and data recording (Figure 10.5), and offer guidance within the area and expertise regarding plants and wildlife, there

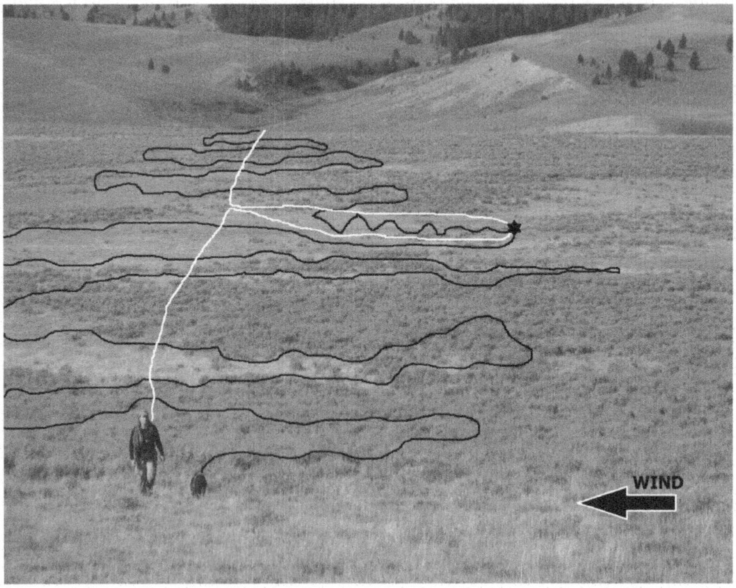

Figure 10.4 A dog working off-leash is able to cover a much greater area than the handler, by moving side to side (i.e., 'quartering'). Here a hypothetical handler and dog track are shown. The dog crosses the plume of scent coming from the target (starred) and weaves in to pinpoint its location, where she briefly meets up with her handler for a reward session, then resumes her main course of travel. Photo courtesy Tyler Roady.

Box 10.5 Factors that impact scent detection

Dogs are fundamentally capable of detecting any target that emits a scent and, as such, hold vast potential as a survey tool. While humans possess 5 million olfactory sensory cells, dogs are armed with an estimated 125 to 220 million such cells, which gives them significantly enhanced scent discrimination ability (Pearsall and Verbruggen, 1982). Even at trace levels, a dog's sense of smell is somewhere between 10 and 100 times greater than a human's, and depending on the odor can be as high as 100,000 to 1 million times more pronounced (Syrotuck, 1972). In addition to their strong sense of smell, dogs possess numerous other qualities that make them well suited to wildlife studies. Dogs have the ability to learn multiple odors and seek them simultaneously (Williams and Johnston, 2002). Furthermore, they can move about freely to locate the source of the odor, and generalize learning from a subset of training samples to being able to recognize samples they've never before encountered (Cablk et al., 2008). A more detailed discussion of canine olfaction can be found in Schoon and Haak (2002).

Another aspect that sets dogs apart from other survey tools is that they are living animals, and therefore individuals, which inherently introduces variation in their detection rates. Overall, the scenting proficiency a given detection dog achieves is far more related to their possessing the requisite traits and skill set than to their breed, as discussed in more detail in Hurt and Smith (2009). Physical fitness, tolerance to heat and cold, and stamina for searching are all influential factors that are best addressed during the selection, training, and conditioning of the candidate dog. As well, the proficiency of a handler and their ability to read the dog may either facilitate or constrain search performance and, thus, detection abilities. Factors within the environment, such as wind speed and direction, precipitation, vegetation density, and terrain, influence the dispersion of scent and, as such, also variably impact how successfully dogs will be able to detect a given target odor. Given the interplay between individual dogs and the environmental conditions they face, it should be anticipated that detection rates may vary from day to day for each dog, and among dogs working on the same target.

Reed et al. (2011) reported that among environmental conditions monitored, time since last significant rainfall was most strongly and positively correlated with the rate of target detection. Similarly, Hunter (2011) showed that detection success decreased with increased rainfall and that detection rates in clear-cut areas, where scat samples were exposed to direct rainfall, were lower than in pine stands, where samples were protected beneath the canopy. By contrast, wind has also been reported to variably affect target detection. While neither wind speed nor direction impacted the rate at which mammalian carnivore scats were detected (Reed et al., 2011), increased wind speed *was* found to increase detection distances of live desert tortoises (Cablk et al., 2008). On the other hand, increased wind has also been correlated with an increase in the amount of time taken by dogs to detect a target (Shivik, 2002).

Many experienced detection dog handlers anecdotally report temperature as an important scent detection variable, but it receives mixed results in peer-reviewed literature. Several studies report no impact on detection rates with increasing temperature (Cablk and Heaton, 2006; Long et al., 2007a; Nussear et al., 2008). Another study reported a weak correlation based on a small sample size; one dog showed *decreased* detection rates with increased temperature while a second dog exhibited *increased* detection rates with increased temperature (Reed et al., 2011). Since dogs pant to cool themselves, the latter example perhaps best illustrates that while increased panting rates can lead to decreased sniffing and scent detection (Gazit and Terkel, 2003), the responses of individual dogs to the prevailing conditions remain variable.

Finally, not all scents are equally potent, thus not equally detectable at the same scale. For example, detection distances of up to 1,500 m were reported for dogs seeking northern right whale scats from boats on the Atlantic Ocean (Rolland et al., 2006), whereas dogs searching for rosy wolf snail (*Euglandina rosea*) on Oahu, Hawaii needed to be within 0.1 m to detect them (Hurt et al., unpublished data). There are without doubt environmental and body size components built in to these variable detection distances. Yet even if conditions were held constant, these two target types would not be equally detectable and would consequently require a different search scale and strategy.

Quantifying the variability in detection rates and measuring environmental impacts can be challenging. When it is material to the study, quantifying inter- or intra-dog variation in detection rates necessitates having some knowledge or appreciation of the number of available targets. This can be accomplished by putting out samples or preemptively finding naturally occurring samples within the testing area. When controlled trials are unnecessary or infeasible, a measure of comparability can be obtained by tracking the number of samples found per unit measure of distance—if the target is considered to be equally dispersed across the study area or all dogs have the opportunity to work the same area.

Figure 10.5 Conservation dog Finny alerts to a burrow—indicating the presence of a black-footed ferret to his handler. Meanwhile, the second human member of the team, also the orienteer, records all relevant data. Photo courtesy Working Dogs for Conservation.

are also many instances when the handler will be required to navigate in the field and collect data/samples solo. Of the numerous detection dog fields, conservation detection work demands some of the highest levels of physical endurance. Long hours in rugged and inhospitable conditions are not uncommon. A conservation dog handler walks within several dozen meters of the dog, and so must essentially traverse the same ground as his or her more agile counterpart. An in-depth consideration of these factors can be found in Hurt and Smith (2009).

Conservation dog handlers must also be trained and able to respond immediately to any harassment or hunting behavior exhibited by the dog. Prior to the start of a project, a handler should apprise him or herself of the behaviors and habits of resident animals and know the best practice to extricate the team from an unintended encounter. In this regard, project partners and local field personnel will have invaluable experience/expertise and should be consulted accordingly.

10.3.3 Incorporating detection dogs in conservation studies: considerations for investigators

It is incumbent upon researchers to draw on their considerable knowledge of the target species and select the most appropriate variant of the target—be it scat, dens or nests, or the living animal. In some cases it will be logistically impossible to obtain training samples (e.g., scat from a species so rare that even zoo specimens are unavailable). This may void plans to use dogs since they must have prior exposure to such samples in order to learn the target scent. Once it has been determined that dogs are a suitable survey or monitoring tool, *all* possible wildlife encounters, including those with non-study species deemed dangerous to the handler or dog, must be scrupulously considered. Here we note that encounters can largely be minimized by avoiding watering holes and refugia or other places where wildlife congregate or are likely to bed down. Being aware of activity patterns and synchronizing surveys with periods of inactivity—when appropriate and logistically feasible—is also advisable.

Ideally, the selected target will offer the least opportunity for interaction with, or adverse impact upon, the focal animal. It would be highly inappropriate, for example, to seek nesting birds of species known to abandon nests if disturbed (Dahlgren et al., 2012). On the other hand, there is evidence that dogs can be used to locate living animals without notable negative consequences. For example, because free-ranging dogs are documented predators of desert tortoises, researchers training dogs to detect tortoises for conservation purposes were explicit about their use of safety evaluations and protocol in written reports, manuscripts, and public presentations (Cablk and Harmon, 2011). A study was conducted in parallel to demonstrate that a tortoise that previously had been located by a dog was no more likely to attract predators, have an increased risk of predation, or alter its movement pattern than one previously located by humans alone (Heaton et al., 2008).

In many instances a non-direct or inanimate target such as scat (discussed in Section 10.2.1) can provide the same or equally valuable data (see Box 10.3) while removing the intent or need to encounter the animal. The endangered San Joaquin kit fox, extensively surveyed using conservation dogs seeking scat, provides a case in point. A similar number of kit fox scats were recovered during initial and follow-up visits for all mark–recapture studies, which required two or four dog-directed searches of precisely the same area. Thus, kit foxes do not appear to be driven from the study site due to repeated dog presence, and all evidence indicates their typical scent-marking behavior continues, unabated (Smith et al., unpublished data).

However, we caution that scat collection carries the potential to be disruptive, particularly for species that are highly susceptible to even the slightest disturbance and contact. In such cases, or when the focal species may present an overt risk to the dog–handler team, it is advisable to formulate a study design that does not require seeking *only* extremely fresh scat (e.g., < 24 hours). This will decrease the likelihood of the focal animal still being nearby when the dog seeks the target. Making noise (e.g., field crew talking and placing bells on dog vests) when moving through an area is an easy and typically effective precaution against chance encounters. Such a momentary element of disturbance is infinitely preferable to risking a face-to-face encounter, which could result from traveling too quietly within an area.

Investigators will also want to consider the potential role of dogs as carrier agents and possible effects to the ecosystem. For example, weed seeds often lodge in dog coats, which must be kept clean to avoid transferring seeds to previously uninfested areas. Though dog hygiene and related post-field body checks are standard handler procedure, their importance in this context should nonetheless be stressed. Dogs may transfer parasites to wildlife and vice versa (Knobel et al., Chapter 6). Minimizing this risk involves stipulating that all dog feces be collected while in the field, requesting up to date vaccinations, and scheduling pre- and post-field zoonotic screenings. This precaution is particularly germane when externally-sourced conservation dogs are brought in to regions where the diseases and parasites exchanged would be novel to either the dog or resident wildlife and hence their resistance and vulnerability is unknown. Sometimes, despite the fact that conservation dogs may constitute the best survey tool, it will simply not be appropriate to employ them (e.g., during an outbreak of canine distemper virus in a Carnivora population).

Having to survey in prohibitively challenging terrain may restrict the practicality and implementation of using detection dogs. Accessibility to study sites will vary, with some requiring extensive overland trekking to reach the starting point where the search will be initiated. The trekking commute distance should be minimized to the extent possible to protect the dog from fatigue and maximize the consistency of the scenting effort over an entire search day. Long commutes also have the potential to increase opportunities for unintentional encounters with wildlife or other dogs.

The distance that can be searched in a single day also depends on factors such as: vegetation density, temperature, length of time to access the search area, strength of target odor, projected number of samples to be found, and duration of the sample collection and data recording procedure. Mesocarnivore and carnivore scat studies (which have relatively large scent sources and require that considerable distances be covered) report daily distances from 2.0 km (Long et al., 2007a) to 7.0 km (Wasser et al., 2004). Therefore, when initiating dog-based searches, we recommend that 'short' search distances be planned with a provision to lengthen them later, if conditions allow.

Finally, not all land management agencies are amenable to granting access permission for dogs, particularly if they are not leashed (see Box 10.4). Consequently, anticipating and securing all required permits and permissions well in advance of the proposed field work is essential to smooth operations.

The best way to plan for these myriad considerations is to work with an experienced professional conservation dog trainer and design the study in consultation with them, discussing any potential areas of concern for either party well in advance of field deployment. When possible, we also recommend using a pilot study to ground-truth the design before large scale implementation.

10.3.4 The value of dogs as conservation ambassadors

While the foremost use of conservation detection dogs is to maximize data collection, they also provide a unique and rather exceptional chance to increase awareness worldwide on the importance of protecting wild species and their habitats. Without question, the involvement of dogs in such studies is still novel enough to be considered unusual, and in any case, dogs tend to pique public interest. Recognizing and harnessing the potential of dogs as ambassadors further magnifies their benefits to conservation. In locations around the globe where dogs and people do not regularly have close relationships, people tend to initially be fearful, yet still curious, about them. In this case, it is helpful for a handler to be able to say 'my dog is friendly' in the local language and create opportunities for local people to see demonstrations of the dog's ability to detect the relevant target, heed a few simple obedience commands, or even just enjoy a bout of play with their handler. A few moments spent with a joyous dog that clearly relishes and excels at his work can break down cultural barriers and misconceptions about dogs in general, and provide a platform for discussing safety around dogs while simultaneously providing a unique chance to engage in a broader dialogue about species conservation (see Figure 10.6).

The majority of people who meet conservation dogs and have the opportunity to watch them work regard the experience positively, sometimes to the extent of wanting to try their hand at it. Members of the public should, however, be discouraged from mimicking researchers and sending their pet dogs on ill-fated conservation pursuits. As discussed

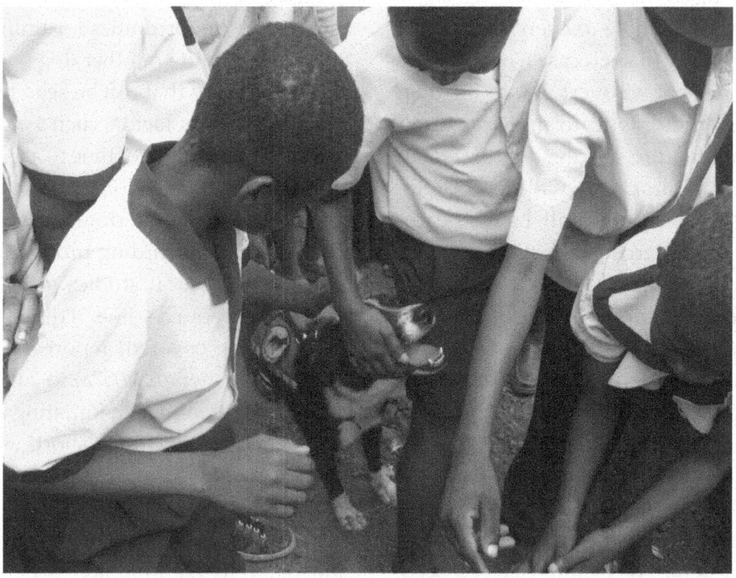

Figure 10.6 Harnessing the role that dogs can play as ambassadors magnifies their conservation impact considerably. Students at a school for the deaf in Cameroon spent a morning learning about rare Cross River gorillas and threats to their survival then had a chance to meet Orbee, one of the dogs working to help them. For many, it was the first time they had ever petted a dog. Photo courtesy Working Dogs for Conservation.

throughout this section, conservation dogs possess a set of qualities and skills that is very rare. More than that, promising detection dog candidates are highly active, energetic, focused, and intense individuals who do not tolerate idle time and need an inordinate amount of structure and supervision. These qualities can be channeled positively, for use in the conservation dog field, but do not typically coincide with the characteristics of agreeable or manageable pets.

10.4 Future role of free-ranging dogs in conservation

Finding wildlife and their sign to obtain demographic information has been the basis of conservation detection dog work for decades, but dog-collected data is increasingly used to examine anthropogenic threats to populations. Detection dog application in studies of wildlife disease is one such area that holds promise. For example, for the first time, dogs were used to collect the carcasses of *Sarcoptes* (mange)-infected wild animals and to identify, separate from their herd, and capture mange-infected wild animals in the Italian Alps (Alasaad et al., 2012). Further development of dogs in such disease application presents an entirely new and exciting role in a wildlife and conservation context.

From determining how human activities impact the stress response of a given species (Wasser et al., 2011) to evaluating presence of biotoxins in the environment (Ayres et al., 2012; Rolland et al., 2006), the use of detection dogs (whether free-ranging or not) often makes it possible for us to increase our understanding without invasive or destructive sampling (e.g., tissue collection for screening). Furthermore, dogs will increasingly be included in enforcing wildlife protection laws by, for example, seeking poisoned bait left out for wildlife (see Fajardo et al., 2011), detecting bushmeat and other wildlife contraband, and tracking poachers immediately following an illegal killing. On the landscape level, we believe that conservation dogs will be increasingly incorporated in efforts to better understand our changing climate and related threats to the integrity of wildlife and plant species as well as ecosystems. As we begin to more fully grasp the consequences of climate change, strategies will be needed to combat threats to native wildlife and ecosystems (e.g., the proliferation of invasive species) in a timely manner. In this regard, detection dogs have the potential to serve as a highly efficient survey tool (see Section 10.2.5) and their application in this context seems both timely and very promising.

10.4.1 Persistent questions and research needs

Over the last decade, many conservation dog-focused articles and theses have highlighted unresolved needs to best understand and use these dogs. When reviewing these suggestions, several themes emerge, and studies that address them will provide welcome information to the field of conservation detection dogs (see Table 10.3).

10.4.2 Quality control

In this chapter we have demonstrated both the versatility of this survey tool and its potential to aid biologists to manage and conserve numerous species. We also discuss the consideration that the training and deployment of conservation dogs requires. Professional training is of critical importance to the successful and safe deployment of free-ranging conservation detection dogs and, as such, it is taken very seriously. Professional conservation dog trainers utilize internal standards to assess a team's field readiness, which at a minimum evaluate its ability to locate a target, communicate, and perform in a directed, controlled manner throughout the search (Hurt and Smith, 2009). In addition, these internal standards may address safety awareness, search focus and endurance, and test complicated (but entirely realistic) search scenarios, ultimately ensuring the dissemination of best practices. All dogs currently participating in wildlife management activities in New Zealand—whether by locating protected species or as part of predator control measures—must undergo formal training and certification in the Department of Conservation's National Conservation Dog Program (MacKay et al., 2008). Although this is not yet the norm in the field of conservation detection, the greater than ever utilization of conservation dogs will likely lead to uniformly accepted industry standards or certification processes. Even

Table 10.3 Emerging themes and study questions regarding conservation dogs.

Theme for exploration	Study question/framework/objective	References
Optimizing the performance of dogs during training and fielding	Is dog motivation reduced by inconsistent delivery of rewards during the search (e.g., brown treesnakes)?	Savidge et al. (2010)
	How does varying target density affect detection rates?	Arnett (2006); Savidge et al. (2010)
	Do dogs have repercussions/an aversive effect on living targets (e.g., Eastern box turtles) thereby reducing probability of detection in subsequent searches?	Kapfer et al. (2012)
	How does the duration of a search impact its efficiency?	Arnett (2006); Goodwin et al. (2010)
	How does handler training and error impact their canine partner?	Dellinger (2011)
Mechanics of olfaction and chemistry/ properties of target scent	What chemical/olfactory components allow dogs to differentiate the scat of one species from that of another?	Smith et al. (2003)
	How long does scent remain detectable?	Harrison (2006)
	Can the physical rates of decay of scats be quantified?; Does the physical rate of decay of scats of different species vary?; How does physical decay affect detection rates/ detectability?	Hunter (2011)
Comparability and analysis of data	How does the efficacy of detection dogs compare with other survey methods?	Long et al. (2007b)
	More modelling required (e.g., to understand scenting conditions/environmental factors)?; More efficacy studies required (e.g., using dogs to determine density estimates)?	Dahlgren et al. (2012)
	How do detection rates differ among dogs?	Goodwin et al. (2010)
	How do detection rates differ across target species?	Reed et al. (2011); Dahlgren et al. (2012)
Call for standardization and/or certification process	Demonstration of the proficiency of a dog/ handler team to detect a given target	Cablk and Heaton (2006); Cablk and Harmon (2011); Reed et al. (2011)
	Fielding teams seeking the same targets using the same protocols and operating conditions, thereby allowing maximum comparability between studies	Arnett (2006); Kapfer et al. (2012)

without such a formalized procedure, fielding only conservation dog teams that have been professionally trained is already viewed by practitioners and research partners as a fundamental component of safety and efficacy.

Over the last 100 years, detection dogs have offered conservationists an efficient means of collecting data to address a multitude of objectives, as outlined throughout this chapter. As such, it is understandable and not unexpected that dogs have assumed a vital role in wildlife and conservation studies. Regardless of whether or not widely accepted quality assurance measures are ever put in place for conservation dogs, it will always be vital that those deploying conservation dog–handler teams maintain the highest standards, for the safety of target and non-target wildlife and for the team. In our view, with careful consideration for risk management, there is virtually limitless potential for the future use of dogs in conservation endeavors.

Acknowledgments

We are especially grateful to Wayne Thogmartin of the United States Geological Survey (USGS, Upper Midwest Environmental Science Center) for his swift response and creation of the map for this chapter. We thank all our colleagues who graciously provided citations relevant to conservation detection dog work, especially Barbara Davenport, Paula MacKay, Todd Steury, and Sam Wasser.

References

Alasaad, S., Permunian, R., Gakuya, F., Mutinda, M., Soriguera, R.C., and Rossi, L. (2012). Sarcoptic-mange detector dogs used to identify infected animals during outbreaks in wildlife. *BMC Veterinary Research*, **8**, 110.

Alexander, B. (2005). Volunteer wildlife detector-dog training for assisted detection of river otter latrines. Ethology Internship Final Evaluation, Texas A&M University.

Arnett, E.B. (2006). A preliminary evaluation of the use of dogs to recovery bat fatalities at wind energy facilities. *Wildlife Society Bulletin*, **34**, 1440–5.

Ayres, K.L., Booth, R.K., Hempelmann, J.A., Koski, K.L., Emmons, C.K., Baird, R.W., Balcomb-Bartok, K., Hanson, M.B., Ford, M.J., and Wasser, S.K. (2012). Distinguishing the impacts of inadequate prey and vessel traffic on an endangered killer whale (*Orcinus orca*) population. PLoS ONE 7(6): e36842.

Beckmann, J.P. (2006). Carnivore conservation and search dogs: The value of a novel, non-invasive technique in the Greater Yellowstone Ecosystem. In A. Wondrak Biel, Ed., *Greater Yellowstone Public Lands: A Century of Discovery, Hard Lessons, and Bright Prospects. Proceedings of the 8th Biennial Scientific Conference on the Greater Yellowstone Ecosystem*. Yellowstone Center for Resources, pp. 20–26.

Boydston, E.E. (2005). Behavior, ecology and detection surveys of mammalian carnivores in the Presidio. Final report. U.S. Geological Survey, Sacramento, CA.

Brooks, S.E., Oi, F.M., and Koehler, P.G. (2003). Ability of canine termite detectors to locate live termites and discriminate them from non-termite material. *Journal of Economic Entomology*, **96**, 1259–66.

Browne, C. and Stafford, K. (2003). The use of dogs in conservation work in New Zealand. *New Zealand Veterinary Association Companion Animal Society Newsletter*, **14**, 58–9.

Browne, C. M. (2005). The use of dogs to detect New Zealand reptile scents. MS thesis, Massey University.

Bulanda, S. (1994). *Ready! Training the Search and Rescue Dog*, Doral Publishing Inc., Sun City, Arizona.

Button, L. (1990). *Practical Scent Dog Training*. Alpine Publications Inc., Loveland, Colorado.

Cablk, M.E. and Harmon, R. (2011). Validation and development of a certification program for using K9s to survey desert tortoises. Final Report. ESTCP Project RC-200609.

Cablk, M.E. and Heaton, J.S. (2006). Accuracy and reliability of dogs in surveying for desert tortoise (*Gopherus agassizii*). *Ecological Applications*, **16**, 1926–35.

Cablk, M.E., Sagebiel, J., Heaton, J., and Valentin, C. (2008). Olfaction-based detection distance: a quantitative analysis of how far away dogs recognize tortoise odor and follow it to source. *Sensors*, **8**, 2208–22.

Chen, M., Daly, M., Williams, N., Williams, S., Williams, C., and Williams, G. (2000). Non-invasive detection of hypoglycaemia using a novel, fully biocompatible and patient friendly alarm system. *British Medical Journal*, **321**, 1565–6.

Christy, M.T., Yackel Adams, A.A., Rodda, G.H., Savidge, J.A., and Tyrrell, C.L. (2010). Modelling detection probabilities to evaluate management and control tools for an invasive species. *Journal of Applied Ecology*, **47**, 106–13.

Cowan, P.E. (1992). The eradication of introduced Australian brushtail possums, *Trichosurus vulpecula*, from Kapiti Island, a New Zealand nature reserve. *Biological Conservation*, **61**, 217–26.

Cross, E. (2006). CBP's four-legged warriors. *Customs and Border Protection Today*, August/September. U.S. Department of Homeland Security, Washington, D.C.

Dahlgren, D.K., Elmore, R.D., Smith, D.A., Hurt, A., Arnett, E.B., and Connelly, J.W. (2012). Use of dogs in wildlife research and management. In N.J. Silvy, Ed., *Wildlife Techniques Manual: Research (Volume I)*. Johns Hopkins University Press, Baltimore, pp. 140–53.

Dean, E.E. (1979). Training of dogs to detect black-footed ferrets. Southwestern Research Institute, New Mexico.

Dellinger, J. A. (2011). Foraging and spatial ecology of red wolves (*Canis rufus*) in northeastern North Carolina, MS thesis, Auburn University.

DeMatteo, K.E., Rinas, M.A., Sede, M.M., Davenport, B., Arguelles, C.F., Lovett, K., and Parker, P.G. (2009). Detection dogs: An effective technique for bush dog surveys. *Journal of Wildlife Management*, **73**, 1436–40.

Duggan, J.M., Heske, E.J., Schooley, R.L., Hurt, A., and Whitelaw, A. (2011). Comparing detection dog and live-trapping surveys for a cryptic rodent. *Journal of Wildlife Management*, **75**, 1209–17.

Eckhardt, L. and Steury, T.D. (2011). Root diseases and timber dogs. *Silviculture Magazine*, Fall, 30–1.

Engeman, R.M., Vice, D.S., Rodriguez, D.V., Gruver, K.S., Santos, W.S., and Pitzler, M.E. (1998a). Effectiveness of the detector dogs used for deterring the dispersal of brown tree snakes. *Pacific Conservation Biology*, **4**, 256–60.

Engeman, R.M., Rodriquez, D.V., Linnell, M.A., and Ptizler, M.E. (1998b). A review of the case histories of the brown tree snakes (*Boiga irregularis*) located by detector dogs on Guam. *International Biodeterioration and Biodegradation*, **42**, 161–5.

Engeman, R. M., Vice, D.S., York, D., and Gruver, K.S. (2002). Sustained evaluation of the effectiveness of detector dogs for locating brown tree snakes in cargo outbound from Guam. *International Biodeterioration and Biodegradation*, **49**, 101–6.

Fajardo, I., Ruiz, A. Zorilla Delgado, I., Valero, A., Fernandez, I., Saez, E., Molino, F.M., and Olivares, J. (2011). Use of specialised canine units to detect poisoned baits and recover forensic evidence in Andalucía (southern Spain). In N.L. Richards, Ed., *Carbofuran and Wildlife Poisoning: Global Perspectives and Forensic Approaches*. John Wiley & Sons, Ltd., Chichester, England, pp. 147–55.

Fukuhara, R., Yamaguchi, T., Ukuta, H., Roy, S., Tanaka, J., and Ogura, G. (2010). Development and introduction of detection dogs in surveying for scats of small Indian mongoose as invasive alien species. *Journal of Veterinary Behavior*, **5**, 101–11.

Gazit, I. and Terkel, J. (2003). Explosives detection by sniffer dogs following strenuous physical activity. *Applied Animal Behaviour Science*, **81**, 149–61.

Goodwin, K.M., Engel, R.E., and Weaver, D.K. (2010). Trained dogs outperform human surveyors in the detection of rare spotted knapweed (*Centaurea stoebe*). *Invasive Plant Science and Management*, **3**, 113–21.

Gsell, A., Innes, J., de Monchy, P., and Brunton, D. (2010). The success of using trained dogs to locate sparse rodents in pest-free sanctuaries. *Wildlife Research*, **37**, 39–46.

Harrison, R.L. (2006). A comparison of survey methods for detecting bobcats. *Wildlife Society Bulletin*, **34**, 548–52.

Heaton, J.S., Cablk, M.E., Nussear, K.E., Esque, T.C., Medica, P.A., Sagebiel, J.C., and Francis, S.S. (2008). Comparison of effects of humans versus wildlife-detector dogs. *Southwestern Naturalist*, **53**, 472–9.

Hill, S. and Hill, J. (1987). *Richard Henry of Resolution Island*, John McIndoe, Dunedin, New Zealand.

Homan, H.J., Linz, G., and Peer, B.D. (2001). Dogs increase recovery of passerine carcasses in dense vegetation. *Wildlife Society Bulletin*, **29**, 292–6.

Hunter, S.A. (2011). Resource partitioning between generalist competitors & factors affecting the detectability of scat by dogs. MS thesis, Auburn University.

Hurt, A. and Smith, D.A. (2009). Conservation dogs. In W.S. Helton, Ed., *Canine Ergonomics: The Science of Working Dogs*. CRC Press, Boca Raton, pp. 175–94.

Kapfer, J.M., Muñoz, D.J., and Tomasek, T. (2012). Use of wildlife detector dogs to study eastern box turtle (*Terrapene carolina caroline*) populations. *Herpetological Conservation and Biology*, **7**, 169–75.

Katz, S.R. and Midkiff, C.R. (1998). Unconfirmed canine accelerant detection: A reliability issue in court. *Journal of Forensic Sciences*, **43**, 329–33.

Kerley, L.L. (2010). Using dogs for tiger conservation and research. *Integrative Zoology*, **5**, 390–6.

Kerley, L.L. and Salkina, G.P. (2007). Using scent-matching dogs to identify individual Amur tigers from scats. *Journal of Wildlife Management*, **71**, 1349–56.

Kretser, H. and Glennon, M. (2011). Evaluating genetic connectivity and re-colonization dynamics of moose in the Northeast. Wildlife Conservation Society, Saranac Lake, New York.

Kohn, M.H. and Wayne, R.K. (1997) Facts from faeces revisited. *Trends in Ecology & Evolution*, **6**, 223–27.

Lin, H.M., Chi, W.L., Lin, C.C., Tseng, Y.C., Chen, W.T., Kung, Y.L., Lien, Y.Y., and Chen, Y.Y. (2011). Fire ant-detecting canines: A complementary method in detecting red imported fire ants. *Journal of Economic Entomology*, **104**, 225–31.

Liu, H., Platt, S.G., and Borg, C.K. (2004). Seed dispersal by the Florida box turtle (*Terrapene carolina bauri*) in pine rockland forests of the lower Florida Keys, United States. *Oecologia*, **138**, 539–46.

Long, R.A., Donovan, T.M., Mackay, P., Zielinski, W.J., and Buzas, J.S. (2007a). Effectiveness of scat detection dogs for detecting forest carnivores. *Journal of Wildlife Management*, **71**, 2007–17.

Long, R.A., Donovan, T.M., Mackay, P., Zielinski, W.J., and Buzas, J.S. (2007b). Comparing scat detection dogs, cameras, and hair snares for surveying carnivores. *Journal of Wildlife Management*, **71**, 2018–25.

Long, R.A., Donovan, T.M., MacKay, P., Zielinski, W.J., and Buzas, J.S. (2011). Predicting carnivore occurrence with noninvasive surveys and occupancy modeling. *Landscape Ecology*, **26**, 327–40.

Lydersen, C. and Gjertz, I. (1986). Studies of the ringed seal *Phoca hispida* in its breeding habitat in Kongsfjorden Svalbard Arctic Ocean. *Polar Research*, **4**, 57–64.

MacKay, P., Smith, D.A., Long, R.A., and Parker, M. (2008). Scat detection dogs. In R.A. Long, P. MacKay, W.J. Zielinski and J.C. Ray, Eds., *Noninvasive Survey Methods for Carnivores*. Island Press, Washington DC, pp. 183–222.

Matre, P. J. (2003). Training, organisation and skills: case studies of practice using mine detection dogs. In I.G. McLean, Ed., *Mine Detection Dogs: Training, Operations, and Odour Detection*. Geneva International Centre for Humanitarian Demining, Geneva, pp. 139–63.

McCulloch, M., Jezierski, T., Broffman, M., Hubbard, A., Turner, K., and Janecki, T. (2006). Diagnostic accuracy of canine scent detection in early- and late-stage lung and breast cancers. *Integrative Cancer Therapies*, **5**, 30–9.

Milne, D. (2007). Controlling an invasive salt marsh grass (*Spartina patens*) in Washington State: A case study of resilience. *Environmental Practice*, **9**, 251–65.

Nussear, K. N., Esque, T.C., Heaton, J.S., Cablk, M.E., Drake, K.K., Valentin, C., Yee, J.L., and Medica, P.A. (2008). Are wildlife detector dogs or people better at finding tortoises (*Gopherus agassizii*)? *Journal of Herpetological Conservation and Biology*, **3**, 103–15.

O'Connor, S., Park, S.J and Goulson, D. (2012). Humans versus dogs; A comparison of methods for the detection of bumble bee nests. *Journal of Apicultural Research*, **51**, 204–11.

Parry-Jones, R. (2009). *The feasibility of using canines to detect wildlife contraband*, TRAFFIC East Asia, Central, Hong Kong.

Paula, J., Leal, M.C., Silva, M.J., Mascarenhas, R., Costa, H., and Mascarenhas, M. (2011). Dogs as a tool to improve bird-strike mortality estimates at wind farms. *Journal for Nature Conservation*, **19**, 202–8.

Pearsall, M.D. and Verbruggen, H. (1982). *Scent: Training to Track, Search, and Rescue*, Alpine Publications Inc., Loveland, Colorado.

Pfiester, M., Koehler, P.G., and Pereira, R.M. (2008). Ability of bed bug-detecting canines to locate live bed bugs and viable bed bug eggs. *Journal of Economic Entomology*, **101**, 1389–96.

Platt, S.G., Ko, W.K., Khaing, L.L, Myo, K.M., Swe, T., Lwin, T., and Rainwater, T.R. (2003). Population status and conservation of the critically endangered Burmese star tortoise *Geochelone platynota* in central Myanmar. *Oryx*, **37**, 464–71.

Putnam, R.J. (1984). Facts from faeces. *Mammal Review*, **14**, 79–97.

Ralls, K. and Smith, D.A. (2004). Latrine use by San Joaquin kit foxes (*Vulpes macrotis mutica*) and coyotes (*Canis latrans*). *Western North American Naturalist*, **64**, 544–7.

Ralls, K., Sharma, S., Smith, D.A., Bremner-Harrison, S., Cypher, B.L., and Maldonado, J.E. (2010). Changes in kit fox defecation patterns during the reproductive season: Implications for noninvasive surveys. *Journal of Wildlife Management*, **74**, 1457–62.

Rebmann, A., David, E., and Sorg, M. (2000). *Cadaver Dog Handbook: Forensic Training and Tactics for the Recovery of Human Remains*, CRC Press, Boca Raton.

Reed, S.E., Bidlack, A.L., Hurt, A., and Getz, W.M. (2011). Detection distance and environmental factors in conservation detection dogs surveys. *Journal of Wildlife Management*, **75**, 243–51.

Reindl-Thompson, S.A., Shivik, J.A., Whitelaw, A., Hurt, A., and Higgins, K.F. (2006). Efficacy of scent dogs in detecting black-footed ferrets at a reintroduction site in South Dakota. *Wildlife Society Bulletin*, **34**, 1435–9.

Robertson, H.A. and Fraser, J.R. (2009). Use of trained dogs to determine the age structure and conservation status of kiwi *Apteryx* spp. populations. *Bird Conservation International*, **19**, 121–9.

Robicheaux, J. and Jons, J.A.R. (1996). *Basic Narcotic Detection Dog Training*. K-9 Concepts, Inc., Broussard, Louisianna.

Rolland, R.M, Hamilton, P.K., Kraus, S.D., Davenport, B., Gillett, R.M., and Wasser, S.K. (2006). Faecal sampling using detection dogs to study reproduction and health in North Atlantic right whales (*Eubalaena glacialis*). *Journal of Cetacean Research and Management*, **8**, 121–5.

Savidge, J.A., Stanford, J.W., Reed, R.N., Haddock, G.R., and Yackel Adams, A.A. (2010). Canine detection of free-ranging brown treesnakes on Guam. *New Zealand Journal of Ecology*, **35**, 174–81.

Schoon, G.A. and Haak, R. (2002). *K9 Suspect Discrimination: Training and Practicing Scent Identification Line-Ups*. Detselig Enterprises, Calgary.

Schwartz, C.W. and Schwartz, E.R. (1974). The three-toed box turtle in central Missouri: Its population, home range and movements. Missouri Department of Conservation, Missouri.

Shivik, J.A. (2002). Odor-adsorptive clothing, environmental factors, and search-dog ability. *Wildlife Society Bulletin*, **30**, 721–7.

Smith, D.A., Ralls, K., Davenport, B., Adams, B., and Maldonado, J.E. (2001). Canine assistants for conservationists. *Science*, **291**, 435.

Smith, D.A., Ralls, K., Hurt, A., Adams, B., Parker, M., Davenport, B., Smith, M.C., and Maldonado, J.E. (2003). Detection and accuracy rates of dogs trained to find scats of San Joaquin kit foxes (*Vulpes macrotis mutica*). *Animal Conservation*, **6**, 339–346.

Smith, D.A., Ralls, K., Cypher, B.L., Clark, Jr., H.O., Kelly, P.A., Williams, D.F., and Maldonado, J.E. (2006). Relative abundance of endangered San Joaquin kit foxes (*Vulpes macrotis mutica*) based on scat-detection dog surveys. *Southwestern Naturalist*, **51**, 210–9.

Steury, T.D. (2011). Distribution of long-tailed weasels in Alabama. Final report to the Alabama Department of Conservation and Natural Resources.

Stevenson, D.J., Ravenscroft, K.R., Zappalorti, R.T., Ravenscroft, M.D., Weigley, S.W., and Jenkins, C.L. (2010). Using a wildlife detector dog for locating eastern indigo snakes (*Drymarchon couperi*). *Herpetological Review*, **41**, 437–42.

Sturdivan, C. (1993). Locating wolves with a dog trained in tracking and scent-discrimination. MS thesis, Evergreen State College.

Syrotuck, W.G. (1972). *Scent and the scenting dog*, Arner Publishing, Rome, New York.

Theobald, S. and Coad, N. (2002). Den control of stoats (*Mustela erminea*) in Trounson Kauri Park, Northland.

DOC Science Internal Series 90. Department of Conservation, Wellington.

Thompson, C.M., Purcell, K., Garner, J., and Green, R. (2010). Kings River Fisher Project. Progress Report 2007-2010. Pacific Southwest Research Center, USDA Forest Service, Pacific Southwest Research Station.

Thompson, C.M., Royle, J.A., and Garner, J.D. (2012). A framework for inference about carnivore density from unstructured spatial sampling of scat using detector dogs. *Journal of Wildlife Management*, **76**, 863–71.

Vice, D.S. and Engeman, R.M. (2000). Brown tree snake discoveries during detector dog inspections following supertyphoon Paka. *Micronesica*, **33**, 105–10.

Vynne, C., Skalski, J.R., Machado, R.B., Groom, M.J., J'Acomo, A.A, Marinho-Filho, J., Ramos Neto, M.B., Pomilla, C., Silveira, L., Smith, H., and Wasser, S.K. (2010). Effectiveness of scat-detection dogs in determining species presence in a tropical Savanna landscape. *Conservation Biology*, **25**, 154–62.

Vynne, C., Keim, J.L., Machado, R.B., Marinho-Filho, J., Silveira, L., Groom, M.J., and Wasser, S.K. (2011). Resource selection and its implications for wide-ranging mammals of the Brazilian Cerrado. PLoS ONE 6(12): e28939.

Wasser, S.K., Davenport, B., Ramage, E.R., Hunt, K.E., Parker, M., Clarke, C., and Stenhouse, G. (2004). Scat detection dogs in wildlife research and management: Application to grizzly and black bears in the Yellowhead Ecosystem, Alberta, Canada. *Canadian Journal of Zoology*, **82**, 475–92.

Wasser, S.K., Smith, H., Madden, L., Marks, N., and Vynne, C. (2009). Scent-matching dogs determine number of unique individuals from scat. *Journal of Wildlife Management*, **73**, 1233–40.

Wasser, S.K., Keim, J.L., Taper, M.L., and Lele, S.R. (2011). The influences of wolf predation, habitat loss, and human activity on caribou and moose in the Alberta oil sands. *Frontiers in Ecology and the Environment*, **9**, 546–51.

Waters, J., O'Connor, S., Park, K.J., and Goulson, D. (2010). Testing a detection dog to locate bumblebee colonies and estimate nest density. *Apidologie*, **42**, 200–5.

Weller, T.J. (2008). Using occupancy estimation to assess the effectiveness of a regional multiple-species conservation plan: Bats in the Pacific Northwest. *Biological Conservation*, **141**, 2279–89.

Williams, M. and Johnston, J.M. (2002). Training and maintaining the performance of dogs (*Canis familiaris*) on an increasing number of odor discriminations in a controlled setting. *Applied Animal Behaviour Science*, **78**, 55–65.

Willis, C.M., Church, S.M., Guest, C.M., Cook, W.A., McCarthy, N., Bransbury, A.J., Church, M.R.T., and Church, J.C.T. (2004). Olfactory detection of human bladder cancer by dogs: Proof of principle study. *British Medical Journal*, **329**, 712–714A.

Winter, W. (1981). Black-footed ferret search dogs. Southwestern Research Institute, New Mexico.

Woodroffe, R. and Ginsberg, J.R. (1998). Edge effects and the extinction of populations inside protected areas. *Science*, **280**, 2126–8.

Wultsch, C. (2009). Molecular scatology: Optimizing noninvasive genetic monitoring of jaguars (*Panthera onca*) and other elusive feline species in the tropics. *Wild Felid Monitor*, **2**, 21.

CHAPTER 11

Hunting dogs and the extraction of wildlife as a resource

Jeremy Koster and Andrew Noss

11.1 Introduction

Dogs fulfill a variety of roles in contemporary societies, and their assistance in hunting may have facilitated the initial domestication and use of dogs (Olsen, 1985). Apparently, few ecosystems exist in which dogs cannot be useful as hunting companions, at least occasionally (Koster, 2009). In some cases, the effects of hunting dogs on wildlife are therefore millennia-old, no doubt contributing selective pressures across and within species that influence patterns found today: game species distribution, communities, and even social or population structures within species. However, the intensification of hunting with dogs, through more numerous human populations and heavier hunting pressure in some regions, and/or more specialist breeding of hunting dogs and prey selection in response to new cultural or market demands, can exacerbate prior tendencies with new implications for conservation. We review the existing literature to describe principal impacts on wildlife, already identified or potentially significant, that result from the contemporary use of dogs for hunting.

The pairing of dogs and human hunters can be a potent combination because of the complementarity of their contributions. Dogs often enhance the probability of successfully harvesting of an animal or reduce the time needed to complete a pursuit (Ikeya, 1994; Lupo, 2012; Milner, 1994). Dogs may detect and indicate as pointers or subsequently corral animals in locations where the human hunters can dispatch the prey (Fitzgerald et al., 1991; Koster, 2008a; White et al., 2003). They are also used by hunters to drive terrestrial animals such as deer (Carr, 1994; Sunde et al., 2009; Sweeney et al., 1971; Torres-Porras et al., 2009; White et al., 2003) or to flush birds (Brøseth and Pedersen, 2010; Guthery and Mecozzi, 2008).

In many cases, prey species exhibit anti-predator strategies that are ill-suited to contend with the combination of dogs and humans. Several species seek refuge in hollow tree trunks (collared peccaries *Pecari tajacu*), burrows (red tegu lizards *Tupinambis rufescens*, nine-banded armadillos *Dasypus novemcinctus*, pacas *Cuniculus paca*, cane rats *Thryonomys* spp.), or bodies of water (lowland and Baird's tapirs *Tapirus terrestris*, *T. bairdii*) when pursued by dogs. Others turn at bay (i.e., stop to face one or more pursuing dogs), allowing hunters to approach them (tapirs, pigs *Sus scrofa*, white-lipped peccaries *Tayassu pecari*, Chacoan peccaries *Catagonus wagneri*, and gemsboks *Oryx gazelle*, among others). The three-banded armadillo *Tolypeutes matacus* rolls into a protective ball upon being located by a dog and is easily collected by the hunter. Some carnivores climb exposed trees to escape pursuit (e.g., bears *Ursus* spp., jaguars *Panthera onca*, pumas *Puma concolor*). Other semi-terrestrial species also take refuge in trees when pursued by hunting dogs, including raccoons *Procyon lotor*, squirrels *Sciurus carolinensis* and *S. vulgaris*, and capercaillie *Tetrao urogallus*. Given such responses, the use of dogs increases the vulnerability of some wildlife species to human hunters who are equipped with sophisticated tools

Free-Ranging Dogs and Wildlife Conservation. Edited by Matthew E. Gompper
© Oxford University Press 2014. Published 2014 by Oxford University Press.

for extracting and dispatching wildlife (Redford and Robinson, 1987; see also Box 11.1).

Readers in affluent, industrialized countries may bear stereotypes of hunting dogs as well-trained, specially-bred dogs that target a specific kind of prey. For example, hounds are used for hunting red foxes (*Vulpes vulpes*) in Europe, North America and Australia (Baker and Harris, 2006; Marvin, 2001; White et al., 2003); mink (*Mustela vison*) in England and Wales (White et al., 2003); raccoons in the USA (Bergman et al., 1995); black bears (*U. americanus*) in the USA (Ryan et al., 2009); brown bears (*U. arctos*) in Sweden (Bischof et al., 2008); pumas in Canada and the USA (Laundré et al., 2000; Ross and Jalkotzy, 1992); wild or feral pigs in Europe, Taiwan, Japan, Brazil, and the USA (Desbiez et al., 2011; Fonseca et al., 2011; Katahira et al., 1993; Maillard and Fournier, 1995; McCann and Garcelon, 2008; Nobayashi, 2006; Ohashi et al., In press; Scillitani et al., 2010; Sodeikat and Pohlmeyer, 2003; Ohashi et al., 2013); moose (*Alces alces*) in Finland (Ruusila and Pesonen 2004); red deer (*Cervus elephas*) and roe deer (*Capreolus capreolus*) in Europe (Bonnot et al., 2013; Torres-Porras et al., 2009; White et al., 2003); and brown hares (*Lepus europaeus*) in Europe, the USA, and New Zealand (White et al., 2003). Pointing dogs are used for willow ptarmigan in Norway (Brøseth and Pedersen, 2010), black grouse (*Tetrao tetrix*) and capercaillie (*T. urogallus*) in Norway (Lande et al., 2010), and northern bobwhite in the USA (Guthery and Mecozzi, 2008), among many other birds hunted with dogs.

The situation in subsistence-based societies is frequently very different, partly because of the overarching motivation to pursue game animals for their edible meat. South American hunters use dogs in the Argentine Chaco for peccaries (Altrichter, 2005), and in the Bolivian Chaco for peccaries, tapirs, and armadillos (Cuéllar, 1999; Fiorello et al., 2006; Noss et al., 2002, 2003, 2004). Hunters in Papua New Guinea use dogs for ring-tailed possums *Pseudocheirus cupreus* and *P. vestilus*, cuscus *Phalanger gymnotis*, rodents, and pigs (Dwyer, 1983; Morren, 1986). Pigs are also hunted with dogs in the Philippines (Estioko-Griffin and Griffin, 1981) and in Borneo (Hoffman, 1986). Hunters in central African forests and savannas use dogs for a variety of mammals and guinea fowl (Gosselin, 1972; Harako, 1981). Poor rural residents may also use hunting dogs for commercial hunting activities, such as harvests of red tegu lizards in the Chaco of Argentina, Paraguay, and Bolivia (Cuéllar et al., 2010; Fitzgerald et al., 1991), or ungulates and banded mongoose *Mungo mungo* in Tanzanian savannas (Caro, 2008; Holmern et al., 2006; Martin et al., 2013). In these settings, dogs are often valuable contributors to household subsistence and livelihoods, although they are often malnourished, haphazardly-trained mongrels receiving little or no veterinary attention (Bronson et al., 2008; Fiorello et al., 2004, 2006; Koster, 2009; Ortega-Pacheco et al., 2007), in sharp contrast to the pampered purebreds mentioned above. In addition, hunters with dogs in

Box 11.1 Local hunting dogs used for biological research

Whereas biologists sometimes use dogs that have specifically been trained for biological research (Woollett et al., Chapter 10), they have also used local hunting dogs in a number of research activities, including telemetry and health studies. Numerous researchers have captured jaguars across their range using local dogs (Crawshaw and Quigley, 1991; Rabinowitz, 1986; Schaller and Crawshaw, 1980; Silveira, 2004; Soisalo and Cavalcanti, 2006; McBride Jr. and McBride, 2007; Azevedo and Murray, 2007; Furtado et al., 2008). Peccary researchers have used local dogs to capture Chacoan peccaries in the Paraguayan Chaco (Taber et al., 1993) and collared peccaries in the United States (Bergman et al., 1995). In the Bolivian Chaco, researchers used local dogs to capture nine-banded and three-banded armadillos, lowland tapir, and collared peccaries for telemetry and health studies (Arambiza and Guerrero, 2000; Ayala, 2000; Deem et al., 2009; Noss et al., 2002). A combination of pointing dogs and hand-held nets was used for willow ptarmigans *Lagopus lagopus* (Brøseth and Pedersen, 2010; Skinner et al., 1998). Finally, local hunting dogs have also been used to census nine-banded and three-banded armadillos in the Bolivian Chaco (Cuéllar, 2002) and to estimate the density of northern bobwhite *Colinus virginianus* in the United States (Guthery and Mecozzi, 2008; Williams et al., 2004).

industrialized countries often exhibit a distinct sense of identity (Boglioli, 2009; Chitwood et al., 2011) and attitudes toward wildlife conservation that differ from the general public, with the latter often being uncomfortable about or even strongly opposed to the use of dogs to pursue and kill animals as a sport and management tool (Agarwala et al., 2010; Liukkonen et al., 2009; Ryan et al., 2009; White et al., 2003). Typically, however, such contrasts in identities and attitudes are less evident in subsistence-based societies, perhaps owing to the reduced differentiation by social class and homogeneity of hunting methods and objectives that characterize such settings (cf. Howe, 1981). Although this review identifies commonalities that transcend socio-ecological differences, conservationists and managers should note that the ramifications of hunting with dogs in any particular locale generally reflect a complex set of variables, including characteristics of both the dogs and the prey species, broader questions of population ecology, and the hunters' motivations and strategies.

In this chapter, we review the literature to evaluate the effects that hunting with dogs may have on wildlife populations and the relevance of these impacts for conservation and wildlife management. We look first at how hunting with dogs, as opposed to other hunting methods, can concentrate hunting pressure on certain species, including vulnerable or endangered species as well as species more resistant to hunting pressure. We subsequently examine scenarios in which the use of dogs can bias harvests towards one sex or the other, or to a particular age category in the prey population. Finally, we briefly consider studies that document the effects of hunting with dogs on the habitat use and ranging behavior of wildlife populations. We explored relevant sources in the primary literature, finding overall a lack of specific attention to hunting dogs even where they are important contributors to hunting activities. Owing to our background in the Neotropics (see Box 11.2 and 11.3), we were better able to locate references on this region in the secondary literature, including unpublished theses and technical reports. This chapter therefore speaks most clearly to issues that are particularly evident in the lowland Neotropics and, by extension, to the subsistence-based societies in other tropical regions. We nevertheless incorporate references from more affluent, sport hunting settings to elucidate generalizable aspects of dog–wildlife interactions, but the different motivations of the hunters and the regulatory management of hunting in these settings often preclude broader comparisons.

Box 11.2 The use of dogs for hunting in lowland Nicaragua

The indigenous Mayangna and Miskito of lowland Nicaragua have a longstanding tradition of hunting with dogs (Figure 11.1). Primarily horticulturalists, the Mayangna and Miskito rely on hunting and fishing as important sources of dietary protein (Stocks, 1996). Typically weighing 11–12 kg, their dogs resemble the generic 'walking hounds' described by Coppinger and Coppinger (2001). Little evidence exists of managed breeding to promote good hunting characteristics or deliberate culling of ineffective dogs. Dogs are usually acquired locally, but some residents purchase multipurpose dogs when traveling in the Mestizo communities beyond the indigenous territories. Locally, puppies typically cost the equivalent of 5 kg of hunted meat, but adults with a reputation for hunting are considerably more expensive. Most households own one or two dogs, but some of the most active hunters own several dogs. There are apparently no un-owned dogs.

Adult dogs vary considerably in their hunting ability. Whereas some dogs are capable of corralling ungulates, others are considered 'house dogs' that do not contribute as hunting companions. Throughout the study, hunters credited dogs with contributions to harvests, and although older dogs and male dogs exhibited statistically greater contributions, the harvests primarily reflected the frequency with which the dogs accompanied their owners beyond the boundaries of the community (Koster and Tankersley, 2012). Hunters with skilled dogs are motivated to bring their dogs on excursions, not only on planned hunting trips but also while collecting firewood and

continued

Box 11.2 *Continued*

Figure 11.1 A Mayangna hunter pursues an agouti (*Dasyprocta punctata*) that fled into a hollow tree trunk upon being chased by the hunter's dogs. (Photograph courtesy of Menuka Scetbon-Didi.)

farming, which are activities that can be interrupted if the dogs locate prey. One prolific hunter, whose dogs were observed to be outside of the community more than 25% of the time during daylight hours, accounted for 31% of the harvested biomass during the study, including half of the tapirs. (The median percentage of observed absences from the community for dogs in other households is 16%. Because observations occurred from 5:30 am to 6:00 pm, this value indicates that a typical dog is beyond the boundary of the community for an average of two hours per day.)

Whereas unaccompanied free-ranging dogs are a concern in other settings, dogs in Arang Dak rarely leave the community without their owners. Although adult dogs are typically unrestrained and free to roam, only 2 of 1,524 randomized observations revealed dogs that were ranging alone beyond the boundaries of the community (see also Lupo, 2012). To some extent, provisioning by their owners and a lack of foraging opportunities in the forest (since dogs can apparently capture few animals without assistance from humans) may dissuade dogs from leaving the community alone.

11.2 Interspecific biases in harvests with dogs

Compared to hunters without dogs, hunters with dogs encounter and capture different game species because the dogs' senses and physical abilities complement those of their human companions. In many cases, the encounter rates with prey species therefore vary considerably depending on the presence or absence of dogs. For example, hunters with dogs in lowland Nicaragua encounter almost 900% more agoutis (*Dasyprocta punctata*) than rifle hunters without dogs (Koster, 2008a). The taxonomic diversity of prey species that are encountered more frequently with dogs reveals the versatility of their potential use, as increased encounter rates when hunting with dogs have been reported for felids (Carvalho Jr. and Pezzuti, 2010), semi-terrestrial primates (Colell et al., 1994; Wright and Priston, 2010), small marsupials (Dwyer, 1983), armadillos (Koster, 2008a), and turtles (Ly et al., 2011). To some extent, these differences reflect the ability of dogs to detect nocturnal species that might otherwise remain unnoticed by human hunters on daytime excursions (Koster, 2008b; Newton et al., 2008) (see Box 11.3).

Box 11.3 Nocturnal hunting with dogs

Hunters in the Bolivian Chaco recorded information on their hunting activities from 1996 to 2000 (Noss et al., 2003, 2004). These hunters use dogs to track armadillos, often extracting them from their burrows. Thus many more nocturnal armadillos are captured by hunters with dogs than without—ratios of between 3:1 and 7:1 (Table 11.1).

These harvest data are displayed visually via a mosaic plot (Figure 11.2). On daytime hunts without dogs, hunters primarily capture diurnal animals. Many such animals are also harvested with dogs during the day, but captures include a much greater representation of nocturnal animals when hunting with dogs—whether at night or during the day. The major limitation of the mosaic plot is that it aggregates across species. For example, the gray brocket deer *Mazama gouazoubira* dominates the diurnal hunting without dogs and the nine-banded armadillo is heavily represented among the nocturnal animals hunted with dogs. Chaco hunters say that dogs can never catch brocket deer; therefore they discourage their dogs from chasing deer at all and prefer to hunt deer by stalking with a firearm. Dogs are especially useful to track and bring to bay species such as tapir and all three peccaries, and to locate armadillos, often driving them to burrows that are subsequently excavated.

To test for a relationship between the use of dogs and harvests of nocturnal prey, we set the presence of dogs as

Table 11.1 Game species captured by subsistence hunters in the Bolivian Chaco (A. Noss, unpublished data). Note: over 700 hunters voluntarily provided written data on over 9,000 hunting events between 1996 and 2000.

Scientific name	English name	Principal activity pattern	Dogs		No dogs		Total
			Day	Night	Day	Night	
Catagonus wagneri	Chacoan peccary	Diurnal	5		5		10
Chaetophractus vellerosus	Small hairy armadillo	Nocturnal	37	56	17	6	116
Chaetophractus villosus	Large hairy armadillo	Diurnal	193	13	234	5	445
Coendou prehensilis	Brazilian porcupine	Nocturnal			7	1	8
Ctenomys conoveri	Conover's tuco tuco	Diurnal	3		18	2	23
Dasyprocta azarae	Agouti	Diurnal	10	2	97	4	113
Dasypus novemcinctus	Nine-banded armadillo	Nocturnal	256	560	70	42	928
Euphractus sexcinctus	Yellow armadillo	Diurnal	87	30	93	12	222
Lagostomus maximus	Plains vizcacha	Nocturnal	1	1	13	6	21
Leopardus geoffroyi	Geoffroy's cat	Nocturnal	6		10		16
Mazama gouazoubira	Grey brocket deer	Diurnal	183	2	1,007	11	1,203
Myrmecophaga tridactyla	Giant anteater	Nocturnal	10		3		13
Pecari tajacu	Collared peccary	Diurnal	401	14	289	7	711
Puma concolor	Puma	Nocturnal	9	1	9		19
Tamandua tetradactyla	Tamandua anteater	Nocturnal	1	2	2	1	6
Tapirus terrestris	Brazilian tapir	Nocturnal	35	4	13	2	54
Tayassu pecari	White-lipped peccary	Diurnal	76	8	98		182
Tolypeutes matacus	Three-banded armadillo	Nocturnal	352	262	96	22	732
Tupinambis rufescens	Red tegu lizard	Diurnal	65	2	84		151
	TOTAL		1,730	957	2,165	121	4,973

continued

Box 11.3 *Continued*

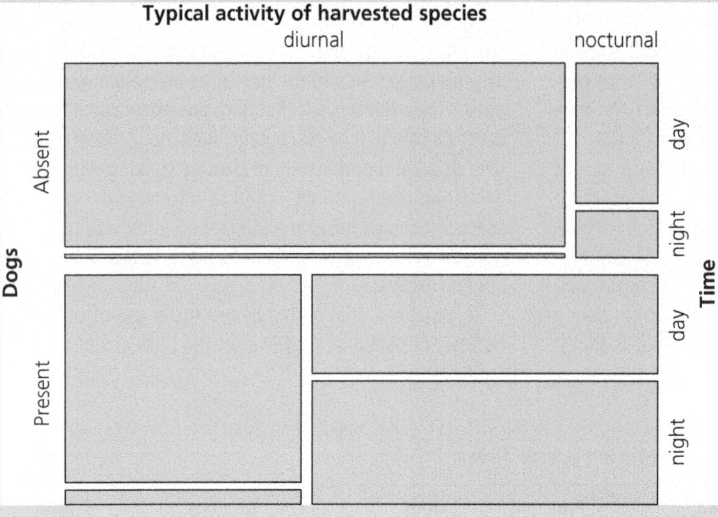

Figure 11.2 Mosaic plot comparing daytime and nighttime captures of nocturnal versus diurnal species by hunters when dogs are present or absent. Note that the self-monitoring records indicate whether dogs accompanied hunters on their outings, but the records do not specify whether dogs played a principal role in catching the game.

a binary outcome variable, analyzed via a generalized linear mixed model (GLMM) with a logit link and a random effect (varying intercept) for the harvested species. Fixed effects in the model include binary variables to denote the typical activity pattern of the species (nocturnal vs. diurnal) and whether the harvest occurred at night, and we include the interaction of these two predictors. Our interest is primarily correlational, and we specify this formula because of the data structure, not to suggest that these variables 'cause' the presence or absence of dogs. As noted in Table 11.2, the results of the model reveal a significant effect of time of day, reflecting the heavy use of dogs for hunting at night. Contrary to our hypothesis, however, the model does not reveal a significant positive association between the presence of dogs and the customary activity pattern of the harvested animals, either during the day (as represented by the main effect) or at night (as reflected by the interaction term). We conclude that, although dogs are clearly assets for hunting some nocturnal species, such as nine-banded armadillos, they are less effective at hunting other nocturnal species (and conversely, dogs are also valuable when hunting some diurnal prey, such as the peccaries). The large estimated variance of the random effect suggests that considerable species-level variation remains unexplained by the predictors in the model.

Table 11.2 Model estimates of the presence of dogs when an animal is harvested ($n = 4{,}973$) in Bolivia. Coefficients are on the log-odds scale.

Variable	Estimate (S.E.)
Fixed effects	
Intercept	−0.69 (0.38)
Harvest at night	0.94 (0.22) ***
Nocturnal species	0.75 (0.54)
Harvest at night * Nocturnal species	0.29 (0.26)
Random effect variance	
Species	1.21

*** $p < 0.001$

Despite the diversity of potential prey species, dogs seem especially well-suited to the hunting of species that are frequent targets of other medium-sized, terrestrial carnivores. Accordingly, reports often indicate that dogs are particularly useful for hunting various ungulates and rodents (Constantino et al., 2008; Corlett, 2007; Crawford and Robinson, 1984; Dei, 1989; Fitzgibbon et al., 1995; Harako, 1981; Lupo, 2012; Morren, 1986; Noss et al., 2004; White et al., 2003; Ruusila and Pesonen 2004; Altrichter, 2005).

By contrast, dogs are not particularly helpful for hunting principally arboreal species, including most primates (Kaplan and Kopischke, 1992; but see Golden, 2009 for reports of dogs used for detecting lemurs).

Among mammals, an evolutionary history of arboreality is associated with extended life history characteristics, apparently because arboreal mammals suffer less predation than their terrestrial counterparts (Shattuck and Williams, 2010). Species that traditionally suffer high predation tend to mature rapidly and have relatively high reproductive rates, which often makes them less vulnerable to overhunting than long-lived species that reproduce slowly (Bodmer et al., 1997). Because the terrestrial ungulates and rodents that are frequently hunted with dogs usually suffer high mortality from other predators, these species may be able to withstand higher rates of off-take than species that incur less predation from non-human predators. By directing hunters' attention and effort toward wildlife species that can endure high levels of off-take, hunting with dogs may reduce the pressure on slowly-reproducing, easily over-hunted species (see Box 11.4).

Box 11.4 Hunting with dogs and life history correlates of prey in Nicaragua

The lead author conducted a year-long study of subsistence hunting in Arang Dak, a community of 35 households along the Lakus River, which flows from the undeveloped core of Nicaragua's Bosawas Biosphere Reserve. Observational data indicate that dogs contribute to hunting primarily by detecting and corralling prey, and hunters with dogs encounter substantially more agoutis, pacas, and nine-banded armadillos than unaided hunters (Koster, 2008a). Dogs also assist in the capture of larger ungulates, including collared peccaries and deer (*Mazama americana*), albeit without comparable increases in encounter rates.

Hunting dogs seem well-suited for capturing fast-breeding species, which can be indexed by the intrinsic rate of natural increase (r_{max}) for the species, as estimated by Robinson and Redford (1991). Although the extent to which wildlife populations can withstand high off-takes depends on several additional considerations, most notably population densities and extrinsic sources of mortality (see Robinson and Redford, 1991), slowly-breeding species are usually more susceptible to over-exploitation (Bodmer et al., 1997). In general, hunting with dogs in Arang Dak is biased toward species whose relatively high reproductive rates permit a greater proportion of individuals to be harvested without over-exploitation (Figure 11.3). The species that are most frequently hunted with the assistance of dogs—namely agoutis, pacas, nine-banded armadillos, and collared peccaries—can typically endure relatively higher harvests than tapirs and primates such as spider (*Ateles* spp.), howler (*Alouatta* spp.), and capuchin (*Cebus* spp.) monkeys. Assuming that hunters would otherwise use firearms while increasing their harvests of these latter species and cracids (e.g., crested guans *Penelope purpurescens*, and great curassows *Crax rubra*), which also reproduce slowly (r_{max} = .15; Alvard, 1998), the use of dogs could lessen the pressure on several of the most vulnerable wildlife populations in the reserve.

The production model of Robinson and Redford (1991) is commonly used to identify unsustainable hunting in the lowland Neotropics. Because the model incorporates numerous assumptions, especially that wildlife population densities correspond to predicted values in un-hunted sites, the maximum harvests in this model are considered over-estimates, and therefore the model cannot be used to infer that observed harvests are sustainable (Bodmer and Robinson, 2005). In Arang Dak, the harvests of two species, Baird's tapir and paca, exceeded the maximum sustainable estimates. Fourteen tapirs were harvested during the study period, including eleven that were detected or bayed by dogs and subsequently killed by the hunters, usually with guns (excepting two occasions when the animals fled to the river, where they were dispatched with machetes and axes). The persistence of tapirs in the core hunting zone around Arang Dak apparently depends on immigration from lightly hunted areas upstream of the settlement (Koster, 2008b). Dogs play a greater role in the harvests of pacas, 92% of which were detected by dogs and subsequently dispatched with machetes and other hand technologies in the burrows, trunks, and waterways where they attempted to hide. Given that the density of pacas shows little evidence of depletion around human settlements (Williams-Guillen et al., 2006), it is possible that the production model underestimates the maximum sustainable harvest of this species in this context, perhaps owing to enhanced foraging opportunities and the

continued

Box 11.4 *Continued*

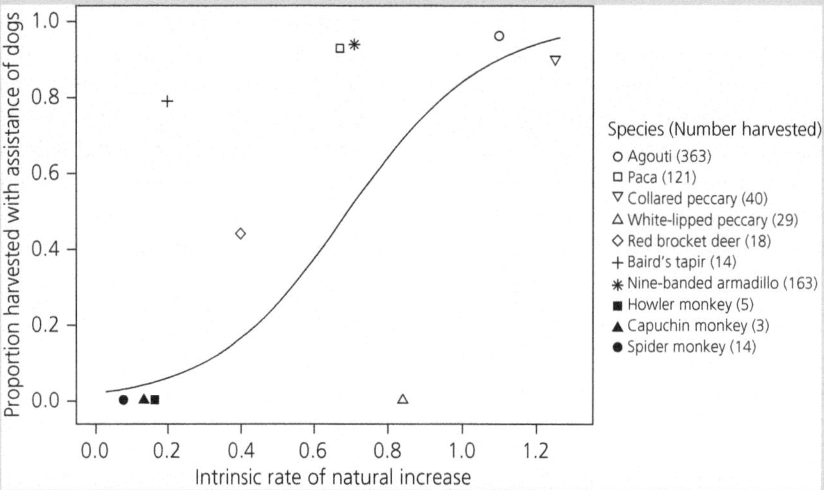

Figure 11.3 The proportion of harvests to which dogs contributed as a function of the intrinsic rate of natural increase (r_{max}) for ten wildlife species in lowland Nicaragua as reported by Robinson and Redford (1991). The fitted line is based on the fixed effects (intercept and r_{max}) from a generalized linear mixed model (GLMM) with a logit link and a random effect (varying intercept) for the ten species. The value in parentheses in the legend indicates the number of animals from each species that were harvested during a year-long study (see Koster, 2008b).

reduced presence of non-human predators in the anthropogenic habitats near the communities (Koster, 2008b).

Overall, hunting with dogs in lowland Nicaragua is directed primarily at fast-reproducing species that adapt well to anthropogenic habitats, where much of the hunting occurs. When paired with firearms, however, dogs apparently increase the rate at which hunters encounter and harvest tapirs.

From a conservationist perspective, however, the ramifications of hunting with dogs can rarely be understood without considering the complementary technologies used by the hunters. Firearms are widely used today, replacing a suite of projectile weapons, sometimes imbued with poison, developed by hunters over millennia, including spears and spear throwers, bows and arrows, crossbows and darts, boomerangs, and blowguns and darts. When equipped with effective projectile weapons, hunters with dogs are better able to harvest wildlife species whose anti-predator strategies would otherwise protect them from being killed. In northern latitudes, for example, harvests of bears that have been treed by dogs almost invariably involve the use of firearms (Bischof et al., 2008). Similarly, in the Neotropics guns are typically needed to harvest jaguars that have fled to trees upon being pursued by dogs (Miller, 1930), and firearms increase the ability of hunters to dispatch tapirs (*Tapirus bairdii*) that have retreated to waterways when chased by dogs (Koster, 2006; see also Noss et al., 2002 and Tobler et al., 2006 for *T. terrestris*). Because such species are relatively susceptible to over-hunting, the use of dogs in such contexts could merit regulatory scrutiny, although the threat might be attenuated by restrictions on firearms if other technologies (e.g., spears or lassos) are ineffective for dispatching or capturing the animals. Although declines in wildlife populations can generally be attributed to habitat destruction and the increase in the number of human hunters, the continued use of dogs in combination with firearms can exacerbate the pressures on wildlife in particular cases and disproportionately affect particular species or classes within species.

In some cases, the hunters' ability to harvest prey depends in part on the number of accompanying dogs. Among small groups of hunters in Finland, for example, the number of harvested moose is positively correlated with the number of dogs that are present on the excursion (Ruusila and Pesonen, 2004). Although few quantitative studies exist, similar correlations might be especially evident when hunting larger prey species, as researchers have reported that the use of multiple dogs facilitates harvests of black bears (Higgins Inman and Vaughan, 2002), tapirs (Noss et al., 2002, 2003) and pigs (Caley and Ottley, 1995; Estioko-Griffin and Griffin 1981; Hoffman, 1986), apparently because multiple dogs are more likely to surround the prey and force it to climb a tree or turn at bay. Larger numbers of dogs can be helpful for hunting relatively smaller prey as well, as seen in the Bolivian Chaco, where hunters who primarily harvest armadillos and peccaries employ significantly more dogs on successful forays (2.8 ± 1.5, n = 2,264) than unsuccessful trips (2.2 ± 1.9, n = 157) (unpaired t test, $p <$ 0.0001) (Fiorello et al., 2006). More noses may increase detection of cryptic prey, such as fossorial armadillos.

Hunting with dogs has been described as an 'unselective' method of hunting, partly because dogs can indiscriminately attack prey, including juvenile animals, females with young, or non-target species, before hunters are able to intervene (Godwin et al., 2013; Grey-Ross et al., 2010). Indeed, observational research indicates that dogs can increase encounters and conflicts with wildlife species that hunters might otherwise disregard. In Nicaragua and Bolivia, hunting with dogs leads to harvests of non-target species including giant anteaters (*Myrmecophaga tridactyla*), tamanduas (*Tamandua mexicana* and *T. tetradactyla*), and greater grisons (*Galictis vittata*). Although these species are not consumed locally, they are killed when corralled or bayed by dogs, largely to protect the dogs from harm (Cuéllar and Noss, 2003; Koster, 2008b, c) (see Box 11.5). Because the carcasses from those hunting interactions are often left at the site of the kill, conventional methods for documenting harvests (hunter self-monitoring, household or market surveys) may underestimate the effect of hunting with dogs on the mortality of non-target species (Noss et al., 2004).

Box 11.5 Hunting dogs and interactions with large predators

In addition to facilitating encounters with prey species, hunting dogs can also attract the attention of large predators. For example, the use of hunting dogs may increase encounters with wolves (Butler et al., Chapter 5), which hunters would ordinarily be inclined to pursue and kill preemptively if not for legal restrictions (Backeryd, 2007; Bisi et al., 2007). Hunters in Finland and Macedonia affirm that wolves seek out dogs to kill and eat them (Kojola et al., 2004; Lescureux and Linnell, 2010). Whereas dogs are used to tree and capture jaguars in the Pantanal and Chaco of South America, in the Atlantic Forest jaguars do not tree when pursued and instead kill the dogs (di Bitetti, pers. comm.). Dogs are also preyed upon by jaguars in Nicaragua, and of the four documented kills of jaguars during a year-long study period, three occurred during attacks on dogs (Koster, 2008b). Occasionally, dogs in Nicaragua also engage in conflicts with smaller carnivores, including ocelots (*Leopardus pardalis*) and tayras (*Eira barbara*). Predictably, hunters seem especially motivated to kill animals that purportedly threaten dogs, even when the meat from the animals is unconsumed.

In subsistence-based societies, the extent to which dogs are trained to stalk and pursue a limited suite of prey species is often unknown. Whereas deliberate breeding, training, and specialization are common among sport hunters, hunting dogs used for subsistence hunting reportedly learn by accompanying and observing experienced dogs (Coppinger and Coppinger, 2001; Koster, 2009). In this way, tendencies for hunting particular prey species may persist, and dogs could struggle to learn to hunt rarely-encountered species, such as tapirs in Bolivia, which require cooperation among multiple dogs to bring them to bay (Noss et al., 2003). On the other hand, there are few reports of dogs in subsistence-based contexts that reliably discriminate between preferred species and non-target species before initiating a pursuit (Koster, 2009).

In addition to arboreal animals, dogs are also ineffective for hunting a variety of terrestrial prey species, including large animals that must be approached stealthily, such as elephants *Loxodonta africana* and buffalo *Syncerus caffer* in the Ituri Forest (Harako, 1981). BaAka hunters in Mossapoula, Central African Republic, rarely use dogs for communal net hunting (their principal hunting method for duikers *Cephalophus spp.* and brush-tailed porcupines *Atherurus africanus*), reportedly because the dogs flush the animals before the nets are in place. Only 3 of 212 Mossapoula residents own dogs, and these residents brought their dogs along on only 8 of 73 net hunts (Noss, 1995). Nor can dogs be used to capture brocket deer in the Bolivian Chaco because the deer do not turn to bay or take refuge as do other ungulates and armadillos in this habitat (Noss et al., 2002). Dogs in these cases are kept to hunt other game species, or for guarding houses rather than hunting.

To summarize, in many environments, dogs enhance the ability of hunters to harvest some species, although dogs are often useless and even counterproductive for other species. For subsistence hunters, the magnitude of these differences largely dictates the merits of hunting with dogs relative to other methods. From a conservationist perspective, concerns about the use of hunting dogs depend on the status of the prey species that can be hunted with dogs. In some settings, dogs are primarily useful for hunting species that can withstand relatively high off-takes, but they can also represent a clear threat to endangered populations (e.g., oribi *Ourebia ourebi* in South Africa, Grey-Ross et al., 2010).

11.3 Intraspecific biases in harvests with dogs

Among wildlife species, females determine recruitment rates and thus the maintenance or growth of populations (Caughley, 1977). The removal of males from a population may therefore have a relatively limited effect on short-term population dynamics if the remaining males are able to impregnate most of the females, as in some species that mate polygynously (Milner et al., 2007). From a management perspective, limitations on the harvests of females may therefore be implemented if the selective take of males enhances the likelihood of sustainable hunting (Ginsberg and Milner-Gulland, 1994). Alternatively, deliberate culling of females could be useful for curtailing the growth of wildlife populations.

Similarly, selective harvesting by age can have varying effects on population dynamics, reflecting individual-level variation in reproductive value (Caughley, 1977). As derived by Fisher (1930), reproductive value indicates the expected number of additional offspring for an individual of age *X*, discounted by the probability of reaching that particular age. Partly because juveniles typically suffer relatively high mortality, reproductive value for most species is moderate at birth, increases until the onset of reproduction, and subsequently declines. The culling of prime, reproductive-aged females therefore has a greater impact on short-term population growth than harvests of juveniles or senesced individuals, which have higher intrinsic risks of mortality. Although these generalizations overlook additional complexities of population dynamics (see Gaillard et al., 1998), hunting strategies that result in disproportionate harvests of juveniles and elderly individuals are more likely to be sustainable.

Because dogs descended from wolves, comparisons might reveal commonalities of predator–prey interactions. As with many predators, research on wolves indicates that they selectively target the most vulnerable members of prey populations, including calves, elderly females, and malnourished individuals (Evans et al., 2006; Sand et al., 2012; Smith et al., 2004; Wright et al., 2006). In several of these studies, the harvests by wolves differed from harvests by humans, which included relatively more prime-aged adults. Armed with projectile weapons, human hunters are typically better-equipped than cursorial predators to hunt healthy adults (Stiner, 1990). When hunting with dogs and hand technologies (e.g., machetes), perhaps the ability of hunters to harvest prime-aged adults is restricted, thus enhancing the sustainability of hunting by directing attention toward individuals with lower reproductive value.

Whereas wolves typically must overtake fleeing ungulates, however, the variable ways in which dogs can be deployed by human hunters mean that the age–sex classes of harvested prey when hunting with dogs do not necessarily match what would be expected from cursorial predators like wolves. On one hand, reports indicate that dogs can overtake and independently kill kangaroos (White, 1972), and endurance running by dogs in Botswana facilitates the hunting of gemsboks, which grow weary from the pursuit until human hunters can catch up and spear these ungulates (Ikeya, 1994). On the other hand, sometimes the primary contribution of dogs is to flush prey toward waiting hunters (e.g., Baker and Harris, 2006; Bonnot et al., 2013; Terashima, 1983), which is reportedly less selective in terms of biasing harvests toward certain age–sex classes (Martínez et al., 2005; Torres-Porras et al., 2009).

Furthermore, it is often not clear that individuals of different ages and sexes vary in their tendency to stop running and seek refuge when chased by dogs. For example, collared peccaries sometimes flee into burrows or hollow trunks when pursued by dogs (Smith, 1976), and it is plausible that juveniles or senesced individuals would be more likely to retreat to such havens if they are less able to elude the dogs by running. The empirical support for this particular hypothesis is mixed, however. Altrichter (2005) reports that in the Argentine Chaco, 70% of harvested collared peccaries were less than 2 years old and therefore more vulnerable to hunting with dogs. A similar study in the Bolivian Chaco reports that juveniles comprise only 36% of harvests by subsistence hunters using dogs (Maffei, 2003; Noss et al., 2003). Meanwhile, except in settings where dogs independently capture and dispatch prey, as in the aforementioned example of kangaroo hunting, the harvest reflects not only the animals that the dogs can corral or bring to bay, but also the willingness and ability of the hunters to pursue the prey when presented with opportunities. For instance, even if dogs disproportionately corral juveniles, the harvest will not necessarily be biased toward juveniles if hunters sometimes refrain from attacking them.

Few studies have explored the impact of hunting with dogs on the demographic composition of harvests (Table 11.3). Methodologically, multiple questions can be addressed. How do observed harvests differ from (1) the expected harvest, given the known age and sex composition of the population, (2) the expected harvest, assuming that different age or sex classes are equally represented, and (3) the composition of harvests when hunting without dogs? The first question is often unanswerable given the general lack of data on the demographic composition of many wildlife species (Bodmer and Robinson, 2005). When such data are available, however, comparisons can be revealing, as when Holmern et al. (2006) show that hunting with dogs results in male-biased and immature-biased harvests of Thomson's gazelles *Gazella thomsoni* (although not of other game species) relative to the observed sex and age ratios from a separate demographic study. In the absence of such comparative data, statistical tests may reveal deviations from equal representation of age or sex classes in the harvest. For example, the harvest of agoutis by indigenous Nicaraguan hunters with dogs includes significantly more females than males (Koster, 2008b), and a review of hunting records in the Bolivian Chaco from 1996 to 2000 also reveals higher proportions of females when hunters use dogs to harvest lowland tapirs and white-lipped peccaries, but similar sex ratios for five armadillo species and collared peccaries (A. Noss, unpublished data). Without data on the sex ratio of these prey populations, however, there is little basis to consider these results to be unexpected given that females may simply be more numerous overall. Furthermore, sex-biased harvests when hunting with dogs may stem primarily from the decision-making of hunters, not sex-related differences in the vulnerability of prey. For example, hunters with dogs reportedly harvest significantly more male bears, which apparently reflects their reluctance to pursue smaller, female bears that have been treed by dogs (Higgins Inman and Vaughan, 2002). The desire for trophy animals by sport hunters contrasts with prey choice decisions in subsistence hunting, with subsistence hunters typically focusing on the expected amount of harvestable meat rather than the age and sex of prey (Alvard, 1998).

Comparisons of the demographic composition of harvests with different methods might be useful as background information for management decisions. For example, Tryjanowski et al. (2009) show that harvests of red foxes are biased toward females and adults when hunting with dogs,

Table 11.3 Reported significant differences in the demographic composition of wildlife harvested with dogs. For sex-related biases, differences can be relative to (A) *Population*: the known demographic composition of the population, (B) *Opposite sex*: an assumed equal ratio of males and females in the population, or (C) *Other hunting*: the observed sex ratio in harvests with other hunting methods (see text for details). Similar distinctions apply to age-related biases.

	Common name	Scientific name	Biased toward	Comparison	Reference
Sex					
	Cougar[a]	*Puma concolor*	Females	Other hunting	Martorello and Beausoleil 2003
	Red fox	*Vulpes vulpes*	Females	Other hunting	Tryjanowski et al. 2009
	Black bear	*Ursus americanus*	Males	Opposite sex	Higgins Inman and Vaughan 2002
	Black bear[b]	*Ursus americanus*	Males	Other hunting	Litvaitis and Kane 1994
	Moose	*Alces alces*	Females	Other hunting	Ball et al. 1999
	Thomson's gazelle	*Gazella thomsoni*	Males	Population	Holmern et al. 2006
	White-lipped peccary	*Tayassu pecari*	Females	Other hunting	A. Noss unpublished data
	White-lipped peccary	*Tayassu pecari*	Males	Opposite sex	Altrichter 2005; Noss et al. 2003
	Collared peccary	*Pecari tajacu*	Males	Opposite sex	Altrichter 2005; Noss et al. 2003
	Agouti	*Dasyprocta punctata*	Females	Opposite sex	Koster 2008b
	Green iguana[c]	*Iguana iguana*	Females	Other hunting	Koster 2008b
	Tegu lizards	*Tupinambis spp.*	Males	Opposite sex	Fitzgerald et al. 1991
Age					
	Black bear (females only)	*Ursus americanus*	Older ages	Other hunting	Malcolm and Van Deelen 2010
	Red fox	*Vulpes vulpes*	Adults	Other hunting	Tryjanowski et al. 2009
	Red deer	*Cervus elaphus*	Juveniles	Other hunting	Martínez et al. 2005
	White-tailed deer	*Odocoileus virginianus*	Juveniles	Other hunting	Godwin et al. 2013
	Collared peccary	*Pecari tajacu*	Juveniles	Adults	Altrichter 2005
	Collared peccary	*Pecari tajacu*	Juveniles	Other hunting	A. Noss unpublished data
	White-lipped peccary	*Tayassu pecari*	Adults	Other hunting	A. Noss unpublished data
	Lowland tapir	*Tapirus terrestris*	Juveniles	Other hunting	A. Noss unpublished data
	Nine-banded armadillo	*Dasypus novemcinctus*	Juveniles	Other hunting	A. Noss unpublished data
	Six-banded armadillo	*Euphractus sexcinctus*	Juveniles	Other hunting	A. Noss unpublished data
	Large hairy armadillo	*Chaetophractus villosus*	Juveniles	Other hunting	A. Noss unpublished data
	Small hairy armadillo	*Chaetophractus vellerosus*	Juveniles	Other hunting	A. Noss unpublished data
	Three-banded armadillo	*Tolypeutes matacus*	Juveniles	Other hunting	A. Noss unpublished data

[a] Harvests in this study were not synchronous, so this difference might also reflect temporal population-level dynamics.
[b] Relative to other hunting methods, Malcolm and Van Deelen (2010) report that the use of dogs has little effect on the sex ratio of harvested bears.
[c] Harvests of iguanas in this study might also include black iguanas (*Ctenosaura similis*).

which indicates that the increased use of dogs may help to limit the population growth of this species at the study site. Seasonal restrictions on the use of dogs could also be needed to protect female iguanas, which are hunted with dogs during the egg-laying season throughout Central America (Koster, 2009).

For a variety of reasons, the use of hunting dogs may alter age and sex ratios of game populations relative to un-hunted populations or populations hunted by other means that do not involve dogs. When detected and pursued by dogs, the escape behavior of males and females may differ, contributing to differences in vulnerability (e.g., moose; Baskin et al., 2004). In social species, males may be more apt to turn and confront the dogs while the females and juveniles flee, which presumably helps to explain the greater number of males in the harvests of white-lipped and collared peccaries in the Argentine and Bolivian Chaco (Altrichter, 2005; Noss et al., 2003). Sexes may differ in the size of their home ranges or dispersal patterns, thus facilitating increased encounters with dogs. In the Chaco of Argentina and Paraguay, for example, male red tegu lizards in the breeding season range over wide areas, leave broader scent trails, and are therefore more vulnerable to hunting with dogs (Fitzgerald et al., 1991). Anecdotal evidence suggests that female brown bears with young are more vulnerable to hunters with dogs—they move more slowly through accessible terrain, leave a wider scent trail, and are more likely to turn and fight to protect cubs. However, a study comparing brown bear harvests across four methods—using dogs, baits, still hunting, stalking—revealed no biases by age or by sex according to hunting method (Bischof et al., 2008). In the case of black bears, Malcolm and Van Deelen (2010) find that hunting with dogs results in a bias toward older females, which typically avoid baits that attract younger females. Additionally, females with young may be more prone than males to climb trees when encountered by hunting dogs, making pinpointing (treeing) those females easier by hunting parties (Hodges et al., 2000).

Based on broader knowledge about the behavioral ecology of wildlife species, researchers could probably make preliminary inferences about the varying vulnerability of demographic subclasses to hunting with dogs, but in most cases there is a clear need for additional empirical research. In addition to monitoring harvests, comparative reports of captures by biologists using both dogs and other methods could reveal demographic sources of vulnerability to hunters with dogs (e.g., biological researchers capturing pacas in Colombia obtained similar proportions of juveniles when using dogs and other methods; Collett, 1981).

11.4 Spatial responses of wildlife to hunting with dogs

A limited number of studies have addressed the ways in which the presence of unrestrained hunting dogs affects the habitat use and ranging behavior of wildlife species. Some species, including non-target species, avoid areas in which the presence of dogs is apparent. In Brazil, for example, evidence from track plots reveals a negative spatial correlation between dogs and maned wolves (*Chrysocyon brachyurus*) in a small national park (Lacerda et al., 2009). Vanak and Gompper (2010) found a significant negative effect of free-ranging dogs on the spatial distribution of the Indian fox *Vulpes bengalensis*. Roe deer in France alter their behavior during the hunting season when hunters use dogs to drive them: they use more forested habitats by day, and more open habitats by night (Bonnot et al., 2013). Wild boars in Japan are more active by night during the hunting season (Ohashi et al., 2013). Upon being disturbed by dogs, species exhibit a variety of responses.[1] In addition to immediately fleeing when chased by dogs, red deer that detect the presence of hunting dogs frequently leave their home range after dark, sometimes staying away for more than a week (Sunde et al., 2009). White-tailed deer *Odocoileus virginianus* may temporally leave established home ranges when pursued by dogs, but usually return within one day (Sweeney et al., 1971). When exposed to recreational hunters with dogs, willow ptarmigan increase their use of habitats with denser vegetation, which helps them to avoid detection and bolsters their chance of escape upon being flushed (Brøseth and Pedersen, 2010; Figure 11.4).

Other species are reluctant to alter their home ranges or use of habitats when exposed to dogs. For example, red fox remained within their home ranges when pursued by dogs in Illinois, USA (Storm, 1965). Raccoons pursued by dogs in Mississippi, USA, quickly treed and did not leave their home ranges (Hodges et al., 2000). Tapir and collared peccary pursued by dogs in the Bolivian Chaco were captured when they took refuge in hollow logs or ponds within the areas that subsequent telemetry tracking confirmed to be the individuals' respective home ranges (Ayala, 2000; R. Miserendino,

[1] In addition to their effects on the movements of wildlife populations, limited research on ungulates suggests that being chased by dogs may compromise the animals' health (Bateson and Bradshaw, 1997; Sforzi and Lovari, 2000).

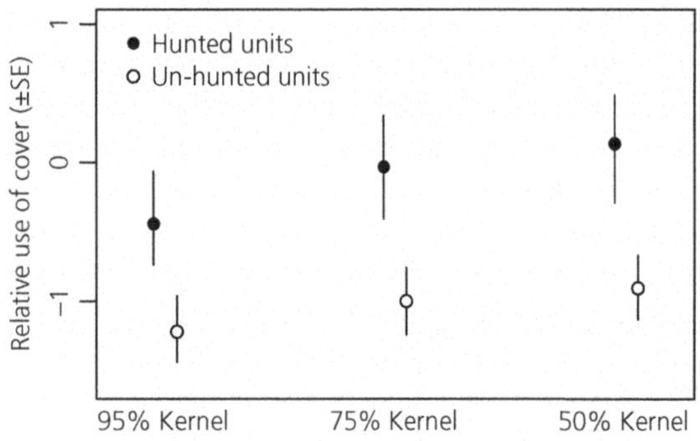

Figure 11.4 Differences in the use of cover habitat by willow ptarmigan (*Lagopus lagopus*) in hunted and un-hunted sites in central Norway. Relative use of cover is based on the difference in log-ratio between use of cover habitat (telemetry locations) and available cover within the home ranges of individual birds. The different kernel estimates (95% Kernel, 75% Kernel, and 50% Kernel) reflect different methods of estimating 'core' home ranges. Redrawn with permission from Brøseth and Pedersen (2010).

unpublished data). A recent study suggests that the use of foxhounds has minimal impact on the spatial behavior of brown hares in Italy (Zaccaroni et al., 2012). Similarly, wild boars typically alter their behavior only when directly pursued by dogs, and only repeated disturbances lead them to relocate outside of their home range (Scillitani et al., 2010; Sodeikat and Pohlmeyer, 2003; Figure 11.5).

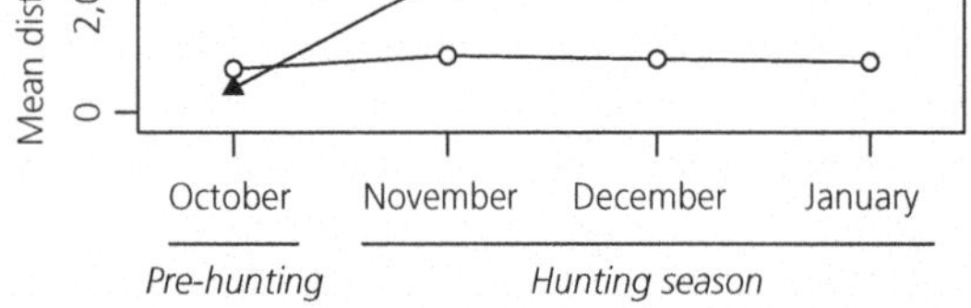

Figure 11.5 Among wild boars subjected to either heavy or light hunting pressure, the mean distances from the capture sites where the animals were first captured and tagged during the pre-hunting season for a subsequent telemetry study. Redrawn with permission from Scillitani et al. (2010).

Few studies compare the effects of hunting with and without dogs, but Grignolio et al. (2011) show that roe deer in Italy expand their home ranges and retreat to undisturbed areas only when hunting involves hounds, not when hunters use rifles. Interestingly, whereas roe deer are the target of hunters with rifles, they are a non-target species when dogs are used for hunting boars and hares, and they are therefore pursued only occasionally by inadequately-trained hounds (Grignolio et al., 2011). This suggests that the use of dogs may elicit general anti-predator behavior among wildlife species that are relatively naïve to the threats posed by unaided human hunters.

The spatial effects of hunting with dogs reflect the influence of hunters, who choose the areas and habitats in which dogs are deployed. Such decisions are frequently designed to maximize hunting success, as when Neotropical hunters deploy dogs in the gallery forest to chase animals into the river, where they can be killed (see Koster, 2009). In regulated areas, hunters with dogs might also aspire to elude detection by authorities, as in Tanzania, where hunters increased the use of dogs in forested habitats following the confiscation of firearms and the construction of a ranger station (Nielsen, 2011).

There is limited evidence from the lowland Neotropics that hunters with dogs remain closer to human settlements than hunters with only firearms (e.g., Sirén et al., 2004). This difference could reflect multiple considerations, including the hunters' reluctance to expose their dogs to being killed by jaguars (Dunn, 2004) and the relative abundance of prey species that can be hunted with dogs in the modified habitats around settlements (Naughton-Treves et al., 2003). Activity budgets of pointing dogs indicate that they spend only half of their time actively ranging in search of prey (Mecozzi and Guthery, 2008; see also Higgins Inman and Vaughan, 2002), and if untrained hounds show similar patterns of time allocation, then their circuitous routes could preclude long-distance forays into the forest. In addition, whereas pursuits with firearms may be virtually instantaneous, pursuits of animals that have been corralled in burrows often require considerable time, leaving less time for hunters and their dogs to travel long distances (Koster, 2008a).

Box 11.6 Questions for future research

How does hunting with dogs affect the composition of wildlife harvests?

This review suggests that wildlife researchers have frequently neglected the contributions of hunting dogs. Although some studies comment briefly on the number of animals that are captured and subdued by dogs (e.g., Ohl-Schacherer et al., 2007), the primary role of dogs in many settings is to detect and corral the prey. Attention to such contributions could reveal both interspecific and intraspecific biases in the composition of harvests, but researchers must simultaneously collect data on harvests and effort with other technologies and, ideally, on the density and demographic structure of the wildlife population itself.

How does variation in hunting ability among dogs affect observable hunting patterns?

There is abundant ethnographic evidence that dogs vary in their hunting ability (Koster and Tankersley, 2012), and there are settings in which virtually all of the harvests with dogs are attributable to a single, highly productive hunter (Alvard, 2000; Caro, 2008; Lee, 1979). Although differences between dogs in subsistence-based societies may be largely idiosyncratic, the specialized breeds and deliberate training that characterize sport hunting can dramatically affect dog–wildlife interactions. Future research could reveal whether regulatory limitations on specific breeds or unusually effective dogs might be a worthwhile management alternative to all-encompassing restrictions on the use of dogs.

What are the broader effects of hunting with dogs on wildlife populations?

Few studies have examined the spatial behavior or the physiological effects of animals after encounters with hunting dogs, partly because such studies typically require researchers to capture live animals, which are fitted with radio collars or comparable devices for monitoring. Considerable potential exists for extending previous research. Instead of twofold distinctions between hunting with dogs and a complete lack of interaction, for example, studies could address the ways in which responses to active hunting dogs differ from encounters with dogs in other contexts (e.g., dogs accompanying hikers or farmers). Also, does exposure to human hunters elicit different responses when they hunt both with and without dogs (e.g., Grignolio et al., 2011)?

11.5 Conclusion

As with other hunting methods, the use of hunting dogs frequently biases harvests toward a distinct suite of prey species and demographic subgroups. Although a few generalizations are currently possible, such as the relative ineffectiveness of dogs for hunting arboreal primates and game birds in the tropics, the consequences of hunting with dogs often depend on the characteristics of prey populations. Both between and within species, some individuals exhibit behavioral tendencies that render them particularly vulnerable to hunters with dogs. In some cases, the vulnerable individuals might belong to endangered species or key demographic subgroups (e.g., adult females), in which case the use of dogs could be incompatible with conservationist agendas. In other cases, the use of dogs might direct pressure away from populations that would be harvested unsustainably with other hunting methods. However, because hunting with dogs can entail other detrimental effects, such as disease spillover and spatial displacement of wildlife populations (including non-target species), considerable empirical research is required to determine whether the use of hunting dogs enhances wildlife conservation relative to other hunting methods (see Box 11.6).

Generalizations about the effects of hunting with dogs are complicated by the mediating influence of their human owners, who typically decide how and where dogs will be deployed. Harvests of wildlife when hunting with dogs frequently depend on the complementary technologies that are employed by the hunters, and the training and care that dogs receive from their owners can shape the interactions between dogs and wildlife. For many subsistence hunters, dogs are an economic investment that can provide a steady supply of bushmeat. Beyond this economic orientation, however, many hunters reportedly enjoy the companionship and the strategic challenges of hunting with their dogs, and emotional attachments to the dogs foster concerns about their welfare. Hunters with dogs may therefore be important constituents in debates about wildlife management, and management plans that accommodate their concerns could generate greater support and compliance than alternatives. Given the aforementioned variation in dog–wildlife interactions, the diversity of hunters' motivations and complementary technologies, and the heterogeneous goals of management programs, conservationist recommendations about hunting with dogs and wildlife management will typically require consideration on a case-by-case basis.

Acknowledgments

Field work in Nicaragua was funded by a Fulbright student grant, the National Science Foundation (Dissertation Improvement Award #0413037), the Hill Foundation, and a William T. Sanders Dissertation Grant. Field work in Bolivia was possible thanks to the support of the United States Agency for International Development (USAID/Bolivia Cooperative Agreement No. 511-A-00-01-00005), the Gordon and Betty Moore Foundation, and the Kaa Iya Foundation. The opinions expressed here are those of the authors and do not necessarily represent the criteria of USAID or other funding agencies. We also thank CABI and the people of the Isoso and Arang Dak for supporting our work. We thank Christine Fiorello, James Perran Ross, and Adrian Treves for suggestions.

References

Agarwala, M., Kumar, S., Treves, A., and Naughton-Treves, L. (2010). Paying for wolves in Solapur, India and Wisconsin, USA: Comparing compensation rules and practice to understand the goals and politics of wolf conservation. *Biological Conservation*, **143**, 2945–55.

Altrichter, M. (2005). Factors Underlying the Interactions Between People and Wildlife in the Argentine Chaco. PhD dissertation, University of Arizona.

Alvard, M. (1998). Indigenous hunting in the Neotropics: Conservation or optimal foraging? In T. Caro, Ed., *Behavioral Ecology and Conservation Biology*. Oxford University Press, Oxford, pp. 474–500.

Alvard, M. S. (2000). The potential for sustainable harvests by traditional Wana hunters in Morowali Nature Reserve, Central Sulawesi, Indonesia. *Human Organization*, **59**, 428–40.

Arambiza S. A. and Guerrero, J. (2000). Métodos de captura y su efectividad en mamíferos en Cerro Cortado, Izozog, Gran Chaco, Santa Cruz, Bolivia. In E. Cabrera, C. Mercolli and R. Resquin, Eds., *Manejo de Fauna Silvestre en Amazonía y Latinoamérica*. CITES Paraguay, Fundación Moises Bertoni, University of Florida, Asunción, Paraguay, pp. 107–12.

Ayala, C. (2000). *Captura de anta (Tapirus terrestris) con perros en Cerro Cortado, Izozog, Santa Cruz, Bolivia*. WCS, CABI, Santa Cruz, Bolivia.

Azevedo, F. C. C. and Murray, D. L. (2007). Spatial organization and food habits of jaguars (*Panthera onca*) in a floodplain forest. *Biological Conservation*, **137**, 391–401.

Backeryd, J. (2007). *Wolf attacks on dogs in Scandinavia 1995–2005*. Institute, Swedish University of Agricultural Sciences, Grimso.

Baker, P. J. and Harris, S. (2006). Does culling reduce fox (*Vulpes vulpes*) density in commercial forests in Wales, UK? *European Journal of Wildlife Research*, **52**, 99–108.

Ball, J. P., Ericsson, G., and Wallin, K. (1999). Climate changes, moose and their human predators. *Ecological Bulletins*, **47**, 178–87.

Baskin, L., Ball, J. P., and Danell, K. (2004). Moose escape behaviour in areas of high hunting pressure. *Alces*, **40**, 123–31.

Bateson, P. and Bradshaw, E. L. (1997). Physiological effects of hunting red deer (*Cervus elaphus*). *Proceedings of the Royal Society, Series B*, **264**, 1707–14.

Bergman, D. L., Bluett, R. D., and Tipton, A. R. (1995). An alternative method for capturing armadillos. *The Southwestern Naturalist*, **40**, 414–6.

Bischof, R., Fujita, R., Zedrosser, A., Söderberg, A., and Swenson, J. E. (2008). Hunting patterns, ban on baiting, and harvest demographics of brown bears in Sweden. *The Journal of Wildlife Management*, **72**, 79–88.

Bisi, J., Kurki, S., Svensberg, M., and Liukkonen, T. (2007). Human dimensions of wolf (*Canis lupus*) conflicts in Finland. *European Journal of Wildlife Research*, **53**, 304–14.

Bodmer, R. E., Eisenberg, J. F., and Redford, K. H. (1997). Hunting and the likelihood of extinction of Amazonian mammals. *Conservation Biology*, **11**, 460–6.

Bodmer, R. E. and Robinson, J. G. (2005). Evaluating the sustainability of hunting in the Neotropics. In K. M. Silvius, R. E. Bodmer and J. M. V. Fragoso, Eds., *People in Nature: Wildlife Conservation in South and Central America*. Columbia University Press, New York, pp. 299–323.

Boglioli, M. (2009). *A Matter of Life and Death: Hunting in Contemporary Vermont*. University of Massachusetts Press, Amherst.

Bonnot, N., Morellet, N., Verheyden, H., Cargnelutti, B., Lourtet, B., Klein, F., and Hewison, A. J. M. (2013).

Habitat use under predation risk: Hunting, roads and human dwellings influence the spatial behaviour of roe deer. *European Journal of Wildlife Research*, **59**, 185–93.

Bronson, E., Emmons, L. H., Murray, S., Dubovi, E. J., and Deem, S. L. (2008). Serosurvey of pathogens in domestic dogs on the border of Noel Kempff Mercado National Park, Bolivia. *Journal of Zoo and Wildlife Medicine*, **39**, 28–36.

Brøseth, H. and Pedersen, H. C. (2010). Disturbance effects of hunting activity in a willow ptarmigan *Lagopus lagopus* population. *Wildlife Biology*, **16**, 241–8.

Caley, P. A. and Ottley, B. (1995). The effectiveness of hunting dogs for removing feral pigs (*Sus scrofa*). *Wildlife Research*, **22**, 147–54.

Caro, T. (2008). Decline of large mammals in the Katavi-Rukwa ecosystem of western Tanzania. *African Zoology*, **43**, 99–116.

Carr, A. (1994). Houndmagic. In A. Carr, Ed., *A Naturalist in Florida: A Celebration of Eden*. Yale University Press, New Haven, pp. 139–47.

Carvalho Jr., E. A. R., and Pezzuti, J. C. B. (2010). Hunting of jaguars and pumas in the Tapajós-Arapiuns Extractive Reserve, Brazilian Amazonia. *Oryx*, **44**, 610–2.

Caughley, G. (1977). *Analysis of Vertebrate Populations*. Wiley & Son, New York.

Chitwood, M. C., Peterson, M. N., and Deperno, C. S. (2011). Assessing dog hunter identity in coastal North Carolina. *Human Dimensions of Wildlife*, **16**, 128–41.

Colell, M., Maté, C., and Fa, J. E. (1994). Hunting among Moka Bubis in Bioko: Dynamics of faunal exploitation at the village level. *Biodiversity and Conservation*, **3**, 939–50.

Collett, S. F. (1981). Population characteristics of *Agouti paca* (Rodentia) in Colombia. *Publications of the Museum, Michigan State University, Biological Series*, **5**, 489–602.

Constantino, P. de A. L., Fortini, L. B., Kaxinawa, F. R. S., Kaxinawa, A. M., Kaxinawa, E. S., Kaxinawa, A. P., Kaxinawa, L. S., Kaxinawa, J. M., and Kaxinawa, J. P. (2008). Indigenous collaborative research for wildlife management in Amazonia: The case of the Kaxinawá, Acre, Brazil. *Biological Conservation*, **141**, 2718–29.

Coppinger, R. and Coppinger, L. (2001). *Dogs: A New Understanding of Canine Origin, Behavior, and Evolution*. University of Chicago Press, Chicago.

Corlett, R. T. (2007). The impact of hunting on the mammalian fauna of tropical Asian forests. *Biotropica*, **39**, 292–303.

Crawford, R. J. M. and Robinson, G. A. (1984). History of the blue duiker *Cephalophus monticola* population in the Tsitsikamma forests, Republic of South Africa. *Koedoe*, **27**, 61–71.

Crawshaw, P. G. and Quigley, H. (1991). Jaguar spacing, activity and habitat use in a seasonally flooded environment in Brazil. *Journal of Zoology*, **223**, 357–70.

Cuéllar, E. (1999). Hunting of armadillos by Izoceño indians of the Gran Chaco, Bolivia. MS thesis, University of Kent.

Cuéllar, E. (2002). Census of the three-banded armadillo *Tolypeutes matacus* using dogs, southern Chaco, Bolivia. *Mammalia*, **66**, 448–51.

Cuéllar, E. and Noss, A. J. (2003). Mamiferos del Chaco y la Chiquitania de Santa Cruz, Bolivia. WCS, FCBS, FAN, Santa Cruz, Bolivia.

Cuéllar, R. L., Fitzgerald, L. A., and Mendoza, F. (2010). Manejo comunitario de peni (*Tupinambis rufescens*) y taitetú (*Tayassu tajacu*) en Isoso: Una iniciativa de conservación a largo plazo. In H.Gómez and A. Llobet, Eds., *Experiencias de Manejo de Fauna Silvestre en Bolivia*. Fundación Amigos de la Naturaleza, Santa Cruz, Bolivia, pp. 57–82.

Deem, S. L., Noss, A. J., Fiorello, C. V., Manharth, A. L., Robbins, R. G., and Karesh, W. B. (2009). Health assessment of free-ranging three-banded (*Tolypeutes matacus*) and nine-banded (*Dasypus novemcinctus*) armadillos in the Gran Chaco, Bolivia. *Journal of Zoo and Wildlife Medicine*, **40**, 245–56.

Dei, G. J. S. (1989). Hunting and gathering in a Ghanaian rain forest community. *Ecology of Food and Nutrition*, **22**, 225–43.

Desbiez, A. L. J., Keuroghlian, A., Piovezan, U., and Bodmer, R. E. (2011). Invasive species and bushmeat hunting contributing to wildlife conservation: The case of feral pigs in a Neotropical wetland. *Oryx*, **45**, 78–83.

Dunn, M. (2004). Re-interpreting the impacts of indigenous hunting: A participatory geographic analysis of Miskito wildlife use in eastern Honduras. MS thesis, Carleton University.

Dwyer, P. D. (1983). Etolo hunting performance and energetics. *Human Ecology*, **11**, 145–74.

Estioko-Griffin, A. and Griffin, P. B. (1981). Woman the hunter: The Agta. In F. Dahlberg, Ed., *Woman the Gatherer*. Yale University Press, New Haven, pp. 121–52.

Evans, S. B., Mech, L. D., White, P. J., and Sargeant, G. A. (2006). Survival of adult female elk in Yellowstone following wolf restoration. *The Journal of Wildlife Management*, **70**, 1372–8.

Fiorello, C. V., Deem, S. L., Gompper, M. E., and Dubovi, E. J. (2004). Seroprevalence of pathogens in domestic carnivores on the border of Madidi National Park, Bolivia. *Animal Conservation*, **7**, 45–54.

Fiorello, C. V., Noss, A. J., and Deem, S. L. (2006). Demography, hunting ecology, and pathogen exposure of domestic dogs in the Isoso of Bolivia. *Conservation Biology*, **20**, 762–71.

Fisher, R. A. (1930). *The Genetical Theorem of Natural Selection*. Clarendon Press, Oxford.

Fitzgerald, L. A., Chani, J. M., and Donadío, O. E. (1991). Tupinambis lizards in Argentina: Implementing management of a traditionally exploited resource. In J. G. Robinson and K. H. Redford, Eds., *Neotropical Wildlife Use and Conservation*. University of Chicago Press, Chicago, pp. 303–16.

Fitzgibbon, C. D., Mogaka, H., and Fanshawe, J. H. (1995). Subsistence hunting in Arabuko-Sokoke, Kenya, and its effects on mammal populations. *Conservation Biology*, **9**, 1116–26.

Fonseca, C., Alves da Silva, A., Alves, J., Vingada, J., and Soares, A. M. V. M. (2011). Reproductive performance of wild boar females in Portugal. *European Journal of Wildlife Research*, **57**, 363–71.

Furtado, M. M., Carrillo-Percastegui, S. E., Jácomo, A. T. A., Powell, G., Silveira, L., Vynne, C., and Sollmann, R. (2008). Studying jaguars in the wild: Past experiences and future perspectives. *Cat News*, **4**, 41–7.

Gaillard, J.-M., Festa-Bianchet, M., and Yoccoz, N. G. (1998). Population dynamics of large herbivores: variable recruitment with constant adult survival. *Trends in Ecology & Evolution*, **13**, 58–63.

Ginsberg, J. R. and Milner-Gulland, E. J. (1994). Sex-biased harvesting and population dynamics in ungulates: Implications for conservation and sustainable use. *Conservation Biology*, **8**, 157–66.

Godwin, C., Schaefer, J. A., Patterson, B. R., and Pond, B. A. (2013). Contribution of dogs to white-tailed deer hunting success. *The Journal of Wildlife Management*, **77**, 290–6.

Golden, C. D. (2009). Bushmeat hunting and use in the Makira Forest, north-eastern Madagascar: A conservation and livelihoods issue. *Oryx*, **43**, 386–92.

Gosselin, G. (1972). *Travail et Changement Social en Pays Gbeya (R.C.A.)*. Labethno, Paris.

Grey-Ross, R., Downs, C. T., and Kirkman, K. (2010). An assessment of illegal hunting on farmland in KwaZulu-Natal, South Africa: Implications for oribi (*Ourebia ourebi*) conservation. *South African Journal of Wildlife Research*, **40**, 43–52.

Grignolio, S., Merli, E., Bongi, P., Ciuti, S., and Apollonio, M. (2011). Effects of hunting with hounds on a non-target species living on the edge of a protected area. *Biological Conservation*, **144**, 641–9.

Guthery, F. S. and Mecozzi, G. E. (2008). Developing the concept of estimating bobwhite density with pointing dogs. *Journal of Wildlife Management*, **72**, 1175–80.

Harako, R. (1981). The cultural ecology of hunting behavior among the Mbuti Pygmies in the Ituri forest, Zaïre. In R. S. O. Harding and G. Teleki, Eds., *Omnivorous Primates: Gathering and Hunting in Human Evolution*. Columbia University, New York, pp. 499–555.

Higgins Inman, K. and Vaughan, M. R. (2002). Hunter effort and success rates of hunting bears with hounds in Virginia. *Ursus*, **13**, 223–30.

Hodges, K. M., Chamberlain, M. J., and Leopold, B. D. (2000). Effects of summer hunting on ranging behavior of adult raccoons in central Mississippi. *Journal of Wildlife Management*, **64**, 194–8.

Hoffman, C. (1986). *The Punan: Hunters and Gatherers of Borneo*. UMI Research, Ann Arbor.

Holmern, T., Mkama, S., Muya, J., and Røskaft, E. (2006). Intraspecific prey choice of bushmeat hunters outside the Serengeti National Park, Tanzania: A preliminary analysis. *African Zoology*, **41**, 81–7.

Howe, J. (1981). Fox hunting as ritual. *American Ethnologist*, **8**, 278–300.

Ikeya, K. (1994). Hunting with dogs among the San in the central Kalahari. *African Study Monographs*, **15**, 119–34.

Kaplan, H. and Kopischke, K. (1992). Resource use, traditional technology, and change among native peoples of Lowland South America. In K. H. Redford and C. Padoch, Eds., *Conservation of Neotropical Forests: Working from Traditional Resource Use*. Columbia University Press, New York, pp. 83–107.

Katahira, L. K., Finnegan, P., and C. P. Stone. (1993). Eradicating feral pigs in montane mesic habitat at Hawaii Volcanoes National Park. *Wildlife Society Bulletin*, **21**, 269–74.

Kojola, I., Ronkainen, S., Hakala, A., Heikkinen, S., and Kokko, S. (2004). Interactions between wolves *Canis lupus* and dogs *C. familiaris* in Finland. *Wildlife Biology*, **10**, 101–5.

Koster, J. M. (2006). Assessing the sustainability of Baird's tapir hunting in the Bosawas Reserve, Nicaragua. *Tapir Conservation*, **15**, 23–28.

Koster, J. M. (2008a). Hunting with dogs in Nicaragua: An optimal foraging approach. *Current Anthropology*, **49**, 935–44.

Koster, J. M. (2008b). The impact of hunting with dogs on wildlife harvests in the Bosawas Reserve, Nicaragua. *Environmental Conservation*, **35**, 211–20.

Koster, J. M. (2008c). Kills of giant anteaters (*Myrmecophaga tridactyla*) by hunters with dogs in the Bosawas Biosphere Reserve, Nicaragua. *The Southwestern Naturalist*, **53**, 414–6.

Koster, J. M. (2009). Hunting dogs in the lowland Neotropics. *Journal of Anthropological Research*, **65**, 575–610.

Koster, J. M. and Tankersley, K. B. (2012). Heterogeneity of hunting ability and nutritional status among domestic dogs in lowland Nicaragua. *Proceedings of the National Academy of Sciences, USA*, **109**, E463–470.

Lacerda, A. C. R., Tomas, W. M., and Marinho-Filho, J. (2009). Domestic dogs as an edge effect in the Brasília

National Park, Brazil: Interactions with native mammals. *Animal Conservation*, **12**, 477–87.

Lande, U. S., Herfindal, I., Finne, M. H., and Kastdalen, L. (2010). Use of hunters in wildlife surveys: Does hunter and forest grouse habitat selection coincide? *European Journal of Wildlife Research*, **56**, 107–15.

Laundré, J. W., Hernández, L., Streubel, D., Altendorf, K., and González, C. L. (2000). Aging mountain lions using gum-line recession. *Wildlife Society Bulletin*, **28**, 963–66.

Lee, R. B. (1979). *The !Kung San: Men, Women, and Work in a Foraging Society*. Cambridge University Press, Cambridge.

Lescureux, N. and Linnell, J. D. C. (2010). Knowledge and perceptions of Macedonian hunters and herders: The influence of species specific ecology of bears, wolves, and lynx. *Human Ecology*, **38**, 389–99.

Litvaitis, J. A. and Kane, D. M. (1994). Relationship of hunting technique and hunter selectivity to composition of black bear harvest. *Wildlife Society Bulletin*, **22**, 604–6.

Liukkonen, T., Mykrä, S., Bisi, J., and Kurki, S. (2009). Conflicts and compromises in lynx *Lynx lynx* conservation and management in Finland. *Wildlife Biology*, **15**, 165–74.

Lupo, K. D. (2012). A dog is for hunting. In U. Albarella, Ed., *Ethnozooarchaeology*. Oxbow Press, Oxford, pp. 4–12.

Ly, T., Hoang, H. D., and Stuart, B. L. (2011). Market turtle mystery solved in Vietnam. *Biological Conservation*, **144**, 1767–71.

Maffei, F. L. (2003). Age structure of two hunted peccary species in the Bolivian Chaco (*Tayassu tajacu* and *T. pecari*). *Mammalia*, **67**, 575–8.

Maillard, D. and Fournier, P. (1995). Effects of shooting with hounds on size of resting range of wild boar (*Sus scrofa* L.) groups in Mediterranean habitat. *Ibex Journal of Mountain Ecology*, **3**, 102–7.

Malcolm, K. D., and Van Deelen, T. R. (2010). Effects of habitat and hunting framework on American black bear harvest structure in Wisconsin. *Ursus*, **21**, 14–22.

Martínez, M., Rodríguez-Vigal, C., Jones, O. R., Coulson, T., and San Miguel, A. (2005). Different hunting strategies select for different weights in red deer. *Biology Letters*, **1**, 353–6.

Martorello, D. A. and Beausoleil, R. A. (2003). Characteristics of cougar harvest with and without dogs. In S.A. Becker, D.D. Bjornlie, F.G. Lindzey and D.S. Moody, Eds., *Proceedings of the Seventh Mountain Lion Workshop*, Lander, Wyoming, 129–35.

Martin, A., Caro, T., and Kiffner, C. (2013). Prey preferences of bushmeat hunters in an East African savannah ecosystem. *European Journal of Wildlife Research*, **59**, 137–45.

Marvin, G. (2001). Cultured killers: Creating and representing foxhounds. *Society & Animals*, **9**, 273–92.

McBride, R. T., Jr. and McBride, R. T. (2007). Safe and selective technique for jaguars in the Paraguayan Chaco. *The Southwestern Naturalist*, **52**, 570–7.

McCann, B. E. and Garcelon, D. K. (2008). The eradication of feral pigs from Pinnacles National Monument. *Journal of Wildlife Management*, **72**, 1287–95.

Mecozzi, G. E. and Guthery, F. S. (2008). Behavior of walk-hunters and pointing dogs during northern bobwhite hunts. *The Journal of Wildlife Management*, **72**, 1399–1404.

Miller, F. W. (1930). Notes on some mammals of southern Matto Grosso, Brazil. *Journal of Mammalogy*, **11**, 10–22.

Milner, J. (1994). Relationships between the forest dwelling people of south-west Mau and tree hyrax, *Dendrohyrax arboreus*. *Journal of East African Natural History*, **83**, 17–29.

Milner, J. M., Nilsen, E. B., and Andreassen, H. P. (2007). Demographic side effects of selective hunting ungulates and carnivores. *Conservation Biology*, **21**, 36–47.

Morren, G. E. B., Jr. (1986). *The Miyanmin: Human Ecology of a Papua New Guinea Society*. UMI Research Press, Ann Arbor.

Naughton-Treves, L., Mena, J. L., Treves, A., Alvarez, N., and Radeloff, V. C. (2003). Wildlife survival beyond park boundaries: The impact of slash-and-burn agriculture and hunting of mammals in Tambopata, Peru. *Conservation Biology*, **17**, 1106–17.

Newton, P., Van Thai, N., Roberton, S., and Bell, D. (2008). Pangolins in peril: Using local hunters' knowledge to conserve elusive species in Vietnam. *Endangered Species Research*, **6**, 41–53.

Nielsen, M. R. (2011). Improving the conservation status of the Udzungwa Mountains, Tanzania? The effect of joint forest management on bushmeat hunting in the Kilombero Nature Reserve. *Conservation and Society*, **9**, 106–18.

Nobayashi, A. (2006). An ethnoarchaeological study of chase hunting with gundogs by the aboriginal peoples of Taiwan. In L. M. Snyder and E. A. Moore, Eds., *Dogs and People in Social, Working, Economic or Symbolic Interaction*. Oxbow Books, Oxford, pp. 77–84.

Noss, A. J. (1995). *Duikers, cables, and nets: The cultural ecology of hunting in a Central African Forest*. PhD dissertation, University of Florida.

Noss, A. J., Cuéllar, E., and Cuéllar, R. L. (2003). Hunter self-monitoring as a basis for biological research: Data from the Bolivian Chaco. *Mastozoología Neotropical*, **10**, 49–67.

Noss, A. J., Cuéllar, E., and Cuéllar, R. L. (2004). An evaluation of hunter self-monitoring in the Bolivian Chaco. *Human Ecology*, **32**, 685–702.

Noss, A. J., Deem, S. L., and Ayala, J. (2002). Use of trained dogs to capture ungulates in the Bolivian Chaco. *Mammalia*, **66**, 297–300.

Ohashi, H., Saito, M., Horie, R., Tsunoda, H., Noba, H., Ishii, H., Kuwabara, T., Hiroshige, Y., Koike, S., Hoshino, Y., Toda, H., and Kaji, K. (2013) Differences in the activity pattern of the wild boar *Sus scrofa* related to human disturbance. *European Journal of Wildlife Research*, **59**, 167–77.

Ohl-Schacherer, J., Shepard, G. H., Jr., Kaplan, H., Peres, C. A., Levi, T., and Yu, D. W. (2007). The sustainability of subsistence hunting by Matsigenka native communities in Manu National Park, Peru. *Conservation Biology*, **21**, 1174–85.

Olsen, S. J. (1985). *Origins of the Domestic Dog: The Fossil Record*. The University of Arizona Press, Tucson.

Ortega-Pacheco, A., Carlos, R.-B. J., Boilo-Gonzalez, M. E., Sauri-Arceo, C. H., Jiménex-Coello, M., and Forsberg, C. L. (2007). A survey of dog populations in urban and rural areas of Yucatan, Mexico. *Anthrozoos*, **20**, 261–74.

Rabinowitz, A. R. (1986). Jaguar predation on domestic livestock in Belize. *Wildlife Society Bulletin*, **14**, 170–4.

Redford, K. H. and Robinson, J. G. (1987). The game of choice: Patterns of Indian and colonist hunting in the Neotropics. *American Anthropologist*, **89**, 650–67.

Robinson, J. G. and Redford, K. H. (1991). Sustainable harvest of Neotropical forest mammals. In J. G. Robinson and K. H. Redford, Eds., *Neotropical Wildlife Use and Conservation*. University of Chicago Press, Chicago, pp. 415–29.

Ross, P. I. and Jalkotzy, M. G. (1992). Characteristics of a hunted population of cougars in southwestern Alberta. *Journal of Wildlife Management*, **56**, 417–26.

Ruusila, V. and Pesonen. (2004). Interspecific cooperation in human (*Homo sapiens*) hunting: The benefits of a barking dog (*Canis familiaris*). *Annales Zoologici Fennici*, **41**, 545–9.

Ryan, C. W., Edwards, J. W., and Duda, M. D. (2009). West Virginia residents' attitudes and opinions toward American black bear hunting. *Ursus*, **20**, 131–42.

Sand, H., Wilkenros, C., Ahlqvist, P., Stromseth, T. H., and Wabakken, P. (2012). Comparing body condition of moose (*Alces alces*) selected by wolves (*Canis lupus*) and human hunters: Consequences for the extent of compensatory mortality. *Canadian Journal of Zoology*, **90**, 403–12.

Schaller, G. B. and Crawshaw, P. G. (1980). Movement patterns of jaguar. *Biotropica*, **12**, 161–8.

Scillitani, L., Monaco, A., and Toso, S. (2010). Do intensive drive hunts affect wild boar (*Sus scrofa*) spatial behaviour in Italy? Some evidences and management implications. *European Journal of Wildlife Research*, **56**, 307–18.

Sforzi, A. and Lovari, S. (2000). Some effects of hunting on wild mammalian populations. *Ibex Journal of Mountain Ecology*, **5**, 165–71.

Shattuck, M. R. and Williams, S. A. (2010). Arboreality has allowed for the evolution of increased longevity in mammals. *Proceedings of the National Academy of Sciences, USA*, **107**, 4635–9.

Silveira, L. (2004). Ecologia comparada e conservação da onça-pintada (*Panthera onca*) e onça-parda (*Puma concolor*) no Cerrado e Pantanal. PhD dissertation, University of Brasilia.

Sirén, A., Hamback, P., and Machoa, J. (2004). Including spatial heterogeneity and animal dispersal when evaluating hunting: A model analysis and an empirical assessment in an Amazonian community. *Conservation Biology*, **18**, 1315–29.

Skinner, J. E., Snow, D. P., and Payne, N. F. (1998). A capture technique for juvenile willow ptarmigan. *Wildlife Society Bulletin*, **26**, 111–2.

Smith, D. W., Drummer, T. D., Murphy, K. M., Guernsey, D. S., and Evans, S. B. (2004). Winter prey selection and estimation of wolf kill rates in Yellowstone National Park, 1995–2000. *The Journal of Wildlife Management*, **68**, 153–66.

Smith, N. J. H. (1976). Utilization of game along Brazil's transamazon highway. *Acta Amazonia*, **6**, 455–66.

Sodeikat, G. and Pohlmeyer, K. (2003). Escape movements of family groups of wild boar *Sus scrofa* influenced by drive hunts in Lower Saxony, Germany. *Wildlife Biology*, **9 (Suppl. 1)**, 43–9.

Soisalo, M. K. and Cavalcanti, S. M. C. (2006). Estimating the density of a jaguar population in the Brazilian Pantanal using camera-traps and capture-recapture sampling in combination with GPS radio-telemetry. *Biological Conservation*, **129**, 487–96.

Stiner, M. C. (1990). The use of mortality patterns in archaeological studies of hominid predatory adaptations. *Journal of Anthropological Archaeology*, **9**, 305–51.

Stocks, A. (1996). The BOSAWAS Natural Reserve and the Mayangna of Nicaragua. In K. H. Redford and J. A. Mansour, Eds., *Traditional Peoples and Biodiversity Conservation in Large Tropical Landscapes*. America Verde Series of The Nature Conservancy, Arlington, Virginia, pp. 1–32.

Storm, G. L. (1965). Movement and activities of foxes as determined by radio-tracking. *Journal of Wildlife Management*, **29**, 1–13.

Sunde, P., Olesen, C. R., Madsen, T. L., and Haugaard, L. (2009). Behavioural responses of GPS-collared female red deer *Cervus elaphus* to driven hunts. *Wildlife Biology*, **15**, 454–60.

Sweeney, J. R., Marchinton, R. L., and Sweeney, J. M. (1971). Responses of radio-monitored white-tailed deer chased by hunting dogs. *Journal of Wildlife Management*, **35**, 707–16.

Taber, A. B., Doncaster, C. P., Neris, N. N., and Colman, F. H. (1993). Ranging behavior and population dynamics of the Chacoan peccary, *Catagonus wagneri*. *Journal of Mammalogy*, **74**, 443–54.

Terashima, H. (1983). Mota and other hunting activities of the Mbuti archers: A socio-ecological study of subsistence technology. *African Study Monographs*, **3**, 71–85.

Tobler, M. W., Naranjo, E. J., and Lira-Torres, I. (2006). Habitat preference, feeding habits and conservation of Baird's tapir in Neotropical montane oak forests. In M. Kappelle, Ed., *Ecology and Conservation of Neotropical Montane Oak Forests*. Springer-Verlag, Berlin, pp. 347–59.

Torres-Porras, J., Carranza, J., and Pérez-González, J. (2009). Selective culling of Iberian red deer stags (*Cervus elaphus hispanicus*) by selective montería in Spain. *European Journal of Wildlife Research*, **55**, 117–23.

Tryjanowski, P., Sparks, T. H., Kamieniarz, R., and Panek, M. (2009). The relationship between hunting methods and sex, age and body weight in a non-trophy animal, the red fox. *Wildlife Research*, **36**, 106–9.

Vanak, A. T. and Gompper, M. E. (2010). Interference competition at the landscape level: The effect of free-ranging dogs on a native mesocarnivore. *Journal of Applied Ecology*, **47**, 1225–32.

White, I. M. (1972). Hunting dogs at Yalata. *Mankind*, **8**, 201–5.

White, P. C. L., Newton-Cross, G. A., Moberly, R. L., Smart, J. C. R., Baker, P. J., and Harris, S. (2003). The current and future management of wild mammals hunted with dogs in England and Wales. *Journal of Environmental Management*, **67**, 187–97.

Williams, C. K., Lutz, R. S., and Applegate, R. D. (2004). Winter survival and additive harvest in northern bobwhite coveys in Kansas. *Journal of Wildlife Management*, **68**, 94–100.

Williams-Guillén, K., Griffith, D., Polisar, J., Camilo, G., and Bauman, K. (2006). Abundancia de animales de caza y características de cacería en el territorio indígena de Kipla Sait Tasbaika, reserva de biósfera BOSAWAS. *Wani*, **23**, 37–61.

Wright, G. J., Peterson, R. O., Smith, D. W., and Lemke, T. O. (2006). Selection of northern Yellowstone elk by gray wolves and hunters. *The Journal of Wildlife Management*, **70**, 1070–78.

Wright, J. H. and Priston, N. E. C. (2010). Hunting and trapping in Lebialem Division, Cameroon: Bushmeat harvesting practices and human reliance. *Endangered Species Research*, **11**, 1–12.

Zaccaroni, M., Biliotti, N., Caroli, M., Sereni, E., and Dessì-Fulgheri, F. (2012). Effect of fox hunting with small hound packs on spatial behaviour of brown hares. *Acta Theriologica*, **57**, 257–60.

CHAPTER 12

The human dimensions of dog–wildlife interactions

Kelly K. Miller, Euan G. Ritchie, and Michael A. Weston

12.1 Introduction

"Science is, however, often one of the first casualties when wildlife management enters the public and political arena" (Thompson et al., 2003).

Throughout history, dogs (*Canis familiaris*) have had a complex and fascinating relationship with humans, with the best-documented domestication records dating back to around 14,000–9,000 years ago, but perhaps occurring as early as ~30,000 years ago (Clutton-Brock, 1995; Napierala and Uerpmann, 2012; Ovodov et al., 2011). The connection between people and dogs is profound, rivaling that of any domesticated animal. There is contention and great uncertainty about how many dogs exist worldwide, but they likely number in the hundreds of millions (Gompper, Chapter 1). Dogs have been, and in many places still are, used for food, companionship, tourism, health aids (e.g., guide dogs), and wildlife management or extraction (e.g., livestock and wildlife guardian dogs, hunting dogs).

The focus of this chapter is on the human dimensions of dog–wildlife interactions, which are far less-studied than the complex and diverse relationships between humans and dogs. The human dimensions of dog–wildlife interactions often constitute what are commonly referred to as human–wildlife conflicts (HWCs; Manfredo and Dayer, 2004). Some well-known examples of HWCs include human conflicts with white-tailed deer (*Odocoileus virginianus*) in North America (Chase et al., 1999), common brushtail possums (*Trichosurus vulpecula*) in Australia (Whiting and Miller, 2008), and elephants (*Loxodonta africana*) in Africa (O'Connell-Rodwell et al., 2000). Manfredo and Dayer (2004) suggest that "despite the diversity of situations and species that spawn HWC, there is one common thread: the thoughts and actions of humans ultimately determine the course and resolution of the conflict." Thus, from a wildlife conservation perspective, understanding the human dimensions of HWC is critical.

The first step in understanding the human dimensions of dog–wildlife interactions is to clearly define the nature of the interaction. In this case, the interactions can be positive or negative depending on the subject (human, dog, or wild animal), and are varied and complex (Table 12.1). The focus of this chapter is on the intersection between humans, dogs, and wildlife and their interactions, and includes their sometimes competing interests.

Since the 1960s, it has been widely acknowledged that it is necessary to investigate the human dimensions of wildlife management issues and problems. Environmental and wildlife management issues almost always arise because of competing human interests and values. If these are not well understood, even the best ecological and biological science may not provide adequate solutions. The field of human dimensions has been defined as "how people value wildlife, how they want wildlife to be managed, and how they affect or are affected by wildlife management decisions" (Decker et al., 2001, p. 3), and is concerned with improving representation in decision-making and with influencing policy and management outcomes (Loker et al., 1998). Human dimensions research is important because, too often, wildlife management decisions

Free-Ranging Dogs and Wildlife Conservation. Edited by Matthew E. Gompper
© Oxford University Press 2014. Published 2014 by Oxford University Press.

Table 12.1 Summary of positive and negative interactions between humans, dogs, and wildlife. The interactions listed in this table are not intended to be exhaustive. These interactions can be positive (i.e., benefit) or negative (i.e., cost) to biodiversity, such as wildlife and/or humans, and some are less evident than others. Perceptions of what constitutes a positive interaction versus a negative interaction can vary.

Human–dog interactions	Human–dog–wildlife interactions	Dog–wildlife interactions
Positive interactions: human–dog bond; companion animals; working dogs (guard dogs, guide dogs, herding dogs, hunting dogs, etc.)	**Positive interactions**: dog walking facilitates human connections with nature	**Positive interactions:** dogs can work as guardians for vulnerable wildlife populations (e.g., Maremma dogs protecting penguins against fox predation)
Negative interactions: threats (perceived and actual) posed by feral dogs (e.g., to humans, livestock, pets); dogs acting as vectors for disease	**Negative interactions:** large carnivores can prey on dogs and this can lead to various human–wildlife conflicts (for example, threats to human safety)	**Negative interactions:** hybridization between dogs and wild canids; dogs directly or indirectly impact wildlife

are based on untested assumptions about people's views and their responses to management programs (Enck and Decker, 1997). This can potentially result in ineffective management and ongoing conflicts (Miller, 2009).

Although there is a significant body of research, particularly in developed countries, on the human dimensions of some large carnivores (involving, for example, wolves, *C. lupus*, and coyotes, *C. latrans*; Lukasik and Alexander, 2011; Meadow et al., 2005; see Boxes 12.1 and 12.2), there has been much less attention paid to the human dimensions of dog–wildlife interactions. This chapter presents an overview of research in the field of the human

Box 12.1 Human dimensions and large carnivores

A significant body of research has been completed on the conflicts that can occur between humans and large carnivores. Studies have focused on attitudes toward large carnivores (for example, gray wolves, bears (*Ursus* spp.), lions (*Panthera leo*)) and views about their conservation and management. This research has revealed several key themes that may inform human dimensions research for dog–wildlife interactions:

- An absence of human dimensions research can contribute to poor management outcomes (Decker et al., 2012; Kellert, 1991). For example, a wolf restoration program in Michigan, USA, in the 1970s failed, in part, because of a lack of information about the social feasibility of the program (see Decker et al., 2001).
- When dealing with human–wildlife conflicts, there will always be a wide range of stakeholder groups and attitudes about the most appropriate management and/or conservation approaches. Given the links between attitudes and behaviors (see Figure 12.1) and the importance of human behaviors in conservation programs (Schultz, 2011), a clear understanding of the full range of attitudes and opinions is essential.
- As large carnivore populations increase and/or human populations increase, human–wildlife conflicts are likely to increase (Bruskotter and Shelby, 2010).
- Level of damage or threat posed by a species will influence attitudes toward that species and management options. Generally speaking, the greater the conflict, the more likely it will be that people accept lethal management techniques (Wittmann et al., 1998). However, this is not a hard and fast rule and the situational context and deep-rooted social identities can also influence judgments about management acceptability (Naughton-Treves et al., 2003).
- Opportunities for public participation in decision-making are essential (Meadow et al., 2005).

Although human dimensions research is now recognized as a critical component of the decision-making process, the "traditional view of wildlife management as limited to biology and ecology still dominates" (Bruskotter and Shelby, 2010).

Box 12.2 A cultural caveat

The role of culture and individual experience of researchers or authors is recognized as a significant influence on the study of people and their attitudes, so much so that authors in the social sciences often declare their backgrounds to enable readers to interpret qualitative human research. The authors of this chapter are not immune to these unintended biases, and acknowledge that they bring a developed world viewpoint to the human dimensions of dog–wildlife interactions. Indeed, there is a clear bias in the available literature towards dogs in the developed world. Such is the diversity of human cultures of which dogs are a part, that all viewpoints and philosophies cannot be detailed in the available space.

However, acknowledging cultural relativities is important (see Ortolani et al., 2009). For example, terms such as 'dog walking' may adequately describe the intentional exercising of a pet, but poorly describe the circumstance whereby a dog accompanies a person going about their daily routine or travels. 'Ownership' of animals such as dogs may be a workable concept in some cultures but not others. The concept of a pet (a companion animal) is more evident in some cultures than others. Managing dogs or any damage they cause is, similarly, not a universal concept and must take the cultural context into account.

dimensions of dog–wildlife interactions and proposes potential solutions for the problems and issues raised.

12.2 Conceptual basis

Human dimensions research focuses largely on human values of wildlife, attitudes toward wildlife and wildlife management issues and approaches, and human behaviors. The field draws from a range of social science disciplines including social psychology, sociology, and economics (Decker et al., 2012). Theories in social psychology often provide the basis for examining the human dimensions of wildlife management and help explain why people hold the values and attitudes that they do and act in particular ways. Social psychological theories can be used to help explain human values, attitudes, behaviors, and knowledge across a range of wildlife management issues and scenarios and can therefore provide a useful framework for understanding the human dimensions of dog–wildlife interactions (Figure 12.1).

The framework presented in Figure 12.1 suggests that values and attitudes are important in predicting subsequent behaviors, but the link is complex and difficult to measure (Fulton et al., 1996). Values can influence attitudes, and attitudes can influence behaviors, but so too can other factors such as availability of time and money, physical barriers (e.g., access to a place), management regulations, cultural context, past experiences, views of others, and so on.

The theories summarized in Figure 12.1 have been further developed and applied within the field of wildlife management. For example, Decker and Purdy (1988) proposed the Wildlife Acceptance Capacity model where there is a "maximum wildlife population level in an area that is acceptable to people." Social factors (what cultural groups and stakeholder groups does a person identify with, and what are their views?), cognitive factors (what beliefs and attitudes does a person hold?), and contextual factors (e.g., the species being controlled; the location of the conflict or issue) all contribute to acceptability judgments (Bruskotter et al., 2009).

Similarly, Wittmann et al. (1998) investigated stakeholder positions on HWCs and wildlife management decisions for specific contexts and animal species. They found that the acceptability of destroying an animal increased as the severity of impact of the human–wildlife interaction on humans increased. Human dimensions studies also report that people view charismatic fauna differently to less charismatic fauna (Kellert, 1996) and that divergent and often polarized views can affect management programs (for example, lethal wolf control in Utah, USA (Bruskotter et al., 2009); wolf restoration in the Rocky Mountains, USA (Meadow et al., 2005)). An understanding of human dimensions aided by such theoretical frameworks can assist managers in developing more socially sustainable and successful long-term management programs.

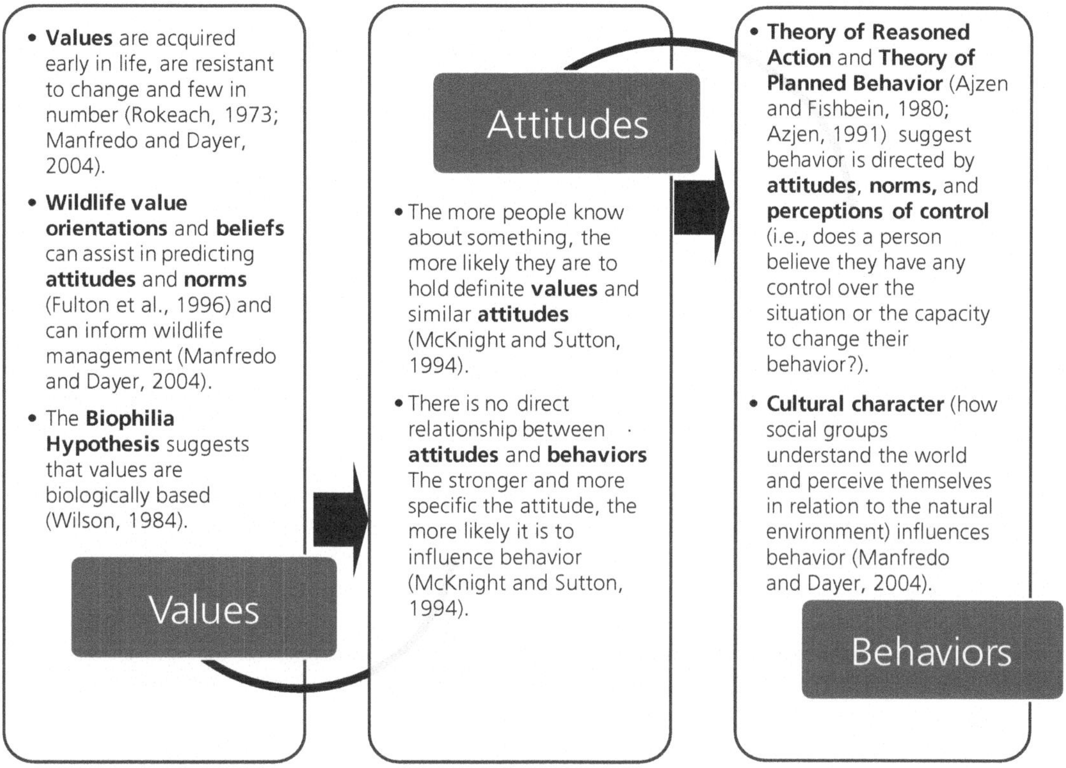

Figure 12.1 Summary of the value–attitude–behavior hierarchy (Fulton et al., 1996) and associated major theories. The list of theories presented here is not exhaustive.

12.3 Human dimensions research on dog–wildlife interactions

It is well established that humans have a longstanding and strong association with dogs (Sillero-Zubiri, 2009) which offers many benefits for both humans (e.g., companionship, mental health; Miller and Howell, 2008) and dogs (e.g., access to food, safety) (see Table 12.1). However, we do not yet fully understand the human dimensions of either the interactions or the conflicts occurring between humans, dogs, and wildlife. This is where we need to draw from human dimensions research as well as ecological and biological science. Given the global and extensive nature of human–dog–wildlife interactions, it is surprising that there has been so little human dimensions research in this field. The human dimensions research completed to date on free-ranging dogs and their interactions with wildlife has focused on pet dogs, working dogs, and feral dogs; and this research is reviewed below.

12.3.1 Pet dogs

It has been estimated that 20–30% of households, globally, have a dog as a companion animal (Ioja et al., 2011), with dog walking being one of the world's most popular recreational activities (Banks and Bryant, 2007). In many areas, dogs are owned and regularly fed but allowed to roam free for much of the time (e.g., Brazil; Torres and Prado, 2010; also see Gompper, Chapter 1); and in many countries dogs are permitted in protected areas with their owners (for example, dogs can accompany human visitors to >96% of protected lands in California, USA, although in most national parks in the USA dogs are only allowed on leashes, near residences and visitor centers, or in campgrounds; Reed and

Merenlender, 2011). In many developed countries, the occurrence of dogs in wildlife habitat mirrors the partitioning of recreational time within societies; peak dog numbers occur during weekends, holidays and outside working hours (Sastre et al., 2009).

To date, studies have focused on behaviors and attitudes relating to 'dog walking' (the act of walking in the presence of a dog) in areas where wildlife occurs. These studies often highlight the spectrum of views on dogs and dog access to natural areas (for example, those who desire unconditional access for dogs versus those who want access limited or controlled; Ioja et al., 2011). There has also been some research documenting threats posed to pet dogs by wildlife and vice versa (e.g., disease, attack) and associated human behaviors (e.g., owner management of dogs) (e.g., Aguirre et al., 1995; Paquet-Durand et al., 2007).

Dog owners place a very high value on being able to walk their dog off-leash (Jenkinson et al., 2009; Maguire et al., 2011; Underhill-Day and Liley, 2007). Therefore, it is not surprising that dog walking studies have reported high rates of non-compliance with dog-leashing laws and other regulations (see Weston and Stankowich, Chapter 4). For example, in a study of wetland buffers in Victoria, Australia, Weston et al. (2009) found that, despite the leashing laws and interpretive and regulatory signs, 68.3% of dogs (n = 104) were unleashed. This rate of non-compliance was similar to that found by Arnberger and Hinterberger (2003) where only one third of visitors to an Austrian national park walked their dog on-leash. Weston et al. (2009) also found walkers and dog walkers were more likely to access a wetland, despite it being closed to the public, than other recreational users (see also, Antos et al., 2007). While the motivation to enter the wetlands when closed to the public is not known, it may be partly because poorly socialized, aggressive dogs, are sometimes deliberately taken to 'no-dog' areas to avoid conflicts with other dogs and their owners (M.A. Weston, pers. obs.).

To understand such non-compliance with leashing laws, Williams et al. (2009) investigated the perceived importance of leashing dogs when visiting beaches. Dog owners expressed moderate support for dog leashing, and were more likely to leash dogs if they thought other beach users expected it or that their dogs would harm wildlife or people. Many participants in their study believed that other dogs caused more damage than their own dog. Williams et al. (2009) also reported that most dogs on the beaches they studied were unleashed, even when their owners were aware of dog regulations and the impact dogs have on nesting shorebirds (see Box 12.3 and Figure 12.2). This was also found by Bridson (2000) when they interviewed dog owners on beaches in New Zealand and found that 42% of dog owners interviewed believed dogs should be allowed on beaches, despite being aware that dogs posed a threat to beach-nesting birds.

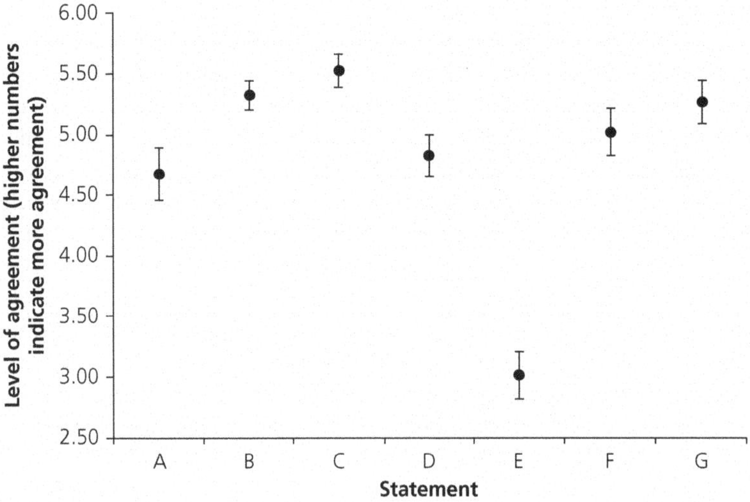

Figure 12.2 Levels of agreement among dog owners (means ± 95% confidence intervals, where 1 is 'strongly disagree' and 7 is 'strongly agree') with the following statements: (A) I feel obliged to leash my dog at the beach, (B) Importance of wildlife protection on beaches, (C) Importance of unleashed exercise for dogs, (D) Unleashed dogs in general have negative consequences, (E) Their own dog has negative consequences, (F) People generally expect dogs to be leashed on beaches and (G) People generally think beaches are a good place for dogs to run around unleashed (after Williams et al., 2009).

Box 12.3 Reining in the roamers: owners' sense of obligation to leash pet dogs

The use of a leash profoundly reduces the roaming behavior of pet dogs in open habitats (i.e., those with structurally simple lower strata through which dogs can move easily) such as wetlands, grasslands, and beaches, and reduces the rate at which dogs disturb wildlife (Weston and Stankowich, Chapter 4). Leashed dogs are less likely to chase or prey upon wildlife, and less likely to interact negatively with other dogs or humans. In many areas, regulations, zoning, and compliance activities seek to maintain a matrix of 'no dogs,' 'leash-only,' 'leash-sometimes' (seasonally or at particular times of day), and 'off-leash' areas, which are designed to accommodate multiple uses including wildlife, and those of dogs and their walkers. A prerequisite for the success of this approach is that compliance with leashing rules is high. However, in countries across the globe where studies are available, which are admittedly biased to the developed world where many dogs live in urban areas, compliance with these leashing regulations is apparently very low (see Weston and Stankowich, Chapter 4).

In some areas inadequate signage and lack of clarity of regulations contribute to low compliance, but even clearly designated 'on-leash' areas are subject to many off-leash dogs. On Victorian (Australia) beaches, while compliance with leashing laws is low, beach goers tend to comply with other management efforts such as temporary beach closures (Weston et al., 2012). Thus, the lack of compliance is not generalized to all regulations, but apparently specifically involves leashing rules. Why compliance with leashing regulations is so low is a critical aspect of the human dimensions of owned dog–wildlife interactions that deserves more attention.

While changing human behavior underpins many potential solutions to environmental problems, it is not well studied, and examples of sustained behavior change are few. Indeed, the available information suggests that many assumptions regarding behavior change (e.g., that education is associated with more pro-environmental attitudes) are questionable and require testing (see Kahan et al., 2012). Similarly, the management solutions to promote behavioral change among humans also need revision in light of new information. For example, a common management response to inappropriate human behavior is to educate the relevant stakeholders. Education, however, rarely changes human behavior by itself (Schultz, 2011). On the other hand, humans are guided strongly by social expectations (norms), and so alternative approaches that seek to change norms hold promise on sustained behavior change (Schultz, 2011).

A case study of underlying human attitudes and perceptions to dog leashing (Williams et al., 2009) involved dog walkers on beaches in southern Victoria, Australia, where unleashed dogs are regarded as a conservation problem for beach-nesting birds, such as the threatened hooded plover *Thinornis rubricollis*. Off-leash dogs disturb, chase, and prey upon eggs, chicks, and breeding adult birds, which occur on beaches heavily used by dog walkers and their dogs. Despite many restrictions involving leashing and 'no dog' zones, compliance is low, particularly with regard to leashing regulations. Thus, enhancing compliance with leashing regulations is considered a conservation management objective.

Williams et al. (2009) surveyed or interviewed 385 dog owners to explore their attitudes to dogs on beaches (see Figure 12.2), and in particular their sense of obligation to leash dogs on beaches. Dog owners were more likely to feel obliged to leash their dog when they believed other people expected dogs to be leashed, and when they believed their dog was a threat to wildlife or people. Dog owners were less likely to feel obliged to leash their dog if they considered unleashed dog recreation was important.

A sense of obligation to leash a dog, as measured by Williams et al. (2009), does not always translate into actual leashing (see Figure 12.2). Nevertheless, a series of recommendations were made that might potentially improve compliance. First, improved compliance may be achieved through community-based approaches to foster social norms (expectations) for dog leashing, and this means that once higher compliance is achieved it may continue to improve. This suggests that work by land managers to improve compliance is not 'pushing against the tide' but may reap ongoing rewards.

For example, on Phillip Island, Victoria, Australia, a concerted management campaign involving awareness, education, and enforcement has led to a reduction in the occurrence of off-leash dogs, especially in hooded plover breeding territories (P. Dann, pers. comm.). Such an approach may not always work; at Mornington Peninsula National Park, adjacent to Phillip Island, compliance slowly improved but remained far from universal (Dowling and Weston, 1999). The differences between these regions remain perplexing, but different management authorities with different approaches dealt with different populations of dog walkers. A second recommendation from the social research conducted was to tailor information products to emphasize the risk that unleashed dogs may pose to wildlife such as

continued

Box 12.3 *Continued*

beach-nesting birds, and raising awareness of designated off-leash dog exercise recreation areas.

Given that the ultimate measure of management success is fewer unleashed dogs where they are not permitted, an understanding of the effectiveness of dog management on compliance rates could inform dog management. Currently, however, little systematically collected data are available on these issues, and this represents a key information gap in managing pet dogs. In particular, if compliance with leashing laws is to be improved, social research will play a critical role in unraveling the complex influences on the decision to leash or not.

These high levels of non-compliance may be related to the low likelihood, in many places, of being caught or fined (Jenkinson et al., 2009; Pelletier, 2006). Alternatively, low compliance may occur where suitable options for off-leash dog exercise are not available or when people place a higher priority on off-leash dog exercise than wildlife conservation. It may also be related to a perception that dogs do not cause significant damage or disturbance in natural environments, or that when they do cause damage it is part of a 'natural process.' Such perceptions may prove to be accurate in some situations. For example, Bekoff and Meaney (1997), in their study of interactions between dogs, people, and the environment in Boulder, Colorado, USA, found that off-leash dogs did not travel far off tracks and that when they did it was for short periods of time. Observations in their study suggested that dogs rarely chased other dogs or wildlife, disturbed people, destroyed vegetation, or entered bodies of water. Their questionnaire found that dog owners and non-dog owners held a view that people were significantly more disruptive to the environment (including wildlife, vegetation, bodies of water) than dogs, and that the quality of the visit would diminish if dogs were required to be leashed.

Yet while dogs may have little impact on the environment in some locations, in other settings the disturbance appears significant (e.g., for shorebirds, Burger et al., 2004; Le Corre et al., 2009; woodland birds, Banks and Bryant, 2007). As science begins to reveal the complex nature of such disturbances (Weston and Stankowich, Chapter 4), these processes need to be fully communicated to the public alongside appropriate behavior change programs if coexistence between dogs and wildlife is to improve. Similarly, an understanding of human perceptions is critical as perceptions can sometimes play as great a role in shaping attitudes as actual experience (Naughton-Treves et al., 2003). Human perceptions of wildlife span from a focus on the individual animal (welfare) to the population (conservation), and even among wildlife scientists, different emphases are evident (Miller and Jones, 2005; Miller and Weston, 2009). It seems prudent, therefore, to define the spectrum of impacts of dogs on wildlife and assess perceptions towards a set of clearly defined impacts.

This recommendation aligns with the work by Sterl et al. (2008) in their study of visitor awareness and assessment of recreational disturbance of wildlife in the Donau-Auen National Park in Austria. Their survey of 271 park visitors found that only 40% of the respondents were aware that wildlife can be disturbed through recreational activities; and only 12% believed that their own visit could have potentially disturbed wildlife. They suggested that this low awareness, coupled with impacting behaviors, is even more problematic given the nature of the area being used. The park studied is in an urban setting, is relatively small in size and lacks buffer zones, is heavily used, has unlimited access, and is highly fragmented by trails. Globally these characteristics are typical of many places visited by people and their dogs. If visitors perceive little or no impact on wildlife when they visit such places, then impacting behaviors are likely to continue (Sterl et al., 2008). It may also be that many (but not all) areas where dog walking occurs tend to be highly degraded as wildlife habitat, and thus most of the impacts are on remnant wildlife species, or on abundant or pest species.

Overall, studies highlight diverse views of visitors to parks, reserves, beaches and protected areas, and these views vary by location and the type of park or open space. For example: Morgan (1999) found that 74.6% of beach users in Wales, UK, wanted dogs banned from the beach; Breton et al. (1996) reported that most beach users would 'forbid' the presence of dogs on the beaches of Barcelona, Spain; Semken et al. (2011) found that 92% of dog walkers and 54% of other users would like the Balcombe Estuary Reserve in Victoria, Australia, to remain open for dog walking; Glover et al. (2011) reported that residents of a major embayment in Victoria supported buffers (exclusion zones) around shorebirds for a variety of recreational activities including dog walking. Therefore, while there are common themes revealed in human dimensions research, management needs to take into account the unique characteristics and contexts of different settings (Manfredo and Dayer, 2004). Studies also need to be contextualized; obtaining representative samples of humans is notoriously difficult, and the perceptions of people will probably vary depending on whether sampling occurs primarily among dog owners, dog walkers, or other user groups.

In addition to human dimensions research on pet dogs accompanying their owners into places important for wildlife, there has also been some human dimensions research focusing on pet dogs and threats posed to them by wildlife. For example, in an analysis of coyote interactions with humans and pets as reported in the Canadian print media from 1995 to 2010 (453 articles reviewed; Alexander and Quinn, 2011), 91 incidents involved dogs (as reported in 108 articles documenting coyote–dog interactions). Coyote-caused dog mortality occurred in 38 cases, many in yards; dogs were off-leash in 92.3% of coyote–dog encounters. While not discussed for the reported coyote attacks on pet dogs, food conditioning was a significant factor in coyote attacks on humans (for example, the person had been feeding the coyote prior to the attack). Thus, Alexander and Quinn suggested that many attacks could be avoided with better waste management and education to deter people from feeding coyotes and to advise dog owners that conflicts can be greater during the coyote pup-rearing season (see also Lukasik and Alexander, 2011).

Dogs can also act as vectors for pathogens that cause disease and this can present a threat to human health (Paquet-Durand et al., 2007). Management regimes in many places are such that dogs are not permitted in parks and protected areas where wildlife occur, or if dogs are permitted they must be leashed. Aguirre et al. (1995), in a survey of individuals representing 179 national parks, 123 state agencies, 103 federal agencies, and 98 colleges and universities in the United States, found that some respondents were concerned because visitors in national parks do not leash their dogs. Their concern was two-fold: dogs posing a health risk to wildlife and dogs acting as vectors for disease transmission to humans. As such, they recommended that parks need comprehensive animal health programs or management plans. Such management programs must take into account the likelihood of compliance by park visitors, and this again points to the need for a better understanding of the human dimensions of dog–wildlife interactions.

12.3.2 Working dogs

The human–dog association is not just one of companionship. Throughout history, dogs have also been placed in working roles to interact with wildlife; for example, as hunting dogs (Chitwood et al., 2011) or livestock guardian dogs (Baker et al., 2007). More recently, dogs have been used as guardians for vulnerable wildlife populations (e.g., Australasian gannets (*Morus serrator*) breeding at Point Danger, Portland, Australia (Peter, 2012); little penguins (*Eudyptula minor*) near Warrnambool, Australia, (Poole, 2010); see VerCauteren et al., Chapter 9; Woollett et al., Chapter 10, and Koster and Noss, Chapter 11 for discussion of how working dogs and wildlife interact).

Such human–dog–wildlife interactions can be complex and difficult to define when diverse stakeholder groups are involved. In some countries it is mandatory to use trained hunting dogs to ensure humane kills (see Chitwood et al., 2011), while in other countries restrictions or bans have been placed on wildlife hunting with dogs (e.g., red fox, *Vulpes vulpes*, hunting in the United Kingdom; Loveridge et al., 2007). Restrictions often come about because of concerns over both dog and wildlife welfare (White et al., 2003) when hunting with dogs and concerns about dog hunting in highly fragmented landscapes (Chitwood et al., 2011). However, many hunters argue that hunting is "woven into the very fabric of personal and social history" (Marks, 1991, p. 5) and that hunting with dogs can help define relationships with family, friends, and nature (Chitwood et al., 2011). Chitwood et al.'s (2011) study on hunter identity in coastal North Carolina, USA, suggested that banning dog hunting "may destabilize rural communities by removing critical elements of community identity and means through which communities cope with challenges to their identity." They also made note of the suggestion by Manfredo et al. (2009) that wildlife managers face a moral imperative to consider the impacts their decisions have on human well-being.

Understanding such hunter perspectives and values is important, as is understanding the perspectives and values of other stakeholders. Chitwood et al. (2011) described hunters' teleological views of animal wellbeing (that they are fulfilling their purpose) and how these contrast with the utilitarian views of other groups (e.g., minimizing stress or pain). They also discussed the role of hunting (with dogs) in facilitating human connections with nature. In order to manage human–dog–wildlife conflicts, such values and perspectives need to be understood fully and common themes across regions identified.

While hunting with dogs is important for some hunters, it does not affect hunter satisfaction for others. For example, in their study of hunter satisfaction when pheasant (*Phasianus colchicus*) hunting in Utah, Frey et al. (2003) found that the presence of dogs did not affect hunter satisfaction or success. Rather, harvest success, relative density of hunters, and the number of cocks seen by hunters predicted hunter satisfaction. Such influences vary in their importance, depending on the location and the hunted species (Frey et al., 2003).

Although there is a significant body of research in the human dimensions field relating to hunting (including with dogs), there has been much less attention paid to the human dimensions of guardian dogs. Livestock guardian dogs (LGDs) have received considerable attention in the literature (Baker et al., 2007; Gehring et al., 2010); indeed, the role of dogs in protecting vulnerable wildlife is gaining more prominence as a wildlife management tool (Ritchie et al., 2012). However very little human dimensions research is documented regarding the role of dogs in assisting with the survey of certain cryptic wildlife, or enabling the use of non-lethal management of pest wildlife (for example, discouraging birds from some airports) (see Weston and Stankowich, Chapter 4).

Van Bommel and Johnson (2012) assessed the role of working dogs for livestock protection in Australia and found them to be an effective management tool with significant reductions in livestock depredation. This, in turn, resulted in less need to control wild predators (often feral dogs and dingoes) through poisoning or shooting. They suggested this was important given the ethical concerns associated with lethal control and the better understanding we now have about the role of top-order predators in ecosystem health (Ritchie et al., 2012). This interaction between feral dogs and working dogs points to the need to understand landholder perceptions of the problem (Ballard and Fleming, 2010) and the likely acceptance of alternative management approaches, such as the use of dogs for livestock protection (see Box 12.4). In a separate study in Namibia (Potgieter, 2011), it was also found that livestock guardian dogs were successful (91% of the LGDs eliminated or reduced livestock losses) at protecting stock against predators, which include black-backed jackal (*C. mesomelas*), chacma baboon (*Papio ursinus*), and cheetahs (*Acinonyx jubatus*). However, farmer satisfaction was more likely to be associated with 'good' LGD behavior than perceptions of a reduction in livestock losses. Common and undesirable LGD behavior included staying at home rather than accompanying livestock and chasing other wildlife. These studies highlight the need to appreciate human perceptions, rather than realities, when assessing the costs and benefits of dogs in mediating human–wildlife conflict.

Box 12.4 Cultures, communities, and canids

Dingoes and wide-ranging dingo–dog hybrids evoke a level of passion and variety of responses from the Australian community that are perhaps unrivaled by any other species. These canids are often reviled by farmers for their impacts on livestock, treasured by others for their inherent beauty and quintessential Australian identity, and spiritually important for many indigenous people as a totemic animal. Others remain largely apathetic. Attitudes towards these wild canids vary as much within groups of the Australian community as they do between. However, it is perhaps in rural and remote Australia where the divide in opinions regarding dingoes and dingo–dog hybrids is most obvious and intense. Such divisions have substantial consequences both for dingoes and dingo–dog hybrids, but also for the people who live within these communities.

Many landholders are aware of the benefits dingoes and dingo–dog hybrids provide in reducing populations of native and introduced herbivores (e.g., kangaroos, goats, pigs, and rabbits) and hence their ability to reduce competition for food with livestock. However, others are non-tolerant of dingoes and hybrid wild dogs and will control them through combinations of poisoning and shooting, in some cases beyond the bounds of their properties. This reduces dingo–dog populations not only in the areas where farmers do not tolerate dogs but also in areas where farmers are tolerant or even encourage these dogs (due to their perceived ecological and management benefits), due to the considerable ranging behavior of dingoes and dogs. Such an effect has been seen for many other carnivores worldwide, where due to their typically wide-ranging behavior, small reserves and their borders afford little protection for carnivores, irrespective of species' population sizes within reserves (Woodroffe and Ginsberg, 1998). The result is animosity between landholders with differing viewpoints about canid management. Indeed, cases exist in which individual landholders who have chosen to encourage wild canid populations on their properties claim to have been harassed by neighbors and ostracized by their local communities. Thus, decisions about the management of wild canids have significant social and potential health costs for individuals living and working in rural and remote Australia, where neighbors may be scores of kilometers away and people are already isolated.

To understand and appreciate just how greatly opinion varies as to the functional roles of dingoes and wild dogs, and what their place should be in the landscape, a study (Kean, 2011) was undertaken in Victoria, Australia, which examined attitudes toward dingo and wild dog management in Victoria. The study had three objectives:

- Identify if attitudes vary between regions where both canid 'species' are present to those where they are absent.
- Identify if attitudes toward dingo and wild dogs vary across stakeholder (community) groups.
- Identify factors influencing attitudes toward dingoes and wild dogs and their management.

There were no significant differences between regions with respect to overall attitudes regarding dingoes and wild dogs. Interestingly, however, differences among stakeholders' viewpoints within regions were apparent. Wildlife managers overall had positive attitudes toward dingoes, while farmers and non-farmers were either indifferent or inconsistent with respect to their attitudes toward dingoes. Farmers and non-farmers in the north-east region of Victoria (where wild dogs are present and common) held stronger negative attitudes toward wild dogs than did farmers in the Grampians region of Victoria (where wild dogs are largely absent). Attitudes toward canids were influenced by both direct (personal stock loss) and indirect experiences (impacts on neighbors' properties or those of other members of the community). In response to wild canids attacking livestock, farmers advocated lethal methods of control whereas non-farmers and wildlife managers of the same communities preferred non-lethal methods. This study demonstrates that the human dimensions of dingo and wild dog management are complex, and preconceived notions of how communities and various stakeholder groups may view canid management do not necessarily hold.

12.3.3 Feral and wide-ranging dogs

Feral dogs, which are sometimes referred to as wild dogs or free-ranging dogs, are widespread canids that occur in the North, Central and South America, Europe, Australia, Africa, and on several remote ocean islands (Green and Gipson, 1994). However, to avoid confusion and due to the importance of how dogs are described in relation to conservation and management actions (Letnic et al., 2012), we make a further distinction. There are truly wild dogs, such as the dingo in Australia, and other dogs, which are free and sometimes wide-ranging, but still rely on human subsidies. The latter type is typical of most dogs in the world, with the dingo being a relatively rare case. Furthermore, sitting between these two groups of dogs (owned/confined vs. feral/wild), are village dogs, which are usually owned or affiliated with a household but that are not easily handled or tame and yet are also heavily dependent on human subsidies. It has been suggested that village dogs are perhaps the most common type of dog in less developed parts of the world (Ortolani et al., 2009).

Populations of free-ranging and feral dogs that inhabit natural areas can present a significant threat to wildlife populations, acting as vectors for disease, competing with, preying upon, or disturbing wildlife, and hybridizing with other wild canids. A 'triple bottom line' analysis in Australia to assess the impact of invasive animals on the environment and agriculture found the feral dog (including the dingo) to be the fifth most significant vertebrate pest (McLeod, 2004). Similarly, a review in the United States found feral dog damage to amount to more than $620 million annually (Bergman et al., 2009). In some situations, however, feral dogs may fulfill an important ecological role and removing them might have unforeseen impacts on biodiversity (Letnic et al., 2012). Thus, management of feral dogs is a key challenge for wildlife managers in some parts of the world.

While there have been numerous studies on the techniques available for feral dog management (for example, repellents, toxicants, fumigants, trapping, shooting, frightening; Green and Gipson, 1994) and human concerns about impacts on livestock, there has been less research on the human dimensions of the interactions between feral dogs and wildlife. So, again, the human–dog interaction is well documented, but our understanding of the human–dog–wildlife interaction is far less developed.

However, we do know that there is a wide spectrum of human views concerning feral dogs and these highlight some important, common themes across wildlife management issues. First, opinions about the problem vary widely and are complicated by the fact that it is difficult to separate a native wild dog, such as the dingo, from a feral dog of more recent ancestry due to hybridization across its range (Fleming et al., 2001; Letnic et al., 2012; Ritchie et al., 2012). The dingo is generally treated as a native Australian species, has a unique place in Aboriginal culture, is protected in many parts of Australia, can be important for tourism, and is considered by many to be an ecologically important predator (Burns and Howard, 2003; Ritchie and Johnson, 2009; Thompson et al., 2003). Feral dogs, on the other hand, are widely perceived to be agricultural pests that derived from animals introduced from Europe or Asia within the past several hundred years.

Second, there are often conflicts between individuals and stakeholder groups over the most appropriate tools for managing the problem. For example, studies on a range of species show that many people are reluctant to accept lethal management techniques, such as shooting or poisoning, and that acceptability of lethal methods varies widely between stakeholder groups (Bruskotter et al., 2009; Nimmo et al., 2007). There is concern about the pain and suffering caused to target and non-target animals (including pet dogs) when managing problem wildlife (McLaren et al., 2007). As noted earlier, people are generally more likely to accept lethal techniques if the problem is severe for them (Wittmann et al., 1998). Such diverse opinions make management challenging and a combination of management approaches is therefore needed.

12.4 Discussion and recommendations

> "When attempting to address human-wildlife conflict, it is well accepted that understanding the role of humans is as important as understanding the ecology of species (Alexander and Quinn, 2011)."

The human dimensions research reviewed above suggests that a combination of management strategies is required to effectively resolve human–dog–wildlife conflicts. Given the complex and varied nature of the conflict, it is important to fully understand both the ecological and human dimensions of the conflict before developing and implementing management programs. This is true of all wildlife management issues (Decker et al., 2012).

Given the long history of human–dog interactions and the strong bond between humans and pet dogs in particular, people and their dogs often need to be managed as a cooperative social unit (Bekoff and Meaney, 1997). Management programs may include:

- education and communication;
- regulations (e.g., leashing laws) and enforcement;
- modifying the environment or managing the wildlife species or dog population (e.g., fences, land use planning, wildlife management tools);
- conflict resolution and public participation.

Management success is likely to result from a combination of the above strategies (see, for example, Dowling and Weston, 1999), and it is important to note that successful outcomes will be influenced by attitudes toward different management options (see Box 12.5).

Box 12.5 Perceptions and awareness of the management of pet dogs

Different management approaches enjoy varying levels of support among the public and, in many areas, managers are sensitive to the views of the public regarding the management of public places. However, public views vary between stakeholder groups (Ioja et al., 2011) and are often assumed rather than documented (but see Glover et al., 2011). Where studies have been conducted, higher levels of support for dog management than expected has been evident.

Christie et al. (2010) documented stakeholder views (n = 295) of shorebird management issues around the 270 km^2 Ramsar-listed Westernport Bay in central southern Victoria, Australia. Cluster analysis of the activities recreationists undertook in the bay revealed a distinct recreational group of 'dog walkers.' The ability to take dogs, or the lack of dogs, did not explain the sites recreationists chose to visit around the bay. Among all recreationists, there were high levels of support for dog management, which was intended to help conserve shorebirds (Figure 12.3). This included support for buffer designations around shorebirds to reduce disturbance caused by dog walking (Glover et al., 2011). There was also at least substantial agreement among dog walkers that the presence of people is detrimental to many bird species (54%), that land and water that provides critical habitat for birds should be protected (92%), that it was good to share the bay with birds (92%), and that they would be upset if bird species went extinct (69%). About half of the dog walkers felt that managers were oversensitive to the needs of birds.

In a related study, dog walkers from a selection of Victorian beaches were highly aware of dog regulations (96%), aware of the consequences of not complying with these regulations (94%), and had seen relevant signage (88%) (Williams et al., 2009). Fewer were aware of their potentially negative influence on beach-nesting birds, had seen enforcement officers, or knew of alternative off-leash zones (Williams et al., 2009).

Like all human studies, the issue of a non-random sample in the two studies mentioned above cannot be absolutely discounted. Nevertheless, levels of awareness and support for dog management should not always be assumed to be low, either among the general public or dog walkers themselves.

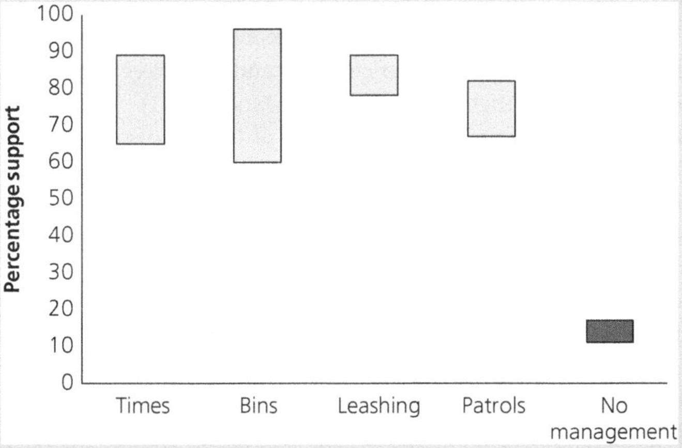

Figure 12.3 Support among 295 stakeholders for dog management (and no management) at Westernport Bay, Victoria, Australia (range of means across four sites shown). The dog managements are restrictions on the times in which dogs were permitted, the installation of dog tidy bins, leashing regulations, ranger patrols, and no management; 14 other managements, not relevant to dogs, are not depicted (after Christie et al., 2010).

12.4.1 Education and communication

> "Neglecting to unpack the intertwined issues wrapped up in a complex wildlife controversy holds the potential pitfall of overlooking important stakes and stakeholders who should be audiences for educational communication (i.e., their concerns might not be addressed and the right channels might not be used to reach them)" (Loker et al., 1998).

Education programs are widely used in wildlife management to raise awareness and foster 'conservation-sympathetic' attitudes in the community. In the absence of human dimensions research, however, education programs are often developed and implemented based on assumptions about the target audience, opinions of the managers and educators, or the views of a vocal minority. As such, the first step in developing any education program is to understand what people already know and the attitudes they hold about the issue (Sterl et al., 2008).

In this case, lessons can be learnt from settings where education programs have already been implemented for resolution of human–dog–wildlife conflicts. For example, Ormsby and Forys (2010) investigated public attitudes toward dog restrictions and the effectiveness of a beach user education program (signs, web pages, print media) in Florida, USA. Their pre-education and post-education survey showed that beach visitors had a similar or more positive attitude after the education program was delivered. They also commented on a need to lessen the focus on the scientific study of birds and to increase the focus on creative education programs that lead to applied conservation.

Managers and educators can use social psychological theories (Figure 12.1) to identify what is already known about the targeted stakeholder groups and to identify gaps. For example, we might have a good understanding of how park visitors behave when walking their dogs but have limited understanding of the underlying reasons and motivations for this behavior (e.g., beliefs, attitudes). Such information is essential in developing educational materials that 'speak the right language' to the intended audience. Social psychological theories have recently expanded into the field of social marketing, which holds promise in longer-term, more widespread, human behavioral change.

The human dimensions literature on dog–wildlife interactions suggests that social norms are an important influence on human behavior (e.g., when people choose to walk their dog off-leash). As such, education campaigns could focus on the development of social norms and community expectations in such contexts. McKenzie-Mohr (2011) suggests that norms should be presented or communicated to the intended audience at the time the targeted behavior is to occur. For example, when entering a park or beach, a sign could display the percentage of visitors who engage in positive environmental behaviors (e.g., dogs on-leash). This gives visitors a sense of what is normal or expected in the environment they are visiting and is likely to be an important influence on subsequent behaviors.

Education campaigns are only part of the solution though. Given that some dog owners know about dog-leashing laws and understand the impacts their dog may have on wildlife (Bridson, 2000; Williams et al., 2009), but still choose to walk their dog off-leash, other behavior change strategies are needed. McKenzie-Mohr (2011) outlines a range of tools that can be used to modify human behavior and they are all based on understanding the barriers and benefits for people in choosing whether or not to participate. Tools available include *communication* (education campaigns), *prompts* (signs, advertisements), *commitment* (seeking a verbal or written commitment from people), *norms* (communication of community expectations), *social diffusion* (early adopters of the behavior spreading the word to others), *convenience* (making it easy for people to change), and *incentives* (financial or otherwise) (McKenzie-Mohr, 2011). Examples of actual implementation of these tools are provided in Table 12.2.

12.4.2 Regulations and enforcement

Regulating human behavior through laws and penalties is another commonly used tool in environmental and wildlife management (Thomas and Murfitt, 2011). The use of such regulations should be considered carefully as it has been found people are

Table 12.2 Examples of management aimed at creating social change to improve dog management for the benefit of wildlife.

Tool	Example	Example Sources
Communication	Campaigns by television, Internet, brochures, rate notices, information with dog registrations, etc.	Dowling and Weston 1999
Prompts	Unambiguous and conspicuous signage in critical areas.	Dowling and Weston 1999
Commitment	Informal agreements fostered through relevant meetings e.g., 'dog breakfasts.'	Maguire 2008
Norms	Expectations to remove dog feces from public areas.	Collier 2011
Social diffusion	Community 'wardens' and 'champions.'	Maguire 2008; Weston et al. 2012
Convenience	Adequate planning for dog zones. Small 'no-dog' wildlife zones which promote coexistence.	McIntosh 1995; Weston et al. 2011
Incentives	Enforcement, free 'leashes' and dog food.	Dowling and Weston 1999; Maguire 2008

more likely to be motivated to change their behavior when they feel they are in control and that their behavior aligns with their values and perceived best interests (DeCaro and Stokes, 2008). That is, approaches that give people control over their decisions (e.g., making a pledge) can work more effectively than the 'fences and fines' approach. As DeCaro and Stokes (2008) note, the "fences and fines approach to conservation has alienated stakeholders from local resources and has undermined intrinsic interest in conservation." Miller and Howell (2008) support this and suggest that social marketing and collaborative management can be more effective than the 'big stick' approach.

Even so, regulations are often needed as one component of an overall management program (Jenkinson et al., 2009). When used, regulations can be most effective when enforced; for example, when a ranger or law enforcement officer is present during peak times of usage or non-compliance (Arnberger and Eder, 2008). However, the application of enforcement is limited; in more rural or underdeveloped settings it may not be feasible.

Before implementing new regulations or enforcing existing regulations, it is important for managers to assess the costs involved and the potential for positive outcomes. That is, are resources being directed where they are most needed? As Reed and Merenlender (2011) point out in their study of dog influences on native carnivores in recreation areas in California, the key factors associated with impacts on wildlife appear to be the presence and number of human visitors. Carnivore abundance and species richness did not differ between recreation areas with different dog policies (i.e., dogs allowed off-leash, allowed on-leash, not allowed), although low leashing compliance may mean that areas with different policies may actually have had a similar presence of dogs. Even so, they suggested that the enforcement of leashing laws may not be the best way to direct limited resources and that restrictions on human visitation (e.g., closing some areas to the public) may be more effective. Clearly, there is a need to better understand the ecological impacts of dog leashing policies in different environments as well as the human dimensions. Where required human behavior change is modest, high compliance (including among dog owners) can be achieved. Weston et al. (2012) describe high compliance among recreationists, including dog walkers, with temporary beach closures that had signs, signs and fences, or signs, fences, and wardens. Thus, coexistence between dogs and threatened wildlife may be possible in some circumstances.

12.4.3 Modifying the environment or managing the species

A commonly used approach in HWCs is to modify the environment in some way. The focus of this may be on changing some aspect of the environment to modify human behavior, for example, fences or boardwalks. Alternatively, and depending on the nature of the conflict, the wildlife population or dog population at the center of the conflict may be managed in some way; for example, through shooting, poisoning, birth control, and other measures.

McKenzie-Mohr (2011) suggests that behavior change strategies focusing on internal barriers for an individual person (e.g., attitudes, opinions, lack of knowledge) are powerful but ineffective if the desired behavior is unpleasant or time-consuming. As a hypothetical example, a person may know that their dog may disturb wildlife and agree with the leashing regulations but find it more convenient to walk through a natural, sensitive, and readily damaged environment to get back to their car with their dog off-leash. As such, the first step in removing external barriers (e.g., the environment, access to an area) is to identify them (McKenzie-Mohr, 2011).

Behavior change strategies focusing on the removal of external barriers are often combined with other strategies. For example, Burger et al. (2004) examined the effect of human activities on migratory shorebirds on Delaware Bay, New Jersey, USA. They documented human recreational behaviors and identified patterns of disturbance to shorebirds. They then developed a management program that included education and communication (through signs), restricted access, wildlife viewing platforms, patrols, and summonses for infractions. These strategies resulted in a significant reduction in the number of recreationists disturbing shorebirds (see also Dowling and Weston, 1999). Although such strategies can be very effective, they also have the potential to create conflict if visitors do not agree with the strategies proposed or implemented (e.g., Kahan et al., 2012). Human dimensions research can provide insights into the acceptability of different management approaches (Bruskotter et al., 2009) and the likely levels of community support once plans are implemented (e.g., Glover et al., 2011).

Strategic land use planning is also increasingly important as human population size increases and pressures mount on parks and open spaces and their biodiversity. If human–dog–wildlife conflicts are considered significant in these areas, planners should consider options for dog walking in other categories of green space (for example, collective house gardens in urban areas; Ioja et al., 2011). Ioja et al. (2011) suggests that such spaces be considered as walking spaces for companion animals to reduce the pressures on urban parks. Options may also exist for establishing designated, fenced, off leash areas, or specific 'dog trails,' which may involve specific facilities such as watering points and 'activity trails' for dogs (Jenkinson, 2011).

12.4.4 Conflict resolution and public participation

No matter what combination of techniques is used to manage human–dog–wildlife conflicts, there will always be an element of disagreement between stakeholders. As such, public and community participation, consensus-building, and conflict resolution strategies are essential for an effective decision-making process (Harding et al., 2009). As with the broader field of environmental management, stakeholder involvement is an important component of wildlife management (Chase et al., 2002; Riley et al., 2003) and there are numerous texts and guides available to inform appropriate participation (e.g., Chase et al., 2002; Harding et al., 2009; Shindler and Cheek, 1999). Loker et al. (1998) suggest that in any wildlife management conflict or problem, a combination of human dimensions and stakeholder approaches can result in decision-making outcomes that can be considered fair, efficient, wise, and stable.

12.5 Future research

This chapter points to a need for further research to examine the human dimensions of dog–wildlife interactions. As this area of research is still emerging in many parts of the world, our understanding of the human values, attitudes, and behaviors relevant for any particular wildlife–human conflict is in its infancy. As Le Corre et al. (2009), in their study of bird disturbance on conservation sites, point out ". . . it is striking to note that the great majority of current studies derive from biology (among the 272 articles selected in our study, there was no social sciences laboratory)." This 'biological bias' is evident in wildlife management globally and may be related to a relative lack of human dimensions training in many university courses (Decker et al., 2001; Miller, 2009).

One of the challenges for human dimensions specialists and wildlife managers is that "human dimensions studies that reveal differences between what managers think stakeholders want and what they really want are not readily embraced by all professionals" (Loker et al., 1998). Nevertheless, if we are to fully understand the nature of human–wildlife conflicts we need to adopt a multi-disciplinary, integrated approach (e.g., Glover et al., 2011).

To date, the scarce human dimensions research on human–dog–wildlife interactions has focused on human behaviors mainly in developed countries; for example, patterns of visitation, the nature of recreational activities, and the number of visitors walking their dog off-leash. While there has been significant research effort in understanding values, attitudes, and motivations of people with regard to working dogs (especially for hunting), there appears to be much less attention paid to the underlying values and attitudes of people living with or near dogs in relation to dog–wildlife interactions (Le Corre et al., 2009). This is an important area for future research.

Acknowledgments

We thank the Environmental Sustainability Research Group at Deakin University and BirdLife Australia's beach-nesting birds project (funded by the Australian Government's Caring For Our Country programme) for supporting this work. Special thanks go to Geoff Wescott, Kasun Ekanayake, Amy Shaw, James Fitzsimons, Thomas Schneider, and Grainne Maguire for their contributions to related projects.

References

Aguirre, A.A., Starkey, E.E., and Hansen, D.E. (1995). Wildlife diseases in national park ecosystems. *Wildlife Society Bulletin*, **23**, 415–9.

Ajzen, I. (1991). The theory of planned behavior. *Organizational Behavior and Human Decision Processes*, **50**, 179–211.

Ajzen, I. and Fishbein, M. (1980). *Understanding Attitudes and Predicting Social Behavior*. Prentice Hall, Englewood Cliffs, New Jersey.

Alexander, S.M. and Quinn, M.S. (2011). Coyote (*Canis latrans*) interactions with humans and pets reported in the Canadian print media (1995-2010). *Human Dimensions of Wildlife*, **16**, 345–59.

Antos, M.J., Ehmke, G.C., Tzaros, C.L., and Weston, M.A. (2007). Unauthorised human use of an urban coastal wetland sanctuary: current and future patterns. *Landscape and Urban Planning*, **80**, 173–83.

Arnberger, A. and Eder, R. (2008). Assessing user interactions on shared recreational trails by long-term video monitoring. *Managing Leisure*, **13**, 36–51.

Arnberger, A. and Hinterberger, B. (2003). Visitor monitoring methods for managing public use pressures in the Donau-Auen National Park, Austria. *Journal for Nature Conservation*, **11**, 260–7.

Baker, S., Singleton, G., and Smith, R. (2007). The nature of the beast: Using biological processes in vertebrate pest management. In D.W.Macdonald and K. Service, Eds., *Key Topics in Conservation Biology*. Blackwell Publishing, Malden, USA, pp. 173–85.

Ballard, G. and Fleming, P.J.S. (2010). Are we barking up the wrong tree? Observations on community co-management of wild dogs in NSW, Australia. *23rd Australasian Wildlife Management Society Annual Conference*, 1-December 3, 2010.

Banks, P.B. and Bryant, J.V. (2007). Four-legged friend or foe? Dog walking displaces native birds from natural areas. *Biology Letters*, **3**, 611–3.

Bekoff, M. and Meaney, C.A. (1997). Interactions among dogs, people, and the environment in Boulder, Colorado: A case study. *Anthrozoos*, **10**, 23–31.

Bergman, D., Breck, S., and Bender, S. (2009). Dogs gone wild: Feral dog damage in the United States. In J.R. Boulanger, Ed., *Proceedings of the 13th Wildlife Damage Management Conference*. Saratoga Springs, NY, pp. 177–83.

Breton, F., Clapes, J., Marques, A., and Priestley, G.K. (1996). The recreational use of beaches and consequences for the development of new trends in management: The case of the beaches of the Metropolitan Region of Barcelona (Catalonia, Spain). *Ocean and Coastal Management*, **32**, 153–80.

Bridson, L. (2000). *Minimising Visitor Impacts on Threatened Shorebirds and their Habitats*. Department of Conservation, Wellington, New Zealand.

Bruskotter, J.T., Vaske, J.J., and Schmidt, R.H. (2009). Social and cognitive correlates of Utah residents' acceptance of the lethal control of wolves. *Human Dimensions of Wildlife*, **14**, 119–32.

Bruskotter, J.T and Shelby, L.B. (2010). Human dimensions of large carnivore conservation and management: Introduction to the special issue. *Human Dimensions of Wildlife*, **15**, 311–4.

Burger, J., Jeitner, C., Clark, K., and Niles, L.J. (2004). The effect of human activities on migrant shorebirds: successful adaptive management. *Environmental Conservation*, **31**, 283–8.

Burns, G.L. and Howard, P. (2003). When wildlife tourism goes wrong: A case study of stakeholder and management issues regarding Dingoes on Fraser Island, Australia. *Tourism Management*, **24**, 699–712.

Chase L.C., Siemer, W.F., and Decker, D.J. (1999). Suburban deer management: A case study in the village of Cayuga Heights, New York. *Human Dimensions of Wildlife*, **4**, 59–60.

Chase L.C., Siemer, W.F., and Decker, D.J. (2002). Designing stakeholder involvement strategies to resolve wildlife management controversies. *Wildlife Society Bulletin*, **30**, 937–950.

Chitwood, M.C., Peterson, M.N., and Deperno, C.S. (2011). Assessing dog hunter identity in coastal North Carolina. *Human Dimensions of Wildlife*, **16**, 128–41.

Christie, B.A., Miller, K.K., and Weston, M.A. (2010). *Western Port welcomes waterbirds: A study of the human dimensions of waterbird conservation in Western Port Bay, Australia*. Deakin University, Burwood.

Clutton-Brock, J. (1995). Origin of the dog: Domestication and early history. In J. Serpell, Ed., *The Domestic Dog: Its Evolution, Behaviour and Interaction with People*. Cambridge University Press, Cambridge, pp. 7–20.

Collier, G. (2011). When's enough enough? Understanding norms. *Water Materials Energy Magazine*, **Dec.**, 35–6.

DeCaro, D., and Stokes, M. (2008). Social-psychological principles of community-based conservation and conservancy motivation: Attaining goals within an autonomy-supportive environment. *Conservation Biology*, **22**, 1443–51.

Decker, D.J., Brown, T.L., and Siemer, W.F. (2001). *Human Dimensions of Wildlife Management in North America*. The Wildlife Society, Bethesda, MD.

Decker, D.J. and Purdy, K.G. (1988). Toward a concept of wildlife acceptance capacity in wildlife management. *Wildlife Society Bulletin*, **16**, 58–62.

Decker, D.J., Riley, S.J., and Siemer, W.F. (2012). *Human Dimensions of Wildlife Management*. 2nd edn. The Johns Hopkins University Press, Baltimore.

Dowling, B. and Weston, M.A. (1999). Managing a breeding population of the Hooded Plover *Thinornis rubricollis* in a high-use recreational environment. *Bird Conservation International*, **9**, 255–70.

Enck, J.W. and Decker, D.J. (1997). Examining assumptions in wildlife management: A contribution of human dimensions inquiry. *Human Dimensions of Wildlife*, **2**, 56–72.

Fleming, P., Corbett, L., Harden, R., and Thomson, P. (2001). *Managing the Impacts of Dingoes and Wild Dogs*. Commonwealth of Australia, Canberra.

Frey, S., Conover, M., Borgo, J., and Messmer, T. (2003). Factors influencing pheasant hunter harvest and satisfaction. *Human Dimensions of Wildlife*, **8**, 277–86.

Fulton, D.C., Manfredo, M.J., and Lipscomb, J. (1996). Wildlife value orientations: A conceptual and measurement approach. *Human Dimensions of Wildlife*, **1**, 24–47.

Gehring, T.M., VerCauteren, K.C., and Landry, J.M. (2010). Livestock protection dogs in the 21st century: Is an ancient tool relevant to modern conservation challenges? *BioScience*, **60**, 299–308.

Glover, H.K., Weston, M.A., Maguire, G.S., Miller, K.K., and Christie, B.A. (2011). Towards ecologically meaningful and socially acceptable buffers: Response distances of shorebirds in Victoria, Australia, to human disturbance. *Landscape and Urban Planning*, **103**, 326–34.

Green, J.S. and Gipson, P.S. (1994). Feral dogs. In S.E. Hygnstrom, R.M. Timm, and G.E. Larson, Eds., *The Handbook: Prevention and Control of Wildlife Damage*. University of Nebraska, Lincoln, pp. 1–7.

Harding, R., Hendriks, C.M., and Faruqi, M. (2009). *Environmental Decision-Making: Exploring Complexity and Context*. Federation Press, Leichhardt, NSW.

Ioja, C.I., Rozylowicz, L., Patroescu, M., Nita, M.R., and Vanau, G.O. (2011). Dog walkers' vs. other park visitors' perceptions: The importance of planning sustainable urban parks in Bucharest, Romania. *Landscape and Urban Planning*, **103**, 74–82.

Jenkinson, S., Hale, J., and Harrop, P. (2009). Influencing walkers with dogs: Three years of progress. *Countryside Recreation*, **17**, 18–22.

Jenkinson, S. (2011). Promoting fitness and fun with your best friend. *Society for Companion Animal Studies Journal*, **Summer**, 12–17.

Kahan, D.M., Peters, E., Wittlin, M., Slovic, P., Ouellette, L.L., Braman, D., and Mandel, G. (2012). The polarizing impact of science literacy and numeracy on perceived climate change risks. *Nature Climate Change*, **2**, 732–5.

Kean, K. (2011). *Human dimensions of dingo and wild dog management in Victoria*. Unpublished Honours Thesis, Deakin University, Melbourne.

Kellert, S.R. (1991). Public views of wolf restoration in Michigan. *Transactions of the North American Wildlife and Natural Resources Conference*, **56**, 152–61.

Kellert, S.R. (1996). *The Value of Life: Biological Diversity and Human Society*. Island Press, Washington, DC.

Le Corre, N., Gelinaud, G., and Brigand, L. (2009). Bird disturbance on conservation sites in Britanny (France): The standpoint of geographers. *Journal of Coastal Conservation*, **13**, 109–18.

Letnic, M., Ritchie, E.G., and Dickman, C.R. (2012). Top predators as biodiversity regulators: The dingo *Canis lupus dingo* as a case study. *Biological Reviews*, **87**, 390–413.

Loker, C.A., Decker, D.J., and Chase, L.C. (1998). Ballot initiatives-antithesis of human dimensions approaches or catalyst for change? *Human Dimensions of Wildlife*, **3**, 8–20.

Loveridge, A.J., Reynolds, J.C., and Milner-Gulland, E.J. (2007). Does sport hunting benefit conservation? In D.W. Macdonald, and K. Service, Eds., *Key Topics in Conservation Biology*. Blackwell Publishing, Malden, USA, pp. 222–38.

Lukasik, V.M. and Alexander, S.M. (2011). Human–coyote interactions in Calgary, Alberta. *Human Dimensions of Wildlife*, **16**, 114–27.

Maguire, G.S. (2008). *A practical guide for managing beach-nesting birds in Australia*. BirdLife Australia, Melbourne.

Maguire, G.S., Miller, K.K., Weston, M.A., and Young, K. (2011). Being beside the seaside: Beach use and preferences among coastal residents of south-eastern Australia. *Ocean and Coastal Management*, **54**, 781–8.

Manfredo, M.J. and Dayer, A.A. (2004). Concepts for exploring the social aspects of human-wildlife conflict in a global context. *Human Dimensions of Wildlife*, **9**, 1–20.

Manfredo, M.J., Vaske, J.J., Brown, P.J., Decker, D.J., and Duke, E.A. (Eds.) (2009). *Wildlife and Society: The Science of Human Dimensions*. Island Press, Washington, DC.

Marks, S.A. (1991). *Southern Hunting in Black and White: Nature, History, and Ritual in a Carolina Community*. Princeton University Press, Princeton, NJ.

McIntosh, R. (1995). The history and multi-purpose management of Kielder Forest. *Forest Ecology and Management*, **79**, 1–11.

McKenzie-Mohr, D. (2011). *Fostering Sustainable Behavior: An Introduction To Community-Based Social Marketing*. 3rd edn. New Society Publishers, Gabriola Island, Canada.

McKnight, J. and Sutton, J. (1994). *Social Psychology*. Prentice Hall, Sydney, Australia.

McLaren G., Bonacic, C., and Rowan, A. (2007). Animal welfare and conservation: Measuring stress in the wild. In D.W. Macdonald, and K. Service, Eds., *Key Topics in Conservation Biology*. Blackwell Publishing, Malden, USA, pp. 220–33.

McLeod, R. (2004). *Counting the Cost: Impact of Invasive Animals in Australia 2004*. Cooperative Research Centre for Pest Animal Control, Canberra.

Meadow, R., Reading, R.P., Phillips, M., Mehringer, M., and Miller, B. (2005). The influence of persuasive arguments on public attitudes toward a proposed wolf restoration in the southern Rockies. *Wildlife Society Bulletin*, **33**, 154–63.

Miller, K.K. (2009). Human dimensions of wildlife population management in Australasia: History, approaches and directions. *Wildlife Research*, **36**, 48–56.

Miller, K.K and Jones, D.N. (2005). Wildlife management in Australasia: Perceptions of objectives and priorities. *Wildlife Research*, **32**, 265–72.

Miller, K.K. and Weston, M.A. (2009). Towards a set of priorities for bird conservation and research in Australia: The perceptions of ornithologists. *Emu*, **109**, 67–74.

Miller, R. and Howell, G.V.J. (2008). Regulating consumption with bite: Building a contemporary framework for urban dog management. *Journal of Business Research*, **61**, 525–31.

Morgan, R. (1999). Preferences and priorities of recreational beach users in Wales, UK. *Journal of Coastal Research*, **15**, 653–67.

Naughton-Treves, L., Grossberg, R., and Treves, R. (2003). Paying for tolerance: Rural citizens' attitudes toward wolf depredation and compensation. *Conservation Biology*, **17**, 1500–11.

Napierala, H., and Uerpmann, H-P. (2012). A "new" Palaeolithic dog from Central Europe. *International Journal of Osteoarchaeology*, **22**, 127–37.

Nimmo, D.G., Miller, K.K., and Adams, R.A. (2007). Managing feral horses in Victoria: A study of community attitudes and perceptions. *Ecological Management & Restoration*, **8**, 237–41.

O'Connell-Rodwell, C.E., Rodwell, T., Rice, M., and Hart, L.A. (2000). Living with the modern conservation paradigm: Can agricultural communities co-exist with elephants? A five-year case study in East Caprivi, Namibia. *Biological Conservation*, **93**, 381–91.

Ormsby, A.A. and Forys, E.A. (2010). The effects of an education campaign on beach user perceptions of beach-nesting birds in Pinellas County, Florida. *Human Dimensions of Wildlife*, **15**, 119–28.

Ortolani, A., Vernooij, H., and Coppinger, R. (2009). Ethiopian village dogs: Behavioural responses to a stranger's approach. *Applied Animal Behaviour Science*, **119**, 210–18.

Ovodov, N.D., Crockford, S.J., Kuzmin, Y.V., Higham, T.F.G., Hodgins G.W.L., and van der Plicht, J. (2011). A 33,000-year-old incipient dog from the Altai Mountains of Siberia: Evidence of the earliest domestication disrupted by the Last Glacial Maximum. *PLoS ONE*, **6**: e22821. doi:10.1371/journal.pone.0022821

Paquet-Durand, I., Hernandez, J., Dolz, G., Romero Zuniga, J.J., Schnieder, T., and Epe, C. (2007). Prevalence of *Toxocara* spp., *Toxascaris leonine* and Ancylostomidae in public parks and beaches in different climate zones of Costa Rica. *ActaTropica*, **104**, 30–7.

Pelletier, F. (2006). Effects of tourist activities on ungulate behaviour in a mountain protected area, *Journal of Mountain Ecology*, **8**, 15–19.

Peter, J. (2012). In dog we trust, *Australian Birdlife*, **1**, 36–9.

Poole, L. (2010). Guardian dogs look after penguins, *ABC Rural*, June 22, 2010, <http://www.abc.net.au/site-archive/rural/content/2010/s2933996.htm>.

Potgieter, C. (2011). *The effectiveness of livestock guarding dogs for livestock production and conservation in Namibia*. Masters Thesis, Nelson Mandela Metropolitan University, South Africa.

Reed, S.E. and Merenlender, A.M. (2011). Effects of management of domestic dogs and recreation on carnivores in protected areas in Northern California. *Conservation Biology*, **25**, 504–13.

Riley, S., Siemer, W., Decker, D., Carpenter, L., Organ, J., and Berchielli, L. (2003). Adaptive impact management: An integrative approach to wildlife management. *Human Dimensions of Wildlife*, **8**, 81–95.

Ritchie, E.G. and Johnson, C.N. (2009). Predator interactions, mesopredator release and biodiversity conservation. *Ecology Letters*, **12**, 982–98.

Ritchie E.G., Elmhagen, B., Glen, A.S., Letnic, M., Ludwig, G., and McDonald, R.A. (2012). Ecosystem restoration with teeth: What role for predators? *Trends in Ecology and Evolution*, **27**, 265–71.

Rokeach, M. (1973). *The Nature of Human Values*. The Free Press: New York.

Sastre, P., Ponce, C., Palacín, C., Martín, C.A., and Alonso, J.C. (2009). Disturbances to great bustards (*Otis tarda*) in central Spain: Human activities, bird responses and management implications. *European Journal of Wildlife Research*, **55**, 425–32.

Schultz, P.W. (2011). Conservation means behavior. *Conservation Biology*, **25**, 1080–3.

Semken, M., Miller, K.K., and James, K. (2011). The value of oral history in natural resource management: the Balcombe Estuary Reserve, Mount Martha, Victoria. *Australasian Journal of Environmental Management*, **18**, 109–22.

Shindler, B. and Cheek, K.A. (1999). Integrating citizens in adaptive management: a propositional analysis. *Conservation Ecology*, **3**, 9.

Sillero-Zubiri, C. (2009). Family Canidae (Dogs). In D.E. Wilson, and R.A. Mittermeier, Eds., *Handbook of the Mammals of the World. Vol 1. Carnivores*. Lynx Edicions, Barcelona, pp. 352–411.

Sterl, P., Brandenburg, C., and Arnberger, A. (2008). Visitors' awareness and assessment of recreational disturbance of wildlife in the Donau-Auen National Park. *Journal for Nature Conservation*, **16**, 135–45.

Thomas, I. and Murfitt, P. (2011). *Environmental Management: Processes and Practices for Australia*. 2nd edn. The Federation Press, Sydney.

Thompson, J., Shirreffs, L., and McPhail, I. (2003). Dingoes on Fraser Island-Tourism Dream or management nightmare? *Human Dimensions of Wildlife*, **8**, 37–47.

Torres, P.C. and Prado, P.I. (2010). Domestic dogs in a fragmented landscape in the Brazilian Atlantic Forest: Abundance, habitat use and caring by owners. *Brazilian Journal of Biology*, **70**, 987–94.

Underhill-Day J.C. and Liley, D. (2007). Visitor patterns on southern heaths: A review of visitor access patterns to heathlands in the UK and the relevance to Annex I bird species. *Ibis*, **149** (Suppl.1), 112–9.

Van Bommel, L. and Johnson, C. (2012). Good dog! Using livestock guardian dogs to protect livestock from predators in Australia's extensive grazing systems. *Wildlife Research*, **39**, 220–29.

Weston, M.A., Antos, M.J., and Glover, H.K. (2009). Birds, buffers and bicycles: A review and case study of wetland buffers. *The Victorian Naturalist*, **126**, 79–86.

Weston, M.A., Dodge, F., Bunce, A., Nimmo, D.G., and Miller, K.K. (2012). Do temporary beach closures assist in the conservation of breeding shorebirds on recreational beaches?*Pacific Conservation Biology*, **18**, 47–55.

Weston, M.A., Ehmke, G.C., and Maguire, G.S. (2011). Nest return times in response to static versus mobile human disturbance. *Journal of Wildlife Management*, **75**, 252–55.

White, P.C.L., Newton-Cross, G.A., Moberly, R.L., Smart, J.C.R., Baker, P.J., and Harris, S. (2003). The current and future management of wild mammals hunted with dogs in England and Wales. *Journal of Environmental Management*, **67**, 187–97.

Whiting, A. and Miller, K. (2008). Examining the living with possums policy in Victoria, Australia: Community knowledge, support and compliance. *Pacific Conservation Biology*, **14**, 169–76.

Williams, K.J.H., Weston, M.A., Henry, S., and Maguire, G.S. (2009). Birds and beaches, dogs and leashes: Dog owners' sense of obligation to leash dogs on beaches in Victoria, Australia. *Human Dimensions of Wildlife*, **14**, 89–101.

Wilson, E.O. (1984). *Biophilia*. Harvard University Press, Cambridge.

Wittmann, K., Vaske, J.J., Zinn, H.C., and Manfredo, M.J. (1998). Standards for lethal response to problem urban wildlife. *Human Dimensions of Wildlife*, **3**, 29–48.

Woodroffe, R. and Ginsberg, J.R. (1998). Edge effects and the extinction of populations inside protected areas. *Science*, **280**, 2126–8.

Index